ELECTRONIC DEVICES

Electron Flow Version

Tenth Edition

ELECTRONIC DEVICES

Electron Flow Version

Tenth Edition

Thomas L. Floyd

330 Hudson Street, NY NY 10013

Vice President, Portfolio Management: Andrew Gilfillan
Portfolio Manager: Tony Webster
Editorial Assistant: Lara Dimmick
Senior Vice President, Marketing: David Gesell
Field Marketing Manager: Thomas Hayward
Marketing Coordinator: Elizabeth MacKenzie-Lamb
Director, Digital Studio and Content Production: Brian Hyland
Managing Producer: Cynthia Zonneveld
Managing Producer: Jennifer Sargunar
Content Producer: Faraz Sharique Ali

Content Producer: Nikhil Rakshit
Manager, Rights Management: Johanna Burke
Operations Specialist: Deidra Smith
Cover Design: Cenveo Publisher Services
Cover Photo: James Steidl/Shutterstock
Full-Service Project Management and Composition:
   Jyotsna Ojha, Cenveo Publisher Services®
Printer/Binder: LSC Communications Willard
Cover Printer: Phoenix Color
Text Font: Times LT Pro

Credits and acknowledgments for materials borrowed from other sources and reproduced, with permission, in this textbook appear on the appropriate page within text.

CIP data on record at the Library of Congress

1   17

ISBN 10:   0-13-442010-1
ISBN 13: 978-0-13-442010-3

PREFACE

This tenth edition of *Electronic Devices* reflects changes recommended by users and reviewers. As in the previous edition, Chapters 1 through 11 are essentially devoted to discrete devices and circuits. Chapters 12 through 17 primarily cover linear integrated circuits. Chapter 18, covers electronic communication devices and methods. Multisim® circuit files in version 14 and LT Spice circuit files are available at the website: www.pearsonhighered.com/careersresources/

New Features

- Chapter covering an introduction to communication devices and methods.

- LT Spice circuit simulation.

- Mutlisim files upgraded to Version 14 and new files added.

- Several new examples.

- Expanded coverage of FETs including JFET limiting parameters, FINFET, UMOSFET, Current source biasing, Cascode dual-gate MOSFET, and tunneling MOSFET.

- Expanded coverage of thyristors including SSRs using SCRs, motor speed control.

- Expanded coverage of switching circuits including interfacing with logic circuits.

- Expanded PLL coverage.

- Many new problems.

Standard Features

- Full-color format.

- Chapter openers include a chapter outline, chapter objectives, introduction, key terms list, Device Application preview, and website reference.

- Introduction and objectives for each section within a chapter.

- Large selection of worked-out examples set off in a graphic box. Each example has a related problem for which the answer can be found at: www.pearsonhighered.com/careersresources/

- Multisim® circuit files for selected examples, troubleshooting, and selected problems are on the companion website.

- LT Spice circuit files for selected examples and problems are on the companion website.

- Section checkup questions are at the end of each section within a chapter. Answers can be found at: www.pearsonhighered.com/careersresources/

- Troubleshooting sections in many chapters.

◆ A Device Application is at the end of most chapters.

◆ A Programmable Analog Technology feature is at the end of selected chapters.

◆ A sectionalized chapter summary, key term glossary, and formula list at the end of each chapter.

◆ True/false quiz, circuit-action quiz, self-test, and categorized problem set with basic and advanced problems at the end of each chapter.

◆ Appendix with answers to odd-numbered problems, glossary, and index are at the end of the book.

◆ Updated PowerPoint® slides, developed by Dave Buchla, are available online. These innovative, interactive slides are coordinated with each text chapter and are an excellent tool to supplement classroom presentations.

◆ A laboratory manual by Dave Buchla and Steve Wetterling coordinated with this textbook is available.

Student Resources

Digital Resources (*www.pearsonhighered.com/careersresources/*)   This section offers students an online study guide that they can check for conceptual understanding of key topics. Also included on the website are tutorials for Multisim® and LT Spice. Answers to Section Checkups, Related Problems for Examples, True/False Quizzes, Circuit-Action Quizzes, and Self-Tests are found on this website.

Circut Simulation (*www.pearsonhighered.com/careersresources/*)   These online files include simulation circuits in Multisim® 14 and LT Spice for selected examples, troubleshooting sections, and selected problems in the text. These circuits were created for use with Multisim® or LT Spice software. These circuit simulation programs are widely regarded as excellent for classroom and laboratory learning. However, no part of your textbook is dependent upon the Multisim® or LT Spice software or provided files.

Laboratory Exercises for Electronic Devices,   Tenth Edition, by Dave Buchla and Steve Wetterling. ISBN: 0-13-442031-4.

Instructor Resources

To access supplementary materials online, instructors need to request an instructor access code. Go to www.pearsonhighered.com/irc to register for an instructor access code. Within 48 hours of registering, you will receive a confirming e-mail including an instructor access code. Once you have received your code, locate your text in the online catalog and click on the Instructor Resources button on the left side of the catalog product page. Select a supplement, and a login page will appear. Once you have logged in, you can access instructor material for all Pearson textbooks. If you have any difficulties accessing the site or downloading a supplement, please contact Customer Service at: http://support.pearson.com/getsupport

Online Instructor's Resource Manual   Includes solutions to chapter problems, Device Application results, summary of Multisim® and LT Spice circuit files, and a test item file. Solutions to the lab manual are also included.

Online Course Support   If your program is offering your electronics course in a distance learning format, please contact your local Pearson sales representative for a list of product solutions.

Online PowerPoint® Slides   This innovative, interactive PowerPoint slide presentation for each chapter in the book provides an effective supplement to classroom lectures.

Online TestGen   This is a test bank of over 800 questions.

Chapter Features

Chapter Opener   Each chapter begins with an opening page, as shown in Figure P–1. The chapter opener includes a chapter introduction, a list of chapter sections, chapter objectives, key terms, an Device Application preview, and a website reference for associated study aids.

Chapter outline

List of performance-based chapter objectives

Key terms

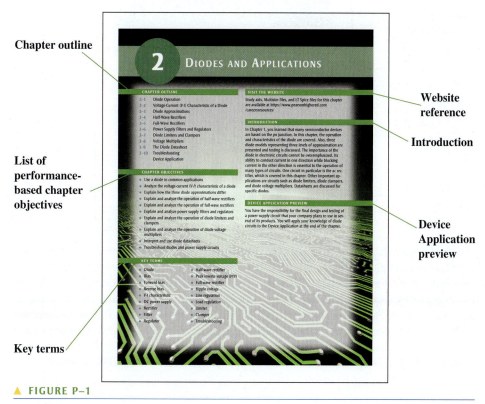

Website reference

Introduction

Device Application preview

▲ FIGURE P–1

A typical chapter opener.

Section Opener   Each section in a chapter begins with a brief introduction and section objectives. An example is shown in Figure P–2.

Section Checkup   Each section in a chapter ends with a list of questions that focus on the main concepts presented in the section. This feature is also illustrated in Figure P–2. The answers to the Section Checkups can be found at: www.pearsonhighered.com/careersresources/

Troubleshooting Sections   Many chapters include a troubleshooting section that relates to the topics covered in the chapter and that illustrates troubleshooting procedures and techniques. The Troubleshooting section also provides Multisim® Troubleshooting exercises.

▶ FIGURE P–2

A typical section opener and section review.

Section checkup ends each section.

Introductory paragraph begins each section.

Performance-based section objectives

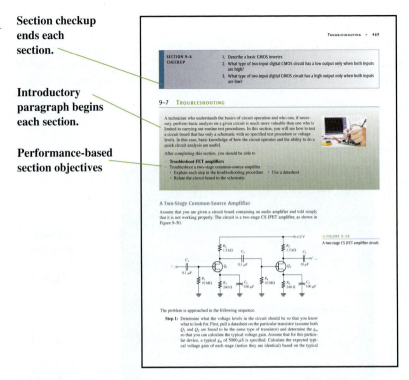

Worked Examples, Related Problems, and Circuit Simulation Exercises Numerous worked-out examples throughout each chapter illustrate and clarify basic concepts or specific procedures. Each example ends with a Related Problem that reinforces or expands on the example by requiring the student to work through a problem similar to the example. Selected examples feature a Multisim® or LT Spice exercise keyed to a file on the companion website that contains the circuit illustrated in the example. A typical example with a Related Problem and a Multisim® or LT Spice exercise are shown in Figure P–3. Answers to Related Problems can be found at: www.pearsonhighered.com/careersresources/

▶ FIGURE P–3

A typical example with a related problem and Multisim®/LT Spice exercise.

Examples are set off from text

Each example contains a related problem relevant to the example.

Selected examples include a Multisim®/LT Spice exercise coordinated with the circuit simulation files on the website.

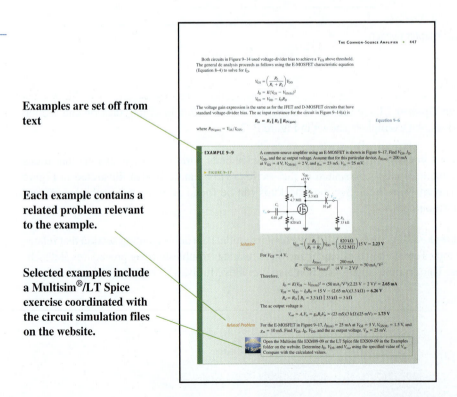

Device Application This feature follows the last section in most chapters and is identified by a special graphic design. A practical application of devices or circuits covered in the chapter is presented. The student learns how the specific device or circuit is used and is taken through the steps of design specification, simulation, prototyping, circuit board implementation, and testing. A typical Device Application is shown in Figure P–4. Device Applications are optional. Results are provided in the Online Instructor's Resource Manual.

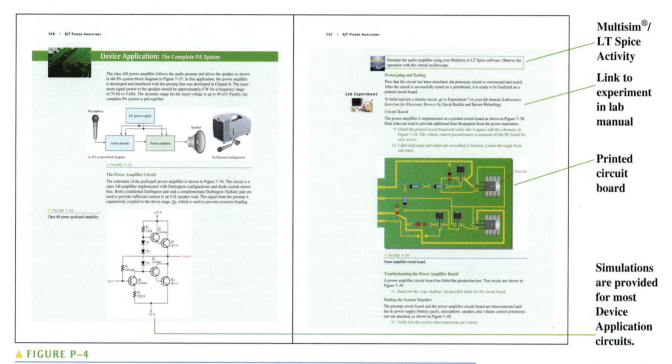

Multisim®/
LT Spice
Activity

Link to
experiment
in lab
manual

Printed
circuit
board

Simulations
are provided
for most
Device
Application
circuits.

▲ FIGURE P–4

Portion of a typical Device Application section.

Chapter End Matter The following pedagogical features are found at the end of most chapters:

- ◆ Summary
- ◆ Key Term Glossary
- ◆ Key Formulas
- ◆ True/False Quiz
- ◆ Circuit-Action Quiz
- ◆ Self-Test
- ◆ Basic Problems
- ◆ Advanced Problems
- ◆ Datasheet Problems (selected chapters)
- ◆ Device Application Problems (many chapters)
- ◆ Multisim® Troubleshooting Problems (most chapters)

Suggestions for Using This Textbook

As mentioned, this book covers discrete devices and circuits in Chapters 1 through 11 and linear integrated circuits in Chapters 12 through 17. Chapter 18 introduces programming concepts for device testing and is linked to Troubleshooting sections.

Option 1 (two terms)   Chapters 1 through 11 can be covered in the first term. Depending on individual preferences and program emphasis, selective coverage may be necessary. Chapters 12 through 17 can be covered in the second term. Again, selective coverage may be necessary.

Option 2 (one term)   By omitting certain topics and by maintaining a rigorous schedule, this book can be used in one-term courses. For example, a course covering only discrete devices and circuits would use Chapters 1 through 11 with, perhaps, some selectivity.

Similarly, a course requiring only linear integrated circuit coverage would use Chapters 12 through 17. Another approach is a very selective coverage of discrete devices and circuits topics followed by a limited coverage of integrated circuits (only op-amps, for example). Also, elements such as the Multisim® and LT Spice exercises, and Device Application can be omitted or selectively used.

To the Student

When studying a particular chapter, study one section until you understand it and only then move on to the next one. Read each section and study the related illustrations carefully; think about the material; work through each example step-by-step, work its Related Problem and check the answer; then answer each question in the Section Checkup, and check your answers. Don't expect each concept to be completely clear after a single reading; you may have to read the material two or even three times. Once you think that you understand the material, review the chapter summary, key formula list, and key term definitions at the end of the chapter. Take the true/false quiz, the circuit-action quiz, and the self-test. Finally, work the assigned problems at the end of the chapter. Working through these problems is perhaps the most important way to check and reinforce your comprehension of the chapter. By working problems, you acquire an additional level of insight and understanding and develop logical thinking that reading or classroom lectures alone do not provide.

Generally, you cannot fully understand a concept or procedure by simply watching or listening to someone else. Only hard work and critical thinking will produce the results you expect and deserve.

Acknowledgments

Many capable people have contributed to the tenth edition of *Electronic Devices*. It has been thoroughly reviewed and checked for both content and accuracy. Those at Pearson who have contributed greatly to this project throughout the many phases of development and production include Faraz Sharique Ali and Rex Davidson. Thanks to Jyotsna Ojha at Cenveo for her management of the art and text programs. Dave Buchla contributed extensively to the content of the book, helping to make this edition the best one yet. Gary Snyder created the circuit files for the Multisim® and LT Spice features in this edition. I wish to express my appreciation to those already mentioned as well as the reviewers who provided many valuable suggestions and constructive criticism that greatly influenced this edition. These reviewers are David Beach, Indiana State University; Mahmoud Chitsazzadeh, Community College of Allegheny County; Wang Ng, Sacramento City College; Almasy Edward, Pennsylvania College of Technology; and Moser Randall, Pennsylvania College of Technology.

Tom Floyd

BRIEF CONTENTS

CONTENTS

ELECTRONIC DEVICES

Electron Flow Version

Tenth Edition

INTRODUCTION TO SEMICONDUCTORS

1

CHAPTER OBJECTIVES

- Describe the structure of an atom
- Discuss insulators, conductors, and semiconductors and how they differ
- Describe how current is produced in a semiconductor
- Describe the properties of *n*-type and *p*-type semiconductors
- Describe how a *pn* junction is formed

KEY TERMS

- Atom
- Proton
- Electron
- Shell
- Valence
- Ionization
- Free electron
- Orbital
- Insulator
- Conductor
- Semiconductor
- Silicon
- Crystal
- Hole
- Metallic bond
- Doping
- *PN* junction
- Barrier potential

VISIT THE WEBSITE

Study aids for this chapter are available at
https://www.pearsonhighered.com/careersresources/

INTRODUCTION

Electronic devices such as diodes, transistors, and integrated circuits are made of a semiconductive material. To understand how these devices work, you should have a basic knowledge of the structure of atoms and the interaction of atomic particles. An important concept introduced in this chapter is that of the *pn* junction that is formed when two different types of semiconductive material are joined. The *pn* junction is fundamental to the operation of devices such as the solar cell, the diode, and certain types of transistors.

1–1 THE ATOM

All matter is composed of atoms; all atoms consist of electrons, protons, and neutrons except normal hydrogen, which does not have a neutron. Each element in the periodic table has a unique atomic structure, and all atoms for a given element have the same number of protons. At first, the atom was thought to be a tiny indivisible sphere. Later it was shown that the atom was not a single particle but was made up of a small, dense nucleus around which electrons orbit at great distances from the nucleus, similar to the way planets orbit the sun. Niels Bohr proposed that the electrons in an atom circle the nucleus in different obits, similar to the way planets orbit the sun in our solar system. The Bohr model is often referred to as the planetary model. Another view of the atom called the *quantum model* is considered a more accurate representation, but it is difficult to visualize. For most practical purposes in electronics, the Bohr model suffices and is commonly used because it is easy to visualize.

After completing this section, you should be able to

❑ **Describe the structure of an atom**
  ◆ Discuss the Bohr model of an atom ◆ Define *electron, proton, neutron,* and *nucleus*
❑ Define *atomic number*
❑ Discuss electron shells and orbits
  ◆ Explain energy levels
❑ Define *valence electron*
❑ Discuss ionization
  ◆ Define *free electron* and *ion*
❑ Discuss the basic concept of the quantum model of the atom

The Bohr Model

An **atom*** is the smallest particle of an element that retains the characteristics of that element. Each of the known 118 elements has atoms that are different from the atoms of all other elements. This gives each element a unique atomic structure. According to the classical Bohr model, atoms have a planetary type of structure that consists of a central nucleus surrounded by orbiting electrons, as illustrated in Figure 1–1. The **nucleus** consists of positively charged particles called **protons** and uncharged particles called **neutrons.** The basic particles of negative charge are called **electrons**.

Each type of atom has a certain number of electrons and protons that distinguishes it from the atoms of all other elements. For example, the simplest atom is that of hydrogen, which has one proton and one electron, as shown in Figure 1–2(a). As another example, the helium atom, shown in Figure 1–2(b), has two protons and two neutrons in the nucleus and two electrons orbiting the nucleus.

Atomic Number

All elements are arranged in the periodic table of the elements in order according to their atomic number. The **atomic number** equals the number of protons in the nucleus, which is the same as the number of electrons in an electrically balanced (neutral) atom. For example, hydrogen has an atomic number of 1 and helium has an atomic number of 2. In their normal (or neutral) state, all atoms of a given element have the same number of electrons as protons; the positive charges cancel the negative charges, and the atom has a net charge of zero.

HISTORY NOTE

Niels Henrik David Bohr (October 7, 1885–November 18, 1962) was a Danish physicist, who made important contributions to understanding the structure of the atom and quantum mechanics by postulating the "planetary" model of the atom. He received the Nobel Prize in physics in 1922. Bohr drew upon the work or collaborated with scientists such as Dalton, Thomson, and Rutherford, among others and has been described as one of the most influential physicists of the 20th century.

---

*All bold terms are in the end-of-book glossary. The bold terms in color are key terms and are also defined at the end of the chapter.

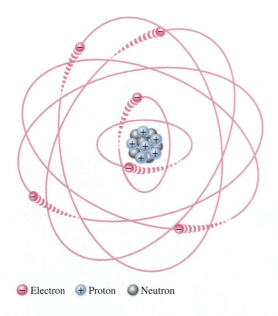

○ Electron   ⊕ Proton   ◉ Neutron

▲ **FIGURE 1–1**

The Bohr model of an atom showing electrons in orbits around the nucleus, which consists of protons and neutrons. The "tails" on the electrons indicate motion.

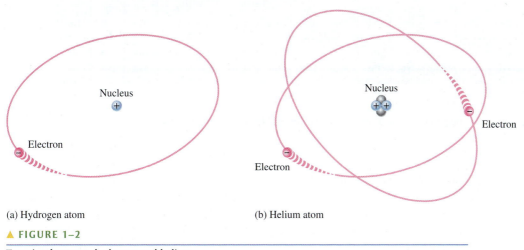

(a) Hydrogen atom   (b) Helium atom

▲ **FIGURE 1–2**

Two simple atoms, hydrogen and helium.

Atomic numbers of all the elements are shown on the periodic table of the elements in Figure 1–3.

Electrons and Shells

Energy Levels   Electrons orbit the nucleus of an atom at certain distances from the nucleus. Electrons near the nucleus have less energy than those in more distant orbits. Only discrete (separate and distinct) values of electron energies exist within atomic structures. Therefore, electrons must orbit only at discrete distances from the nucleus.

Each discrete distance (**orbit**) from the nucleus corresponds to a certain energy level. In an atom, the orbits are grouped into energy levels known as shells. A given atom has a fixed number of shells. Each shell has a fixed maximum number of electrons. The shells (energy levels) are designated 1, 2, 3, and so on, with 1 being closest to the nucleus. The Bohr model of the silicon atom is shown in Figure 1–4. Notice that there are 14 electrons surrounding the nucleus with exactly 14 protons, and usually 14 neutrons in the nucleus.

Helium
Atomic number = 2

1 H																	2 He
3 Li	4 Be											5 B	6 C	7 N	8 O	9 F	10 Ne
11 Na	12 Mg											13 Al	14 Si	15 P	16 S	17 Cl	18 Ar
19 K	20 Ca	21 Sc	22 Ti	23 V	24 Cr	25 Mn	26 Fe	27 Co	28 Ni	29 Cu	30 Zn	31 Ga	32 Ge	33 As	34 Se	35 Br	36 Kr
37 Rb	38 Sr	39 Y	40 Zr	41 Nb	42 Mo	43 Tc	44 Ru	45 Rh	46 Pd	47 Ag	48 Cd	49 In	50 Sn	51 Sb	52 Te	53 I	54 Xe
55 Cs	56 Ba	*	72 Hf	73 Ta	74 W	75 Re	76 Os	77 Ir	78 Pt	79 Au	80 Hg	81 Tl	82 Pb	83 Bi	84 Po	85 At	86 Rn
87 Fr	88 Ra	**	104 Rf	105 Db	106 Sg	107 Bh	108 Hs	109 Mt	110 Ds	111 Rg	112 Cp	113 Uut	114 Uuq	115 Uup	116 Uuh	117 Uus	118 Uuo

Silicon
Atomic number = 14

57 La	58 Ce	59 Pr	60 Nd	61 Pm	62 Sm	63 Eu	64 Gd	65 Tb	66 Dy	67 Ho	68 Er	69 Tm	70 Yb	71 Lu
89 Ac	90 Th	91 Pa	92 U	93 Np	94 Pu	95 Am	96 Cm	97 Bk	98 Cf	99 Es	100 Fm	101 Md	102 No	103 Lr

▲ **FIGURE 1–3**

The periodic table of the elements. Some tables also show atomic mass.

▶ **FIGURE 1–4**

Illustration of the Bohr model of the silicon atom.

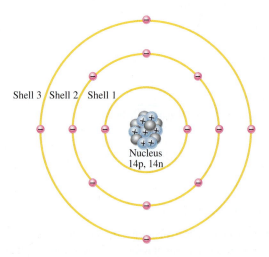

Shell 3 Shell 2 Shell 1

Nucleus
14p, 14n

The Maximum Number of Electrons in Each Shell The maximum number of electrons (N_e) that can exist in each shell of an atom is a fact of nature and can be calculated by the formula,

Equation 1–1

$$N_e = 2n^2$$

where n is the number of the shell. The maximum number of electrons that can exist in the innermost shell (shell 1) is

$$N_e = 2n^2 = 2(1)^2 = 2$$

The maximum number of electrons that can exist in shell 2 is

$$N_e = 2n^2 = 2(2)^2 = 2(4) = 8$$

The maximum number of electrons that can exist in shell 3 is

$$N_e = 2n^2 = 2(3)^2 = 2(9) = 18$$

The maximum number of electrons that can exist in shell 4 is

$$N_e = 2n^2 = 2(4)^2 = 2(16) = 32$$

Valence Electrons

Electrons that are in orbits farther from the nucleus have higher energy and are less tightly bound to the atom than those closer to the nucleus. This is because the force of attraction between the positively charged nucleus and the negatively charged electron decreases with increasing distance from the nucleus. Electrons with the highest energy exist in the outermost shell of an atom and are relatively loosely bound to the atom. This outermost shell is known as the **valence** shell, and electrons in this shell are called *valence electrons*. These valence electrons contribute to chemical reactions and bonding within the structure of a material and determine its electrical properties. When a valence electron gains sufficient energy from an external source, it can break free from its atom. This is the basis for conduction in materials.

Ionization

When an atom absorbs energy, the valence electrons can easily jump to higher energy shells. If a valence electron acquires a sufficient amount of energy, called *ionization energy,* it can actually escape from the outer shell and the atom's influence. The departure of a valence electron leaves a previously neutral atom with an excess of positive charge (more protons than electrons). The process of losing a valence electron is known as **ionization**, and the resulting positively charged atom is called a *positive ion.* For example, the chemical symbol for hydrogen is H. When a neutral hydrogen atom loses its valence electron and becomes a positive ion, it is designated H^+. The escaped valence electron is called a **free electron**.

The reverse process can occur in certain atoms when a free electron collides with the atom and is captured, releasing energy. The atom that has acquired the extra electron is called a *negative ion.* The ionization process is not restricted to single atoms. In many chemical reactions, a group of atoms that are bonded together can lose or acquire one or more electrons.

For some nonmetallic materials such as chlorine, a free electron can be captured by the neutral atom, forming a negative ion. In the case of chlorine, the ion is more stable than the neutral atom because it has a filled outer shell. The chlorine ion is designated as Cl^-.

The Quantum Model

Although the Bohr model of an atom is widely used because of its simplicity and ease of visualization, it is not a complete model. The quantum model is considered to be more accurate. The quantum model is a statistical model and very difficult to understand or visualize. Like the Bohr model, the quantum model has a nucleus of protons and neutrons surrounded by electrons. Unlike the Bohr model, the electrons in the quantum model do not exist in precise circular orbits as particles. Three important principles underlie the quantum model: the wave-particle duality principle, the uncertainty principle, and the superposition principle.

◆ *Wave-particle duality.* Just as light can be thought of as exhibiting both a wave and a particle **(photon),** electrons are thought to exhibit a wave-particle duality. The velocity of an orbiting electron is related to its wavelength, which interferes with neighboring electron wavelengths by amplifying or canceling each other.

◆ *Uncertainty principle.* As you know, a wave is characterized by peaks and valleys; therefore, electrons acting as waves cannot be precisely identified in terms of their position. According to a principle ascribed to Heisenberg, it is impossible to determine simultaneously both the position and velocity of an electron with any degree of accuracy or certainty. The result of this principle produces a concept of the atom with *probability clouds,* which are mathematical descriptions of where electrons in an atom are most likely to be located.

◆ *Superposition.* A principle of quantum theory that describes a challenging concept about the behavior of matter and forces at the subatomic level. Basically, the principle states that although the state of any object is unknown, it is actually in all possible states simultaneously as long as an observation is not attempted. An analogy known as Schrödinger's cat is often used to illustrate in an oversimplified way quantum superposition. The analogy goes as follows: A living cat is placed in a metal box with a vial of hydrocyanic acid and a very small amount of a radioactive substance. Should even a single atom of the radioactive substance decay during a test period, a relay mechanism will be activated and will cause a hammer to break the vial and kill the cat. An observer cannot know whether or not this sequence has occurred. According to quantum theory, the cat exists in a superposition of both the alive and dead states simultaneously.

In the quantum model, each shell or energy level consists of up to four subshells called **orbitals**, which are designated *s, p, d,* and *f.* Orbital *s* can hold a maximum of two electrons, orbital *p* can hold six electrons, orbital *d* can hold 10 electrons, and orbital *f* can hold 14 electrons. Each atom can be described by an electron configuration table that shows the shells or energy levels, the orbitals, and the number of electrons in each orbital. For example, the electron configuration table for the nitrogen atom is given in Table 1–1. The first full-size number is the shell or energy level, the letter is the orbital, and the exponent is the number of electrons in the orbital.

▶ **TABLE 1–1**

Electron configuration table for nitrogen.

NOTATION	EXPLANATION
$1s^2$	2 electrons in shell 1, orbital s
$2s^2 \quad 2p^3$	5 electrons in shell 2: 2 in orbital s, 3 in orbital p

Atomic orbitals do not resemble a discrete circular path for the electron as depicted in Bohr's planetary model. In the quantum picture, each shell in the Bohr model is a three-dimensional space surrounding the atom that represents the mean (average) energy of the electron cloud. The term **electron cloud** (probability cloud) is used to describe the area around an atom's nucleus where an electron will probably be found.

EXAMPLE 1–1

Using the atomic number from the periodic table in Figure 1–3, describe a silicon (Si) atom using an electron configuration table.

Solution

The atomic number of silicon is 14. This means that there are 14 protons in the nucleus. Since there is always the same number of electrons as protons in a neutral atom, there are also 14 electrons. As you know, there can be up to two electrons in shell 1, eight in shell 2, and eighteen in shell 3. Therefore, in silicon there are two electrons in shell 1, eight electrons in shell 2, and four electrons in shell 3 for a total of 14 electrons. The electron configuration table for silicon is shown in Table 1–2.

▶ **TABLE 1–2**

NOTATION	EXPLANATION
$1s^2$	2 electrons in shell 1, orbital s
$2s^2 \quad 2p^6$	8 electrons in shell 2: 2 in orbital s, 6 in orbital p
$3s^2 \quad 3p^2$	4 electrons in shell 3: 2 in orbital s, 2 in orbital p

*Related Problem**

Develop an electron configuration table for the germanium (Ge) atom in the periodic table.

*Answers can be found at www.pearsonhighered.com/floyd.

In a three-dimensional representation of the quantum model of an atom, the *s*-orbitals are shaped like spheres with the nucleus in the center. For energy level 1, the sphere is a single sphere, but for energy levels 2 or more, each single *s*-orbital is composed of nested spherical shells. A *p*-orbital for shell 2 has the form of two ellipsoidal lobes with a point of tangency at the nucleus (sometimes referred to as a dumbbell shape.) The three *p*-orbitals in each energy level are oriented at right angles to each other. One is oriented on the *x*-axis, one on the *y*-axis, and one on the *z*-axis. For example, a view of the quantum model of a sodium atom (Na) that has 11 electrons is shown in Figure 1–5. The three axes are shown to give you a 3-D perspective.

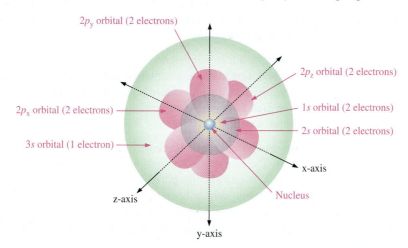

◀ **FIGURE 1–5**

Three-dimensional quantum model of the sodium atom, showing the orbitals and number of electrons in each orbital.

SECTION 1–1 CHECKUP
Answers can be found at www.pearsonhighered.com/floyd.

1. Describe the Bohr model of the atom.
2. Define *electron*.
3. What is the nucleus of an atom composed of? Define each component.
4. Define *atomic number*.
5. Discuss electron shells and orbits and their energy levels.
6. What is a valence electron?
7. What is a free electron?
8. Discuss the difference between positive and negative ionization.
9. Name three principles that distinguish the quantum model.

1–2 MATERIALS USED IN ELECTRONIC DEVICES

In terms of their electrical properties, materials can be classified into three groups: conductors, semiconductors, and insulators. When atoms combine to form a solid, crystalline material, they arrange themselves in a symmetrical pattern. The atoms within a semiconductor crystal structure are held together by covalent bonds, which are created by the interaction of the valence electrons of the atoms. Silicon is a crystalline material.

After completing this section, you should be able to

❏ **Discuss insulators, conductors, and semiconductors and how they differ**
  ◆ Define the *core* of an atom ◆ Describe the carbon atom ◆ Name two types each of semiconductors, conductors, and insulators
❏ Explain the band gap
  ◆ Define *valence band* and *conduction band* ◆ Compare a semiconductor atom to a conductor atom
❏ Discuss silicon and gemanium atoms
❏ Explain covalent bonds
  ◆ Define *crystal*

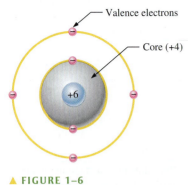

Valence electrons

Core (+4)

+6

▲ FIGURE 1–6

Diagram of a carbon atom.

Insulators, Conductors, and Semiconductors

All materials are made up of atoms. These atoms contribute to the electrical properties of a material, including its ability to conduct electrical current.

For purposes of discussing electrical properties, an atom can be represented by the valence shell and a **core** that consists of all the inner shells and the nucleus. This concept is illustrated in Figure 1–6 for a carbon atom. Carbon is used in some types of electrical resistors. Notice that the carbon atom has four electrons in the valence shell and two electrons in the inner shell. The nucleus consists of six protons and six neutrons, so the +6 indicates the positive charge of the six protons. The core has a net charge of +4 (+6 for the nucleus and −2 for the two inner-shell electrons).

Insulators   An insulator is a material that does not conduct electrical current under normal conditions. Most good insulators are compounds rather than single-element materials and have very high resistivities. Valence electrons are tightly bound to the atoms; therefore, there are very few free electrons in an insulator. Examples of insulators are rubber, plastics, glass, mica, and quartz.

Conductors   A conductor is a material that easily conducts electrical current. Most metals are good conductors. The best conductors are single-element materials, such as copper (Cu), silver (Ag), gold (Au), and aluminum (Al), which are characterized by atoms with only one valence electron very loosely bound to the atom. These loosely bound valence electrons can become free electrons with the addition of a small amount of energy to free them from the atom. Therefore, in a conductive material the free electrons are available to carry current.

Semiconductors   A semiconductor is a material that is between conductors and insulators in its ability to conduct electrical current. A semiconductor in its pure (intrinsic) state is neither a good conductor nor a good insulator. Single-element semiconductors are antimony (Sb), arsenic (As), astatine (At), boron (B), polonium (Po), tellurium (Te), silicon (Si), and germanium (Ge). Compound semiconductors such as gallium arsenide, indium phosphide, gallium nitride, silicon carbide, and silicon germanium are also commonly used. The single-element semiconductors are characterized by atoms with four valence electrons. Silicon is the most commonly used semiconductor.

Band Gap

In solid materials, interactions between atoms "smear" the valence shell into a band of energy levels called the *valence band.* Valence electrons are confined to that band. When an electron acquires enough additional energy, it can leave the valence shell, become a *free electron,* and exist in what is known as the *conduction band.*

The difference in energy between the valence band and the conduction band is called an *energy gap* or **band gap.** This is the amount of energy that a valence electron must have in order to jump from the valence band to the conduction band. Once in the conduction band, the electron is free to move throughout the material and is not tied to any given atom.

Figure 1–7 shows energy diagrams for insulators, semiconductors, and conductors. The energy gap or band gap is the difference between two energy levels and electrons are "not allowed" in this energy gap based on quantum theory. Although an electron may not exist in this region, it can "jump" across it under certain conditions. For insulators, the gap can be crossed only when breakdown conditions occur—as when a very high voltage is applied across the material. The band gap is illustrated in Figure 1–7(a) for insulators. In semiconductors the band gap is smaller, allowing an electron in the valence band to jump into the conduction band if it absorbs a photon. The band gap depends on the semiconductor material. This is illustrated in Figure 1–7(b). In conductors, the conduction band and valence band overlap, so there is no gap, as shown in Figure 1–7(c). This means that electrons in the valence band move freely into the conduction band, so there are always electrons available as free electrons.

F Y I

Next to silicon, the second most common semiconductive material is gallium arsenide, GaAs. This is a crystalline compound, not an element. Its properties can be controlled by varying the relative amount of gallium and arsenic.

GaAs has the advantage of making semiconductor devices that respond very quickly to electrical signals. It is widely used in high-frequency applications and in light-emitting diodes and solar cells.

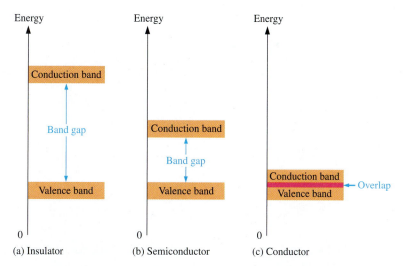

◀ FIGURE 1–7

Energy diagrams for the three types of materials.

Comparison of a Semiconductor Atom to a Conductor Atom

Silicon is a semiconductor and copper is a conductor. Bohr diagrams of the silicon atom and the copper atom are shown in Figure 1–8. Notice that the core of the silicon atom has a net charge of +4 (14 protons – 10 electrons) and the core of the copper atom has a net charge of +1 (29 protons – 28 electrons). Recall that the core includes everything except the valence electrons.

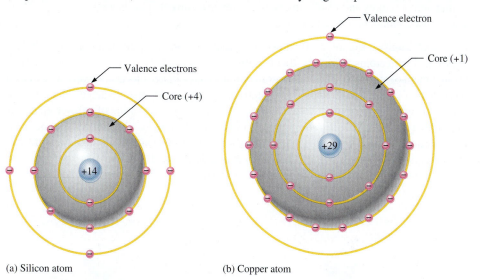

(a) Silicon atom   (b) Copper atom

◀ FIGURE 1–8

Bohr diagrams of the silicon and copper atoms.

The valence electron in the copper atom "feels" an attractive force of +1 compared to a valence electron in the silicon atom which "feels" an attractive force of +4. Therefore, there is more force trying to hold a valence electron to the atom in silicon than in copper. The copper's valence electron is in the fourth shell, which is a greater distance from its nucleus than the silicon's valence electron in the third shell. Recall that, electrons farthest from the nucleus have the most energy. The valence electron in copper has more energy than the valence electron in silicon. This means that it is easier for valence electrons in copper to acquire enough additional energy to escape from their atoms and become free electrons than it is in silicon. In fact, large numbers of valence electrons in copper already have sufficient energy to be free electrons at normal room temperature.

Silicon and Germanium

The atomic structures of silicon and germanium are compared in Figure 1–9. **Silicon** is used in diodes, transistors, integrated circuits, and other semiconductor devices. Notice that both silicon and **germanium** have the characteristic four valence electrons.

▶ **FIGURE 1–9**

Diagrams of the silicon and germanium atoms.

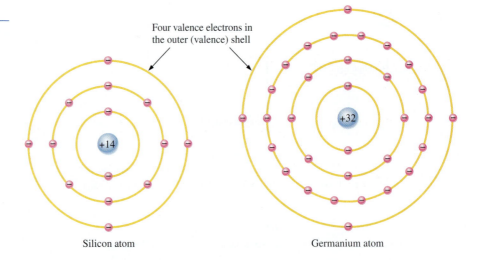

Four valence electrons in the outer (valence) shell

Silicon atom   Germanium atom

The valence electrons in germanium are in the fourth shell while those in silicon are in the third shell, closer to the nucleus. This means that the germanium valence electrons are at higher energy levels than those in silicon and, therefore, require a smaller amount of additional energy to escape from the atom. This property makes germanium more unstable at high temperatures and results in excessive reverse current. This is why silicon is a more widely used semiconductive material.

Covalent Bonds   Figure 1–10 shows how each silicon atom positions itself with four adjacent silicon atoms to form a silicon **crystal**, which is a three-dimensional symmetrical arrangement of atoms. A silicon (Si) atom with its four valence electrons shares an electron with each of its four neighbors. This effectively creates eight shared valence electrons for each atom and produces a state of chemical stability. Also, this sharing of valence electrons produces a strong **covalent bond** that hold the atoms together; each valence electron is attracted equally by the two adjacent atoms which share it. Covalent bonding in an intrinsic silicon crystal is shown in Figure 1–11. An **intrinsic** crystal is one that has no impurities. Covalent bonding for germanium is similar because it also has four valence electrons.

▶ **FIGURE 1–10**

Illustration of covalent bonds in silicon.

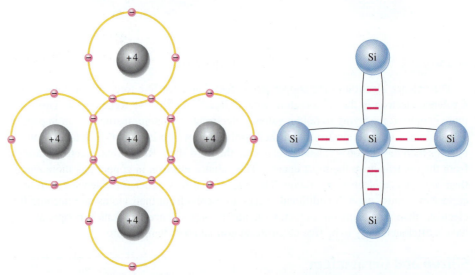

(a) The center silicon atom shares an electron with each of the four surrounding silicon atoms, creating a covalent bond with each. The surrounding atoms are in turn bonded to other atoms, and so on.

(b) Bonding diagram. The red negative signs represent the shared valence electrons.

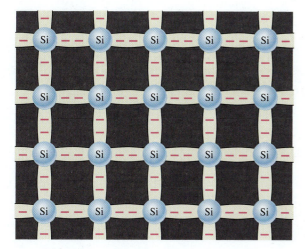

◀ **FIGURE 1–11**

Covalent bonds in a silicon crystal.

SECTION 1–2 CHECKUP

1. What is the basic difference between conductors and insulators?
2. How do semiconductors differ from conductors and insulators?
3. How many valence electrons does a conductor such as copper have?
4. How many valence electrons does a semiconductor have?
5. Name three of the best conductive materials.
6. What is the most widely used semiconductive material?
7. Why does a semiconductor have fewer free electrons than a conductor?
8. How are covalent bonds formed?
9. What is meant by the term *intrinsic*?
10. What is a crystal?

1–3 CURRENT IN SEMICONDUCTORS

The way a material conducts electrical current is important in understanding how electronic devices operate. You can't really understand the operation of a device such as a diode or transistor without knowing something about current in semiconductors.

After completing this section, you should be able to

❑ **Describe how current is produced in a semiconductor**
❑ Discuss conduction electrons and holes
  ◆ Explain an electron-hole pair ◆ Discuss recombination
❑ Explain electron and hole current

As you have learned, the electrons in a solid can exist only within prescribed energy bands. Each shell corresponds to a certain energy band and is separated from adjacent shells by band gaps, in which no electrons can exist. Figure 1–12 shows the energy band diagram for the atoms in a pure silicon crystal at its lowest energy level. There are no electrons shown in the conduction band, a condition that occurs *only* at a temperature of absolute 0 Kelvin.

▶ FIGURE 1–12

Energy band diagram for an atom in a pure (intrinsic) silicon crystal at it's lowest energy state. There are no electrons in the conduction band at a temperature of 0 K.

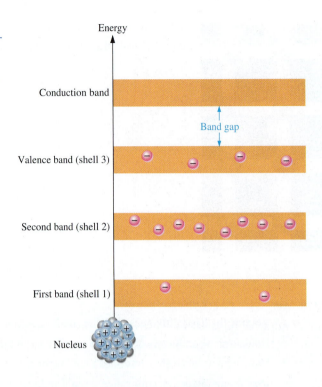

Conduction Electrons and Holes

An intrinsic (pure) silicon crystal at room temperature has sufficient heat (thermal) energy for some valence electrons to jump the gap from the valence band into the conduction band, becoming free electrons. Free electrons are also called **conduction electrons.** This is illustrated in the energy diagram of Figure 1–13(a) and in the bonding diagram of Figure 1–13(b).

▶ FIGURE 1–13

Creation of electron-hole pairs in a silicon crystal. Electrons in the conduction band are free electrons.

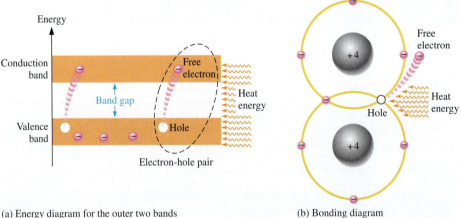

(a) Energy diagram for the outer two bands   (b) Bonding diagram

When an electron jumps to the conduction band, a vacancy is left in the valence band within the crystal. This vacancy is called a hole. For every electron raised to the conduction band by external energy, there is one hole left in the valence band, creating what is called an **electron-hole pair. Recombination** occurs when a conduction-band electron loses energy and falls back into a hole in the valence band.

To summarize, a piece of intrinsic silicon at room temperature has, at any instant, a number of conduction-band (free) electrons that are unattached to any atom and are essentially drifting randomly throughout the material. There is also an equal number of holes in the valence band created when these electrons jump into the conduction band. This is illustrated in Figure 1–14.

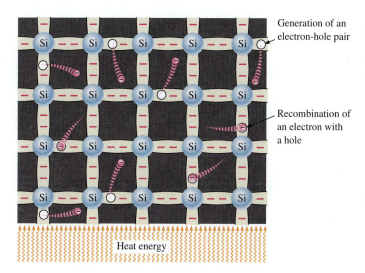

◀ FIGURE 1–14

Electron-hole pairs in a silicon crystal. Free electrons are being generated continuously while some recombine with holes.

Electron and Hole Current

When a voltage is applied across a piece of intrinsic silicon, as shown in Figure 1–15, the thermally generated free electrons in the conduction band, which are free to move randomly in the crystal structure, are now easily attracted toward the positive end. This movement of free electrons is one type of **current** in a semiconductive material and is called *electron current*.

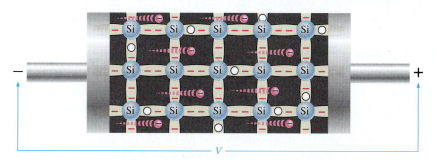

◀ FIGURE 1–15

Electron current in intrinsic silicon is produced by the movement of thermally generated free electrons.

Another type of current occurs in the valence band, where the holes created by the free electrons exist. Electrons remaining in the valence band are still attached to their atoms and are not free to move randomly in the crystal structure as are the free electrons. However, a valence electron can move into a nearby hole with little change in its energy level, thus leaving another hole where it came from. Effectively the hole has moved from one place to another in the crystal structure, as illustrated in Figure 1–16. Although current in the valence band is produced by valence electrons, it is called *hole current* to distinguish it from electron current in the conduction band.

As you have seen, conduction in semiconductors is considered to be either the movement of free electrons in the conduction band or the movement of holes in the valence band, which is actually the movement of valence electrons to nearby atoms, creating hole current in the opposite direction.

It is interesting to contrast the two types of charge movement in a semiconductor with the charge movement in a metallic conductor, such as copper. Copper atoms form a different type of crystal in which the atoms are not covalently bonded to each other but consist of a "sea" of positive ion cores, which are atoms stripped of their valence electrons. The valence electrons are attracted to the positive ions, keeping the positive ions together and forming the **metallic bond**. The valence electrons do not belong to a given atom, but to the crystal as a whole. Since the valence electrons in copper are free to move, the application of a voltage results in current. There is only one type of current—the movement of free electrons—because there are no "holes" in the metallic crystal structure.

► FIGURE 1–16

Hole current in intrinsic silicon.

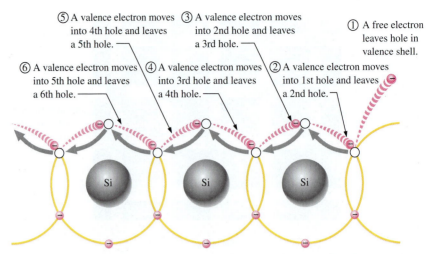

⑤ A valence electron moves into 4th hole and leaves a 5th hole.

③ A valence electron moves into 2nd hole and leaves a 3rd hole.

① A free electron leaves hole in valence shell.

⑥ A valence electron moves into 5th hole and leaves a 6th hole.

④ A valence electron moves into 3rd hole and leaves a 4th hole.

② A valence electron moves into 1st hole and leaves a 2nd hole.

When a valence electron moves left to right to fill a hole while leaving another hole behind, the hole has effectively moved from right to left. Gray arrows indicate effective movement of a hole.

SECTION 1–3 CHECKUP

1. Are free electrons in the valence band or in the conduction band?
2. Which electrons are responsible for electron current in silicon?
3. What is a hole?
4. At what energy level does hole current occur?

1–4 *N*-TYPE AND *P*-TYPE SEMICONDUCTORS

Semiconductive materials do not conduct current well and are of limited value in their intrinsic state. This is because of the limited number of free electrons in the conduction band and holes in the valence band. Intrinsic silicon (or germanium) must be modified by increasing the number of free electrons or holes to increase its conductivity and make it useful in electronic devices. This is done by adding impurities to the intrinsic material. Two types of extrinsic (impure) semiconductive materials, *n*-type and *p*-type, are the key building blocks for most types of electronic devices.

After completing this section, you should be able to

❑ **Describe the properties of *n*-type and *p*-type semiconductors**
   ◆ Define *doping*
❑ Explain how *n*-type semiconductors are formed
   ◆ Describe a majority carrier and minority carrier in *n*-type material
❑ Explain how *p*-type semiconductors are formed
   ◆ Describe a majority carrier and minority carrier in *p*-type material

Since semiconductors are generally poor conductors, their conductivity can be drastically increased by the controlled addition of impurities to the intrinsic (pure) semiconductive material. This process, called **doping**, increases the number of current carriers (electrons or holes). The two categories of impurities are *n*-type and *p*-type.

N-Type Semiconductor

To increase the number of conduction-band electrons in intrinsic silicon, **pentavalent** impurity atoms are added. These are atoms with five valence electrons such as arsenic (As), phosphorus (P), bismuth (Bi), and antimony (Sb).

As illustrated in Figure 1–17, each pentavalent atom (antimony, in this case) forms co-valent bonds with four adjacent silicon atoms. Four of the antimony atom's valence elec-trons are used to form the covalent bonds with silicon atoms, leaving one extra electron. This extra electron becomes a conduction electron because it is not involved in bonding. Because the pentavalent atom gives up an electron, it is often called a *donor atom.* The number of conduction electrons can be carefully controlled by the number of impurity atoms added to the silicon. A conduction electron created by this doping process does not leave a hole in the valence band because it is in excess of the number required to fill the valence band.

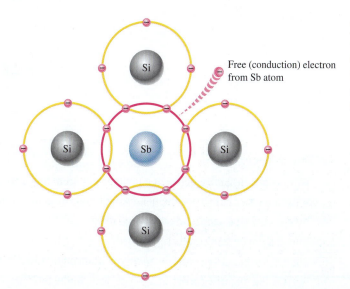

Free (conduction) electron from Sb atom

◀ **FIGURE 1–17**

Pentavalent impurity atom in a sili-con crystal structure. An antimony (Sb) impurity atom is shown in the center. The extra electron from the Sb atom becomes a free electron.

Majority and Minority Carriers   Since most of the current carriers are electrons, silicon (or germanium) doped with pentavalent atoms is an *n*-type semiconductor (the *n* stands for the negative charge on an electron). The electrons are called the **majority carriers** in *n*-type material. Although the majority of current carriers in *n*-type material are electrons, there are also a few holes that are created when electron-hole pairs are thermally generated. These holes are *not* produced by the addition of the pentavalent impurity atoms. Holes in an *n*-type material are called **minority carriers.**

P-Type Semiconductor

To increase the number of holes in intrinsic silicon, **trivalent** impurity atoms are added. These are atoms with three valence electrons such as boron (B), indium (In), and gallium (Ga). As illustrated in Figure 1–18, each trivalent atom (boron, in this case) forms covalent bonds with four adjacent silicon atoms. All three of the boron atom's valence electrons are used in the covalent bonds; and, since four electrons are required, a hole results when each trivalent atom is added. Because the trivalent atom can take an electron, it is often referred to as an *acceptor atom.* The number of holes can be carefully controlled by the number of trivalent impurity atoms added to the silicon. A hole created by this doping process is *not* accompanied by a conduction (free) electron.

Majority and Minority Carriers   Since most of the current carriers are holes, silicon (or germanium) doped with trivalent atoms is called a *p*-type semiconductor. The holes are the majority carriers in *p*-type material. Although the majority of current carriers in *p*-type material are holes, there are also a few conduction-band electrons that are created when electron-hole pairs are thermally generated. These conduction-band electrons are *not* pro-duced by the addition of the trivalent impurity atoms. Conduction-band electrons in *p*-type material are the minority carriers.

▶ **FIGURE 1–18**

Trivalent impurity atom in a silicon crystal structure. A boron (B) impurity atom is shown in the center.

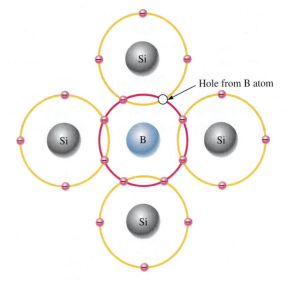

Hole from B atom

SECTION 1–4 CHECKUP

1. Define *doping*.
2. What is the difference between a pentavalent atom and a trivalent atom?
3. What are other names for the pentavalent and trivalent atoms?
4. How is an *n*-type semiconductor formed?
5. How is a *p*-type semiconductor formed?
6. What is the majority carrier in an *n*-type semiconductor?
7. What is the majority carrier in a *p*-type semiconductor?
8. By what process are the majority carriers produced?
9. By what process are the minority carriers produced?
10. What is the difference between intrinsic and extrinsic semiconductors?

1–5 THE *PN* JUNCTION

When you take a block of silicon and dope part of it with a trivalent impurity and the other part with a pentavalent impurity, a boundary called the *pn* junction is formed between the resulting *p*-type and *n*-type portions. The *pn* junction is the basis for diodes, certain transistors, solar cells, and other devices, as you will learn later.

After completing this section, you should be able to

❑ **Describe how a *pn* junction is formed**
  ◆ Discuss diffusion across a *pn* junction
❑ Explain the formation of the depletion region
  ◆ Define *barrier potential* and discuss its significance ◆ State the values of barrier potential in silicon and germanium
❑ Discuss energy diagrams
  ◆ Define *energy hill*

A *p*-type material consists of silicon atoms and trivalent impurity atoms such as boron. The boron atom adds a hole when it bonds with the silicon atoms. However, since the number of protons and the number of electrons are equal throughout the material, there is no net charge in the material and so it is neutral.

An *n*-type silicon material consists of silicon atoms and pentavalent impurity atoms such as antimony. As you have seen, an impurity atom releases an electron when it bonds with four silicon atoms. Since there is still an equal number of protons and electrons (including the free electrons) throughout the material, there is no net charge in the material and so it is neutral.

If a piece of intrinsic silicon is doped so that part is *n*-type and the other part is *p*-type, a ***pn* junction** forms at the boundary between the two regions and a diode is created, as indicated in Figure 1–19(a). The *p* region has many holes (majority carriers) from the impurity atoms and only a few thermally generated free electrons (minority carriers). The *n* region has many free electrons (majority carriers) from the impurity atoms and only a few thermally generated holes (minority carriers).

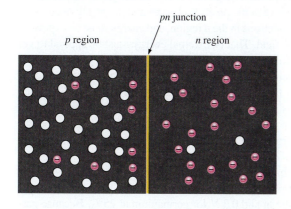

(a) The basic silicon structure at the instant of junction formation showing only the majority and minority carriers. Free electrons in the *n* region near the *pn* junction begin to diffuse across the junction and fall into holes near the junction in the *p* region.

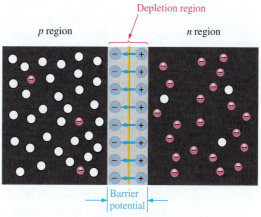

(b) For every electron that diffuses across the junction and combines with a hole, a positive charge is left in the *n* region and a negative charge is created in the *p* region, forming a barrier potential. This action continues until the voltage of the barrier repels further diffusion. The blue arrows between the positive and negative charges in the depletion region represent the electric field.

▲ **FIGURE 1–19**

Formation of the depletion region. The width of the depletion region is exaggerated for illustration purposes.

Formation of the Depletion Region

The free electrons in the *n* region are randomly drifting in all directions. At the instant of the *pn* junction formation, the free electrons near the junction in the *n* region begin to diffuse across the junction into the *p* region where they combine with holes near the junction, as shown in Figure 1–19(b).

Before the *pn* junction is formed, recall that there are as many electrons as protons in the *n*-type material, making the material neutral in terms of net charge. The same is true for the *p*-type material.

When the *pn* junction is formed, the *n* region loses free electrons as they diffuse across the junction. This creates a layer of positive charges (pentavalent ions) near the junction. As the electrons move across the junction, the *p* region loses holes as the electrons and holes combine. This creates a layer of negative charges (trivalent ions) near the junction. These two layers of positive and negative charges form the **depletion region,** as shown in Figure 1–19(b). The term *depletion* refers to the fact that the region near the *pn* junction is depleted of charge carriers (electrons and holes) due to diffusion across the junction. Keep in mind that the depletion region is formed very quickly and is very thin compared to the *n* region and *p* region.

After the initial surge of free electrons across the *pn* junction, the depletion region has expanded to a point where equilibrium is established and there is no further diffusion of

HISTORY NOTE

Russell Ohl, working at Bell Labs in 1940, stumbled on the semiconductor *pn* junction. Ohl was working with a silicon sample that had an accidental crack down its middle. He was using an ohmmeter to test the electrical resistance of the sample when he noted that when the sample was exposed to light, the current between the two sides of the crack made a significant jump. This discovery was fundamental to the work of the team that invented the transistor in 1947.

electrons across the junction. This occurs as follows: As electrons continue to diffuse across the junction, more and more positive and negative charges are created near the junction as the depletion region is formed. A point is reached where the total negative charge in the depletion region repels any further diffusion of electrons (negatively charged particles) into the *p* region (like charges repel) and the diffusion stops. In other words, the depletion region acts as a barrier to the further movement of electrons across the junction.

Barrier Potential Any time there is a positive charge and a negative charge near each other, there is a force acting on the charges as described by Coulomb's law. In the depletion region there are many positive charges and many negative charges on opposite sides of the *pn* junction. The forces between the opposite charges form an *electric field,* as illustrated in Figure 1–19(b) by the blue arrows between the positive charges and the negative charges. This electric field is a barrier to the free electrons in the *n* region, and energy must be expended to move an electron through the electric field. That is, external energy must be applied to get the electrons to move across the barrier of the electric field in the depletion region.

The potential difference of the electric field across the depletion region is the amount of voltage required to move electrons through the electric field. This potential difference is called the barrier potential and is expressed in volts. Stated another way, a certain amount of voltage equal to the barrier potential and with the proper polarity must be applied across a *pn* junction before electrons will begin to flow across the junction. You will learn more about this when we discuss *biasing* in Chapter 2.

The barrier potential of a *pn* junction depends on several factors, including the type of semiconductive material, the amount of doping, and the temperature. The typical barrier potential is approximately 0.7 V for silicon and 0.3 V for germanium at 25°C. Because germanium devices are not widely used, silicon will be assumed throughout the rest of the book.

Energy Diagrams of the *PN* Junction and Depletion Region

The valence and conduction bands in an *n*-type material are at slightly lower energy levels than the valence and conduction bands in a *p*-type material. Recall that *p*-type material has trivalent impurities and *n*-type material has pentavalent impurities. The trivalent impurities exert lower forces on the outer-shell electrons than the pentavalent impurities. The lower forces in *p*-type materials mean that the electron orbits are slightly larger and hence have greater energy than the electron orbits in the *n*-type materials.

An energy diagram for a *pn* junction at the instant of formation is shown in Figure 1–20(a). As you can see, the valence and conduction bands in the *n* region are at lower energy levels than those in the *p* region, but there is a significant amount of overlapping.

The free electrons in the *n* region that occupy the upper part of the conduction band in terms of their energy can easily diffuse across the junction (they do not have to gain additional energy) and temporarily become free electrons in the lower part of the *p*-region conduction band. After crossing the junction, the electrons quickly lose energy and fall into the holes in the *p*-region valence band as indicated in Figure 1–20(a).

As the diffusion continues, the depletion region begins to form and the energy level of the *n*-region conduction band decreases. The decrease in the energy level of the conduction band in the *n* region is due to the loss of the higher-energy electrons that have diffused across the junction to the *p* region. Soon, there are no electrons left in the *n*-region conduction band with enough energy to get across the junction to the *p*-region conduction band, as indicated by the alignment of the top of the *n*-region conduction band and the bottom of the *p*-region conduction band in Figure 1–20(b). At this point, the junction is at equilibrium; and the depletion region is complete because diffusion has ceased. There is an energy gradient across the depletion region which acts as an "energy hill" that an *n*-region electron must climb to get to the *p* region.

Notice that as the energy level of the *n*-region conduction band has shifted downward, the energy level of the valence band has also shifted downward. It still takes the same amount of energy for a valence electron to become a free electron. In other words, the energy gap between the valence band and the conduction band remains the same.

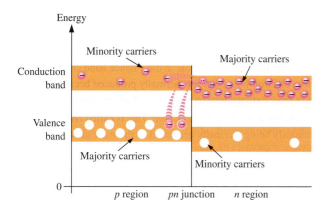

(a) At the instant of junction formation

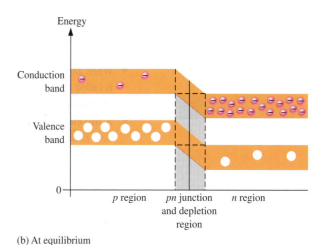

(b) At equilibrium

▲ FIGURE 1–20

Energy diagrams illustrating the formation of the *pn* junction and depletion region.

SECTION 1–5 CHECKUP	1. What is a *pn* junction?
	2. Explain diffusion.
	3. Describe the depletion region.
	4. Explain what the barrier potential is and how it is created.
	5. What is the typical value of the barrier potential for a silicon diode?
	6. What is the typical value of the barrier potential for a germanium diode?

SUMMARY

Section 1–1
◆ According to the classical Bohr model, the atom is viewed as having a planetary-type structure with electrons orbiting at various distances around the central nucleus.

◆ According to the quantum model, electrons do not exist in precise circular orbits as particles as in the Bohr model. The electrons can be waves or particles and precise location at any time is uncertain.

◆ The nucleus of an atom consists of protons and neutrons. The protons have a positive charge and the neutrons are uncharged. The number of protons is the atomic number of the atom.

◆ Electrons have a negative charge and orbit around the nucleus at distances that depend on their energy level. An atom has discrete bands of energy called *shells* in which the electrons orbit. Atomic structure allows a certain maximum number of electrons in each shell. In their natural state, all atoms are neutral because they have an equal number of protons and electrons.

◆ The outermost shell or band of an atom is called the *valence band,* and electrons that orbit in this band are called *valence electrons.* These electrons have the highest energy of all those in the atom. If a valence electron acquires enough energy from an outside source, it can jump out of the valence band and break away from its atom.

Section 1–2
◆ Insulating materials have very few free electrons and do not conduct current under normal circumstances.

◆ Materials that are conductors have a large number of free electrons and conduct current very well.

◆ Semiconductive materials fall in between conductors and insulators in their ability to conduct current.

◆ Semiconductor atoms have four valence electrons. Silicon is the most widely used semiconductive material.

◆ Semiconductor atoms bond together in a symmetrical pattern to form a solid material called a *crystal.* The bonds that hold the type of crystal used in semiconductors are called *covalent bonds.*

9. In a semiconductor crystal, the atoms are held together by

   (a) the interaction of valence electrons   (b) forces of attraction

   (c) covalent bonds   (d) answers (a), (b), and (c)

10. The atomic number of silicon is

    (a) 8   (b) 2   (c) 4   (d) 14

11. The atomic number of germanium is

    (a) 8   (b) 2   (c) 4   (d) 32

12. The valence shell in a silicon atom has the number designation of

    (a) 0   (b) 1   (c) 2   (d) 3

13. Each atom in a silicon crystal has

    (a) four valence electrons

    (b) four conduction electrons

    (c) eight valence electrons, four of its own and four shared

    (d) no valence electrons because all are shared with other atoms

Section 1–3   14. Electron-hole pairs are produced by

    (a) recombination   (b) thermal energy   (c) ionization   (d) doping

15. Recombination is when

    (a) an electron falls into a hole

    (b) a positive and a negative ion bond together

    (c) a valence electron becomes a conduction electron

    (d) a crystal is formed

16. The current in a semiconductor is produced by

    (a) electrons only   (b) holes only   (c) negative ions   (d) both electrons and holes

Section 1–4   17. In an intrinsic semiconductor,

    (a) there are no free electrons

    (b) the free electrons are thermally produced

    (c) there are only holes

    (d) there are as many electrons as there are holes

    (e) answers (b) and (d)

18. The process of adding an impurity to an intrinsic semiconductor is called

    (a) doping   (b) recombination   (c) atomic modification   (d) ionization

19. A trivalent impurity is added to silicon to create

    (a) germanium   (b) a *p*-type semiconductor

    (c) an *n*-type semiconductor   (d) a depletion region

20. The purpose of a pentavalent impurity is to

    (a) reduce the conductivity of silicon   (b) increase the number of holes

    (c) increase the number of free electrons   (d) create minority carriers

21. The majority carriers in an *n*-type semiconductor are

    (a) holes   (b) valence electrons   (c) conduction electrons   (d) protons

22. Holes in an *n*-type semiconductor are

    (a) minority carriers that are thermally produced

    (b) minority carriers that are produced by doping

    (c) majority carriers that are thermally produced

    (d) majority carriers that are produced by doping

Section 1–5   23. A *pn* junction is formed by

    (a) the recombination of electrons and holes

    (b) ionization

    (c) the boundary of a *p*-type and an *n*-type material

    (d) the collision of a proton and a neutron

24. The depletion region is created by

    (a) ionization   **(b)** diffusion   **(c)** recombination   **(d)** answers (a), (b), and (c)

25. The depletion region consists of

    (a) nothing but minority carriers   **(b)** positive and negative ions

    (c) no majority carriers   **(d)** answers (b) and (c)

PROBLEMS

Answers to all odd-numbered problems are at the end of the book.

BASIC PROBLEMS

Section 1–1   **The Atom**

   1. Describe the Bohr model of the atom.

   2. What is a shell in the atomic structure?

   3. If the atomic number of a neutral atom is 6, how many electrons does the atom have? How many protons?

   4. What is the maximum number of electrons that can exist in the third shell of an atom?

Section 1–2   **Materials Used in Electronic Devices**

   5. For each of the energy diagrams in Figure 1–21, determine the class of material based on relative comparisons.

   6. A certain atom has four valence electrons. What type of atom is it?

   7. In a silicon crystal, how many covalent bonds does a single atom form?

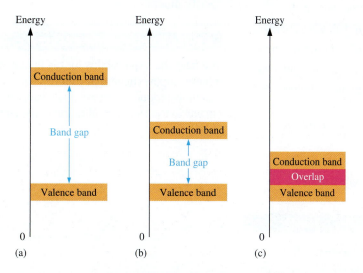

◀ **FIGURE 1–21**

Section 1–3   **Current in Semiconductors**

   8. What happens when heat is added to silicon?

   9. Name the two energy bands at which current is produced in silicon.

   10. Why can't there be hole current in the conduction band?

   11. What is a metallic bond?

Section 1–4   ***N*-Type and *P*-Type Semiconductors**

   12. Describe the process of doping and explain how it alters the atomic structure of silicon.

   13. What is antimony? What is boron?

Section 1–5   **The *PN* Junction**

   14. How is the electric field across the *pn* junction created?

   15. Because of its barrier potential, can a diode be used as a voltage source? Explain.

2

DIODES AND APPLICATIONS

CHAPTER OBJECTIVES

◆ Use a diode in common applications
◆ Analyze the voltage-current (*V-I*) characteristic of a diode
◆ Explain how the three diode approximations differ
◆ Explain and analyze the operation of half-wave rectifiers
◆ Explain and analyze the operation of full-wave rectifiers
◆ Explain and analyze power supply filters and regulators
◆ Explain and analyze the operation of diode limiters and clampers
◆ Explain and analyze the operation of diode voltage multipliers
◆ Interpret and use diode datasheets
◆ Troubleshoot diodes and power supply circuits

KEY TERMS

◆ Diode
◆ Bias
◆ Forward bias
◆ Reverse bias
◆ *V-I* characteristic
◆ DC power supply
◆ Rectifier
◆ Filter
◆ Regulator

◆ Half-wave rectifier
◆ Peak inverse voltage (PIV)
◆ Full-wave rectifier
◆ Ripple voltage
◆ Line regulation
◆ Load regulation
◆ Limiter
◆ Clamper
◆ Troubleshooting

VISIT THE WEBSITE

Study aids, Multisim files, and LT Spice files for this chapter are available at https://www.pearsonhighered.com /careersresources/

INTRODUCTION

In Chapter 1, you learned that many semiconductor devices are based on the *pn* junction. In this chapter, the operation and characteristics of the diode are covered. Also, three diode models representing three levels of approximation are presented and testing is discussed. The importance of the diode in electronic circuits cannot be overemphasized. Its ability to conduct current in one direction while blocking current in the other direction is essential to the operation of many types of circuits. One circuit in particular is the ac rectifier, which is covered in this chapter. Other important applications are circuits such as diode limiters, diode clampers, and diode voltage multipliers. Datasheets are discussed for specific diodes.

DEVICE APPLICATION PREVIEW

You have the responsibility for the final design and testing of a power supply circuit that your company plans to use in several of its products. You will apply your knowledge of diode circuits to the Device Application at the end of the chapter.

2–1 DIODE OPERATION

A modern diode is a two-terminal semiconductor device formed by two doped regions of silicon separated by a *pn* junction. In this chapter, the most common category of diode, known as the general-purpose diode, is covered. Other descriptors, such as rectifier diode or signal diode, are used depending on the particular application for which the diode was designed. You will learn how to use a voltage to cause the diode to conduct current in one direction and block it in the other direction. This process is called *biasing*.

After completing this section, you should be able to

❑ Recognize the electrical symbol for a diode and several diode package configurations
❑ Apply forward bias to a diode
  ◆ Define *forward bias* and state the required conditions   ◆ Discuss the effect of forward bias on the depletion region   ◆ Explain how the barrier potential affects the forward bias.
❑ Reverse-bias a diode
  ◆ Define *reverse bias* and state the required conditions   ◆ Discuss reverse current and reverse breakdown

The Diode

As mentioned, a **diode** is made from a small piece of semiconductor material, usually silicon, in which half is doped as a *p* region and half is doped as an *n* region with a *pn* junction and depletion region in between. The *p* region is called the **anode** and is connected to a conductive terminal. The *n* region is called the **cathode** and is connected to a second conductive terminal. The basic diode structure and schematic symbol are shown in Figure 2–1.

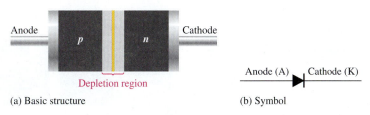

(a) Basic structure   (b) Symbol

◀ FIGURE 2–1

The diode.

Typical Diode Packages   Several common physical configurations of through-hole mounted diodes are illustrated in Figure 2–2(a). The anode (A) and cathode (K) are indicated on a diode in several ways, depending on the type of package. The cathode is usually marked by a band, a tab, or some other feature. On those packages where one lead is connected to the case, the case is the cathode.

Surface-Mount Diode Packages   Figure 2–2(b) shows typical diode packages for surface mounting on a printed circuit board. The SOD and SOT packages have gull-wing shaped leads. The SMA package has L-shaped leads that bend under the package. The SOD and SMA types have a band on one end to indicate the cathode. The SOT type is a three-terminal package in which there are either one or two diodes. In a single-diode SOT package, pin 1 is usually the anode and pin 3 is the cathode. In a dual-diode SOT package, pin 3 is the common terminal and can be either the anode or the cathode. Always check the datasheet for the particular diode to verify the pin configurations.

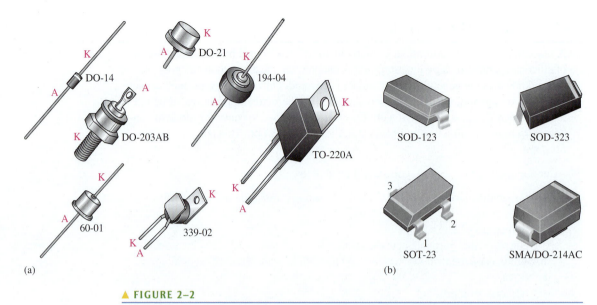

(a)　　　　　　　　　　　　　　　　　　　　(b)

▲ FIGURE 2–2

Typical diode packages with terminal identification. The letter K is used for cathode to avoid confusion with certain electrical quantities that are represented by C. Case type numbers are indicated for each diode.

Forward Bias

To **bias** a diode, you apply a dc voltage across it. **Forward bias** is the condition that allows current through the *pn* junction. Figure 2–3 shows a dc voltage source connected by conductive material (contacts and wire) across a diode in the direction to produce forward bias. This external bias voltage is designated as V_{BIAS}. The resistor limits the forward current to a value that will not damage the diode. Notice that the negative side of V_{BIAS} is connected to the *n* region of the diode and the positive side is connected to the *p* region. This is one requirement for forward bias. A second requirement is that the bias voltage, V_{BIAS}, must be greater than the barrier potential (V_B).

▶ FIGURE 2–3

A diode connected for forward bias.

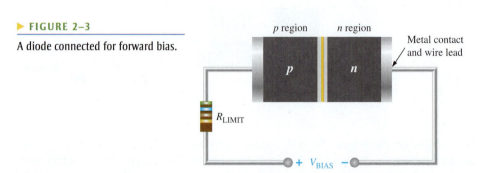

A fundamental picture of what happens when a diode is forward-biased is shown in Figure 2–4. Because like charges repel, the negative side of the bias-voltage source "pushes" the free electrons, which are the majority carriers in the *n* region, toward the *pn* junction. This flow of free electrons is called *electron current*. The negative side of the source also provides a continuous flow of electrons through the external connection (conductor) and into the *n* region as shown.

The bias-voltage source imparts sufficient energy to the free electrons for them to overcome the barrier potential of the depletion region and move on through into the *p* region. Once in the *p* region, these conduction electrons have lost enough energy to immediately combine with holes in the valence band.

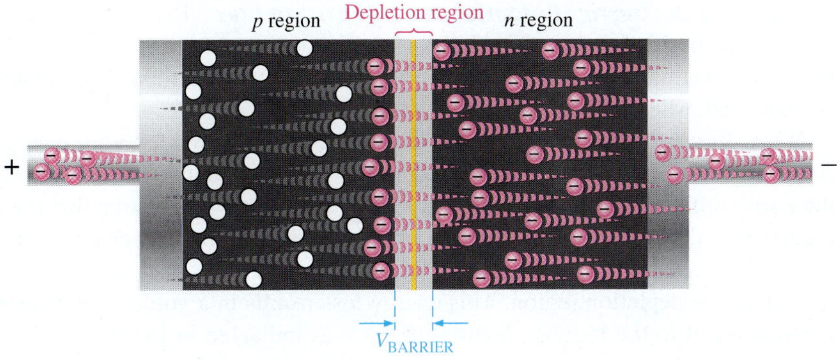

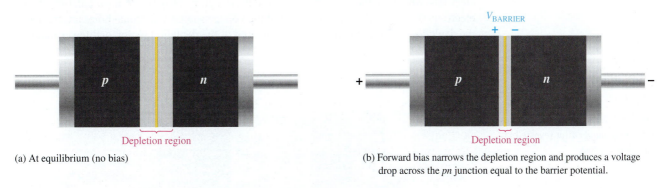

▶ FIGURE 2–4

A forward-biased diode showing the flow of majority carriers and the voltage due to the barrier potential across the depletion region.

Now, the electrons are in the valence band in the *p* region, simply because they have lost too much energy overcoming the barrier potential to remain in the conduction band. Since unlike charges attract, the positive side of the bias-voltage source attracts the valence electrons toward the left end of the *p* region. The holes in the *p* region provide the medium or "pathway" for these valence electrons to move through the *p* region. The valence electrons move from one hole to the next toward the left. The holes, which are the majority carriers in the *p* region, effectively (not actually) move to the right toward the junction, as you can see in Figure 2–4. This *effective* flow of holes is the hole current. You can also view the hole current as being created by the flow of valence electrons through the *p* region, with the holes providing the only means for these electrons to flow.

As the electrons flow out of the *p* region through the external connection (conductor) and to the positive side of the bias-voltage source, they leave holes behind in the *p* region; at the same time, these electrons become conduction electrons in the metal conductor. Recall that the conduction band in a conductor overlaps the valence band so that it takes much less energy for an electron to be a free electron in a conductor than in a semiconductor and that metallic conductors do not have holes in their structure. There is a continuous availability of holes effectively moving toward the *pn* junction to combine with the continuous stream of electrons as they come across the junction into the *p* region.

The Effect of Forward Bias on the Depletion Region

As electrons from the *n* side are pushed into the depletion region, they combine with holes on the *p* side, effectively reducing the depletion region. This process during forward bias causes the depletion region to narrow, as indicated in Figure 2–5(b).

(a) At equilibrium (no bias)

(b) Forward bias narrows the depletion region and produces a voltage drop across the *pn* junction equal to the barrier potential.

▲ **FIGURE 2–5**

The depletion region narrows and a voltage drop is produced across the *pn* junction when the diode is forward-biased.

The Effect of the Barrier Potential During Forward Bias   Recall that the electric field between the positive and negative sides of the junction creates an "energy hill" that prevents free electrons from diffusing across the junction at equilibrium. This creates the barrier potential, which in silicon is approximately 0.7 V.

When forward bias is applied, the free electrons are provided with enough energy from the bias-voltage source to overcome the barrier potential and effectively "climb the energy hill" and cross the depletion region. The energy per charge that the electrons require in order to cross the depletion region is equal to the barrier potential. In other words, the electrons give up an amount of energy equivalent to the barrier potential when they cross the depletion region. This energy loss results in a voltage drop across the *pn* junction equal to the barrier potential (0.7 V), as indicated in Figure 2–5(b). An additional small voltage drop occurs across the *p* and *n* regions due to the internal resistance of the material. For doped semiconductive material, this resistance, called the **dynamic resistance,** is very small and can usually be neglected. This is discussed in more detail in Section 2–2.

Reverse Bias

Reverse bias is the condition that essentially prevents current through the diode. Figure 2–6 shows a dc voltage source connected across a diode in the direction to produce reverse bias. This external bias voltage is designated as V_{BIAS} just as it was for forward bias. Notice that the positive side of V_{BIAS} is connected to the *n* region of the diode and the negative side is connected to the *p* region. Also note that the depletion region is shown much wider than in forward bias or equilibrium.

▶ **FIGURE 2–6**

A diode connected for reverse bias. A limiting resistor is shown although it is not important in reverse bias because there is essentially no current.

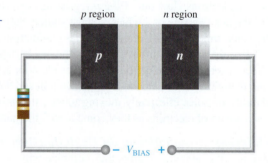

An illustration of what happens when a diode is reverse-biased is shown in Figure 2–7. Because unlike charges attract, the positive side of the bias-voltage source "pulls" the free electrons, which are the majority carriers in the *n* region, away from the *pn* junction. As the electrons flow toward the positive side of the voltage source, additional holes are created at the depletion region. This results in a widening of the depletion region and fewer majority carriers.

▶ **FIGURE 2–7**

The diode during the short transition time immediately after reverse-bias voltage is applied.

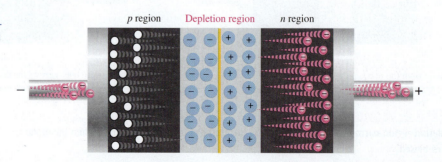

In the p region, electrons from the negative side of the voltage source enter as valence electrons and move from hole to hole toward the depletion region where they create additional negative charge. This results in a widening of the depletion region and a depletion of majority carriers. The flow of valence electrons can be viewed as holes being "pulled" toward the positive side.

The initial flow of charge carriers is transitional and lasts for only a very short time after the reverse-bias voltage is applied. As the depletion region widens, the availability of majority carriers decreases. As more of the n and p regions become depleted of majority carriers, the electric field increases in strength until the potential across the depletion region equals the bias voltage, V_{BIAS}. At this point, the transition current essentially ceases except for a very small reverse current that can usually be neglected.

Reverse Current   The extremely small current that exists in reverse bias after the transition current dies out is caused by the minority carriers in the n and p regions that are produced by thermally generated electron-hole pairs. The small number of free minority electrons in the p region are "pushed" toward the pn junction by the negative bias voltage. When these electrons reach the wide depletion region, they "fall down the energy hill" and combine with the minority holes in the n region as valence electrons and flow toward the positive bias voltage, creating a small hole current.

The conduction band in the p region is at a higher energy level than the conduction band in the n region. Therefore, the minority electrons easily pass through the depletion region because they require no additional energy. Reverse current is illustrated in Figure 2–8.

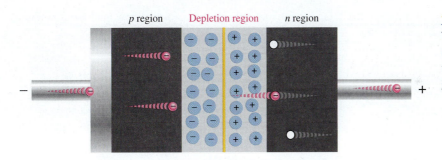

p region   Depletion region   n region

◄ **FIGURE 2–8**

The extremely small reverse current in a reverse-biased diode is due to the minority carriers from thermally generated electron-hole pairs.

Reverse Breakdown   Normally, the reverse current is so small that it can be neglected. However, if the external reverse-bias voltage is increased to a value called the *breakdown voltage,* the reverse current will drastically increase.

This is what happens. The high reverse-bias voltage imparts energy to the free minority electrons so that as they speed through the p region, they collide with atoms with enough energy to knock valence electrons into the conduction band. The newly created conduction electrons are also high in energy and repeat the process. If one electron knocks only two others out of their valence orbit during its travel through the p region, the numbers quickly multiply. As these high-energy electrons go through the depletion region, they have enough energy to go through the n region as conduction electrons, rather than combining with holes.

The multiplication of conduction electrons just discussed is known as the **avalanche effect,** and reverse current can increase dramatically if steps are not taken to limit the current. When the reverse current is not limited, the resulting heating will permanently damage the diode. Most diodes are not operated in reverse breakdown, but if the current is limited (by adding a series-limiting resistor for example), there is no permanent damage to the diode.

SECTION 2–1
CHECK-UP
Answers can be found at
www.pearsonhighered.com/
floyd.

1. Describe forward bias of a diode.
2. Explain how to forward-bias a diode.
3. Describe reverse bias of a diode.
4. Explain how to reverse-bias a diode.
5. Compare the depletion regions in forward bias and reverse bias.
6. Which bias condition produces majority carrier current?
7. How is reverse current in a diode produced?
8. When does reverse breakdown occur in a diode?
9. Define *avalanche effect* as applied to diodes.

2–2 VOLTAGE-CURRENT CHARACTERISTIC OF A DIODE

As you have learned, forward bias produces current through a diode and reverse bias essentially prevents current, except for a negligible reverse current. Reverse bias prevents current as long as the reverse-bias voltage does not equal or exceed the breakdown voltage of the junction. In this section, we will examine the relationship between the voltage and the current in a diode on a graphical basis.

After completing this section, you should be able to

❏ **Analyze the voltage-current (*V-I*) characteristic of a diode**
❏ Explain the *V-I* characteristic for forward bias
  ◆ Graph the *V-I* curve for forward bias ◆ Describe how the barrier potential affects the *V-I* curve ◆ Define *dynamic resistance*
❏ Explain the *V-I* characteristic for reverse bias
  ◆ Graph the *V-I* curve for reverse bias
❏ Discuss the complete *V-I* characteristic curve
  ◆ Describe the effects of temperature on the diode characteristic

V-I Characteristic for Forward Bias

When a forward-bias voltage is applied across a diode, there is current. This current is called the *forward current* and is designated I_F. Figure 2–9 illustrates what happens as the forward-bias voltage is increased positively from 0 V. The resistor is used to limit the forward current to a value that will not overheat the diode and cause damage.

With 0 V across the diode, there is no forward current. As you gradually increase the forward-bias voltage, the forward current *and* the voltage across the diode gradually increase, as shown in Figure 2–9(a). A portion of the forward-bias voltage is dropped across the limiting resistor. When the forward-bias voltage is increased to a value where the voltage across the diode reaches approximately 0.7 V (barrier potential), the forward current begins to increase rapidly, as illustrated in Figure 2–9(b).

As you continue to increase the forward-bias voltage, the current continues to increase very rapidly, but the voltage across the diode increases only gradually above 0.7 V. This small increase in the diode voltage above the barrier potential is due to the voltage drop across the internal dynamic resistance of the semiconductive material.

Graphing the V-I Curve If you plot the results of the type of measurements shown in Figure 2–9 on a graph, you get the ***V-I characteristic*** curve for a forward-biased diode, as shown in Figure 2–10(a). The diode forward voltage (V_F) increases to the right along the horizontal axis, and the forward current (I_F) increases upward along the vertical axis.

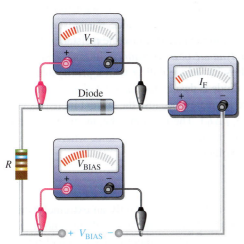

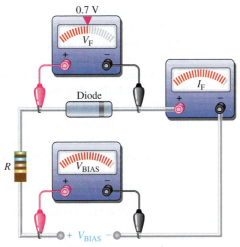

(a) Small forward-bias voltage ($V_F < 0.7$ V), very small forward current.

(b) Forward voltage reaches and remains nearly constant at approximately 0.7 V. Forward current continues to increase as the bias voltage is increased.

▲ **FIGURE 2–9**

Forward-bias measurements show general changes in V_F and I_F as V_{BIAS} is increased.

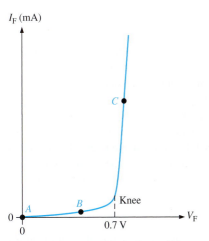

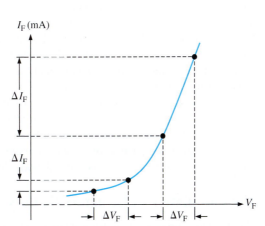

◀ **FIGURE 2–10**

Relationship of voltage and current in a forward-biased diode.

(a) *V-I* characteristic curve for forward bias.

(b) Expanded view of a portion of the curve in part (a). The dynamic resistance r'_d decreases as you move up the curve, as indicated by the decrease in the value of $\Delta V_F / \Delta I_F$.

As you can see in Figure 2–10(a), the forward current increases very little until the forward voltage across the *pn* junction reaches approximately 0.7 V at the knee of the curve. After this point, the forward voltage remains nearly constant at approximately 0.7 V, but I_F increases rapidly. As previously mentioned, there is a slight increase in V_F above 0.7 V as the current increases due mainly to the voltage drop across the dynamic resistance. The I_F scale is typically in mA, as indicated.

Three points *A*, *B*, and *C* are shown on the curve in Figure 2–10(a). Point *A* corresponds to a zero-bias condition. Point *B* corresponds to Figure 2–10(a) where the forward voltage is less than the barrier potential of 0.7 V. Point *C* corresponds to Figure 2–10(a) where the forward voltage *approximately* equals the barrier potential. As the external bias voltage and forward current continue to increase above the knee, the forward voltage will increase slightly above 0.7 V. In reality, the forward voltage can be as much as approximately 1 V, depending on the forward current.

Bias Connections

Forward-Bias Recall that a diode is forward-biased when a voltage source is connected as shown in Figure 2–14(a). The positive terminal of the source is connected to the anode through a current-limiting resistor. The negative terminal of the source is connected to the cathode. The forward electron flow current (I_F) is from cathode to anode as indicated. The forward voltage drop (V_F) due to the barrier potential is from positive at the anode to negative at the cathode.

▶ **FIGURE 2–14**

Forward-bias and reverse-bias connections showing the diode symbol.

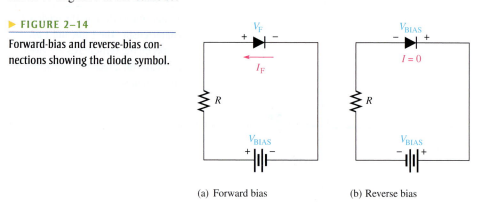

(a) Forward bias (b) Reverse bias

Reverse-Bias Connection A diode is reverse-biased when a voltage source is connected as shown in Figure 2–14(b). The negative terminal of the source is connected to the anode side of the circuit, and the positive terminal is connected to the cathode side. A resistor is not necessary in reverse bias but it is shown for circuit consistency. The reverse current is extremely small and can be considered to be zero. Notice that the entire bias voltage (V_{BIAS}) appears across the diode.

Diode Approximations

The Ideal Diode Model The ideal model of a diode is the least accurate approximation and can be represented by a simple switch. When the diode is forward-biased, it ideally acts like a closed (on) switch, as shown in Figure 2–15(a). When the diode is reverse-biased, it

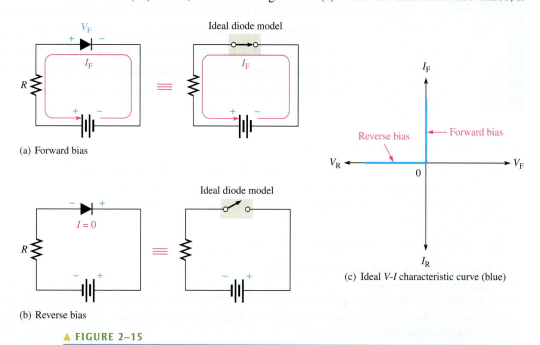

(a) Forward bias

(b) Reverse bias

(c) Ideal *V-I* characteristic curve (blue)

▲ **FIGURE 2–15**

The ideal model of a diode.

ideally acts like an open (off) switch, as shown in part (b). Although the barrier potential, the forward dynamic resistance, and the reverse current are all neglected, this model is adequate for most troubleshooting when you are trying to determine if the diode is working properly.

In Figure 2–15(c), the ideal *V-I* characteristic curve graphically depicts the ideal diode operation. Since the barrier potential and the forward dynamic resistance are neglected, the diode is assumed to have a zero voltage across it when forward-biased, as indicated by the portion of the curve on the positive vertical axis.

$$V_F = 0 \text{ V}$$

The forward current is determined by the bias voltage and the limiting resistor using Ohm's law.

$$I_F = \frac{V_{BIAS}}{R_{LIMIT}}$$

Equation 2–1

Since the reverse current is neglected, its value is assumed to be zero, as indicated in Figure 2–15(c) by the portion of the curve on the negative horizontal axis.

$$I_R = 0 \text{ A}$$

The reverse voltage equals the bias voltage.

$$V_R = V_{BIAS}$$

You may want to use the ideal model when you are troubleshooting or trying to figure out the operation of a circuit and are not concerned with more exact values of voltage or current.

The Practical Diode Model   The practical model includes the barrier potential. When the diode is forward-biased, it is equivalent to a closed switch in series with a small equivalent voltage source (V_F) equal to the barrier potential (0.7 V) with the positive side toward the anode, as indicated in Figure 2–16(a). This equivalent voltage source represents the barrier potential that must be exceeded by the bias voltage before the diode will conduct and is not an active source of voltage. When conducting, a voltage drop of 0.7 V appears across the diode.

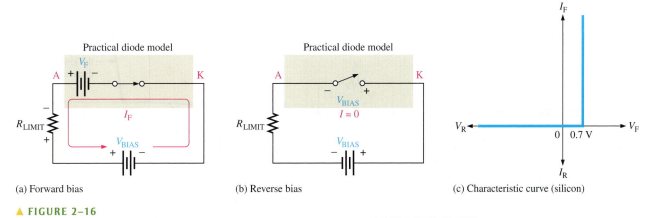

(a) Forward bias   (b) Reverse bias   (c) Characteristic curve (silicon)

▲ **FIGURE 2–16**

The practical model of a diode.

When the diode is reverse-biased, it is equivalent to an open switch just as in the ideal model, as shown in Figure 2–16(b). The barrier potential does not affect reverse bias, so it is not a factor.

The characteristic curve for the practical diode model is shown in Figure 2–16(c). Since the barrier potential is included and the dynamic resistance is neglected, the diode is assumed to have a voltage across it when forward-biased, as indicated by the portion of the curve to the right of the origin.

$$V_F = 0.7 \text{ V}$$

(b) Ideal model:

$$I_R = 0\text{ A}$$
$$V_R = V_{BIAS} = 10\text{ V}$$
$$V_{R_{LIMIT}} = 0\text{ V}$$

Practical model:

$$I_R = 0\text{ A}$$
$$V_R = V_{BIAS} = 10\text{ V}$$
$$V_{R_{LIMIT}} = 0\text{ V}$$

Complete model:

$$I_R = 1\ \mu\text{A}$$
$$V_{R_{LIMIT}} = I_R R_{LIMIT} = (1\ \mu\text{A})\,(1.0\text{ k}\Omega) = 1\text{ mV}$$
$$V_R = V_{BIAS} - V_{R_{LIMIT}} = 10\text{ V} - 1\text{ mV} = 9.999\text{ V}$$

Related Problem*   Assume that the diode in Figure 2–18(a) fails open. What is the voltage across the diode and the voltage across the limiting resistor?

* Answers can be found at www.pearsonhighered.com/floyd.

 Open the Multisim file EXM02-01 or LT Spice file EXS02-01 in the Examples folder on the website. Measure the voltages across the diode and the resistor in both circuits and compare with the calculated results in this example.

SECTION 2–3 CHECKUP

1. What are the two conditions under which a diode is operated?
2. Under what condition is a general-purpose diode never intentionally operated?
3. What is the simplest way to visualize a diode?
4. To more accurately represent a diode, what factors must be included?
5. Which diode model represents the most accurate approximation?

2–4   HALF-WAVE RECTIFIERS

Because of their ability to conduct current in one direction and block current in the other direction, diodes are used in circuits called rectifiers that convert ac voltage into dc voltage. Rectifiers are found in all dc power supplies that operate from an ac voltage source. A power supply is an essential part of each electronic system from the simplest to the most complex.

After completing this section, you should be able to

❑ **Explain and analyze the operation of half-wave rectifiers**
❑ Describe a basic dc power supply
❑ Discuss half-wave rectification
   ◆ Determine the average value of a half-wave voltage
❑ Explain how the barrier potential affects a half-wave rectifier output
   ◆ Calculate the output voltage
❑ Define *peak inverse voltage*
❑ Explain the operation of a transformer-coupled rectifier

The Basic DC Power Supply

All active electronic devices require a source of constant dc that can be supplied by a battery or a dc power supply. The **dc power supply** converts the North American standard 120 V, 60 Hz ac voltage available at wall outlets into a constant dc voltage. The dc power supply is one of the most common circuits you will find, so it is important to understand how it works. The output dc voltage is used to power most electronic circuits, including consumer electronics, computers, industrial controllers, and laboratory instrumentation systems and equipment. The dc voltage level required depends on the application, but most applications require relatively low dc voltages.

A basic block diagram of a complete power supply is shown in Figure 2–19(a). Generally the ac input line voltage is stepped down to a lower ac voltage with a transformer (although it may be stepped up when higher voltages are needed or there may be no transformer at all in rare instances). As you learned in your dc/ac course, a **transformer** changes ac voltages based on the turns ratio between the primary and secondary. If the secondary has more turns than the primary, the output voltage across the secondary will be higher and the current will be smaller. If the secondary has fewer turns than the primary, the output voltage across the secondary will be lower and the current will be higher. The rectifier can be either a half-wave rectifier or a full-wave rectifier (covered in Section 2–5). The **rectifier** converts the ac input voltage to a pulsating dc voltage, called a half-wave rectified voltage, as shown in Figure 2–19(b). The **filter** eliminates the fluctuations in the rectified voltage and produces a relatively smooth dc voltage. The power supply filter is covered in Section 2–6. The **regulator** is a circuit that maintains a constant dc voltage for variations in the input line voltage or in the load. Regulators vary from a single semiconductor device to more complex integrated circuits. The load is a circuit or device connected to the output of the power supply and operates from the power supply voltage and current.

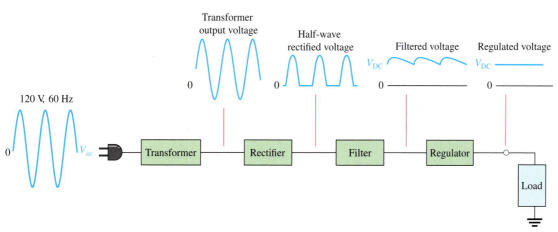

(a) Complete power supply with transformer, rectifier, filter, and regulator

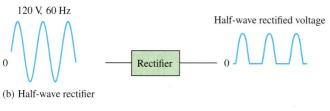

(b) Half-wave rectifier

▲ **FIGURE 2–19**

Block diagram of a dc power supply with a load and a rectifier.

Center-Tapped Full-Wave Rectifier Operation

A **center-tapped rectifier** is a type of full-wave rectifier that uses two diodes connected to the secondary of a center-tapped transformer, as shown in Figure 2–31. The input voltage is coupled through the transformer to the center-tapped secondary. Half of the total secondary voltage appears between the center tap and each end of the secondary winding as shown.

▶ **FIGURE 2–31**

A center-tapped full-wave rectifier.

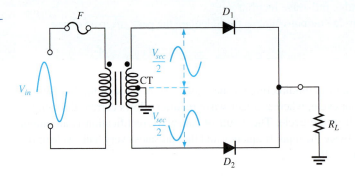

For a positive half-cycle of the input voltage, the polarities of the secondary voltages are as shown in Figure 2–32(a). This condition forward-biases diode D_1 and reverse-biases diode D_2. The current path is through D_1 and the load resistor R_L, as indicated. For a negative half-cycle of the input voltage, the voltage polarities on the secondary are as shown in Figure 2–32(b). This condition reverse-biases D_1 and forward-biases D_2. The current path is through D_2 and R_L, as indicated. Because the output current during both the positive and negative portions of the input cycle is in the same direction through the load, the output voltage developed across the load resistor is a full-wave rectified dc voltage, as shown.

▶ **FIGURE 2–32**

Basic operation of a center-tapped full-wave rectifier. Note that the current through the load resistor is in the same direction during the entire input cycle, so the output voltage always has the same polarity.

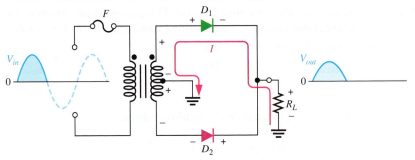

(a) During positive half-cycles, D_1 is forward-biased and D_2 is reverse-biased.

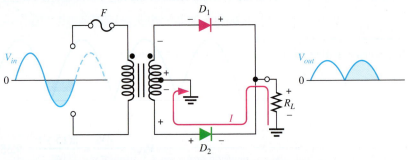

(b) During negative half-cycles, D_2 is forward-biased and D_1 is reverse-biased.

Effect of the Turns Ratio on the Output Voltage If the transformer's turns ratio is 1, the peak value of the rectified output voltage equals half the peak value of the primary input voltage less the barrier potential, as illustrated in Figure 2–33. Half of the primary voltage

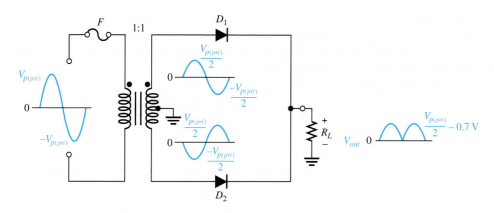

Center-tapped full-wave rectifier with a transformer turns ratio of 1. $V_{p(pri)}$ is the peak value of the primary voltage.

appears across each half of the secondary winding ($V_{p(sec)} = V_{p(pri)}$). We will begin referring to the forward voltage due to the barrier potential as the **diode drop.**

In order to obtain an output voltage with a peak equal to the input peak (less the diode drop), a step-up transformer with a turns ratio of $n = 2$ must be used, as shown in Figure 2–34. In this case, the total secondary voltage (V_{sec}) is twice the primary voltage ($2V_{pri}$), so the voltage across each half of the secondary is equal to V_{pri}.

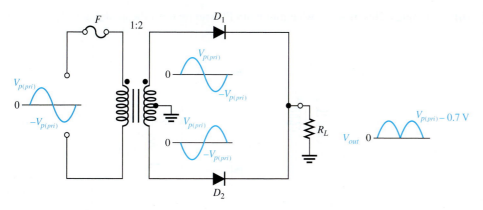

Center-tapped full-wave rectifier with a transformer turns ratio of 2.

In any case, the output voltage of a center-tapped full-wave rectifier is always one-half of the total secondary voltage less the diode drop, no matter what the turns ratio.

$$V_{out} = \frac{V_{sec}}{2} - 0.7 \text{ V}$$

Equation 2–7

Peak Inverse Voltage Each diode in the full-wave rectifier is alternately forward-biased and then reverse-biased. The maximum reverse voltage that each diode must withstand is the peak secondary voltage $V_{p(sec)}$. This is shown in Figure 2–35 where D_2 is assumed to be reverse-biased (red) and D_1 is assumed to be forward-biased (green) to illustrate the concept.

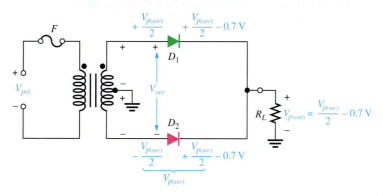

Diode reverse voltage (D_2 shown reverse-biased and D_1 shown forward-biased).

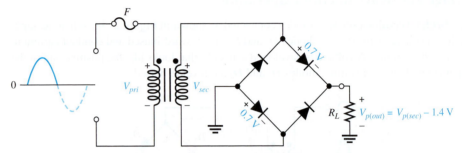

(a) Ideal diodes

(b) Practical diodes (Diode drops included)

▲ FIGURE 2–39

Bridge operation during a positive half-cycle of the primary and secondary voltages.

Peak Inverse Voltage Let's assume that D_1 and D_2 are forward-biased and examine the reverse voltage across D_3 and D_4. Visualizing D_1 and D_2 as shorts (ideal model), as in Figure 2–40(a), you can see that D_3 and D_4 have a peak inverse voltage equal to the peak secondary voltage. Since the output voltage is *ideally* equal to the secondary voltage,

$$\text{PIV} = V_{p(out)}$$

If the diode drops of the forward-biased diodes are included as shown in Figure 2–40(b), the peak inverse voltage across each reverse-biased diode in terms of $V_{p(out)}$ is

Equation 2–10

$$\textbf{PIV} = V_{p(out)} + \textbf{0.7 V}$$

The PIV rating of the bridge diodes is less than that required for the center-tapped configuration. If the diode drop is neglected, the bridge rectifier requires diodes with half the PIV rating of those in a center-tapped rectifier for the same output voltage.

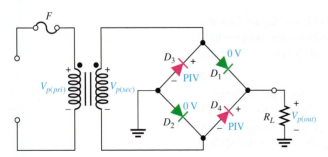

(a) For the ideal diode model (forward-biased diodes D_1 and D_2 are shown in green), PIV = $V_{p(out)}$.

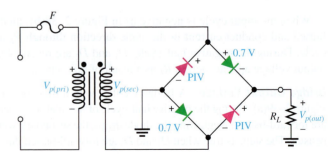

(b) For the practical diode model (forward-biased diodes D_1 and D_2 are shown in green), PIV = $V_{p(out)}$ + 0.7 V.

▲ FIGURE 2–40

Peak inverse voltages across diodes D_3 and D_4 in a bridge rectifier during the positive half-cycle of the secondary voltage.

EXAMPLE 2–7

Determine the peak output voltage for the bridge rectifier in Figure 2–41. Assuming the practical model, what PIV rating is required for the diodes? The transformer is specified to have a 12 V rms secondary voltage for the standard 120 V across the primary.

▶ **FIGURE 2–41**

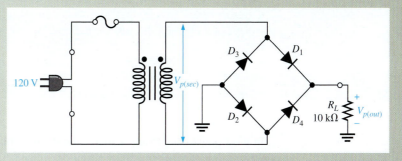

Solution

The peak output voltage (taking into account the two diode drops) is

$$V_{p(sec)} = 1.414 V_{rms} = 1.414(12 \text{ V}) \cong 17 \text{ V}$$
$$V_{p(out)} = V_{p(sec)} - 1.4 \text{ V} = 17 \text{ V} - 1.4 \text{ V} = \mathbf{15.6 \text{ V}}$$

The PIV for each diode is

$$\text{PIV} = V_{p(out)} + 0.7 \text{ V} = 15.6 \text{ V} + 0.7 \text{ V} = \mathbf{16.3 \text{ V}}$$

The PIV rating must exceed this value.

Related Problem

Determine the peak output voltage for the bridge rectifier in Figure 2–41 if the transformer produces an rms secondary voltage of 30 V. What is the minimum PIV rating for the diodes?

Open the Multisim file EXM02-07 or LT Spice file EXS02-07 in the Examples folder on the website. Measure the output voltage and compare to the calculated value.

SECTION 2–5 CHECKUP

1. How does a full-wave voltage differ from a half-wave voltage?

2. What is the average value of a full-wave rectified voltage with a peak value of 60 V?

3. Which type of full-wave rectifier has the greater output voltage for the same input voltage and transformer turns ratio?

4. For a peak output voltage of 45 V, in which type of rectifier would you use diodes with a PIV rating of 50 V?

5. What PIV rating is required for diodes used in the type of rectifier that was not selected in Question 4?

2–6 POWER SUPPLY FILTERS AND REGULATORS

A power supply filter ideally eliminates the fluctuations in the output voltage of a half-wave or full-wave rectifier and produces a constant-level dc voltage. Filtering is necessary because electronic circuits require a constant source of dc voltage and current to provide power and biasing for proper operation. Filters are implemented with capacitors, as you will see in this section. Voltage regulation in power supplies is usually done with integrated circuit voltage regulators. A voltage regulator prevents changes in the filtered dc voltage due to variations in input voltage or load.

In most power supply applications, the standard 60 Hz ac power line voltage must be converted to an approximately constant dc voltage. The 60 Hz pulsating dc output of a half-wave rectifier or the 120 Hz pulsating output of a full-wave rectifier must be filtered to reduce the large voltage variations. Figure 2–42 illustrates the filtering concept showing a nearly smooth dc output voltage from the filter. The small amount of fluctuation in the filter output voltage is called *ripple*.

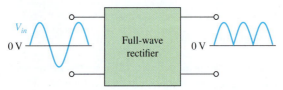

(a) Rectifier without a filter

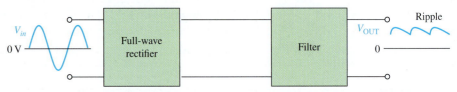

(b) Rectifier with a filter (output ripple is exaggerated)

▲ **FIGURE 2–42**

Power supply filtering.

Capacitor-Input Filter

A half-wave rectifier with a capacitor-input filter is shown in Figure 2–43. The filter is simply a capacitor connected from the rectifier output to ground. R_L represents the equivalent resistance of a load. We will use the half-wave rectifier to illustrate the basic principle and then expand the concept to full-wave rectification.

During the positive first quarter-cycle of the input, the diode is forward-biased, allowing the capacitor to charge to within 0.7 V of the input peak, as illustrated in Figure 2–43(a). When the input begins to decrease below its peak, as shown in part (b), the capacitor retains its charge and the diode becomes reverse-biased because the cathode is more positive than the anode. During the remaining part of the cycle, the capacitor can discharge only through the load resistance at a rate determined by the $R_L C$ time constant, which is normally long compared to the period of the input. The larger the time constant, the less the capacitor will discharge. During the first quarter of the next cycle, as illustrated in part (c), the diode will again become forward-biased when the input voltage exceeds the capacitor voltage by approximately 0.7 V.

SAFETY NOTE

When installing polarized capacitors in a circuit, be sure to observe the proper polarity. The positive lead always connects to the more positive side of the circuit. An incorrectly connected polarized capacitor can explode.

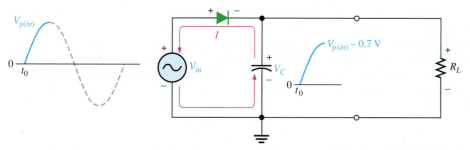

(a) Initial charging of the capacitor (diode is forward-biased) happens only once when power is turned on.

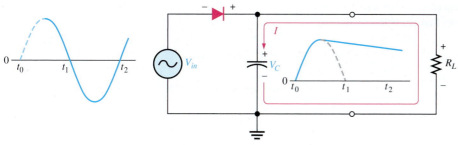

(b) The capacitor discharges through R_L after peak of positive alternation when the diode is reverse-biased. This discharging occurs during the portion of the input voltage indicated by the solid dark blue curve.

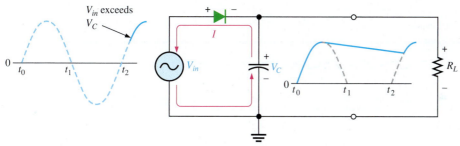

(c) The capacitor charges back to peak of input when the diode becomes forward-biased. This charging occurs during the portion of the input voltage indicated by the solid dark blue curve.

▲ **FIGURE 2–43**

Operation of a half-wave rectifier with a capacitor-input filter. The current indicates charging or discharging of the capacitor.

Ripple Voltage As you have seen, the capacitor quickly charges at the beginning of a cycle and slowly discharges through R_L after the positive peak of the input voltage (when the diode is reverse-biased). The variation in the capacitor voltage due to the charging and discharging is called the **ripple voltage**. Generally, ripple is undesirable; thus, the smaller the ripple, the better the filtering action, as illustrated in Figure 2–44.

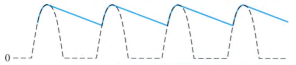

(a) Larger ripple (blue) means less effective filtering.

(b) Smaller ripple means more effective filtering. Generally, the larger the capacitor value, the smaller the ripple for the same input and load.

▲ **FIGURE 2–44**

Half-wave ripple voltage (blue line).

For a given input frequency, the output frequency of a full-wave rectifier is twice that of a half-wave rectifier, as illustrated in Figure 2–45. This makes a full-wave rectifier easier to filter because of the shorter time between peaks. When filtered, the full-wave rectified voltage has a smaller ripple than does a half-wave voltage for the same load resistance and capacitor values. The capacitor discharges less during the shorter interval between full-wave pulses, as shown in Figure 2–46.

▶ **FIGURE 2–45**

The period of a full-wave rectified voltage is half that of a half-wave rectified voltage. The output frequency of a full-wave rectifier is twice that of a half-wave rectifier.

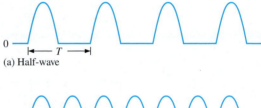

(a) Half-wave

(b) Full-wave

▶ **FIGURE 2–46**

Comparison of ripple voltages for half-wave and full-wave rectified voltages with the same filter capacitor and load and derived from the same sinusoidal input voltage.

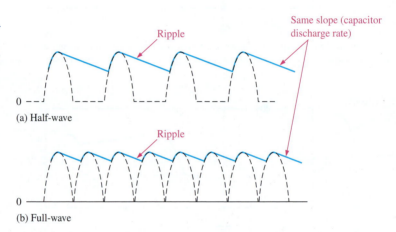

(a) Half-wave

(b) Full-wave

Ripple Factor The **ripple factor (r)** is an indication of the effectiveness of the filter and is defined as

Equation 2–11

$$r = \frac{V_{r(pp)}}{V_{DC}}$$

where $V_{r(pp)}$ is the peak-to-peak ripple voltage and V_{DC} is the dc (average) value of the filter's output voltage, as illustrated in Figure 2–47. The lower the ripple factor, the better the filter. The ripple factor can be lowered by increasing the value of the filter capacitor or increasing the load resistance.

▶ **FIGURE 2–47**

V_r and V_{DC} determine the ripple factor.

For a full-wave rectifier with a capacitor-input filter, approximations for the peak-to-peak ripple voltage, $V_{r(pp)}$, and the dc value of the filter output voltage, V_{DC}, are given in the following equations. The variable $V_{p(rect)}$ is the unfiltered peak rectified voltage. Notice that if R_L or C increases, the ripple voltage decreases and the dc voltage increases.

$$V_{r(pp)} \cong \left(\frac{1}{fR_LC}\right)V_{p(rect)}$$

Equation 2–12

$$V_{DC} \cong \left(1 - \frac{1}{2fR_LC}\right)V_{p(rect)}$$

Equation 2–13

The derivations for these equations can be found in "Derivations of Selected Equations" at www.pearsonhighered.com/floyd.

EXAMPLE 2–8

Determine the ripple factor for the filtered bridge rectifier with a load as indicated in Figure 2–48.

▶ **FIGURE 2–48**

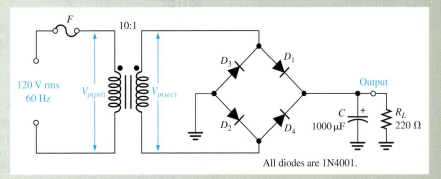

All diodes are 1N4001.

Solution

The transformer turns ratio is $n = 0.1$. The peak primary voltage is

$$V_{p(pri)} = 1.414V_{rms} = 1.414(120\ V) = 170\ V$$

The peak secondary voltage is

$$V_{p(sec)} = nV_{p(pri)} = 0.1(170\ V) = 17.0\ V$$

The unfiltered peak full-wave rectified voltage is

$$V_{p(rect)} = V_{p(sec)} - 1.4\ V = 17.0\ V - 1.4\ V = 15.6\ V$$

The frequency of a full-wave rectified voltage is 120 Hz. The approximate peak-to-peak ripple voltage at the output is

$$V_{r(pp)} \cong \left(\frac{1}{fR_LC}\right)V_{p(rect)} = \left(\frac{1}{(120\ Hz)(220\ \Omega)(1000\ \mu F)}\right)15.6\ V = 0.591\ V$$

The approximate dc value of the output voltage is determined as follows:

$$V_{DC} = \left(1 - \frac{1}{2fR_LC}\right)V_{p(rect)} = \left(1 - \frac{1}{(240\ Hz)(220\ \Omega)(1000\ \mu F)}\right)15.6\ V = 15.3\ V$$

The resulting ripple factor is

$$r = \frac{V_{r(pp)}}{V_{DC}} = \frac{0.591\ V}{15.3\ V} = \mathbf{0.039}$$

The percent ripple is 3.9%.

Related Problem

Determine the peak-to-peak ripple voltage if the filter capacitor in Figure 2–48 is increased to 2200 μF and the load resistance changes to 2.2 kΩ.

Open the Multisim file EXM02-08 or LT Spice file EXS02-08 in the Examples folder on the website. For the specified input voltage, measure the peak-to-peak ripple voltage and the dc value at the output. Do the results agree closely with the calculated values? If not, can you explain why?

3. What causes the ripple voltage on the output of a capacitor-input filter?

4. If the load resistance connected to a filtered power supply is decreased, what happens to the ripple voltage?

5. Define *ripple factor*.

6. What is the difference between input (line) regulation and load regulation?

2–7 DIODE LIMITERS AND CLAMPERS

Diode circuits, called limiters or clippers, are sometimes used to clip off portions of signal voltages above or below certain levels. Another type of diode circuit, called a clamper, is used to add or restore a dc level to an electrical signal. Both limiter and clamper diode circuits will be examined in this section.

After completing this section, you should be able to

❑ **Explain and analyze the operation of diode limiters and clampers**
❑ Describe the operation of a diode limiter
  ◆ Discuss biased limiters ◆ Discuss voltage-divider bias ◆ Describe an application
❑ Describe the operation of a diode clamper

Diode Limiters

Figure 2–52(a) shows a diode positive **limiter** (also called **clipper**) that limits or clips the positive part of the input voltage. As the input voltage goes positive, the diode becomes forward-biased and conducts current. Point *A* is limited to +0.7 V when the input voltage exceeds this

▶ **FIGURE 2–52**

Examples of diode limiters (clippers).

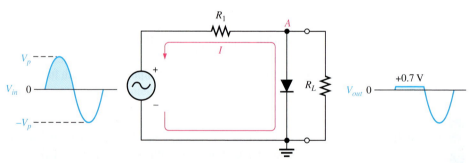

(a) Limiting of the positive alternation. The diode is forward-biased during the positive alternation (above 0.7 V) and reverse-biased during the negative alternation.

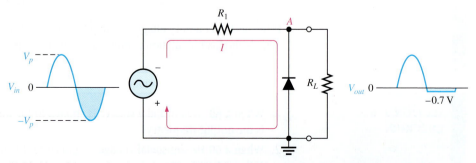

(b) Limiting of the negative alternation. The diode is forward-biased during the negative alternation (below −0.7 V) and reverse-biased during the positive alternation.

value. When the input voltage goes back below 0.7 V, the diode is reverse-biased and appears as an open. The output voltage looks like the negative part of the input voltage, but with a magnitude determined by the voltage divider formed by R_1 and the load resistor, R_L, as follows:

$$V_{out} = \left(\frac{R_L}{R_1 + R_L} \right) V_{in}$$

If R_1 is small compared to R_L, then $V_{out} \cong V_{in}$.

   If the diode is turned around, as in Figure 2–52(b), the negative part of the input voltage is clipped off. When the diode is forward-biased during the negative part of the input voltage, point A is held at -0.7 V by the diode drop. When the input voltage goes above -0.7 V, the diode is no longer forward-biased; and a voltage appears across R_L proportional to the input voltage.

EXAMPLE 2–10

What would you expect to see displayed on an oscilloscope connected across R_L in the limiter shown in Figure 2–53?

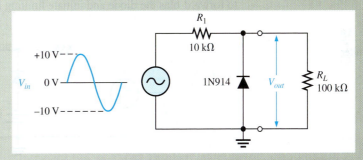

▲ FIGURE 2–53

Solution   The diode is forward-biased and conducts when the input voltage goes below -0.7 V. So, for the negative limiter, determine the peak output voltage across R_L by the following equation:

$$V_{p(out)} = \left(\frac{R_L}{R_1 + R_L} \right) V_{p(in)} = \left(\frac{100 \text{ k}\Omega}{110 \text{ k}\Omega} \right) 10 \text{ V} = 9.09 \text{ V}$$

The scope will display an output waveform as shown in Figure 2–54.

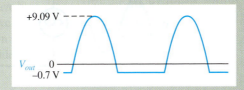

▲ FIGURE 2–54

Output voltage waveform for Figure 2–53.

Related Problem   Describe the output waveform for Figure 2–53 if R_1 is changed to 1 kΩ.

   Open the Multisim file EXM02-10 or LT Spice file EXS02-10 in the Examples folder on the website. For the specified input, measure the resulting output waveform. Compare with the waveform shown in the example.

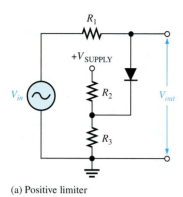

(a) Positive limiter

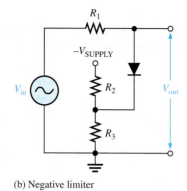

(b) Negative limiter

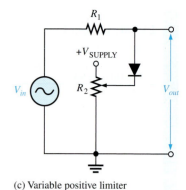

(c) Variable positive limiter

▲ FIGURE 2–60

Diode limiters implemented with voltage-divider bias.

EXAMPLE 2–12

Describe the output voltage waveform for the diode limiter in Figure 2–61.

▶ FIGURE 2–61

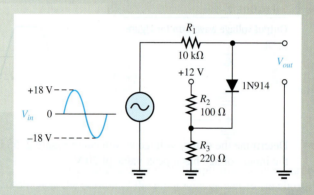

Solution   The circuit is a positive limiter. Use the voltage-divider formula to determine the bias voltage.

$$V_{BIAS} = \left(\frac{R_3}{R_2 + R_3}\right)V_{SUPPLY} = \left(\frac{220\ \Omega}{100\ \Omega + 220\ \Omega}\right)12\ V = 8.25\ V$$

The output voltage waveform is shown in Figure 2–62. The positive part of the output voltage waveform is limited to $V_{BIAS} + 0.7\ V$.

▶ FIGURE 2–62

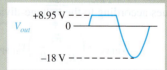

Related Problem   How would you change the voltage divider in Figure 2–61 to limit the output voltage to +6.7 V?

Open the Multisim file EXM02-12 or LT Spice file EXS02-12 in the Examples folder on the website. Observe the output voltage on the oscilloscope and compare to the calculated result.

Diode Clampers

A clamper adds a dc level to an ac voltage. **Clampers** are sometimes known as *dc restorers*. Figure 2–63 shows a diode clamper that inserts a positive dc level in the output waveform. The operation of this circuit can be seen by considering the first negative half-cycle of the input voltage. When the input voltage initially goes negative, the diode is forward-biased, allowing the capacitor to charge to near the peak of the input ($V_{p(in)} - 0.7$ V), as shown in Figure 2–63(a). Just after the negative peak, the diode is reverse-biased. This is because the cathode is held near $V_{p(in)} - 0.7$ V by the charge on the capacitor. The capacitor can only discharge through the high resistance of R_L. So, from the peak of one negative half-cycle to the next, the capacitor discharges very little. The amount that is discharged, of course, depends on the value of R_L.

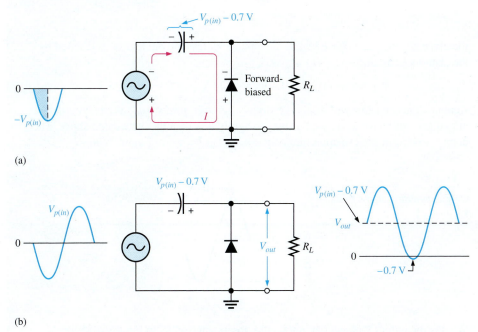

◀ **FIGURE 2–63**

Positive clamper operation.

(a)

(b)

If the capacitor discharges during the period of the input wave, clamping action is affected. If the *RC* time constant is 100 times the period, the clamping action is excellent. An *RC* time constant of ten times the period will have a small amount of distortion at the ground level due to the charging current.

The net effect of the clamping action is that the capacitor retains a charge approximately equal to the peak value of the input less the diode drop. The capacitor voltage acts essentially as a battery in series with the input voltage. The dc voltage of the capacitor adds to the input voltage by superposition, as in Figure 2–63(b).

If the diode is turned around, a negative dc voltage is added to the input voltage to produce the output voltage as shown in Figure 2–64.

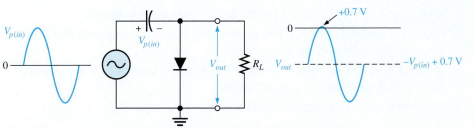

◀ **FIGURE 2–64**

Negative clamper.

EXAMPLE 2–13

What is the output voltage that you would expect to observe across R_L in the clamping circuit of Figure 2–65? Assume that RC is long compared to the period to prevent significant capacitor discharge.

▶ FIGURE 2–65

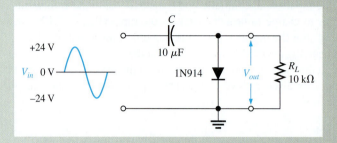

Solution   Ideally, a negative dc value equal to the input peak less the diode drop is inserted by the clamping circuit.

$$V_{DC} \cong -(V_{p(in)} - 0.7 \text{ V}) = -(24 \text{ V} - 0.7 \text{ V}) = \mathbf{-23.3\ V}$$

Actually, the capacitor will discharge slightly between peaks, and, as a result, the output voltage will have an average value of slightly less than that calculated above. The output waveform goes to approximately +0.7 V, as shown in Figure 2–66.

▶ FIGURE 2–66

Output waveform across R_L for Figure 2–65.

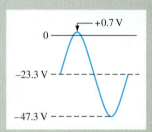

Related Problem   What is the output voltage that you would observe across R_L in Figure 2–65 for $C = 22\ \mu\text{F}$ and $R_L = 18\ \text{k}\Omega$?

Open the Multisim file EXM02-13 or LT Spice file EXS02-13 in the Examples folder on the website. For the specified input, measure the output waveform. Compare with the waveform shown in the example.

SECTION 2–7 CHECKUP

1. Discuss how diode limiters and diode clampers differ in terms of their function.

2. What is the difference between a positive limiter and a negative limiter?

3. What is the maximum voltage across an unbiased positive silicon diode limiter during the positive alternation of the input voltage?

4. To limit the output voltage of a positive limiter to 5 V when a 10 V peak input is applied, what value must the bias voltage be?

5. What component in a clamping circuit effectively acts as a battery?

2–8 VOLTAGE MULTIPLIERS

Voltage multipliers use clamping action to increase peak rectified voltages without the necessity of increasing the transformer's voltage rating. Multiplication factors of two, three, and four are common. Voltage multipliers are used in high-voltage, low-current applications such as cathode-ray tubes (CRTs) and particle accelerators.

After completing this section, you should be able to

❑ **Explain and analyze the operation of diode voltage multipliers**
  ❑ Discuss voltage doublers
    ◆ Explain the half-wave voltage doubler ◆ Explain the full-wave voltage doubler
  ❑ Discuss voltage triplers
  ❑ Discuss voltage quadruplers

Voltage Doubler

Half-Wave Voltage Doubler A voltage doubler is a **voltage multiplier** with a multiplication factor of two. A half-wave voltage doubler is shown in Figure 2–67. During the positive half-cycle of the secondary voltage, diode D_1 is forward-biased and D_2 is reverse-biased. Capacitor C_1 is charged to the peak of the secondary voltage (V_p) less the diode drop with the polarity shown in part (a). During the negative half-cycle, diode D_2 is forward-biased and D_1 is reverse-biased, as shown in part (b). Since C_1 can't discharge, the peak voltage on C_1 adds to the secondary voltage to charge C_2 to approximately $2V_p$. Applying Kirchhoff's law around the loop as shown in part (b), the voltage across C_2 is

$$V_{C1} - V_{C2} + V_p = 0$$
$$V_{C2} = V_p + V_{C1}$$

Neglecting the diode drop of D_2, $V_{C1} = V_p$. Therefore,

$$V_{C2} = V_p + V_p = 2V_p$$

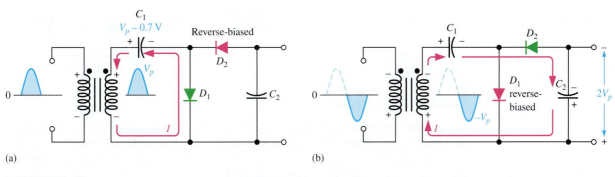

(a)   (b)

▲ **FIGURE 2–67**

Half-wave voltage doubler operation. V_p is the peak secondary voltage.

Under a no-load condition, C_2 remains charged to approximately $2V_p$. If a load resistance is connected across the output, C_2 discharges slightly through the load on the next positive half-cycle and is again recharged to $2V_p$ on the following negative half-cycle. The resulting output is a half-wave, capacitor-filtered voltage. The peak inverse voltage across each diode is $2V_p$. If the diode were reversed, the output voltage across C_2 would have the opposite polarity. In this case, polarized capacitors should also be reversed.

Full-Wave Voltage Doubler   A full-wave doubler is shown in Figure 2–68. When the secondary voltage is positive, D_1 is forward-biased and C_1 charges to approximately V_p, as shown in part (a). During the negative half-cycle, D_2 is forward-biased and C_2 charges to approximately V_p, as shown in part (b). The output voltage, $2V_p$, is taken across the two capacitors in series.

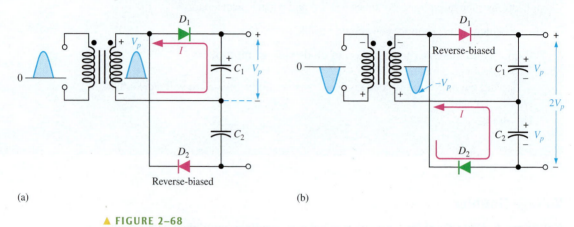

(a)

(b)

▲ **FIGURE 2–68**

Full-wave voltage doubler operation.

Voltage Tripler

The addition of another diode-capacitor section to the half-wave voltage doubler creates a voltage tripler, as shown in Figure 2–69. The operation is as follows: On the positive half-cycle of the secondary voltage, C_1 charges to V_p through D_1. During the negative half-cycle, C_2 charges to $2V_p$ through D_2, as described for the doubler. During the next positive half-cycle, C_3 charges to $2V_p$ through D_3. The tripler output is taken across C_1 and C_3, as shown in the figure.

▶ **FIGURE 2–69**

Voltage tripler.

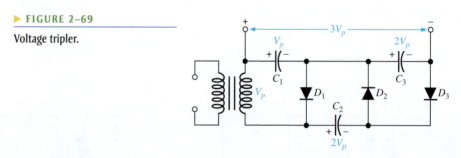

Voltage Quadrupler

The addition of still another diode-capacitor section, as shown in Figure 2–70, produces an output four times the peak secondary voltage. C_4 charges to $2V_p$ through D_4 on a negative half-cycle. The $4V_p$ output is taken across C_2 and C_4, as shown. In both the tripler and quadrupler circuits, the PIV of each diode is $2V_p$.

▶ **FIGURE 2–70**

Voltage quadrupler.

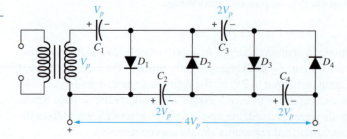

SECTION 2–8
CHECKUP

1. What must be the peak voltage rating of the transformer secondary for a voltage doubler that produces an output of 200 V?
2. The output voltage of a quadrupler is 620 V. What minimum PIV rating must each diode have?

2–9 THE DIODE DATASHEET

A manufacturer's datasheet gives detailed information on a device so that it can be used properly in a given application. A typical datasheet provides maximum ratings, electrical characteristics, mechanical data, and graphs of various parameters.

After completing this section, you should be able to

❏ **Interpret and use diode datasheets**
   ◆ Define several absolute maximum ratings   ◆ Define diode thermal characteristics   ◆ Define several electrical characteristics   ◆ Interpret the forward current derating curve   ◆ Interpret the forward characteristic curve   ◆ Discuss nonrepetitive surge current   ◆ Discuss the reverse characteristics

Figure 2–71 shows a typical rectifier diode datasheet. The presentation of information on datasheets may vary from one manufacturer to another, but they basically all convey the same information. The mechanical information, such as package dimensions, are not shown on this particular datasheet but are generally available from the manufacturer. Notice on this datasheet that there are three categories of data given in table form and four types of characteristics shown in graphical form.

Data Categories

Absolute Maximum Ratings   The absolute maximum ratings indicate the maximum values of the several parameters under which the diode can be operated without damage or degradation. For greatest reliability and longer life, the diode should be operated well under these maximums. Generally, the maximum ratings are specified for an operating ambient temperature (T_A) of 25°C unless otherwise stated. Ambient temperature is the temperature of the air surrounding the device. The parameters given in Figure 2–71 are as follows:

V_{RRM} The peak reverse voltage that can be applied repetitively across the diode. Notice that it is 50 V for the 1N4001 and 1000 V for the 1N4007. This rating is the same as the PIV.

$I_{F(AV)}$ The maximum average value of a 60 Hz half-wave rectified forward current. This current parameter is 1.0 A for all of the diode types and is specified for an ambient temperature of 75°C.

I_{FSM} The maximum peak value of nonrepetitive single half-sine-wave forward surge current with a duration of 8.3 ms. This current parameter is 30 A for all of the diode types.

T_{stg} The allowable range of temperatures at which the device can be kept when not operating or connected to a circuit.

T_J The allowable range of temperatures for the *pn* junction when the diode is operated in a circuit.

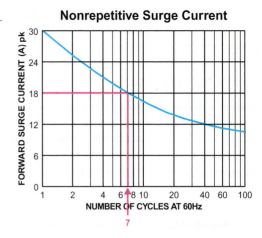

◀ FIGURE 2–74

Reverse Characteristics   This graph from the datasheet shows how the reverse current varies with the reverse voltage for three different junction temperatures. The horizontal axis is the percentage of maximum reverse voltage, V_{RRM}. For example, at 25°C, a 1N4001 has a reverse current of approximately 0.04 μA at 20% of its maximum V_{RRM} or 10 V. If the V_{RRM} is increased to 90%, the reverse current increases to approximately 0.11 μA, as shown in Figure 2–75.

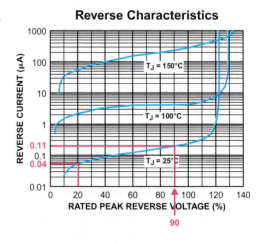

◀ FIGURE 2–75

SECTION 2–9 CHECKUP	
	1. Determine the peak repetitive reverse voltage for each of the following diodes: 1N4002, 1N4003, 1N4004, 1N4005, 1N4006.
	2. If the forward current is 800 mA and the forward voltage is 0.75 V in a 1N4005, is the power rating exceeded?
	3. What is $I_{F(AV)}$ for a 1N4001 at an ambient temperature of 100°C?
	4. What is I_{FSM} for a 1N4003 if the surge is repeated 40 times at 60 Hz?

2–10 TROUBLESHOOTING

This section provides a general overview and application of an approach to troubleshooting. Specific troubleshooting examples of the power supply and diode circuits are covered.

After completing this section, you should be able to

❑ **Troubleshoot diodes and power supply circuits**
❑ Test a diode with a DMM
  ◆ Use the diode test position ◆ Determine if the diode is good or bad
  ◆ Use the Ohms function to check a diode
❑ Troubleshoot a dc power supply by analysis, planning, and measurement
  ◆ Use the half-splitting method
❑ Perform fault analysis
  ◆ Isolate fault to a single component

Testing a Diode

A multimeter can be used as a fast and simple way to check a diode out of the circuit. A good diode will show an extremely high resistance (ideally an open) with reverse bias and a very low resistance with forward bias. A defective open diode will show an extremely high resistance (or open) for both forward and reverse bias. A defective shorted or resistive diode will show zero or a low resistance for both forward and reverse bias. An open diode is the most common type of failure.

The DMM Diode Test Position Many digital multimeters (DMMs) have a diode test function that provides a convenient way to test a diode. A typical DMM, as shown in Figure 2–76, has a small diode symbol to mark the position of the function switch. When set to *diode test*, the meter provides an internal voltage sufficient to forward-bias and reverse-bias a diode. This internal voltage may vary among different makes of DMM, but 2.5 V to 3.5 V is a typical range of values. The meter provides a voltage reading or other indication to show the condition of the diode under test.

When the Diode Is Working In Figure 2–76(a), the red (positive) lead of the meter is connected to the anode and the black (negative) lead is connected to the cathode to forward-bias the diode. If the diode is good, you will get a reading of between approximately 0.5 V and 0.9 V, with 0.7 V being typical for forward bias.

In Figure 2–76(b), the diode is turned around for reverse bias as shown. If the diode is working properly, you will typically get a reading of "OL." Some DMMs may display the internal voltage for a reverse-bias condition.

When the Diode Is Defective When a diode has failed open, you get an out-of-range "OL" indication for both the forward-bias and the reverse-bias conditions, as illustrated in Figure 2–76(c). If a diode is shorted, the meter reads 0 V in both forward- and reverse-bias tests, as indicated in part (d).

Checking a Diode with the OHMs Function DMMs that do not have a diode test position can be used to check a diode by setting the function switch on an OHMs range. For a forward-bias check of a good diode, you will get a resistance reading that can vary depending on the meter's internal battery. Many meters do not have sufficient voltage on the OHMs setting to fully forward-bias a diode and you may get a reading of from several hundred to several thousand ohms. For the reverse-bias check of a good diode, you will get an out-of-range indication such as "OL" on most DMMs because the reverse resistance is too high for the meter to measure.

Even though you may not get accurate forward- and reverse-resistance readings on a DMM, the relative readings indicate that a diode is functioning properly, and that is usually all you need to know. The out-of-range indication shows that the reverse resistance is extremely high, as you expect. The reading of a few hundred to a few thousand ohms

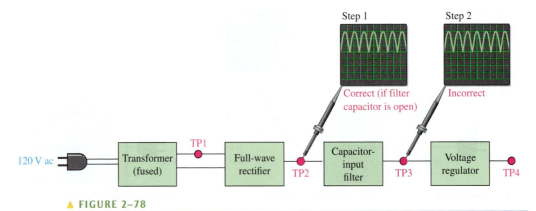

▲ FIGURE 2–78

Example of the half-splitting approach. An open filter capacitor is indicated.

are working properly. This measurement also indicates that the filter capacitor is open, which is verified by the full-wave voltage at TP3. If the filter were working properly, you would measure a dc voltage at both TP2 and TP3. If the filter capacitor were shorted, you would observe no voltage at all of the test points because the fuse would most likely be blown. A short anywhere in the system is very difficult to isolate because, if the system is properly fused, the fuse will blow immediately when a short to ground develops.

For the case illustrated in Figure 2–78, the half-splitting method took two measurements to isolate the fault to the open filter capacitor. If you had started from the transformer output, it would have taken three measurements; and if you had started at the final output, it would have also taken three measurements, as illustrated in Figure 2–79.

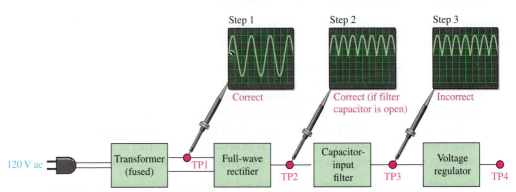

(a) Measurements starting at the transformer output

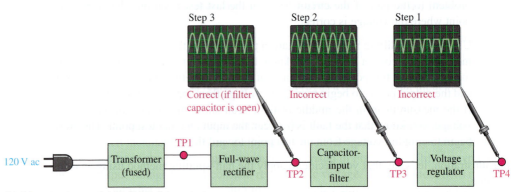

(b) Measurements starting at the regulator output

▲ FIGURE 2–79

In this particular case, the two other approaches require more oscilloscope measurements than the half-splitting approach in Figure 2–78.

Fault Analysis

In some cases, after isolating a fault to a particular circuit, it may be necessary to isolate the problem to a single component in the circuit. In this event, you have to apply logical thinking and your knowledge of the symptoms caused by certain component failures. Some typical component failures and the symptoms they produce are now discussed.

Effect of an Open Diode in a Half-Wave Rectifier A half-wave filtered rectifier with an open diode is shown in Figure 2–80. The resulting symptom is zero output voltage as indicated. This is obvious because the open diode breaks the current path from the transformer secondary winding to the filter and load resistor and there is no load current.

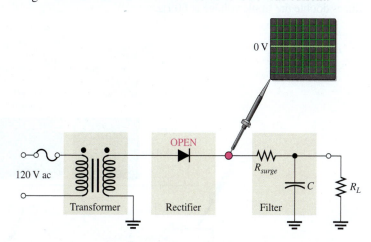

◀ FIGURE 2–80

The effect of an open diode in a half-wave rectifier is an output of 0 V.

Other faults that will cause the same symptom in this circuit are an open transformer winding, an open fuse, or no input voltage.

Effect of an Open Diode in a Full-Wave Rectifier A full-wave center-tapped filtered rectifier is shown in Figure 2–81. If either of the two diodes is open, the output voltage will have twice the normal ripple voltage at 60 Hz rather than at 120 Hz, as indicated.

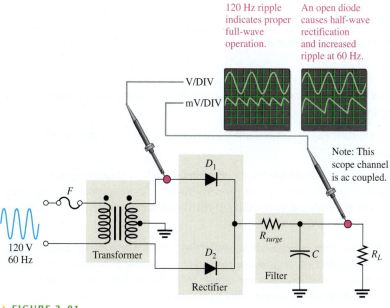

▲ FIGURE 2–81

The effect of an open diode in a center-tapped rectifier is half-wave rectification and twice the ripple voltage at 60 Hz.

Another fault that will cause the same symptom is an open in the transformer secondary winding.

The reason for the increased ripple at 60 Hz rather than at 120 Hz is as follows: If one of the diodes in Figure 2–81 is open, there is current through R_L only during one half-cycle of the input voltage. During the other half-cycle of the input, the open path caused by the open diode prevents current through R_L. The result is half-wave rectification, as shown in Figure 2–81, which produces the larger ripple voltage with a frequency of 60 Hz.

An open diode in a full-wave bridge rectifier will produce the same symptom as in the center-tapped circuit, as shown in Figure 2–82. The open diode prevents current through R_L during half of the input voltage cycle. The result is half-wave rectification, which produces double the ripple voltage at 60 Hz.

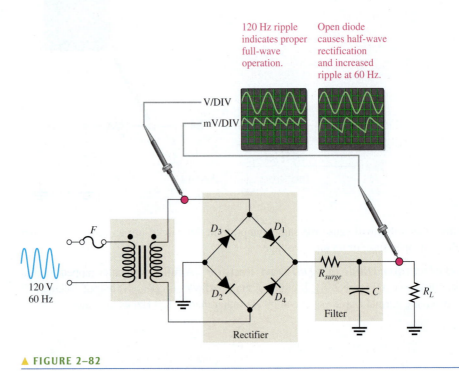

▲ **FIGURE 2–82**

Effect of an open diode in a bridge rectifier.

Effects of a Faulty Filter Capacitor   Three types of defects of a filter capacitor are illustrated in Figure 2–83.

◆ *Open*   If the filter capacitor for a full-wave rectifier opens, the output is a full-wave rectified voltage.

◆ *Shorted*   If the filter capacitor shorts, the output is 0 V. A shorted capacitor should cause the fuse to blow open. If not properly fused, a shorted capacitor may cause some or all of the diodes in the rectifier to burn open due to excessive current. In any event, the output is 0 V.

◆ *Leaky*   A leaky filter capacitor is equivalent to a capacitor with a parallel leakage resistance. The effect of the leakage resistance is to reduce the time constant and allow the capacitor to discharge more rapidly than normal. This results in an increase in the ripple voltage on the output. This fault is rare.

Effects of a Faulty Transformer   An open primary or secondary winding of a power supply transformer results in an output of 0 V, as mentioned before.

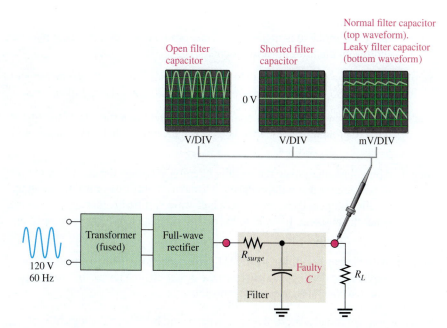

Effects of a faulty filter capacitor.

EXAMPLE 2–14

You are troubleshooting the power supply shown in the block diagram of Figure 2–84. You have found in the analysis phase that there is no output voltage from the regulator, as indicated. Also, you have found that the unit is plugged into the outlet and have verified the input to the transformer with a DMM. You decide to use the half-splitting method using the scope. What is the problem?

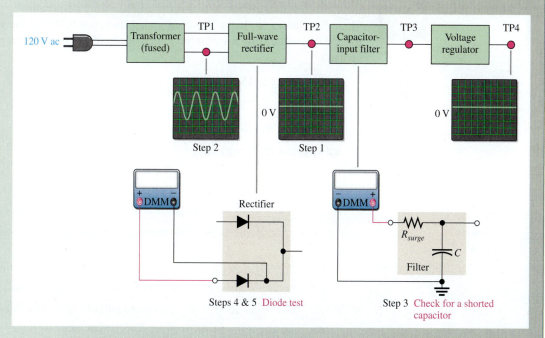

▲ FIGURE 2–84

Solution   The step-by-step measurement procedure is illustrated in the figure and described as follows:

Step 1:   There is no voltage at test point 2 (TP2). This indicates that the fault is between the input to the transformer and the output of the rectifier. Most likely, the problem is in the transformer or in the rectifier, but there may be a short from the filter input to ground.

Step 2: The voltage at test point 1 (TP1) is correct, indicating that the transformer is working. So, the problem must be in the rectifier or a shorted filter input.

Step 3: With the power turned off, use a DMM to check for a short from the filter input to ground. Assume that the DMM indicates no short. The fault is now isolated to the rectifier.

Step 4: Apply fault analysis to the rectifier circuit. Determine the component failure in the rectifier that will produce a 0 V input. If only one of the diodes in the rectifier is open, there should be a half-wave rectified output voltage, so this is not the problem. In order to have a 0 V output, there must be an open in the rectifier circuit.

Step 5: With the power off, use the DMM in the diode test mode to check each diode. Replace the defective diodes, turn the power on, and check for proper operation. Assume this corrects the problem.

Related Problem   Suppose you had found a short in Step 3, what would have been the logical next step?

Multisim Troubleshooting Exercises

These file circuits are in the Troubleshooting Exercises folder on the website. Open each file and determine if the circuit is working properly. If it is not working properly, determine the fault.

   1. Multisim file TSM02-01

   2. Multisim file TSM02-02

   3. Multisim file TSM02-03

   4. Multisim file TSM02-04

SECTION 2–10 CHECKUP	

SECTION 2–10 CHECKUP

1. A properly functioning diode will produce a reading in what range when forward-biased?

2. What reading might an ohmmeter produce when it reverse-biases a diode?

3. What effect does an open diode have on the output voltage of a half-wave rectifier?

4. What effect does an open diode have on the output voltage of a full-wave rectifier?

5. If one of the diodes in a bridge rectifier shorts, what are some possible consequences?

6. What happens to the output voltage of a rectifier if the filter capacitor becomes very leaky?

7. The primary winding of the transformer in a power supply opens. What will you observe on the rectifier output?

8. The dc output voltage of a filtered rectifier is less than it should be. What may be the problem?

Device Application: *DC Power Supply*

Assume that you are working for a company that designs, tests, manufactures, and markets various electronic instruments including dc power supplies. Your first assignment is to develop and test a basic unregulated power supply using the knowledge that you have acquired so far. Later modifications will include the addition of a regulator. The power supply must meet or exceed the following specifications:

* Input voltage: 120 V rms @60 Hz
* Output voltage: 16 V dc ±10%
* Ripple factor (max): 3.00%
* Load current (max): 250 mA

Design of the Power Supply

The Rectifier Circuit   A full-wave rectifier has less ripple for a given filter capacitor than a half-wave rectifier. A full-wave bridge rectifier is probably the best choice because it provides the most output voltage for a given input voltage and the PIV is less than for a center-tapped rectifier. Also, the full-wave bridge does not require a center-tapped transformer.

1. Compare Equations 2–7 and 2–9 for output voltages.
2. Compare Equations 2–8 and 2–10 for PIV.

The full-wave bridge rectifier circuit is shown in Figure 2–85.

▶ FIGURE 2–85

Power supply with full-wave bridge rectifier and capacitor filter.

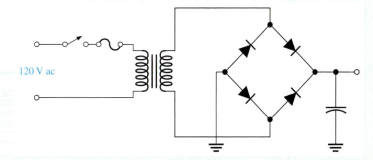

120 V ac

The Rectifier Diodes   There are two approaches for implementing the full-wave bridge: Four individual diodes, as shown in Figure 2–86(a) or a single IC package containing four diodes connected as a bridge rectifier, as shown in part (b).

▶ FIGURE 2–86

Rectifier components.

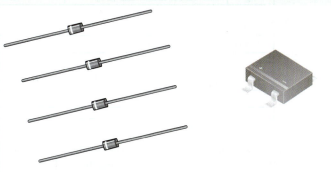

(a) Separate rectifier diodes

(b) Full-wave bridge rectifier

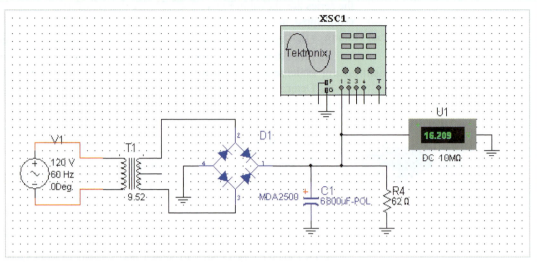

(a) Multisim circuit screen

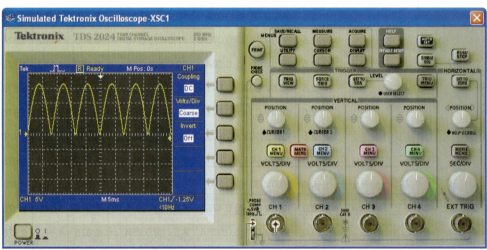

(b) Output voltage without the filter capacitor

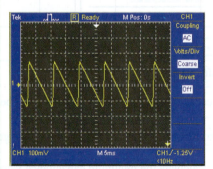

(c) Ripple voltage is less than 300 mV pp

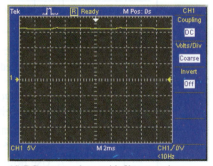

(d) DC output voltage with filter capacitor
(near top of screen)

▲ **FIGURE 2–89**

Power supply simulation.

supply circuit with a load connected and scope displays of the output voltage with and without the filter capacitor connected. The filter capacitor value of 6800 μF is the next highest standard value closest to the minimum calculated value required. A load resistor value was chosen to draw a current equal to or greater than the specified maximum load current.

$$R_L = \frac{16 \text{ V}}{250 \text{ mA}} = 64 \text{ }\Omega$$

The closest standard value is 62 Ω, which draws 258 mA at 16 V and which meets and exceeds the load current specification.

    8. Determine the power rating for the load resistor.

To produce a dc output of 16 V, a peak secondary voltage of 16 V + 1.4 V = 17.4 V is required. The rms secondary voltage must be

$$V_{rms(sec)} = 0.707V_{p(sec)} = 0.707(16 \text{ V} + 1.4 \text{ V}) = 12.3 \text{ V}$$

A standard transformer rms output voltage is 12.6 V. The transformer specification required by Multisim is

$$120 \text{ V}:12.6 \text{ V} = 9.52:1$$

The dc voltmeter in Figure 2–89(a) indicates an output voltage of 16.209 V, which is well within the 16 V ± 10% requirement. In part (c), the scope is AC coupled and set at 100 mV/division. You can see that the peak-to-peak ripple voltage is less than 300 mV, which is less than 480 mV, corresponding to the specified maximum ripple factor of 3%.

Build and simulate the circuit using your Multisim or LT Spice software. Observe the operation with the virtual oscilloscope and voltmeter.

Prototyping and Testing

Now that all the components have been selected, the prototype circuit is constructed and tested. After the circuit is successfully tested, it is ready to be finalized on a printed circuit board.

Lab Experiment

To build and test a similar circuit, go to Experiment 2 in your lab manual (*Laboratory Exercises for Electronic Devices* by David Buchla and Steven Wetterling).

The Printed Circuit Board

The circuit board is shown in Figure 2–90. There are additional traces and connection points on the board for expansion to a regulated power supply, which will be done in Chapter 3. The circuit board is connected to the ac voltage and to a power load resistor via a cable. The power switch shown in the original schematic will be on the PC board housing and is not shown for the test setup. A DMM measurement of the output voltage indicates a correct value. Oscilloscope measurement of the ripple shows that it is within specifications.

19. If the value of R_3 in Figure 2–61 is decreased, the positive output voltage will

(a) increase   (b) decrease   (c) not change

20. If the input voltage in Figure 2–65 is increased, the peak negative value of the output voltage will

(a) increase   (b) decrease   (c) not change

SELF-TEST

Answers can be found at www.pearsonhighered.com/floyd.

Section 2–1   1. The term *bias* means

(a) the ratio of majority carriers to minority carriers

(b) the amount of current across a diode

(c) a dc voltage that is applied to control the operation of a device

(d) neither (a), (b), nor (c)

2. To forward-bias a diode,

(a) an external voltage is applied that is positive at the anode and negative at the cathode

(b) an external voltage is applied that is negative at the anode and positive at the cathode

(c) an external voltage is applied that is positive at the *p* region and negative at the *n* region

(d) answers (a) and (c)

3. When a diode is forward-biased,

(a) the only current is hole current

(b) the only current is electron current

(c) the only current is produced by majority carriers

(d) the current is produced by both holes and electrons

4. Although current is blocked in reverse bias,

(a) there is some current due to majority carriers

(b) there is a very small current due to minority carriers

(c) there is an avalanche current

5. For a silicon diode, the value of the forward-bias voltage typically

(a) must be greater than 0.3 V

(b) must be greater than 0.7 V

(c) depends on the width of the depletion region

(d) depends on the concentration of majority carriers

6. When forward-biased, a diode

(a) blocks current   (b) conducts current

(c) has a high resistance   (d) drops a large voltage

Section 2–2   7. A diode is normally operated in

(a) reverse breakdown   (b) the forward-bias region

(c) the reverse-bias region   (d) either (b) or (c)

8. The dynamic resistance can be important when a diode is

(a) reverse-biased   (b) forward-biased

(c) in reverse breakdown   (d) unbiased

9. The *V-I* curve for a diode shows

(a) the voltage across the diode for a given current

(b) the amount of current for a given bias voltage

(c) the power dissipation

(d) none of these

Section 2–3   10. Ideally, a diode can be represented by a

(a) voltage source   (b) resistance   (c) switch   (d) all of these

11. In the practical diode model,

    (a) the barrier potential is taken into account

    (b) the forward dynamic resistance is taken into account

    (c) none of these

    (d) both (a) and (b)

12. In the complete diode model,

    (a) the barrier potential is taken into account

    (b) the forward dynamic resistance is taken into account

    (c) the reverse resistance is taken into account

    (d) all of these

Section 2–4   **13.** The average value of a half-wave rectified voltage with a peak value of 200 V is

    (a) 63.7 V   **(b)** 127.2 V   **(c)** 141 V   **(d)** 0 V

14. When a 60 Hz sinusoidal voltage is applied to the input of a half-wave rectifier, the output frequency is

    (a) 120 Hz   **(b)** 30 Hz   **(c)** 60 Hz   **(d)** 0 Hz

15. The peak value of the input to a half-wave rectifier is 10 V. The approximate peak value of the output is

    (a) 10 V   **(b)** 3.18 V   **(c)** 10.7 V   **(b)** 9.3 V

16. For the circuit in Question 15, the diode must be able to withstand a reverse voltage of

    (a) 10 V   **(b)** 5 V   **(c)** 20 V   **(d)** 3.18 V

Section 2–5   **17.** The average value of a full-wave rectified voltage with a peak value of 75 V is

    (a) 53 V   **(b)** 47.8 V   **(c)** 37.5 V   **(d)** 23.9 V

18. When a 60 Hz sinusoidal voltage is applied to the input of a full-wave rectifier, the output frequency is

    (a) 120 Hz   **(b)** 60 Hz   **(c)** 240 Hz   **(d)** 0 Hz

19. The total secondary voltage in a center-tapped full-wave rectifier is 125 V rms. Neglecting the diode drop, the rms output voltage is

    (a) 125 V   **(b)** 177 V   **(c)** 100 V   **(d)** 62.5 V

20. When the peak output voltage is 100 V, the PIV for each diode in a center-tapped full-wave rectifier is (neglecting the diode drop)

    (a) 100 V   **(b)** 200 V   **(c)** 141 V   **(d)** 50 V

21. When the rms output voltage of a bridge full-wave rectifier is 20 V, the peak inverse voltage across the diodes is (neglecting the diode drop)

    (a) 20 V   **(b)** 40 V   **(c)** 28.3 V   **(d)** 56.6 V

Section 2–6   **22.** The ideal dc output voltage of a capacitor-input filter is equal to

    (a) the peak value of the rectified voltage

    (b) the average value of the rectified voltage

    (c) the rms value of the rectified voltage

23. A certain power-supply filter produces an output with a ripple of 100 mV peak-to-peak and a dc value of 20 V. The ripple factor is

    (a) 0.05   **(b)** 0.005   **(c)** 0.00005   **(d)** 0.02

24. A 60 V peak full-wave rectified voltage is applied to a capacitor-input filter. If $f = 120$ Hz, $R_L = 10$ kΩ, and $C = 10$ μF, the ripple voltage is

    (a) 0.6 V   **(b)** 6 mV   **(c)** 5.0 V   **(d)** 2.88 V

25. If the load resistance of a capacitor-filtered full-wave rectifier is reduced, the ripple voltage

    (a) increases   **(b)** decreases   **(c)** is not affected   **(d)** has a different frequency

26. Line regulation is determined by

    (a) load current

    (b) zener current and load current

Section 2–5 **Full-Wave Rectifiers**

17. Find the average value of each voltage in Figure 2–95.

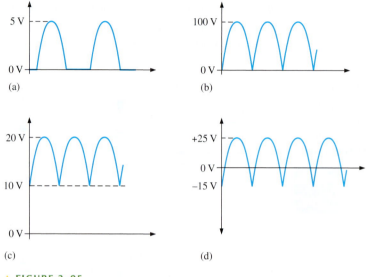

(a)

(b)

(c)

(d)

▲ **FIGURE 2–95**

18. Consider the circuit in Figure 2–96.

    (a) What type of circuit is this?

    (b) What is the total peak secondary voltage?

    (c) Find the peak voltage across each half of the secondary.

    (d) Sketch the voltage waveform across R_L.

    (e) What is the peak current through each diode?

    (f) What is the PIV for each diode?

▶ **FIGURE 2–96**

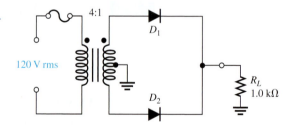

19. Calculate the peak voltage across each half of a center-tapped transformer used in a full-wave rectifier that has an average output voltage of 120 V.

20. Show how to connect the diodes in a center-tapped rectifier in order to produce a negative-going full-wave voltage across the load resistor.

21. What PIV rating is required for the diodes in a bridge rectifier that produces an average output voltage of 50 V?

22. The rms output voltage of a bridge rectifier is 20 V. What is the peak inverse voltage across the diodes?

23. Draw the output voltage waveform for the bridge rectifier in Figure 2–97. Notice that all the diodes are reversed from circuits shown earlier in the chapter.

◢ FIGURE 2–97

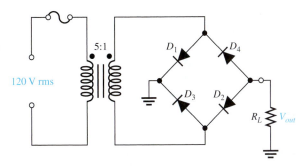

Section 2–6 Power Supply Filters and Regulators

24. A certain rectifier filter produces a dc output voltage of 75 V with a peak-to-peak ripple voltage of 0.5 V. Calculate the ripple factor.

25. A certain full-wave rectifier has a peak output voltage of 30 V. A 50 μF capacitor-input filter is connected to the rectifier. Calculate the peak-to-peak ripple and the dc output voltage developed across a 600 Ω load resistance.

26. What is the percentage of ripple for the rectifier filter in Problem 25?

27. What value of filter capacitor is required to produce a 1% ripple factor for a full-wave rectifier having a load resistance of 1.5 kΩ? Assume the rectifier produces a peak output of 18 V.

28. A full-wave rectifier produces an 80 V peak rectified voltage from a 60 Hz ac source. If a 10 μF filter capacitor is used, determine the ripple factor for a load resistance of 10 kΩ.

29. Determine the peak-to-peak ripple and dc output voltages in Figure 2–98. The transformer has a 36 V rms secondary voltage rating, and the line voltage has a frequency of 60 Hz.

30. Refer to Figure 2–98 and draw the following voltage waveforms in relationship to the input waveforms: V_{AB}, V_{AD}, and V_{CD}. A double letter subscript indicates a voltage from one point to another.

31. If the no-load output voltage of a regulator is 15.5 V and the full-load output is 14.9 V, what is the percent load regulation?

32. Assume a regulator has a percent load regulation of 0.5%. What is the output voltage at full-load if the unloaded output is 12.0 V?

◢ FIGURE 2–98

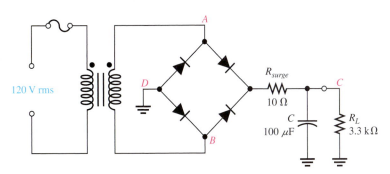

Section 2–7 Diode Limiters and Clampers

33. Determine the output waveform for the circuit of Figure 2–99.

◢ FIGURE 2–99

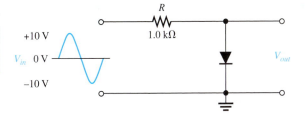

Section 2–8 **Voltage Multipliers**

43. A certain voltage doubler has 20 V rms on its input. What is the output voltage? Draw the circuit, indicating the output terminals and PIV rating for the diode.

44. Repeat Problem 43 for a voltage tripler and quadrupler.

Section 2–9 **The Diode Datasheet**

45. From the datasheet in Figure 2–71, determine how much peak inverse voltage a 1N4002 diode can withstand.

46. Repeat Problem 45 for a 1N4007.

47. If the peak output voltage of a bridge full-wave rectifier is 50 V, determine the minimum value of the load resistance that can be used when 1N4002 diodes are used.

Section 2–10 **Troubleshooting**

48. Consider the meter indications in each circuit of Figure 2–106, and determine whether the diode is functioning properly, or whether it is open or shorted. Assume the ideal model.

◀ FIGURE 2–106

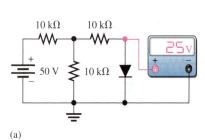
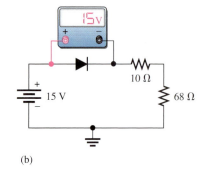

(a)   (b)

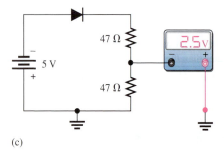
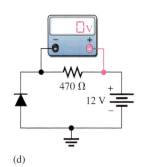

(c)   (d)

49. Determine the voltage with respect to ground at each point in Figure 2–107. Assume the practical model.

50. If one of the diodes in a bridge rectifier opens, what happens to the output?

◀ FIGURE 2–107

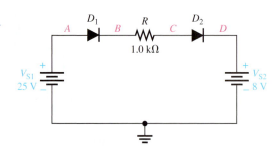

51. From the meter readings in Figure 2–108, determine if the rectifier is functioning properly. If it is not, determine the most likely failure(s).

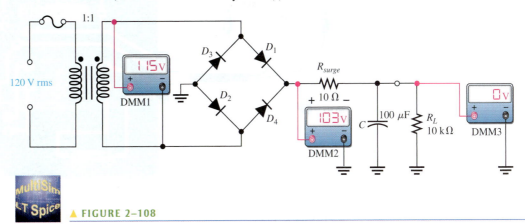

▲ FIGURE 2–108

52. Each part of Figure 2–109 shows oscilloscope displays of various rectifier output voltages. In each case, determine whether or not the rectifier is functioning properly and if it is not, determine the most likely failure(s).

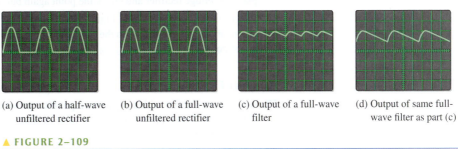

(a) Output of a half-wave unfiltered rectifier

(b) Output of a full-wave unfiltered rectifier

(c) Output of a full-wave filter

(d) Output of same full-wave filter as part (c)

▲ FIGURE 2–109

53. Based on the values given, would you expect the circuit in Figure 2–110 to fail? If so, why?

► FIGURE 2–110

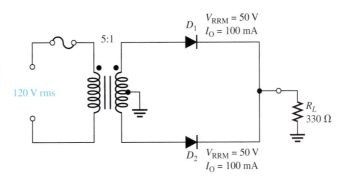

DEVICE APPLICATION PROBLEMS

54. Determine the most likely failure(s) in the circuit of Figure 2–111 for each of the following symptoms. State the corrective action you would take in each case. The transformer has a rated output of 10 V rms.

(a) No voltage from test point 1 to test point 2

(b) No voltage from test point 3 to test point 4

(c) 8 V rms from test point 3 to test point 4

(d) Excessive 120 Hz ripple voltage at test point 6

(e) There is a 60 Hz ripple voltage at test point 6

(f) No voltage at test point 6

3

SPECIAL-PURPOSE DIODES

CHAPTER OBJECTIVES

◆ Describe the characteristics of a zener diode and analyze its operation

◆ Apply a zener diode in voltage regulation

◆ Describe the varactor diode characteristic and analyze its operation

◆ Discuss the characteristics, operation, and applications of LEDs, quantum dots, and photodiodes

◆ Explain the basic operation of a solar cell

◆ Discuss the basic characteristics of several types of diodes

◆ Troubleshoot zener diode regulators

KEY TERMS

◆ Zener diode
◆ Zener breakdown
◆ Varactor
◆ Light-emitting diode (LED)
◆ Electroluminescence
◆ Pixel
◆ Photodiode
◆ PV cell
◆ Laser

VISIT THE WEBSITE

Study aids, Multisim files, and LT Spice files for this chapter are available at https://www.pearsonhighered.com/careersresources/

INTRODUCTION

Chapter 2 was devoted to general-purpose and rectifier diodes, which are the most widely used types. In this chapter, we will cover several other types of diodes that are designed for specific applications, including the zener, varactor (variable-capacitance), light-emitting, photo, solar cell, laser, Schottky, tunnel, *pin,* step-recovery, and current regulator diodes.

DEVICE APPLICATION PREVIEW

The Device Application in this chapter is the expansion of the 16 V power supply developed in Chapter 2 into a 12 V regulated power supply with an LED power-on indicator. The new circuit will incorporate a voltage regulator IC, which is introduced in this chapter.

3–1 THE ZENER DIODE

A major application for zener diodes is to provide a stable reference voltage for use in power supplies, voltmeters, and other instruments and they can be used as voltage regulators in certain applications. In this section, you will see how the zener diode maintains a nearly constant dc voltage under the proper operating conditions. You will learn the conditions and limitations for properly using the zener diode and the factors that affect its performance.

After completing this section, you should be able to

❑ **Describe the characteristics of a zener diode and analyze its operation**
❑ Recognize a zener diode by its schematic symbol
❑ Discuss zener breakdown
  ◆ Define *avalanche breakdown*
❑ Explain zener breakdown characteristics
  ◆ Describe zener regulation
❑ Discuss zener equivalent circuits
❑ Define *temperature coefficient*
  ◆ Analyze zener voltage as a function of temperature
❑ Discuss zener power dissipation and derating
  ◆ Apply power derating to a zener diode
❑ Interpret zener diode datasheets

The symbol for a zener diode is shown in Figure 3–1. Instead of a straight line representing the cathode, the zener diode has a bent line that reminds you of the letter Z (for zener). A **zener diode** is a silicon *pn* junction device that is designed for operation in the reverse-breakdown region. The breakdown voltage of a zener diode is set by carefully controlling the doping level during manufacture. Recall, from the discussion of the diode characteristic curve in Chapter 2, that when a diode reaches reverse breakdown, its voltage remains almost constant even though the current changes drastically, and this is the key to zener diode operation. This volt-ampere characteristic is shown again in Figure 3–2 with the normal operating region for zener diodes shown as a shaded area.

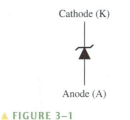

Cathode (K)

Anode (A)

▲ **FIGURE 3–1**

Zener diode symbol.

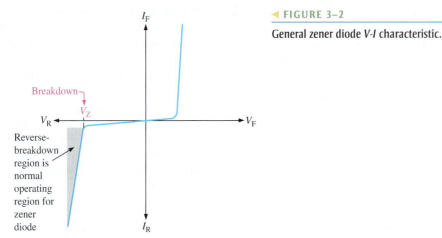

◀ **FIGURE 3–2**

General zener diode *V-I* characteristic.

Zener Breakdown

Zener diodes are designed to operate in reverse breakdown. Two types of reverse breakdown in a zener diode are *avalanche* and *zener*. The avalanche effect, discussed in Chapter 2, occurs in both rectifier and zener diodes at a sufficiently high reverse voltage. **Zener breakdown** occurs in a zener diode at low reverse voltages. A zener diode is heavily doped to reduce the breakdown voltage. This causes a very thin depletion region. As a result, an intense electric field exists within the depletion region. Near the zener breakdown voltage (V_Z), the field is intense enough to pull electrons from their valence bands and create current.

Zener diodes with breakdown voltages of less than approximately 5 V operate predominately in zener breakdown. Those with breakdown voltages greater than approximately 5 V operate predominately in **avalanche breakdown.** Both types, however, are called *zener diodes.* Zeners are commercially available with breakdown voltages from less than 1 V to more than 250 V with specified tolerances from 1% to 20%.

Breakdown Characteristics

Figure 3–3 shows the reverse portion of a zener diode's characteristic curve. Notice that as the reverse voltage (V_R) is increased, the reverse current (I_R) remains extremely small up to the "knee" of the curve. The reverse current is also called the zener current, I_Z. At this point, the breakdown effect begins; the internal zener resistance, also called zener impedance (Z_Z), begins to decrease as the reverse current increases rapidly. From the bottom of the knee, the zener breakdown voltage (V_Z) remains essentially constant although it increases slightly as the zener current, I_Z, increases.

▶ **FIGURE 3–3**

Reverse characteristic of a zener diode. V_Z is usually specified at a value of the zener current known as the test current.

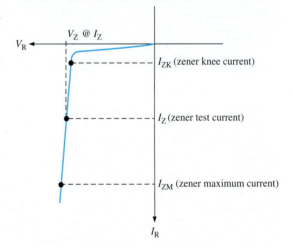

Zener Regulation   The ability to keep the reverse voltage across its terminals essentially constant is the key feature of the zener diode. A zener diode operating in breakdown can act as a low-current voltage regulator because it maintains a nearly constant voltage across its terminals over a specified range of reverse-current values.

A minimum value of reverse current, I_{ZK}, must be maintained in order to keep the diode in breakdown for voltage regulation. You can see on the curve in Figure 3–3 that when the reverse current is reduced below the knee of the curve, the voltage decreases drastically and regulation is lost. Also, there is a maximum current, I_{ZM}, above which the diode may be damaged due to excessive power dissipation. So, basically, the zener diode maintains a nearly constant voltage across its terminals for values of reverse current ranging from I_{ZK} to I_{ZM}. A nominal zener voltage, V_Z, is usually specified on a datasheet at a value of reverse current called the *zener test current.*

Zener Equivalent Circuits

Figure 3–4 shows the ideal model (first approximation) of a zener diode in reverse breakdown and its ideal characteristic curve. It has a constant voltage drop equal to the nominal zener voltage. This constant voltage drop across the zener diode produced by reverse breakdown is represented by a dc voltage symbol even though the zener diode does not produce a voltage.

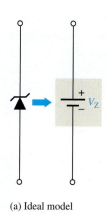
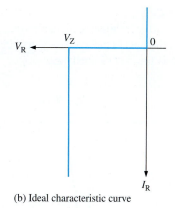

(a) Ideal model (b) Ideal characteristic curve

◀ FIGURE 3–4

Ideal zener diode equivalent circuit model and the characteristic curve.

Figure 3–5(a) represents the practical model (second approximation) of a zener diode, where the zener impedance (resistance), Z_Z, is included. Since the actual voltage curve is not ideally vertical, a change in zener current (ΔI_Z) produces a small change in zener voltage (ΔV_Z), as illustrated in Figure 3–5(b). By Ohm's law, the ratio of ΔV_Z to ΔI_Z is the impedance, as expressed in the following equation:

$$Z_Z = \frac{\Delta V_Z}{\Delta I_Z}$$

Equation 3–1

Because Z_Z is defined as a change in voltage over a change in current, it is a dynamic (or ac) resistance. Normally, Z_Z is specified at the zener test current. In most cases, you can assume that Z_Z is a small constant over the full range of zener current values and is purely resistive. It is best to avoid operating a zener diode near the knee of the curve because the impedance changes dramatically in that area.

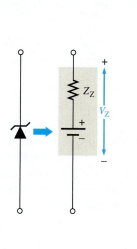
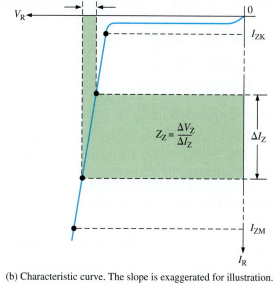

(a) Practical model (b) Characteristic curve. The slope is exaggerated for illustration.

◀ FIGURE 3–5

Practical zener diode equivalent circuit and the characteristic curve illustrating Z_Z.

For most circuit analysis and troubleshooting work, the ideal model will give very good results and is much easier to use than more complicated models. When a zener diode is operating normally, it will be in reverse breakdown and you should observe the nominal breakdown voltage across it. Most **schematics** will indicate on the drawing what this voltage should be.

EXAMPLE 3–1

A zener diode exhibits a certain change in V_Z for a certain change in I_Z on a portion of the linear characteristic curve between I_{ZK} and I_{ZM} as illustrated in Figure 3–6. What is the zener impedance?

▶ **FIGURE 3–6**

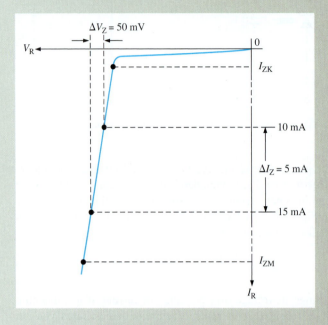

Solution

$$Z_Z = \frac{\Delta V_Z}{\Delta I_Z} = \frac{50 \text{ mV}}{5 \text{ mA}} = \mathbf{10 \ \Omega}$$

Related Problem*   Calculate the zener impedance if the change in zener voltage is 100 mV for a 20 mA change in zener current on the linear portion of the characteristic curve.

* Answers can be found at www.pearsonhighered.com/floyd.

Temperature Coefficient

The temperature coefficient specifies the percent change in zener voltage for each degree Celsius change in temperature. For example, a 12 V zener diode with a positive temperature coefficient of 0.01%/°C will exhibit a 1.2 mV increase in V_Z when the junction temperature increases one degree Celsius. The formula for calculating the change in zener voltage for a given junction temperature change, for a specified temperature coefficient, is

Equation 3–2

$$\Delta V_Z = V_Z \times TC \times \Delta T$$

where V_Z is the nominal zener voltage at the reference temperature of 25°C, TC is the temperature coefficient, and ΔT is the change in temperature from the reference temperature. A positive TC means that the zener voltage increases with an increase in temperature or decreases with a decrease in temperature. A negative TC means that the zener voltage decreases with an increase in temperature or increases with a decrease in temperature.

In some cases, the temperature coefficient is expressed in mV/°C rather than as %/°C. For these cases, ΔV_Z is calculated as:

Equation 3–3

$$\Delta V_Z = TC \times \Delta T$$

EXAMPLE 3–2

An 8.2 V zener diode (8.2 V at 25°C) has a positive temperature coefficient of 0.05%/°C. What is the zener voltage at 60°C?

Solution

The change in zener voltage is

$$\Delta V_Z = V_Z \times TC \times \Delta T = (8.2 \text{ V})(0.05\%/°C)(60°C - 25°C)$$
$$= (8.2 \text{ V})(0.0005/°C)(35°C) = 144 \text{ mV}$$

Notice that 0.05%/°C was converted to 0.0005/°C. The zener voltage at 60°C is

$$V_Z + \Delta V_Z = 8.2 \text{ V} + 144 \text{ mV} = \textbf{8.34 V}$$

Related Problem

A 12 V zener has a positive temperature coefficient of 0.075%/°C. How much will the zener voltage change when the junction temperature decreases 50 degrees Celsius?

Zener Power Dissipation and Derating

All diodes are rated by the manufacturer for absolute maximum ratings. Recall that dc power dissipation is the product of the voltage drop and the current. For a zener, the voltage drop is V_Z and the current in the device is I_Z. The power dissipated is simply $P = V_Z I_Z$. The maximum power dissipation is designated P_D by the manufacturer. For example, the 1N746 zener is rated for a maximum power of 500 mW and the 1N3305 is rated for a P_D of 50 W.

Power Derating   The maximum power dissipation of a zener diode is typically specified for temperatures at or below a certain value (e.g., 50°C). Above the specified temperature, the maximum power dissipation is reduced according to a derating factor. The derating factor is expressed in mW/°C. The maximum derated power can be determined with the following formula:

$$P_{D(derated)} = P_D - (mW/°C)\Delta T$$

EXAMPLE 3–3

A certain zener diode has a maximum power rating of 400 mW at 50°C and a derating factor of 3.2 mW/°C. Determine the maximum power the zener can dissipate at a temperature of 90°C.

Solution

$$P_{D(derated)} = P_D - (mW/°C)\Delta T$$
$$= 400 \text{ mW} - (3.2 \text{ mW/°C})(90°C - 50°C)$$
$$= 400 \text{ mW} - 128 \text{ mW} = \textbf{272 mW}$$

Related Problem

A certain 50 W zener diode must be derated with a derating factor of 0.5W/°C above 75°C. Determine the maximum power it can dissipate at 160°C.

Zener Diode Datasheet Information

The amount and type of information found on datasheets for zener diodes (or any category of electronic device) varies from one type of diode to the next. The datasheet for some zeners contains more information than for others. Figure 3–7 gives an example of the type of information you have studied that can be found on a typical datasheet. This particular information is for a zener series, the 1N4728A–1N4764A.

1N4728A - 1N4764A

Zeners

DO-41 Glass case
COLOR BAND DENOTES CATHODE

Absolute Maximum Ratings * T_a = 25°C unless otherwise noted

Symbol	Parameter	Value	Units
P_D	Power Dissipation @ TL ≤ 50°C, Lead Length = 3/8"	1.0	W
	Derate above 50°C	6.67	mW/°C
T_J, T_{STG}	Operating and Storage Temperature Range	-65 to +200	°C

* These ratings are limiting values above which the serviceability of the diode may be impaired.

Electrical Characteristics T_a = 25°C unless otherwise noted

Device	V_Z (V) @ I_Z (Note 1)			Test Current I_Z (mA)	Max. Zener Impedance			Leakage Current	
	Min.	Typ.	Max.		Z_Z @ I_Z (Ω)	Z_{ZK} @ I_{ZK} (Ω)	I_{ZK} (mA)	I_R (μA)	V_R (V)
1N4728A	3.315	3.3	3.465	76	10	400	1	100	1
1N4729A	3.42	3.6	3.78	69	10	400	1	100	1
1N4730A	3.705	3.9	4.095	64	9	400	1	50	1
1N4731A	4.085	4.3	4.515	58	9	400	1	10	1
1N4732A	4.465	4.7	4.935	53	8	500	1	10	1
1N4733A	4.845	5.1	5.355	49	7	550	1	10	1
1N4734A	5.32	5.6	5.88	45	5	600	1	10	2
1N4735A	5.89	6.2	6.51	41	2	700	1	10	3
1N4736A	6.46	6.8	7.14	37	3.5	700	1	10	4
1N4737A	7.125	7.5	7.875	34	4	700	0.5	10	5
1N4738A	7.79	8.2	8.61	31	4.5	700	0.5	10	6
1N4739A	8.645	9.1	9.555	28	5	700	0.5	10	7
1N4740A	9.5	10	10.5	25	7	700	0.25	10	7.6
1N4741A	10.45	11	11.55	23	8	700	0.25	5	8.4
1N4742A	11.4	12	12.6	21	9	700	0.25	5	9.1
1N4743A	12.35	13	13.65	19	10	700	0.25	5	9.9
1N4744A	14.25	15	15.75	17	14	700	0.25	5	11.4
1N4745A	15.2	16	16.8	15.5	16	700	0.25	5	12.2
1N4746A	17.1	18	18.9	14	20	750	0.25	5	13.7
1N4747A	19	20	21	12.5	22	750	0.25	5	15.2
1N4748A	20.9	22	23.1	11.5	23	750	0.25	5	16.7
1N4749A	22.8	24	25.2	10.5	25	750	0.25	5	18.2
1N4750A	25.65	27	28.35	9.5	35	750	0.25	5	20.6
1N4751A	28.5	30	31.5	8.5	40	1000	0.25	5	22.8
1N4752A	31.35	33	34.65	7.5	45	1000	0.25	5	25.1
1N4753A	34.2	36	37.8	7	50	1000	0.25	5	27.4
1N4754A	37.05	39	40.95	6.5	60	1000	0.25	5	29.7
1N4755A	40.85	43	45.15	6	70	1500	0.25	5	32.7
1N4756A	44.65	47	49.35	5.5	80	1500	0.25	5	35.8
1N4757A	48.45	51	53.55	5	95	1500	0.25	5	38.8
1N4758A	53.2	56	58.8	4.5	110	2000	0.25	5	42.6
1N4759A	58.9	62	65.1	4	125	2000	0.25	5	47.1
1N4760A	64.6	68	71.4	3.7	150	2000	0.25	5	51.7
1N4761A	71.25	75	78.75	3.3	175	2000	0.25	5	56
1N4762A	77.9	82	86.1	3	200	3000	0.25	5	62.2
1N4763A	86.45	91	95.55	2.8	250	3000	0.25	5	69.2
1N4764A	95	100	105	2.5	350	3000	0.25	5	76

Notes:

1. Zener Voltage (V_Z)
   The zener voltage is measured with the device junction in the thermal equilibrium at the lead temperature(T_L) at 30°C ± 1°C and 3/8" lead length.

▲ FIGURE 3–7

Partial datasheet for the 1N4728A–1N4764A series 1 W zener diodes. Copyright Fairchild Semiconductor Corporation. Used by permission. Datasheets are available at www.fairchildsemi.com.

Absolute Maximum Ratings   The absolute maximum power dissipation, P_D, is specified as 1.0 W up to 50°C on the data sheet in Figure 3–7. Generally, the zener diode should be operated at least 20% below this maximum to assure reliability and longer life. The power dissipation is derated as shown on the datasheet at 6.67 mW for each degree above 50°C. For example, using the procedure illustrated in Example 3–3, the maximum derated power dissipation at 60°C is

$$P_{D(derated)} = 1 \text{ W} - 10°C(6.67 \text{ mW/°C}) = 1 \text{ W} - 66.7 \text{ mW} = 0.9933 \text{ W}$$

At 125°C, the maximum power dissipation is

$$P_{D(derated)} = 1 \text{ W} - 75°C(6.67 \text{ mW/°C}) = 1 \text{ W} - 500.25 \text{ mW} = 0.4998 \text{ W}$$

Notice that a maximum reverse current is not specified but can be determined from the maximum derated power dissipation for a given value of V_Z. For example, at 50°C, the maximum zener current for a zener voltage of 3.3 V is

$$I_{ZM} = \frac{P_D}{V_Z} = \frac{1 \text{ W}}{3.3 \text{ V}} = 303 \text{ mA}$$

The operating junction temperature, T_J, and the storage temperature, T_{STG}, have a range of from −65°C to 200°C.

Electrical Characteristics   The first column in the datasheet lists the zener type numbers, 1N4728A through 1N4764A.

Zener voltage, V_Z, and zener test current, I_Z   For each device type, the minimum, typical, and maximum zener voltages are listed. V_Z is measured at the specified zener test current, I_Z. For example, the zener voltage for a 1N4728A can range from 3.315 V to 3.465 V with a typical value of 3.3 V at a test current of 76 mA.

Maximum zener impedance Z_Z is the maximum zener impedance at the specified test current, I_Z. For example, for a 1N4728A, Z_Z is 10 Ω at 76 mA. The maximum zener impedance, Z_{ZK}, at the knee of the characteristic curve is specified at I_{ZK}, which is the current at the knee of the curve. For example, Z_{ZK} is 400 Ω at 1 mA for a 1N4728A.

Leakage current   Reverse leakage current is specified for a reverse voltage that is less than the knee voltage. This means that the zener is not in reverse breakdown for these measurements. For example I_R is 100 μA for a reverse voltage of 1 V in a 1N4728A.

EXAMPLE 3–4

From the datasheet in Figure 3–7, a 1N4736A zener diode has a Z_Z of 3.5 Ω. The datasheet gives $V_Z = 6.8$ V at a test current, I_Z, of 37 mA. What is the voltage across the zener terminals when the current is 50 mA? When the current is 25 mA? Figure 3–8 represents the zener diode.

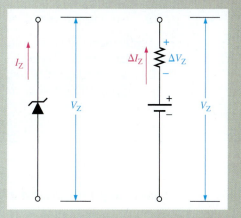

▲ FIGURE 3–8

Solution   For $I_Z = 50$ mA: The 50 mA current is a 13 mA increase above the test current, I_Z, of 37 mA.

$$\Delta I_Z = I_Z - 37\ \text{mA} = 50\ \text{mA} - 37\ \text{mA} = +13\ \text{mA}$$
$$\Delta V_Z = \Delta I_Z Z_Z = (13\ \text{mA})(3.5\ \Omega) = +45.5\ \text{mV}$$

The change in voltage due to the increase in current above the I_Z value causes the zener terminal voltage to increase. The zener voltage for $I_Z = 50$ mA is

$$V_Z = 6.8\ \text{V} + \Delta V_Z = 6.8\ \text{V} + 45.5\ \text{mV} = \mathbf{6.85\ V}$$

For $I_Z = 25$ mA: The 25 mA current is a 12 mA decrease below the test current, I_Z, of 37 mA.

$$\Delta I_Z = -12\ \text{mA}$$
$$\Delta V_Z = \Delta I_Z Z_Z = (-12\ \text{mA})(3.5\ \Omega) = -42\ \text{mV}$$

The change in voltage due to the decrease in current below the test current causes the zener terminal voltage to decrease. The zener voltage for $I_Z = 25$ mA is

$$V_Z = 6.8\ \text{V} - \Delta V_Z = 6.8\ \text{V} - 42\ \text{mV} = \mathbf{6.76\ V}$$

Related Problem   Repeat the analysis for $I_Z = 10$ mA and for $I_Z = 30$ mA using a 1N4742A zener with $V_Z = 12$ V at $I_Z = 21$ mA and $Z_Z = 9\ \Omega$.

**SECTION 3–1
CHECKUP**
Answers can be found at www
.pearsonhighered.com/floyd.

1. In what region of their characteristic curve are zener diodes operated?
2. At what value of zener current is the zener voltage normally specified?
3. How does the zener impedance affect the voltage across the terminals of the device?
4. What does a positive temperature coefficient of **0.05%/°C** mean?
5. Explain power derating.

3–2 ZENER DIODE APPLICATIONS

The zener diode can be used as a type of voltage regulator for providing stable reference voltages. In this section, you will see how zeners can be used as voltage references, regulators, and as simple limiters or clippers.

After completing this section, you should be able to

❏ **Apply a zener diode in voltage regulation**
 ❏ Analyze zener regulation with a variable input voltage
 ❏ Discuss zener regulation with a variable load
 ❏ Describe zener regulation from no load to full load
 ❏ Discuss zener limiting

Zener Regulation with a Variable Input Voltage

Zener diode regulators can provide a reasonably constant dc level at the output, but they are not particularly efficient. For this reason, they are limited to applications that require only low current to the load. Figure 3–9 illustrates how a zener diode can be used to regulate a

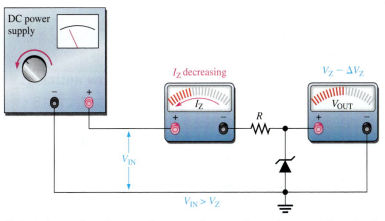

Zener regulation of a varying input voltage.

(a) As the input voltage increases, the output voltage remains nearly constant ($I_{ZK} < I_Z < I_{ZM}$).

(b) As the input voltage decreases, the output voltage remains nearly constant ($I_{ZK} < I_Z < I_{ZM}$).

dc voltage. As the input voltage varies (within limits), the zener diode maintains a nearly constant output voltage across its terminals. However, as V_{IN} changes, I_Z will change proportionally so that the limitations on the input voltage variation are set by the minimum and maximum current values (I_{ZK} and I_{ZM}) with which the zener can operate. Resistor R is the series current-limiting resistor. The meters indicate the relative values and trends.

To illustrate regulation, let's use the ideal model of the 1N4740A zener diode (ignoring the zener resistance) in the circuit of Figure 3–10. The absolute lowest current that will maintain regulation is specified at I_{ZK}, which for the 1N4740A is 0.25 mA and represents the no-load current. The maximum current is not given on the datasheet but can be calculated from the power specification of 1 W, which is given on the datasheet. Keep in mind that both the minimum and maximum values are at the operating extremes and represent worst-case operation.

$$I_{ZM} = \frac{P_D}{V_Z} = \frac{1\ W}{10\ V} = 100\ mA$$

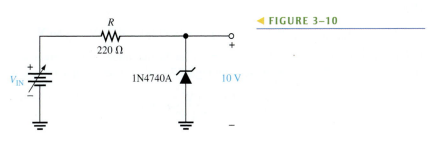

▶ FIGURE 3–14

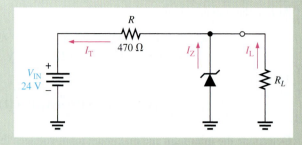

Solution   When $I_L = 0$ A ($R_L = \infty$), I_Z is maximum and equal to the total circuit current I_T.

$$I_{Z(max)} = I_T = \frac{V_{IN} - V_Z}{R} = \frac{24\text{ V} - 12\text{ V}}{470\ \Omega} = 25.5\text{ mA}$$

If R_L is removed from the circuit, the load current is 0 A. Since $I_{Z(max)}$ is less than I_{ZM}, 0 A is an acceptable minimum value for I_L because the zener can handle all of the 25.5 mA.

$$I_{L(min)} = \mathbf{0\ A}$$

The maximum value of I_L occurs when I_Z is minimum ($I_Z = I_{ZK}$), so

$$I_{L(max)} = I_T - I_{ZK} = 25.5\text{ mA} - 1\text{ mA} = \mathbf{24.5\ mA}$$

The minimum value of R_L is

$$R_{L(min)} = \frac{V_Z}{I_{L(max)}} = \frac{12\text{ V}}{24.5\text{ mA}} = \mathbf{490\ \Omega}$$

Therefore, if R_L is less than 490 Ω, R_L will draw more of the total current away from the zener and I_Z will be reduced below I_{ZK}. This will cause the zener to lose regulation. Regulation is maintained for any value of R_L between 490 Ω and infinity.

Related Problem   Find the minimum and maximum load currents for which the circuit in Figure 3–14 will maintain regulation. Determine the minimum value of R_L that can be used. $V_Z = 3.3$ V (constant), $I_{ZK} = 1$ mA, and $I_{ZM} = 150$ mA. Assume an ideal zener.

 Open the Multisim file EXM03-06 or LT Spice file EXS03-06 in the Examples folder on the website. For the calculated minimum value of load resistance, verify that regulation occurs.

In the last example, we assumed that Z_Z was zero and, therefore, the zener voltage remained constant over the range of currents. We made this assumption to demonstrate the concept of how the regulator works with a varying load. Such an assumption is often acceptable and in many cases produces results that are reasonably accurate. In Example 3–7, we will take the zener impedance into account.

EXAMPLE 3–7   For the circuit in Figure 3–15:

(a) Determine V_{OUT} at I_{ZK} and at I_{ZM}.

(b) Calculate the value of R that should be used.

(c) Determine the minimum value of R_L that can be used.

▶ FIGURE 3–15

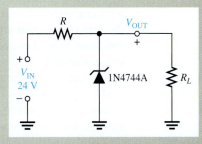

Solution   The 1N4744A zener used in the regulator circuit of Figure 3–15 is a 15 V diode. The datasheet in Figure 3–7 gives the following information:
$V_Z = 15$ V @ $I_Z = 17$ mA, $I_{ZK} = 0.25$ mA, and $Z_Z = 14$ Ω.

(a) For I_{ZK}:

$$V_{OUT} = V_Z - \Delta I_Z Z_Z = 15 \text{ V} - \Delta I_Z Z_Z = 15 \text{ V} - (I_Z - I_{ZK})Z_Z$$
$$= 15 \text{ V} - (16.75 \text{ mA})(14 \text{ Ω}) = 15 \text{ V} - 0.235 \text{ V} = \mathbf{14.76 \text{ V}}$$

Calculate the zener maximum current. The maximum power dissipation is 1 W.

$$I_{ZM} = \frac{P_D}{V_Z} = \frac{1 \text{ W}}{15 \text{ V}} = 66.7 \text{ mA}$$

For I_{ZM}:

$$V_{OUT} = V_Z + \Delta I_Z Z_Z = 15 \text{ V} + \Delta I_Z Z_Z$$
$$= 15 \text{ V} + (I_{ZM} - I_Z)Z_Z = 15 \text{ V} + (49.7 \text{ mA})(14 \text{ Ω}) = \mathbf{15.7 \text{ V}}$$

(b) Calculate the value of R for the maximum zener current that occurs when there is no load as shown in Figure 3–16(a).

$$R = \frac{V_{IN} - V_{OUT}}{I_{ZK}} = \frac{24 \text{ V} - 15.7 \text{ V}}{66.7 \text{ mA}} = 124 \text{ Ω}$$

$R = \mathbf{130 \text{ Ω}}$ (nearest larger standard value), which reduces I_{ZM} to 63.8 mA.

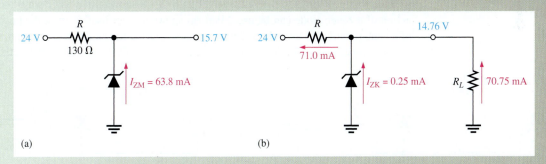

(a)   (b)

▲ FIGURE 3–16

(c) For the minimum load resistance (maximum load current), the zener current is minimum ($I_{ZK} = 0.25$ mA) as shown in Figure 3–16(b).

$$I_T = \frac{V_{IN} - V_{OUT}}{R} = \frac{24 \text{ V} - 14.76 \text{ V}}{130 \text{ Ω}} = 71.0 \text{ mA}$$
$$I_L = I_T - I_{ZK} = 71.0 \text{ mA} - 0.25 \text{ mA} = 70.75 \text{ mA}$$
$$R_{L(min)} = \frac{V_{OUT}}{I_L} = \frac{14.76 \text{ V}}{70.75 \text{ mA}} = \mathbf{209 \text{ Ω}}$$

Related Problem   Repeat each part of the preceding analysis if the zener is changed to a 1N4742A 12 V device.

You have seen how the zener diode regulates voltage. Its regulating ability is somewhat limited by the change in zener voltage over a range of current values, which restricts the load current that it can handle. To achieve better regulation and provide for greater variations in load current, the zener diode is combined as a key element with other circuit components to create a three-terminal linear voltage regulator. Three-terminal voltage regulators that were introduced in Chapter 2 are IC devices that use an internal zener diode to provide a reference voltage for an amplifier. For a given dc input voltage, the three-terminal regulator maintains an essentially constant dc voltage over a range of input voltages and load currents. The dc output voltage is always less than the input voltage. The details of this type of regulator are covered in Chapter 17. Figure 3–17 illustrates a basic three-terminal regulator showing where the zener diode is used.

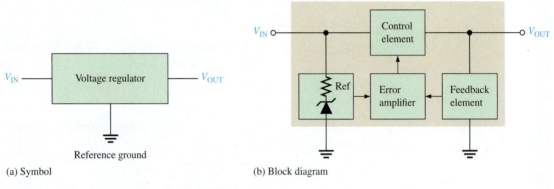

(a) Symbol   (b) Block diagram

▲ FIGURE 3–17

Three-terminal voltage regulators.

Zener Limiter

In addition to voltage regulation applications, zener diodes can be used in ac applications to limit voltage swings to desired levels. Figure 3–18 shows three basic ways the limiting action of a zener diode can be used. Part (a) shows a zener used to limit the positive peak of a signal voltage to the selected zener voltage. During the negative alternation, the zener acts as a forward-biased diode and limits the negative voltage to −0.7V. When the zener is

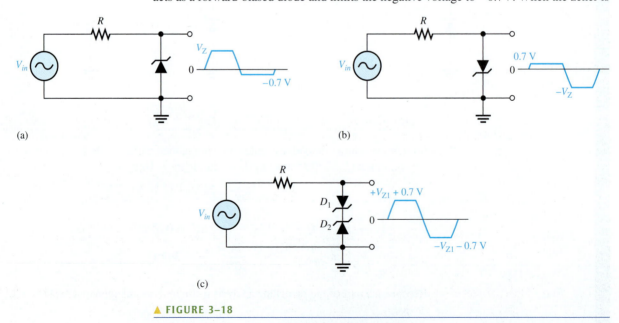

(a)   (b)

(c)

▲ FIGURE 3–18

Basic zener limiting action with a sinusoidal input voltage.

turned around, as in part (b), the negative peak is limited by zener action and the positive voltage is limited to +0.7 V. Two back-to-back zeners limit both peaks to the zener voltage ±0.7 V, as shown in part (c). During the positive alternation, D_2 is functioning as the zener limiter and D_1 is functioning as a forward-biased diode. During the negative alternation, the roles are reversed.

EXAMPLE 3–8

Determine the output voltage for each zener limiting circuit in Figure 3–19.

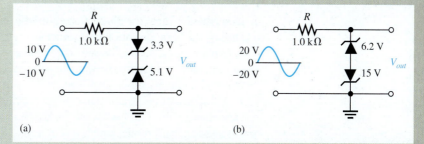

▲ FIGURE 3–19

Solution See Figure 3–20 for the resulting output voltages. Remember, when one zener is operating in breakdown, the other one is forward-biased with approximately 0.7 V across it.

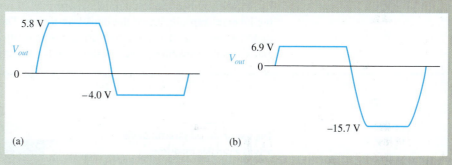

▲ FIGURE 3–20

Related Problem **(a)** What is the output in Figure 3–19(a) if the input voltage is increased to a peak value of 20 V?

(b) What is the output in Figure 3–19(b) if the input voltage is decreased to a peak value of 5 V?

 Open the Multisim file EXM03-08 or LT Spice file EXS03-08 in the Examples folder on the website. For the specified input voltages, measure the resulting output waveforms. Compare with the waveforms shown in the example.

SECTION 3–2 CHECKUP

1. In a zener diode regulator, what value of load resistance results in the maximum zener current?

2. Explain the terms *no load* and *full load*.

3. How much voltage appears across a zener diode when it is forward-biased?

4. What voltage is across the series resistor in a zener limiter circuit?

3–3 VARACTOR DIODES

The junction capacitance of diodes varies with the amount of reverse bias. Varactor diodes are specially designed to take advantage of this characteristic and are used as voltage-controlled capacitors rather than traditional diodes. These devices are commonly used in communication systems. Varactor diodes are also referred to as *varicaps* or *tuning diodes*.

After completing this section, you should be able to

❑ **Describe the varactor diode characteristic and analyze its operation**
❑ Discuss the basic operation of a varactor
   ◆ Explain why a reverse-biased varactor acts as a capacitor   ◆ Calculate varactor capacitance   ◆ Identify the varactor schematic symbol
❑ Interpret a varactor diode datasheet
   ◆ Define and discuss capacitance tolerance range   ◆ Define and discuss capacitance ratio   ◆ Discuss the back-to-back configuration
❑ Discuss and analyze the application of a varactor in a resonant band-pass filter

A **varactor** is a diode that always operates in reverse bias and is doped to maximize the inherent capacitance of the depletion region. The depletion region acts as a capacitor dielectric because of its nonconductive characteristic. The *p* and *n* regions are conductive and act as the capacitor plates, as illustrated in Figure 3–21.

▶ **FIGURE 3–21**

The reverse-biased varactor diode acts as a variable capacitor.

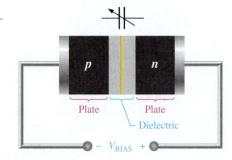

Basic Operation

Recall that capacitance is determined by the parameters of plate area (*A*), dielectric constant (ϵ), and plate separation (*d*), as expressed in the following formula:

$$C = \frac{A\epsilon}{d}$$

As the reverse-bias voltage increases, the depletion region widens, effectively increasing the plate separation, thus decreasing the capacitance. When the reverse-bias voltage decreases, the depletion region narrows, thus increasing the capacitance. This action is shown in Figure 3–22(a) and (b).

In a varactor diode, these capacitance parameters are controlled by the method of doping near the *pn* junction and the size and geometry of the diode's construction. Nominal

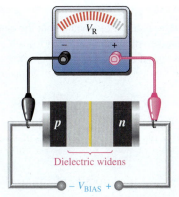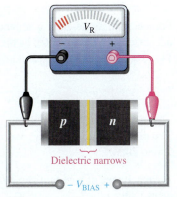

◄ FIGURE 3–22

Varactor diode capacitance varies with reverse voltage.

(a) Greater reverse bias, less capacitance   (b) Less reverse bias, greater capacitance

varactor capacitances are typically available from a few picofarads to several hundred picofarads. Figure 3–23 shows a common symbol for a varactor.

Varactor Capacitance Ratio

The varactor **capacitance ratio**, CR, is also known as the *tuning ratio*. It is the ratio of the diode capacitance at a maximum reverse voltage to the diode capacitance at a minimum reverse voltage. For the varactor diode represented by the graph in Figure 3–24, the capacitance ratio is the ratio of capacitance measured at a reverse voltage of 1.4 V divided by capacitance measured at a reverse voltage of 25 V.

▲ FIGURE 3–23

Varactor diode symbol.

$$CR = \frac{C_{MAX}}{C_{MIN}}$$

The doping in the *n* and *p* regions is made uniform so that at the *pn* junction there is a very abrupt change from *n* to *p* instead of the more gradual change found in the rectifier diodes. The abruptness of the *pn* junction determines the capacitance ratio.

A graph of diode capacitance (C_T) versus reverse voltage for a certain varactor is shown in Figure 3–24. For this particular device, C_T varies from 30 pF to slightly less than 4 pF as V_R varies from 1 V to 30 V.

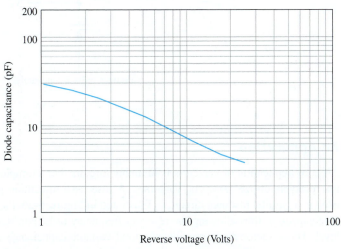

◄ FIGURE 3–24

Diode capacitance as a function of reverse voltage; typical values.

Example of a diode capacitance versus reverse voltage graph

EXAMPLE 3–9

For a certain diode, the capacitance ratio is 6.0. This means that the capacitance value decreases by a factor of 6.0 as the reverse voltage is increased from $V_{MIN} = 2$ V to $V_{MAX} = 20$ V. Find the capacitance range, if $C_{MAX} = 22$ pF.

Solution

$$C_{MIN} = \frac{C_{MAX}}{CR} = \frac{22 \text{ pF}}{6.0} = 3.7 \text{ pF}$$

The diode capacitance range is from 22 pF to 3.7 pF when V_R is increased from 2 V to 20 V.

Related Problem

A capacitance of a certain varactor can be varied from 15 pF to 100 pF. What is the capacitance ratio?

Back-to-Back Configuration   One of the drawbacks of using just a single varactor diode in certain applications, such as rf tuning, is that if the diode is forward-biased by the rf signal during part of the ac cycle, its reverse leakage will increase momentarily. Also, a type of distortion called *harmonic distortion* is produced if the varactor is alternately biased positively and negatively. To avoid harmonic distortion, you will often see two varactor diodes back to back, as shown in Figure 3–25(a) with the reverse dc voltage applied to both devices simultaneously. The two tuning diodes will be driven alternately into high and low capacitance, and the net capacitance will remain constant and is unaffected by the rf signal amplitude. The Zetex 832A varactor diode is available in a back-to-back configuration in an SOT23 surface mount package or as a single diode in an SOD523 surface mount package, as shown in Figure 3–25(b). Although the cathodes in the back-to-back configuration are connected to a common pin, each diode can also be used individually.

▶ **FIGURE 3–25**

Varactor diodes and typical packages.

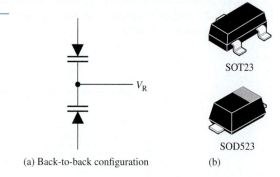

(a) Back-to-back configuration   (b)

An Application

A major application of varactors is in tuning circuits. For example, VHF, UHF, and satellite receivers utilize varactors. Varactors are also used in cellular communications. When used in a parallel resonant circuit, as illustrated in Figure 3–26, the varactor acts as a variable capacitor, thus allowing the resonant frequency to be adjusted by a variable voltage level. The varactor diode provides the total variable capacitance in the parallel resonant band-pass filter. The varactor diode and the inductor form a parallel resonant circuit from

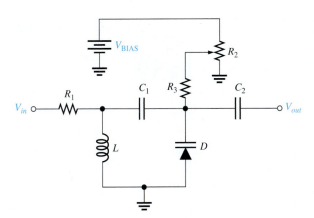

A resonant band-pass filter using a varactor diode for adjusting the resonant frequency over a specified range.

the output to ac ground. The capacitors C_1 and C_2 have no effect on the filter's frequency response because their reactances are negligible at the resonant frequencies. C_1 prevents a dc path from the potentiometer wiper back to the ac source through the inductor and R_1. C_2 prevents a dc path from the wiper of the potentiometer to a load on the output. The potentiometer R_2 forms a variable dc voltage for biasing the varactor. The reverse-bias voltage across the varactor can be varied with the potentiometer. The bias voltage must possess good voltage and temperature stability to avoid frequency drift and provide a constant capacitance.

Recall that the parallel resonant frequency is

$$f_r \cong \frac{1}{2\pi\sqrt{LC}}$$

EXAMPLE 3–10

(a) Given that the capacitance of a certain varactor is approximately 40 pF at 0 V bias and that the capacitance at a 2 V reverse bias is 22 pF, determine the capacitance at a reverse bias of 20 V using a minimum capacitance ratio of 5.0.

(b) Using the capacitances at bias voltages of 0 V and 20 V, calculate the resonant frequencies at the bias extremes for the circuit in Figure 3–26 if $L = 2$ mH.

(c) Verify the frequency calculations by simulating the circuit in Figure 3–26 for the following component values: $R_1 = 47$ kΩ, $R_2 = 10$ kΩ, $R_3 = 5.1$ MΩ, $C_1 = 10$ nF, $C_2 = 10$ nF, $L = 2$ mH, and $V_{BIAS} = 20$ V.

Solution

(a) $C_{20} = \dfrac{C_2}{CR} = \dfrac{22 \text{ pF}}{5.0} = \mathbf{4.4 \text{ pF}}$

(b) $f_0 = \dfrac{1}{2\pi\sqrt{LC}} = \dfrac{1}{2\pi\sqrt{(2 \text{ mH})(40 \text{ pF})}} = \mathbf{563 \text{ kHz}}$

$f_{20} = \dfrac{1}{2\pi\sqrt{LC}} = \dfrac{1}{2\pi\sqrt{(2 \text{ mH})(4.4 \text{ pF})}} = \mathbf{1.7 \text{ MHz}}$

(c) The Multsim simulation of the circuit is shown in Figure 3–27. The Bode plotters show the frequency responses at 0 V and 20 V reverse bias. The center of the 0 V bias response curve is at 558 kHz and the center of the 20 V bias response curve is at 1.69 MHz. These results agree reasonably well with the calculated values.

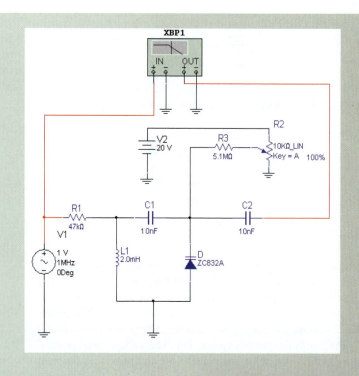

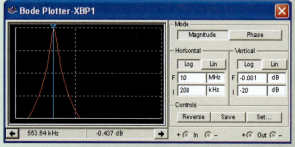
Frequency response for 0 V varactor bias

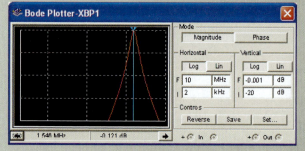
Frequency response for 20 V reverse varactor bias

▲ **FIGURE 3–27**

Multisim simulation.

These results show that this circuit can be tuned over most of the AM broadcast band.

Related Problem   How could you increase the tuning range of the circuit?

SECTION 3–3 CHECKUP

1. What is the key feature of a varactor diode?
2. Under what bias condition is a varactor operated?
3. What part of the varactor produces the capacitance?
4. Based on the graph in Figure 3–22(c), what happens to the diode capacitance when the reverse voltage is increased?
5. Define *capacitance ratio*.

3–4   OPTICAL DIODES

In this section, three types of optoelectronic devices are introduced: the light-emitting diode, quantum dots, and the photodiode. As the name implies, the light-emitting diode is a light emitter. Quantum dots are very tiny light emitters made from silicon with great promise for various devices, including light-emitting diodes. On the other hand, the photodiode is a light detector.

After completing this section, you should be able to

❑ **Discuss the basic characteristics, operation, and applications of LEDs, quantum dots, and photodiodes**
❑ Describe the light-emitting diode (LED)
   ◆ Identify the LED schematic symbol   ◆ Discuss the process of electroluminescence   ◆ List some LED semiconductor materials   ◆ Discuss LED biasing
   ◆ Discuss light emission
❑ Interpret an LED datasheet
   ◆ Define and discuss radiant intensity and irradiance
❑ Describe some LED applications
❑ Discuss high-intensity LEDs and applications
   ◆ Explain how high-intensity LEDs are used in traffic lights   ◆ Explain how high-intensity LEDs are used in displays
❑ Describe the organic LED (OLED)
❑ Discuss quantum dots and their application
❑ Describe the photodiode and interpret a typical datasheet
   ◆ Discuss photodiode sensitivity

The Light-Emitting Diode (LED)

The symbol for an LED is shown in Figure 3–28.

The basic operation of the **light-emitting diode (LED)** is as follows. When the device is forward-biased, electrons cross the *pn* junction from the *n*-type material and recombine with holes in the *p*-type material. Recall from Chapter 1 that these free electrons are in the conduction band and at a higher energy than the holes in the valence band. The difference in energy between the electrons and the holes corresponds to the energy of visible light. When recombination takes place, the recombining electrons release energy in the form of **photons.** The emitted light tends to be monochromatic (one color) that depends on the band gap (and other factors). A large exposed surface area on one layer of the semiconductive material permits the photons to be emitted as visible light. This process, called **electroluminescence**, is illustrated in Figure 3–29. Various impurities are added during the doping process to establish the **wavelength** of the emitted light. The wavelength determines the color of visible light. Some LEDs emit photons that are not part of the visible spectrum but have longer wavelengths and are in the **infrared** (IR) portion of the spectrum.

▲ **FIGURE 3–28**

Symbol for an LED. When forward-biased, it emits light.

LED Semiconductor Materials   The semiconductor gallium arsenide (GaAs) was used in early LEDs and emits IR radiation, which is invisible. The first visible red LEDs were produced using gallium arsenide phosphide (GaAsP) on a GaAs substrate. The efficiency was increased using a gallium phosphide (GaP) substrate, resulting in brighter red LEDs and also allowing orange LEDs.

Later, GaP was used as the light-emitter to achieve pale green light. By using a red and a green chip, LEDs were able to produce yellow light. The first super-bright red, yellow, and green LEDs were produced using gallium aluminum arsenide phosphide (GaAlAsP). By the early 1990s ultrabright LEDs using indium gallium aluminum phosphide (InGaAlP) were available in red, orange, yellow, and green.

▶ FIGURE 3–29

Electroluminescence in a forward-biased LED.

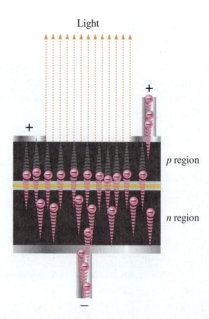

Light

p region

n region

Blue LEDs using silicon carbide (SiC) and ultrabright blue LEDs made of gallium nitride (GaN) became available. High intensity LEDs that produce green and blue are also made using indium gallium nitride (InGaN). High-intensity white LEDs are formed using ultrabright blue GaN coated with fluorescent phosphors that absorb the blue light and reemit it as white light.

LED Biasing   The forward voltage across an LED is considerably greater than for a silicon diode. Typically, the maximum V_F for LEDs is between 1.2 V and 3.2 V, depending on the material. Reverse breakdown for an LED is much less than for a silicon rectifier diode (3 V to 10 V is typical).

The LED emits light in response to a sufficient forward current, as shown in Figure 3–30(a). The amount of power output translated into light is directly proportional to the forward current, as indicated in Figure 3–30(b). An increase in I_F corresponds proportionally to an increase in light output. The light output (both intensity and color) is also dependent on temperature. Light intensity goes down with higher temperature as indicated in the figure.

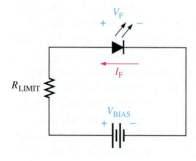

(a) Forward-biased operation

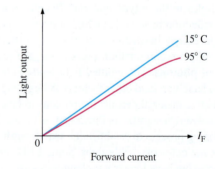

(b) General light output versus forward current for two temperatures

▲ FIGURE 3–30

Basic operation of an LED.

Light Emission   An LED emits light over a specified range of wavelengths as indicated by the **spectral** output curves in Figure 3–31. The curves in part (a) represent the light output versus wavelength for typical visible LEDs, and the curve in part (b) is for a typical infrared LED. The wavelength (λ) is expressed in nanometers (nm). The normalized output of the visible red LED peaks at 660 nm, the yellow at 590 nm, green at 540 nm, and blue at 460 nm. The output for the infrared LED peaks at 940 nm.

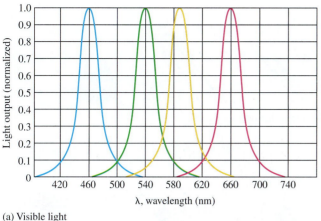

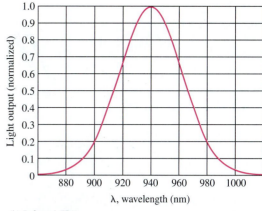

(a) Visible light

(b) Infrared (IR)

▲ **FIGURE 3–31**

Examples of typical spectral output curves for LEDs.

The graphs in Figure 3–32 show typical **radiation** patterns for small LEDs. LEDs are directional light sources (unlike filament or fluorescent bulbs). The radiation pattern is generally perpendicular to the emitting surface; however, it can be altered by the shape of the emitter surface and by lenses and diffusion films to favor a specific direction. Directional patterns can be an advantage for certain applications, such as traffic lights, where the light is intended to be seen only by certain drivers. Figure 3–32(a) shows the pattern for a forward-directed LED such as those used in small panel indicators. Figure 3–32(b) shows the pattern for a wider viewing angle such as that produced by many super-bright LEDs. A wide variety of patterns are available from manufacturers; one variation is to design the LED to emit nearly all the light to the side in two lobes.

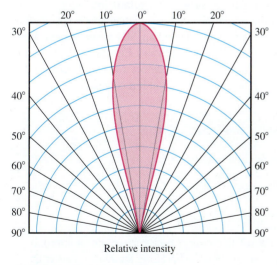

(a) A narrow viewing angle LED

(b) A wide viewing angle LED

▲ **FIGURE 3–32**

Radiation patterns for two different LEDs.

Typical small LEDs for indicators are shown in Figure 3–33(a). In addition to small LEDs for indicators, bright LEDs are becoming popular for lighting because of their superior efficiency and long life. A typical LED for lighting can deliver 50–60 lumens per watt, which is approximately five times greater efficiency than a standard incandescent bulb. LEDs for lighting are available in a variety of configurations, including low-wattage bulbs

► FIGURE 3-33

Typical LEDs.

(a) Typical small LEDs for indicators

| Helion 12 V overhead light with socket and module | 120 V, 3.5 W screw base for low-level illumination | 120 V, 1 W small screw base candelabra style | 6 V, bayonet base for flashlights, etc. |

(b) Typical LEDs for lighting applications

for outdoor walkways and gardens. Many LED lamps are designed to work in 120 V standard fixtures. A few representative configurations are shown in Figure 3–33(b).

LED Datasheet Information

A partial datasheet for an TSMF1000 infrared (IR) light-emitting diode is shown in Figure 3–34. Notice that the maximum reverse voltage is only 5 V, the maximum forward current is 100 mA, and the forward voltage drop is approximately 1.3 V for $I_F = 20$ mA.

From the graph in part (c), you can see that the peak power output for this device occurs at a wavelength of 870 nm; its radiation pattern is shown in part (d).

Radiant Intensity and Irradiance   In Figure 3–34(a), the **radiant intensity,** I_e (symbol not to be confused with current), is the output power per steradian and is specified as 5 mW/sr at $I_F = 20$ mA. The steradian (sr) is the unit of solid angular measurement. **Irradiance,** E, is the power per unit area at a given distance from an LED source expressed in mW/cm². Irradiance is important because the response of a detector (photodiode) used in conjunction with an LED depends on the irradiance of the light it receives.

EXAMPLE 3–11

From the LED datasheet in Figure 3–34 determine the following:

(a) The radiant power at 910 nm if the maximum output is 35 mW.

(b) The forward voltage drop for $I_F = 20$ mA.

(c) The radiant intensity for $I_F = 40$ mA.

Solution   **(a)** From the graph in Figure 3–34(c), the relative radiant power at 910 nm is approximately 0.25 and the peak radiant power is 35 mW. Therefore, the radiant power at 910 nm is

$$\phi_e = 0.25(35 \text{ mW}) = \textbf{8.75 mW}$$

(b) From the graph in part (b), $V_F \cong \textbf{1.25 V}$ for $I_F = 20$ mA.

(c) From the graph in part (e), $I_e \cong \textbf{10 mW/sr}$ for $I_F = 40$ mA.

Related Problem   Determine the relative radiant power at 850 nm.

Absolute Maximum Ratings

T_{amb} = 25°C, unless otherwise specified

Parameter	Test condition	Symbol	Value	Unit
Reverse Voltage		V_R	5	V
Forward current		I_F	100	mA
Peak Forward Current	t_p/T = 0.5, t_p = 100 μs	I_{FM}	200	mA
Surge Forward Current	t_p = 100 μs	I_{FSM}	0.8	A
Power Dissipation		P_V	190	mW
Junction Temperature		T_j	100	°C
Operating Temperature Range		T_{amb}	- 40 to + 85	°C

Basic Characteristics

T_{amb} = 25°C, unless otherwise specified
T_{amb} = 25°C, unless otherwise specified

Parameter	Test condition	Symbol	Min	Typ.	Max	Unit
Forward Voltage	I_F = 20 mA	V_F		1.3	1.5	V
	I_F = 1 A, t_p = 100 μs	V_F		2.4		V
Temp. Coefficient of V_F	I_F = 1.0 mA	TK_{VF}		- 1.7		mV/K
Reverse Current	V_R = 5 V	I_R			10	μA
Junction capacitance	V_R = 0 V, f = 1 MHz, E = 0	C_j		160		pF
Radiant Intensity	I_F = 20 mA	I_e	2.5	5	13	mW/sr
	I_F = 100 mA, t_p = 100 μs	I_e		25		mW/sr
Radiant Power	I_F = 100 mA, t_p = 20 ms	Φ_e		35		mW
Temp. Coefficient of $\cdot \Phi_e$	I_F = 20 mA	$TK\Phi_e$		- 0.6		%/K
Angle of Half Intensity		ϕ		± 17		deg
Peak Wavelength	I_F = 20 mA	λ_p		870		nm
Spectral Bandwidth	I_F = 20 mA	$\Delta\lambda$		40		nm
Temp. Coefficient of λ_p	I_F = 20 mA	$TK\lambda_p$		0.2		nm/K
Rise Time	I_F = 20 mA	t_r		30		ns
Fall Time	I_F = 20 mA	t_f		30		ns
Virtual Source Diameter		$\varnothing$		1.2		mm

(a)

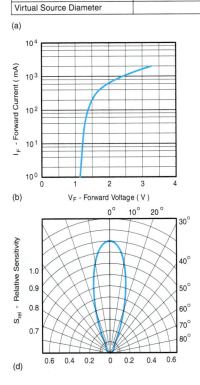

(b)

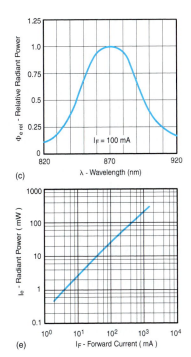

(c)

(d)

(e)

▲ **FIGURE 3–34**

Partial datasheet for an TSMF1000 IR light-emitting diode. Datasheet courtesy of Vishay Intertechnology, Inc. Datasheets are available at www.vishay.com.

▶ FIGURE 3–38

The lens directs the light emitted from the LED to optimize visibility.

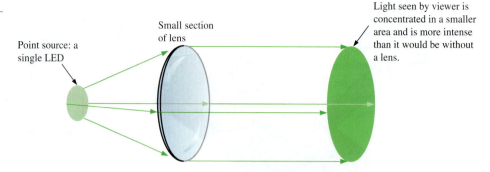

and control the current to the LEDs. This is a more efficient way to control the current to the LEDs and can save power while reducing heat dissipation.

Some LED traffic arrays use small reflectors for each LED to help maximize the effect of the light output. Also, an optical lens covers the front of the array to direct the light from each individual diode to prevent improper dispersion of light and to optimize the visibility. Figure 3–38 illustrates how a lens is used to direct the light toward the viewer.

The particular LED circuit configuration depends on the voltage and the color of the LED. Different color LEDs require different forward voltages to operate. Red LEDs take the least; and as the color moves up the color spectrum toward blue, the voltage requirement increases. Typically, a red LED requires about 2 V, while blue LEDs require between 3 V and 4 V. Generally, LEDs, however, need 20 mA to 30 mA of current, regardless of their voltage requirements. Typical V-I curves for red, yellow, green, and blue LEDs are shown in Figure 3–39.

▶ FIGURE 3–39

V-I characteristic curves for visible-light LEDs.

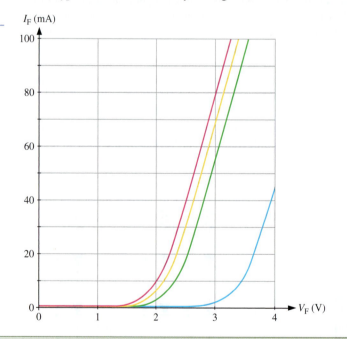

EXAMPLE 3–12

Using the graph in Figure 3–39, determine the green LED forward voltage for a current of 20 mA. Design a 12 V LED circuit to minimize the number of limiting resistors for an array of 60 diodes.

Solution From the graph, a green LED has a forward voltage of approximately 2.5 V for a forward current of 20 mA. The maximum number of series LEDs is 3. The total voltage across three LEDs is

$$V = 3 \times 2.5 \text{ V} = 7.5 \text{ V}$$

The voltage drop across the series-limiting resistor is

$$V = 12 \text{ V} - 7.5 \text{ V} = 4.5 \text{ V}$$

The value of the limiting resistor is

$$R_{LIMIT} = \frac{4.5 \text{ V}}{20 \text{ mA}} = 225 \ \Omega$$

The LED array has 20 parallel branches each with a limiting resistor and three LEDs, as shown in Figure 3–40.

► **FIGURE 3–40**

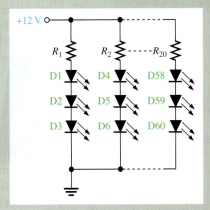

Related Problem   Design a 12 V red LED array with minimum limiting resistors, a forward current of 30 mA, and containing 64 diodes.

LED Displays   LEDs are widely used in large and small signs and message boards for both indoor and outdoor uses, including large-screen television. Signs can be single-color, multicolor, or full-color. Full-color screens use a tiny grouping of high-intensity red, green, and blue LEDs to form a **pixel**. A typical screen is made of thousands of RGB pixels with the exact number determined by the sizes of the screen and the pixel.

Red, green, and blue (RGB) are primary colors and when mixed together in varying amounts, can be used to produce any color in the visible spectrum. A basic pixel formed by three LEDs is shown in Figure 3–41. The light emission from each of the three diodes can be varied independently by varying the amount of forward current. Yellow is added to the three primary colors (RGBY) in some TV screen applications.

Other Applications   High-intensity LEDs are becoming more widely used in automotive lighting for taillights, brakelights, turn signals, back-up lights, and interior applications. LED arrays are expected to replace most incandescent bulbs in automotive lighting. Eventually, headlights may also be replaced by white LED arrays. LEDs can be seen better in poor weather and can last 100 times longer than an incandescent bulb.

LEDs are also finding their way into interior home and business lighting applications. Arrays of white LEDs may eventually replace incandescent light bulbs and flourescent lighting in interior living and work areas. As previously mentioned, most white LEDs use a blue GaN (gallium nitride) LED covered by a yellowish phosphor coating made of a certain type of crystals that have been powdered and bound in a type of viscous adhesive. Since yellow light stimulates the red and green receptors of the eye, the resulting mix of blue and yellow light gives the appearance of white.

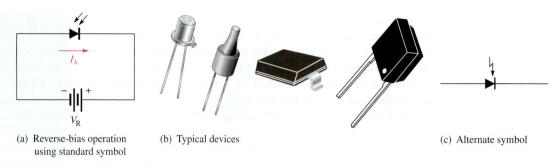

(a) Reverse-bias operation using standard symbol

(b) Typical devices

(c) Alternate symbol

▲ FIGURE 3–43

Photodiode.

Recall that when reverse-biased, a rectifier diode has a very small reverse leakage current. The same is true for a photodiode. The reverse-biased current is produced by thermally generated electron-hole pairs in the depletion region, which are swept across the *pn* junction by the electric field created by the reverse voltage. In a rectifier diode, the reverse leakage current increases with temperature due to an increase in the number of electron-hole pairs.

A photodiode differs from a rectifier diode in that when its *pn* junction is exposed to light, the reverse current increases with the light intensity. When there is no incident light, the reverse current, I_λ, is almost negligible and is called the **dark current.** An increase in the amount of light intensity, expressed as irradiance (mW/cm^2), produces an increase in the reverse current, as shown by the graph in Figure 3–44(a).

From the graph in Figure 3–44(b), you can see that the reverse current for this particular device is approximately 1.4 μA at a reverse-bias voltage of 10 V with an irradiance of 0.5 mW/cm^2. Therefore, the resistance of the device is

$$R_R = \frac{V_R}{I_\lambda} = \frac{10 \text{ V}}{1.4 \text{ } \mu A} = 7.14 \text{ M}\Omega$$

At 20 mW/cm^2, the current is approximately 55 μA at $V_R = 10$ V. The resistance under this condition is

$$R_R = \frac{V_R}{I_\lambda} = \frac{10 \text{ V}}{55 \text{ } \mu A} = 182 \text{ k}\Omega$$

These calculations show that the photodiode can be used as a variable-resistance device controlled by light intensity.

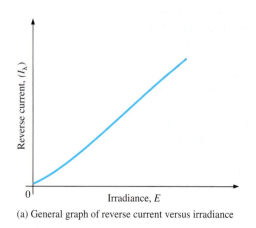

(a) General graph of reverse current versus irradiance

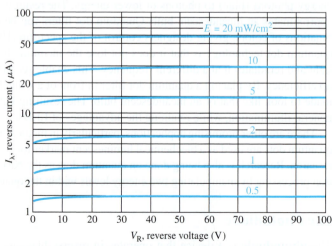

(b) Example of a graph of reverse current versus reverse voltage for several values of irradiance

▲ FIGURE 3–44

Typical photodiode characteristics.

Figure 3–45 illustrates that the photodiode allows essentially no reverse current (except for a very small dark current) when there is no incident light. When a light beam strikes the photodiode, it conducts an amount of reverse current that is proportional to the light intensity (irradiance).

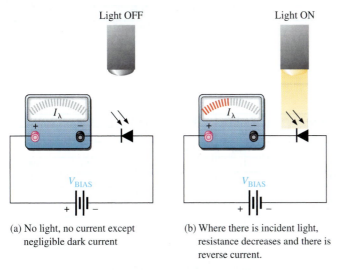

(a) No light, no current except negligible dark current

(b) Where there is incident light, resistance decreases and there is reverse current.

▲ **FIGURE 3–45**

Operation of a photodiode showing current measurements.

Photodiode Datasheet Information

A partial datasheet for an TEMD1000 photodiode is shown in Figure 3–46. Notice that the maximum reverse voltage is 60 V and the dark current (reverse current with no light) is typically 1 nA for a reverse voltage of 10 V. The dark current increases with an increase in reverse voltage and also with an increase in temperature.

Sensitivity From the graph in part (b), you can see that the maximum sensitivity for this device occurs at a wavelength of 950 nm. The angular response graph in part (c) shows an area of response measured as relative sensitivity. At 10° on either side of the maximum orientation, the sensitivity drops to approximately 82% of maximum.

EXAMPLE 3–13

For a TEMD1000 photodiode,

(a) Determine the maximum dark current for $V_R = 10$ V.

(b) Determine the reverse light current for an irradiance of 1 mW/cm^2 at a wavelength of 850 nm if the device angle is oriented at 10° with respect to the maximum irradiance and the reverse voltage is 5 V.

Solution (a) From Figure 3–46(a), the maximum dark current $I_{ro} = $ **10 nA.**

(b) From the graph in Figure 3–46(d), the reverse light current is 12 μA at 950 nm. From Figure 3–46(b), the relative sensitivity is 0.6 at 850 nm. Therefore, the reverse light current is

$$I_\lambda = I_{ra} = 0.6(12 \ \mu A) = 72 \ \mu A$$

For an angle of 10°, the relative sensitivity is reduced to 0.92 of its value at 0°.

$$I_\lambda = I_{ra} = 0.92(7.2 \ \mu A) = \textbf{6.62} \ \boldsymbol{\mu}\textbf{A}$$

Related Problem What is the reverse current if the wavelength is 1050 nm and the angle is 0°?

A complete PV solar cell.

The conductive grid across the top of the cell is necessary so that the electrons have a shorter distance to travel through the silicon when an external load is connected. The farther electrons travel through the silicon material, the greater the energy loss due to resistance. A solid contact covering all of the bottom of the wafer is then added, as indicated in the figure. Thickness of the solar cell compared to the surface area is greatly exaggerated for purposes of illustration.

After the contacts are incorporated, an antireflective coating is placed on top the contact grid and *n* region, as shown in Figure 3–47(c). This allows the solar cell to absorb as much of the sun's energy as possible by reducing the amount of light energy reflected away from the surface of the cell. Finally, a glass or transparent plastic layer is attached to the top of the cell with transparent adhesive to protect it from the weather. Figure 3–48 shows a completed solar cell.

Operation of a Solar Cell   As indicated before, sunlight is composed of photons, or "packets" of energy. The sun produces an astounding amount of energy. The small fraction of the sun's total energy that reaches the earth is enough to meet all of our power needs many times over. There is sufficient solar energy striking the earth each hour to meet worldwide demands for an entire year.

The *n*-type layer is very thin compared to the *p* region to allow light penetration into the *p* region. The thickness of the entire cell is actually about the thickness of an eggshell. When a photon penetrates either the *n* region or the *p*-type region and strikes a silicon atom near the *pn* junction with sufficient energy to knock an electron out of the valence band, the electron becomes a free electron and leaves a hole in the valence band, creating an *electron-hole pair.* The amount of energy required to free an electron from the valence band of a silicon atom is called the band-gap energy and is 1.12 eV (electron volts). In the *p* region, the free electron is swept across the depletion region by the electric field into the *n* region. In the *n* region, the hole is swept across the depletion region by the electric field into the *p* region. Electrons accumulate in the *n* region, creating a negative charge; and holes accumulate in the *p* region, creating a positive charge. A voltage is developed between the *n* region and *p* region contacts, as shown in Figure 3–49.

Basic operation of a solar cell with incident sunlight.

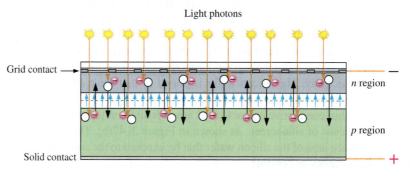

When a load is connected to a solar cell via the top and bottom contacts, the free electrons flow out of the *n* region to the grid contacts on the top surface, through the negative contact, through the load and back into the positive contact on the bottom surface, and into the *p* region where they can recombine with holes. The sunlight energy continues to create new electron-hole pairs and the process goes on, as illustrated in Figure 3–50.

A solar cell producing voltage and current through a load under incident sunlight.

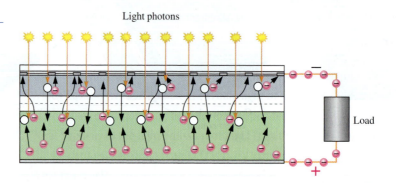

Solar Cell Characteristics

Solar cells are typically 100 cm² to 225 cm² in size. The usable voltage from silicon solar cells is approximately 0.5 V to 0.6 V. Terminal voltage is only slightly dependent on the intensity of light radiation, but the current increases with light intensity. For example, a 100 cm² silicon cell reaches a maximum current of approximately 2 A when radiated by 1000 W/m² of light.

Figure 3–51 shows the *V-I* characteristic curves for a typical solar cell for various light intensities. Higher light intensity produces more current. The operating point for maximum power output for a given light intensity should be in the "knee" area of the curve, as indicated by the dashed line. The load on the solar cell controls this operating point ($R_L = V/I$).

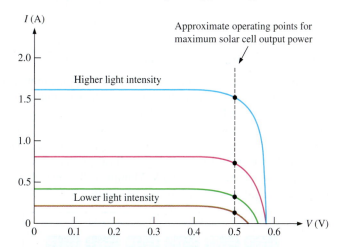

◀ FIGURE 3–51

V-I characteristic for a typical single solar cell from increasing light intensities.

In a solar power system, the cell is generally loaded by a charge controller or an inverter. A special method called *maximum power point tracking* will sense the operating point and adjust the load resistance to keep it in the knee region. For example, assume the solar cell is operating on the highest intensity curve (blue) shown in Figure 3–51. For maximum power (dashed line), the voltage is 0.5 V and the current is 1.5 A. For this condition, the load is

$$R_L = \frac{V}{I} = \frac{0.5 \text{ V}}{1.5 \text{ A}} = 0.33 \ \Omega$$

Now, if the light intensity falls to where the cell is operating on the red curve, the current is less and the load resistance will have to change to maintain maximum power output as follows:

$$R_L = \frac{V}{I} = \frac{0.5 \text{ V}}{0.8 \text{ A}} = 0.625 \ \Omega$$

If the resistance did not change, the voltage output would drop to

$$V = IR = (0.8 \text{ A})(0.33 \text{ W}) = 0.264 \text{ V}$$

resulting in less than maximum power output for the red curve. Of course, the power will still be less on the red curve than on the blue curve because the current is less.

The output voltage and current of a solar cell is also temperature dependent. Notice in Figure 3–52 that for a constant light intensity the output voltage decreases as the temperature increases but the current is affected only by a small amount.

Solar Cell Panels

Currently, the problem is in harnessing solar energy in sufficient amounts and at a reasonable cost to meet our requirements. It takes approximately a square meter solar panel to produce 100 W in a sunny climate. Some energy can be harvested even if cloud cover exists, but no energy can be obtained during the night.

The basic construction of a laser diode is shown in Figure 3–54(b). A *pn* junction is formed by two layers of doped gallium arsenide. The length of the *pn* junction bears a precise relationship with the wavelength of the light to be emitted. There is a highly reflective surface at one end of the *pn* junction and a partially reflective surface at the other end, forming a resonant cavity for the photons. External leads provide the anode and cathode connections.

The basic operation is as follows. The laser diode is forward-biased by an external voltage source. As electrons move through the junction, recombination occurs just as in an ordinary diode. As electrons fall into holes to recombine, photons are released. A released photon can strike an atom, causing another photon to be released. As the forward current is increased, more electrons enter the depletion region and cause more photons to be emitted. Eventually some of the photons that are randomly drifting within the depletion region strike the reflected surfaces perpendicularly. These reflected photons move along the depletion region, striking atoms and releasing additional photons due to the avalanche effect. This back-and-forth movement of photons increases as the generation of photons "snowballs" until a very intense beam of laser light is formed by the photons that pass through the partially reflective end of the *pn* junction.

Each photon produced in this process is identical to the other photons in energy level, phase relationship, and frequency. So a single wavelength of intense light emerges from the laser diode, as indicated in Figure 3–54(c). Laser diodes have a threshold level of current above which the laser action occurs and below which the diode behaves essentially as an LED, emitting incoherent light.

An Application   Laser diodes and photodiodes are used in the pick-up systems of compact disc (CD) players, DVD players, and Blu-ray discs. All three work in a similar manner. In the case of CD players, audio information (sound) is digitally recorded in stereo on the surface of a compact disc in the form of microscopic "pits" and "flats." A lens arrangement focuses the laser beam from the diode onto the CD surface. As the CD rotates, the lens and beam follow the track under control of a servomotor. The laser light, which is altered by the pits and flats along the recorded track, is reflected back from the track through a lens and optical system to infrared photodiodes. The signal from the photodiodes is then used to reproduce the digitally recorded sound. In the case of the DVD player, a shorter wavelength is used and the tracks are denser, so they hold more data and the data is sampled faster; most computers use a DVD drive instead of the older CD drive. DVDs can be encoded with audio, video, or other digital data. As in the case of the CD player, the pits and flats represent the data, but in this case they are closer together and smaller, to squeeze more data into an equivalent space. The latest technology that uses this idea is that of Blu-ray discs, which have even smaller pits and lands and can hold considerably more data than a DVD. Blu-rays are named for the shorter-wavelength blue laser that can write smaller than the red laser used in DVDs. The net result is the ability to store higher-quality movies (HD) or just more data than on a DVD.

Laser diodes are also used in many other applications, such as laser printers and fiber-optic systems, including sensors used in a wide variety of products such as bar code readers. Light from the laser diode is moved across the bar code, and the reflected light is turned into a digital code depending on whether it detects a dark line or a white space.

The Schottky Diode

Schottky diodes are high-current diodes used primarily in high-frequency and fast-switching applications. They are also known as *hot-carrier diodes*. The term *hot-carrier* is derived from the higher energy level of electrons in the *n* region compared to those in the metal region. A Schottky diode symbol is shown in Figure 3–55. A Schottky diode is formed by joining a doped semiconductor region (usually *n*-type) with a metal such as gold, silver, or platinum. Rather than a *pn* junction, there is a metal-to-semiconductor junction, as shown in Figure 3–56. The forward voltage drop is typically around 0.3 V because there is no depletion region as in a *pn* junction diode.

▲ **FIGURE 3–55**

Schottky diode symbol.

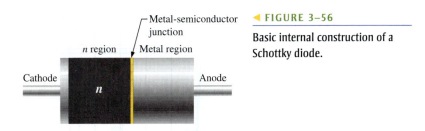

◀ FIGURE 3–56

Basic internal construction of a Schottky diode.

The Schottky diode operates only with majority carriers. There are no minority carriers and thus no reverse leakage current as in other types of diodes. The metal region is heavily occupied with conduction-band electrons, and the *n*-type semiconductor region is lightly doped. When forward-biased, the higher-energy electrons in the *n* region are injected into the metal region where they give up their excess energy very rapidly. Since there are no minority carriers, as in a conventional rectifier diode, there is a very rapid response to a change in bias. The Schottky is a fast-switching diode, and most of its applications make use of this property. It can be used in high-frequency applications and in many digital circuits to decrease switching times. The LS family of TTL logic (LS stands for low-power Schottky) is one type of digital integrated circuit that uses the Schottky diode.

The *PIN* Diode

The *pin* diode consists of heavily doped *p* and *n* regions separated by an intrinsic (*i*) region, as shown in Figure 3–57(a). When reverse-biased, the *pin* diode acts like a nearly constant capacitance. When forward-biased, it acts like a current-controlled variable resistance. This is shown in Figure 3–57(b) and (c). The low forward resistance of the intrinsic region decreases with increasing current.

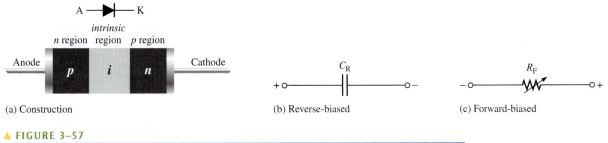

(a) Construction

(b) Reverse-biased

(c) Forward-biased

▲ FIGURE 3–57

PIN diode.

The forward series resistance characteristic and the reverse capacitance characteristic are shown graphically in Figure 3–58 for a typical *pin* diode.

The *pin* diode is used as a dc-controlled microwave switch operated by rapid changes in bias or as a modulating device that takes advantage of the variable forward-resistance characteristic. Since no rectification occurs at the *pn* junction, a high-frequency signal can be modulated (varied) by a lower-frequency bias variation. A *pin* diode can also be used in attenuator applications because its resistance can be controlled by the amount of current. Certain types of *pin* diodes are used as photodetectors in fiber-optic systems.

The Step-Recovery Diode

The step-recovery diode uses graded doping where the doping level of the semiconductive materials is reduced as the *pn* junction is approached. This produces an abrupt turn-off time by allowing a fast release of stored charge when switching from forward to reverse bias. It also allows a rapid re-establishment of forward current when switching from reverse to forward bias. This diode is used in very high frequency (VHF) and fast-switching applications.

Electrical Characteristics (MC7812E)

(Refer to test circuit ,0°C < T_J < 125°C, I_O = 500mA, V_I =19V, C_I= 0.33μF, C_O=0.1μF, unless otherwise specified)

Parameter	Symbol	Conditions		MC7812E			Unit
				Min.	Typ.	Max.	
Output Voltage	V_O	T_J = +25°C		11.5	12	12.5	V
		5.0mA ≤ I_O ≤ 1.0A, P_O ≤15W V_I = 14.5V to 27V		11.4	12	12.6	
Line Regulation (Note1)	Regline	T_J = +25°C	V_I = 14.5V to 30V	-	10	240	mV
			V_I = 16V to 22V	-	3.0	120	
Load Regulation (Note1)	Regload	T_J = +25°C	I_O = 5mA to 1.5A	-	11	240	mV
			I_O = 250mA to 750mA	-	5.0	120	
Quiescent Current	I_Q	T_J = +25°C		-	5.1	8.0	mA
Quiescent Current Change	ΔI_Q	I_O = 5mA to 1.0A		-	0.1	0.5	mA
		V_I = 14.5V to 30V		-	0.5	1.0	
Output Voltage Drift (Note2)	ΔV_O/ΔT	I_O = 5mA		-	-1	-	mV/°C
Output Noise Voltage	V_N	f = 10Hz to 100kHz, T_A = +25 °C		-	76	-	μV/V_o
Ripple Rejection (Note2)	RR	f = 120Hz V_I = 15V to 25V		55	71	-	dB
Dropout Voltage	V_Drop	I_O = 1A, T_J = +25°C		-	2	-	V
Output Resistance (Note2)	r_O	f = 1kHz		-	18	-	mΩ
Short Circuit Current	I_SC	V_I = 35V, T_A= +25°C		-	230	-	mA
Peak Current (Note2)	I_PK	T_J = +25°C		-	2.2	-	A

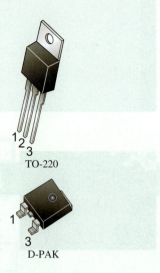

1
2
3
TO-220

1

3
D-PAK

(a)

(b) 1—input, 2—ground, 3—output

▲ FIGURE 3–69

Partial datasheet and packages for a 7812 regulator. You can view an entire datasheet at *www.fairchildsemi.com*. Copyright Fairchild Semiconductor Corporation. Used by permission.

A partial datasheet for a 7812 is shown in Figure 3–69(a) Notice that there is a range of nominal output voltages, but it is typically 12 V. The line and load regulation specify how much the output can vary about the nominal output value. For example, the typical 12 V output will change no more than 11 mV (typical) as the load current changes from 5 mA to 1. 5 A. Package configurations are shown in part (b).

1. From the datasheet, determine the maximum output voltage if the input voltage to the regulator increases to 22 V, assuming a nominal output of 12 V.
2. From the datasheet, determine how much the typical output voltage changes when the load current changes from 250 mA to 750 mA.

The LED A typical partial datasheet for a visible red LED is shown in Figure 3–70. As the datasheet shows, a forward current of 10 mA to 20 mA is used for the test data.

Optical and Electrical Characteristics

T_amb = 25 °C, unless otherwise specified

Red

TLHK51..

Parameter	Test condition	Part	Symbol	Min	Typ.	Max	Unit
Luminous intensity [1]	I_F = 20 mA	TLHK5100	I_V	320			mcd
Dominant wavelength	I_F = 10 mA		λ_d	626	630	639	nm
Peak wavelength	I_F = 10 mA		λ_p		643		nm
Angle of half intensity	I_F = 10 mA		φ		± 9		deg
Forward voltage	I_F = 20 mA		V_F		1.9	2.6	V
Reverse voltage	I_R = 10 μA		V_R	5			V
Junction capacitance	V_R = 0, f = 1 MHz		C_j		15		pF

[1] in one Packing Unit I_Vmin/I_Vmax ≤ 0.5

▲ FIGURE 3–70

Partial datasheet and package for a typical red LED. To view a complete datasheet, go to *www.vishay.com*. Datasheet courtesy of Vishay Intertechnology, Inc.

3. Determine the value of the resistor shown in Figure 3–68 for limiting the LED current to 20 mA and use the next higher standard value. Also specify the power rating of the limiting resistor.

The Fuse The fuse will be in series with the primary winding of the transformer, as shown in Figure 3–68. The fuse should be calculated based on the maximum allowable primary current. Recall from your dc/ac circuits course that if the voltage is stepped down, the current is stepped up. From the specifications for the unregulated power supply, the maximum load current is 250 mA. The current required for the power-on LED indicator is 15 mA. So, the total secondary current is 265 mA. The primary current will be the secondary current divided by the turns ratio.

4. Calculate the primary current and use this value to select a fuse rating.

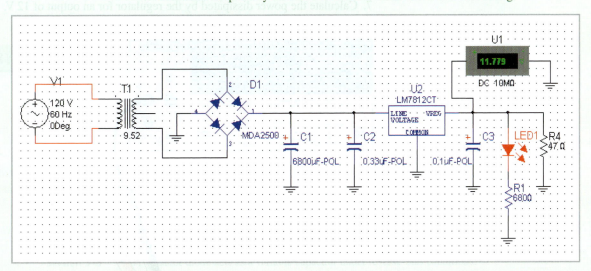

▲ **FIGURE 3–71**

Simulation of the regulated 12 V power supply circuit.

Simulation

In the development of a new circuit, it is helpful to simulate the circuit using a software program before actually building it and committing it to hardware. We will use Multisim to simulate this power supply circuit. Figure 3–71 shows the simulated regulated power supply circuit. The unregulated power supply was previously tested, so you need only to verify that the regulated output is correct. A load resistor value is chosen to draw a current equal to or greater than the specified maximum load current.

$$R_L = \frac{12 \text{ V}}{250 \text{ mA}} = 48 \ \Omega$$

The closest standard value is 47 Ω, which draws 255 mA at 12 V.

5. Determine the power rating for the load resistor.

Simulate the circuit using your Multisim or LT Spice software. Verify the operation with the virtual voltmeter.

Prototyping and Testing

Now that all the components have been selected and the circuit has been simulated, the new components are added to the power supply protoboard from Experiment 2 and the circuit is tested.

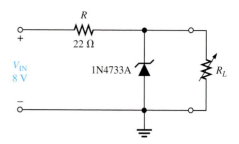

▶ FIGURE 3–76

Multisim and LT Spice file circuits are identified with a logo and are in the Problems folder on the website. Filenames correspond to figure numbers (e.g., FGM03-76 or FGS03-76).

16. In a certain zener regulator, the output voltage changes 0.2 V when the input voltage goes from 5 V to 10 V. What is the input regulation expressed as a percentage? Refer to Chapter 2, Equation 2–14.

17. The output voltage of a zener regulator is 3.6 V at no load and 3.4 V at full load. Determine the load regulation expressed as a percentage. Refer to Chapter 2, Equation 2–15.

Section 3–3   The Varactor Diode

18. Figure 3–77 is a curve of reverse voltage versus capacitance for a certain varactor. Determine the change in capacitance if V_R varies from 5 V to 20 V.

▶ FIGURE 3–77

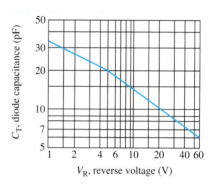

19. Refer to Figure 3–77 and determine the approximate value of V_R that produces 25 pF.

20. What capacitance value is required for each of the varactors in Figure 3–78 to produce a resonant frequency of 1 MHz?

▶ FIGURE 3–78

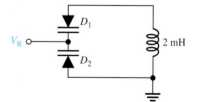

21. At what value must the voltage V_R be set in Problem 20 if the varactors have the characteristic curve in Figure 3–78?

Section 3–4   Optical Diodes

22. The LED in Figure 3–79(a) has a light-producing characteristic as shown in part (b). Neglecting the forward voltage drop of the LED, determine the amount of radiant (light) power produced in mW.

23. Determine how to connect the seven-segment display in Figure 3–80 to display "5." The maximum continuous forward current for each LED is 30 mA and a +5 V dc source is to be used.

24. Specify the number of limiting resistors and their value for a series-parallel array of 48 red LEDs using a 9 V dc source for a forward current of 20 mA.

25. Develop a yellow LED traffic-light array using a minimum number of limiting resistors that operates from a 24 V supply and consists of 100 LEDs with $I_F = 30$ mA and an equal number of LEDs in each parallel branch. Show the circuit and the resistor values.

▶ FIGURE 3–79

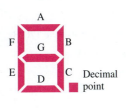

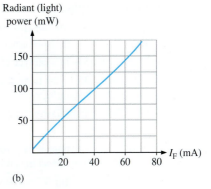

(a)

(b)

▶ FIGURE 3–80

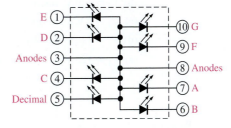

26. For a certain photodiode at a given irradiance, the reverse resistance is 200 kΩ and the reverse voltage is 10 V. What is the current through the device?

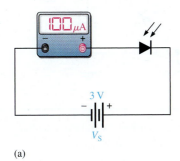

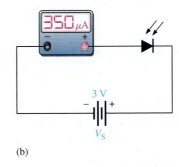

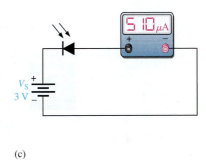

(a)

(b)

(c)

▲ FIGURE 3–81

27. What is the resistance of each photodiode in Figure 3–81?

28. When the switch in Figure 3–82 is closed, will the microammeter reading increase or decrease? Assume D_1 and D_2 are optically coupled.

▶ FIGURE 3–82

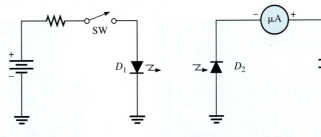

Section 3–5 The Solar Cell

29. List five parts of a typical solar cell.

30. Determine the number and type of connection for PV cells each with a nominal output voltage of 0.5 V to produce a total output of 15 V.

31. For the connection in Problem 30, how much current is supplied to a 10 kΩ load?

32. Determine how to modify the connection in Problem 30 to achieve a load current capacity of 10 mA.

Section 3–6 Other Types of Diodes

33. The *V-I* characteristic of a certain tunnel diode shows that the current changes from 0.25 mA to 0.15 mA when the voltage changes from 125 mV to 200 mV. What is the resistance?

34. In what type of circuit are tunnel diodes commonly used?

35. What purpose do the reflective surfaces in the laser diode serve? Why is one end only partially reflective?

Section 3–7 Troubleshooting

36. For each set of measured voltages at the points (1, 2, and 3) indicated in Figure 3–83, determine if they are correct and if not, identify the most likely fault(s). State what you would do to correct the problem once it is isolated. The zener is rated at 12 V.

(a) $V_1 = 120$ V rms, $V_2 = 30$ V dc, $V_3 = 12$ V dc

(b) $V_1 = 120$ V rms, $V_2 = 30$ V dc, $V_3 = 30$ V dc

(c) $V_1 = 0$ V, $V_2 = 0$ V, $V_3 = 0$ V

(d) $V_1 = 120$ V rms, $V_2 = 30$ V peak full-wave 120 Hz, $V_3 = 12$ V, 120 Hz pulsating voltage

(e) $V_1 = 120$ V rms, $V_2 = 9$ V, $V_3 = 0$ V

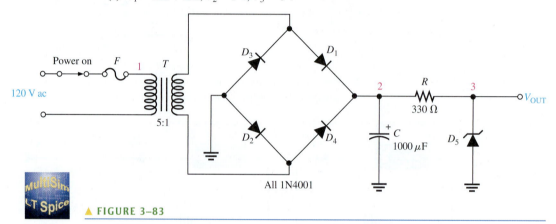

▲ **FIGURE 3–83**

37. What is the output voltage in Figure 3–83 for each of the following faults?

(a) D_5 open (b) *R* open (c) *C* leaky (d) *C* open

(e) D_3 open (f) D_2 open (g) *T* open (h) *F* open

DEVICE APPLICATION PROBLEMS

38. Based on the indicated voltage measurements with respect to ground in Figure 3–84(a), determine the probable fault(s).

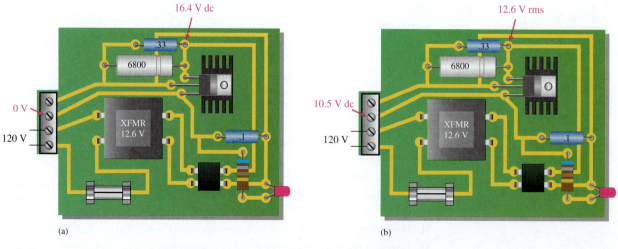

(a)   (b)

▲ **FIGURE 3–84**

39. Determine the probable fault(s) indicated by the voltage measurements in Figure 3–84(b).

40. List the possible reasons for the LED in Figure 3–84 not emitting light when the power supply is plugged in.

41. If a 1k Ω load resistor is connected from the output pin to ground on a properly operating power supply circuit like shown in Figure 3–84, how much power will the 7812 regulator dissipate?

DATASHEET PROBLEMS

42. Refer to the zener diode datasheet in Figure 3–7.

(a) What is the maximum dc power dissipation at 25°C for a 1N4738A?

(b) Determine the maximum power dissipation at 70°C and at 100°C for a 1N4751A.

(c) What is the minimum current required by the 1N4738A for regulation?

(d) What is the maximum current for the 1N4750A at 25°C?

(e) The current through a 1N4740A changes from 25 mA to 0.25 mA. How much does the zener impedance change?

43. Refer to the varactor diode datasheet in Figure 3–24.

(a) What is the maximum forward current for the 832A?

(b) What is the maximum capacitance of an 830A at a reverse voltage of 2 V?

(c) What is the maximum capacitance range of an 836A?

44. Refer to the LED datasheet in Figure 3–34.

(a) Can 9 V be applied in reverse across an TSMF1000 LED?

(b) Determine the typical value of series resistor for the TSMF1000 when a voltage of 5.1 V is used to forward-bias the diode with $I_F = 20$ mA.

(c) Assume the forward current is 50 mA and the forward voltage drop is 1.5 V at an ambient temperature of 15°C. Is the maximum power rating exceeded?

(d) Determine the radiant intensity for a forward current of 40 mA.

(e) What is the radiant intensity at an angle of 20° from the axis if the forward current is 100 mA?

45. Refer to the photodiode datasheet in Figure 3–46.

(a) A TEMD1000 is connected in series with a 1k Ω resistor and a reverse-bias voltage source. There is no incident light on the diode. What is the maximum voltage drop across the resistor?

(b) At what wavelength will the reverse current be the greatest for a given irradiance?

(c) At what wavelength is relative spectral sensitivity of the TEMD1000 equal to 0.4?

ADVANCED PROBLEMS

46. Develop the schematic for the circuit board in Figure 3–85 and determine what type of circuit it is.

47. If a 30 V rms, 60 Hz input voltage is connected to the ac inputs, determine the output voltages on the circuit board in Figure 3–85.

48. If each output of the board in Figure 3–85 is loaded with 10k Ω, what fuse rating should be used?

49. Design a zener voltage regulator to meet the following specifications: The input voltage is 24 V dc, the load current is 35 mA, and the load voltage is 8.2 V.

50. The varactor-tuned band-pass filter in Figure 3–27 is to be redesigned to produce a bandwidth of from 350 kHz to 850 kHz within a 10% tolerance. Specify what change you would have to make using the graph in Figure 3–86.

51. Design a seven-segment red LED display circuit in which any of the ten digits can be displayed using a set of switches. Each LED segment is to have a current of 20 mA ± 10% from a 12 V source and the circuit must be designed with a minimum number of switches.

52. If you used a common-anode seven-segment display in Problem 51, redesign it for a common-cathode display or vice versa.

▶ FIGURE 3–85

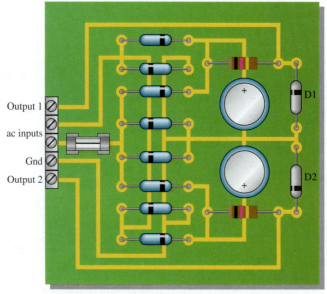

Rectifier diodes: 1N4001A
Zener diodes: D1-1N4736A, D2-1N4749A
Filter capacitors: 100 μF

▶ FIGURE 3–86

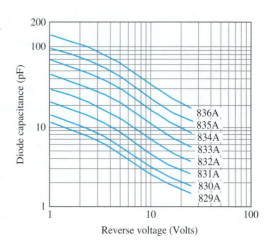

MULTISIM TROUBLESHOOTING PROBLEMS

These file circuits are in the Troubleshooting Problems folder on the website.

53. Open file TPM03-53 and determine the fault.

54. Open file TPM03-54 and determine the fault.

55. Open file TPM03-55 and determine the fault.

56. Open file TPM03-56 and determine the fault.

BIPOLAR JUNCTION TRANSISTORS

4

CHAPTER OBJECTIVES

◆ Describe the structure of the BJT
◆ Discuss basic BJT operation
◆ Discuss important BJT parameters and characteristics and analyze transistor circuits
◆ Discuss how a BJT is used as a voltage amplifier
◆ Discuss how a BJT is used as a switch
◆ Discuss the phototransistor and its operation
◆ Identify various types of transistor packages
◆ Troubleshoot faults in transistor circuits

KEY TERMS

◆ BJT
◆ Emitter
◆ Base
◆ Collector
◆ Gain
◆ Beta
◆ Saturation
◆ Linear
◆ Cutoff
◆ Load Line
◆ AND gate
◆ OR gate
◆ Amplification
◆ Phototransistor

VISIT THE WEBSITE

Study aids, Multisim files, and LT Spice files for this chapter are available at https://www.pearsonhighered.com /careersresources/

INTRODUCTION

The invention of the transistor was the beginning of a technological revolution that is still continuing. All of the complex electronic devices and systems today are an outgrowth of early developments in semiconductor transistors.

Two basic types of transistors are the bipolar junction transistor (BJT), which we will begin to study in this chapter, and the field-effect transistor (FET), which we will cover in later chapters. The BJT is used in two broad areas—as a linear amplifier to boost or amplify an electrical signal and as an electronic switch. Both of these applications are introduced in this chapter.

DEVICE APPLICATION PREVIEW

Suppose you work for a company that makes a security alarm system for protecting homes and businesses against illegal entry. You are given the responsibility for final development and for testing each system before it is shipped out. The first step is to learn all you can about transistor operation. You will then apply your knowledge to the Device Application at the end of the chapter.

4–1 BIPOLAR JUNCTION TRANSISTOR (BJT) STRUCTURE

The structure of the bipolar junction transistor (BJT) determines its operating characteristics. In this section, you will see how semiconductive materials are used to form a BJT, and you will learn the standard BJT symbols.

After completing this section, you should be able to

❑ **Describe the structure of the BJT**
  ◆ Explain the difference between the structure of an *npn* and a *pnp* transistor
  ◆ Identify the symbols for *npn* and *pnp* transistors ◆ Name the three regions of a BJT and their labels

The **BJT** is constructed with three doped semiconductor regions separated by two *pn* junctions, as shown in the epitaxial planar structure in Figure 4–1(a). The three regions are called **emitter**, **base**, and **collector**. Physical representations of the two types of BJTs are shown in Figure 4–1(b) and (c). One type consists of two *n* regions separated by a *p* region (*npn*), and the other type consists of two *p* regions separated by an *n* region (*pnp*). The term **bipolar** refers to the use of both holes and electrons as current carriers in the transistor structure.

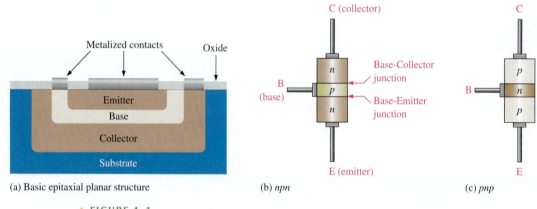

(a) Basic epitaxial planar structure   (b) *npn*   (c) *pnp*

▲ **FIGURE 4–1**

BJT construction. The substrate is a physical supporting material for the transistor.

The *pn* junction joining the base region and the emitter region is called the *base-emitter junction*. The *pn* junction joining the base region and the collector region is called the *base-collector junction,* as indicated in Figure 4–1(b). A lead connects to each of the three regions, as shown. These leads are labeled E, B, and C for emitter, base, and collector, respectively. The base region is lightly doped and very thin compared to the heavily doped emitter and the moderately doped collector regions. Because of this difference in doping levels, the emitter and collector are not interchangeable. (The reason for this is discussed in the next section.) Figure 4–2 shows the schematic symbols for the *npn* and *pnp* bipolar junction transistors.

▶ **FIGURE 4–2**

Standard BJT (bipolar junction transistor) symbols.

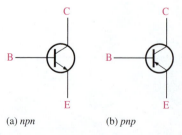

(a) *npn*   (b) *pnp*

1. Name the two types of BJTs according to their structure.
2. The BJT is a three-terminal device. Name the three terminals.
3. What separates the three regions in a BJT?
4. Why aren't the collector and emitter interchangeable on a BJT?

4–2   BASIC BJT OPERATION

In order for a BJT to operate properly as an amplifier, the two *pn* junctions must be correctly biased with external dc voltages. In this section, we mainly use the *npn* transistor for illustration. The operation of the *pnp* is the same as for the *npn* except that the roles of the electrons and holes, the bias voltage polarities, and the current directions are all reversed.

After completing this section, you should be able to

- ❑ **Discuss basic BJT operation**
- ❑ Describe forward-reverse bias
  - ◆ Show how to bias *pnp* and *npn* BJTs with dc sources
- ❑ Explain the internal operation of a BJT
  - ◆ Discuss the hole and electron movement
- ❑ Discuss transistor currents
  - ◆ Calculate any of the transistor currents if the other two are known

Biasing

Figure 4–3 shows a bias arrangement for both *npn* and *pnp* BJTs for operation as an **amplifier.** Notice that in both cases the base-emitter (BE) junction is forward-biased and the base-collector (BC) junction is reverse-biased. This condition is called *forward-reverse bias.*

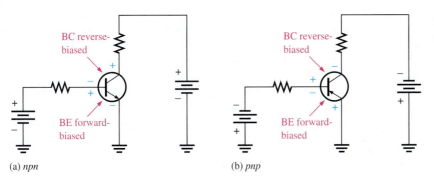

◄ FIGURE 4–3

Forward-reverse bias of a BJT.

(a) *npn*

(b) *pnp*

Operation

To understand how a transistor operates, let's examine what happens inside the *npn* structure. The heavily doped *n*-type emitter region has a very high density of conduction-band (free) electrons, as indicated in Figure 4–4. These free electrons easily diffuse through the forward-based BE junction into the lightly doped and very thin *p*-type base region, as indicated by the wide arrow. The base has a low density of holes, which are the majority carriers, as represented by the white circles. A small percentage of the total number of free electrons injected into the base region recombine with holes and move as valence electrons through the base region and into the emitter region as hole current, indicated by the red arrows.

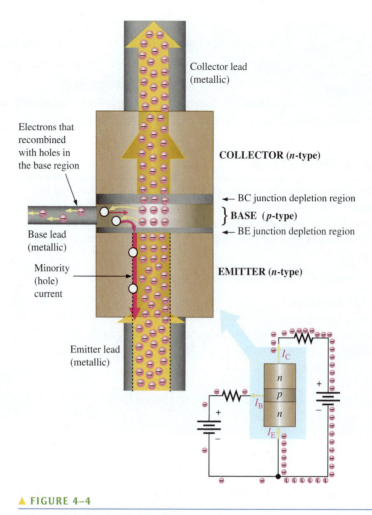

▲ FIGURE 4–4

BJT operation showing electron flow.

When the electrons that have recombined with holes as valence electrons leave the crystalline structure of the base, they become free electrons in the metallic base lead and produce the external base current. Most of the free electrons that have entered the base do not recombine with holes because the base is very thin. As the free electrons move toward the reverse-biased BC junction, they are swept across into the collector region by the attraction of the positive collector supply voltage. The free electrons move through the collector region, into the external circuit, and then return into the emitter region along with the base current, as indicated. The emitter current is slightly greater than the collector current because of the small base current that splits off from the total current injected into the base region from the emitter.

Transistor Currents

The directions of the currents in an *npn* transistor and its schematic symbol are as shown in Figure 4–5(a); those for a *pnp* transistor are shown in Figure 4–5(b). Notice that the arrow on the emitter inside the transistor symbols points in the direction of conventional current which is opposite electron flow. These diagrams show that the emitter current (I_E) is the sum of the collector current (I_C) and the base current (I_B), expressed as follows:

Equation 4–1

$$I_E = I_C + I_B$$

As mentioned before, I_B is very small compared to I_E or I_C. The capital-letter subscripts indicate dc values.

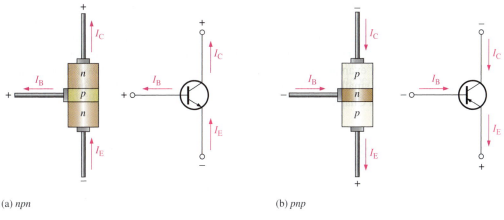

(a) *npn*   (b) *pnp*

▲ **FIGURE 4–5**

Transistor currents.

SECTION 4–2
CHECKUP

1. What are the bias conditions of the base-emitter and base-collector junctions for a transistor to operate as an amplifier?
2. Which is the largest of the three transistor currents?
3. Is the base current smaller or larger than the emitter current?
4. Is the base region much thinner or much wider than the collector and emitter regions?
5. If the collector current is 1 mA and the base current is 10 μA, what is the emitter current?

4–3 BJT CHARACTERISTICS AND PARAMETERS

Two important parameters, β_{DC} (dc current gain) and α_{DC} (ratio of collector current to emitter current) are introduced and used to analyze a BJT circuit. Also, transistor characteristic curves are covered, and you will learn how a BJT's operation can be determined from these curves. Finally, maximum ratings of a BJT are discussed.

After completing this section, you should be able to

❑ **Discuss basic BJT parameters and characteristics and analyze transistor circuits**
❑ Define *dc beta* (β_{DC}) and *dc alpha* (α_{DC})
   ◆ Calculate (β_{DC}) and (α_{DC}) based on transistor current
❑ Describe a basic dc model of a BJT
❑ Analyze BJT circuits
   ◆ Identify transistor currents and voltages   ◆ Calculate each transistor current
   ◆ Calculate each transistor voltage
❑ Interpret collector characteristic curves
   ◆ Discuss the linear region   ◆ Explain saturation and cutoff in relation to the curves
❑ Describe the cutoff condition in a BJT circuit
❑ Describe the saturation condition in a BJT circuit
❑ Discuss the dc load line and apply it to circuit analysis
❑ Discuss how β_{DC} changes with temperature
❑ Explain and apply maximum transistor ratings
❑ Derate a transistor for power dissipation
❑ Interpret a BJT datasheet

Also, by Ohm's law,

$$V_{R_B} = I_B R_B$$

Substituting for V_{R_B} yields

$$I_B R_B = V_{BB} - V_{BE}$$

Solving for I_B,

Equation 4–4

$$I_B = \frac{V_{BB} - V_{BE}}{R_B}$$

The voltage at the collector with respect to the grounded emitter is

$$V_{CE} = V_{CC} - V_{R_C}$$

Since the drop across R_C is

$$V_{R_C} = I_C R_C$$

the voltage at the collector with respect to the emitter can be written as

Equation 4–5

$$V_{CE} = V_{CC} - I_C R_C$$

where $I_C = \beta_{DC} I_B$.

The voltage across the reverse-biased collector-base junction is

Equation 4–6

$$V_{CB} = V_{CE} - V_{BE}$$

EXAMPLE 4–2

Determine I_B, I_C, I_E, V_{BE}, V_{CE}, and V_{CB} in the circuit of Figure 4–9. The transistor has a $\beta_{DC} = 150$.

▶ **FIGURE 4–9**

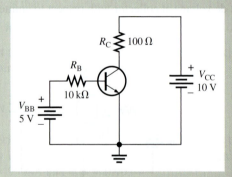

Solution   From Equation 4–3, $V_{BE} \cong$ **0.7 V**. Calculate the base, collector, and emitter currents as follows:

$$I_B = \frac{V_{BB} - V_{BE}}{R_B} = \frac{5\ \text{V} - 0.7\ \text{V}}{10\ \text{k}\Omega} = \textbf{430}\ \boldsymbol{\mu}\textbf{A}$$

$$I_C = \beta_{DC} I_B = (150)(430\ \mu\text{A}) = \textbf{64.5 mA}$$

$$I_E = I_C + I_B = 64.5\ \text{mA} + 430\ \mu\text{A} = \textbf{64.9 mA}$$

Solve for V_{CE} and V_{CB}.

$$V_{CE} = V_{CC} - I_C R_C = 10\ \text{V} - (64.5\ \text{mA})(100\ \Omega) = 10\ \text{V} - 6.45\ \text{V} = \textbf{3.55 V}$$

$$V_{CB} = V_{CE} - V_{BE} = 3.55\ \text{V} - 0.7\ \text{V} = \textbf{2.85 V}$$

Since the collector is at a higher voltage than the base, the collector-base junction is reverse-biased.

Related Problem   Determine I_B, I_C, I_E, V_{CE}, and V_{CB} in Figure 4–9 for the following values: $R_B = 22$ kΩ, $R_C = 220$ Ω, $V_{BB} = 6$ V, $V_{CC} = 9$ V, and $\beta_{DC} = 90$.

Open the Multisim file EXM04-02 or LT Spice file EXS04-02 in the Examples folder on the website. Measure each current and voltage and compare with the calculated values.

Collector Characteristic Curves

Using a circuit like that shown in Figure 4–10(a), a set of *collector characteristic curves* can be generated that show how the collector current, I_C, varies with the collector-to-emitter voltage, V_{CE}, for specified values of base current, I_B. Notice in the circuit diagram that both V_{BB} and V_{CC} are variable sources of voltage.

Assume that V_{BB} is set to produce a certain value of I_B and V_{CC} is zero. For this condition, both the base-emitter junction and the base-collector junction are forward-biased because the base is at approximately 0.7 V while the emitter is at 0 V and the collector is near 0 V. The base current is primarily through the base-emitter junction because of the low

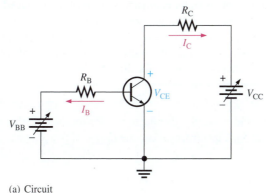

(a) Circuit

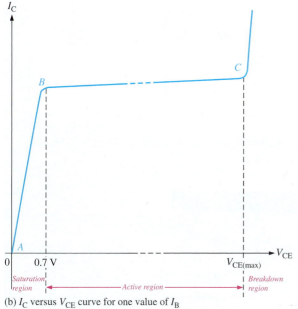

(b) I_C versus V_{CE} curve for one value of I_B

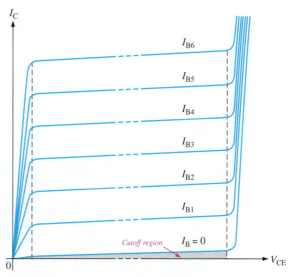

(c) Family of I_C versus V_{CE} curves for several values of I_B ($I_{B1} < I_{B2} < I_{B3}$, etc.)

▲ **FIGURE 4–10**

Collector characteristic curves.

impedance path to ground and, therefore, I_C will be approximately zero and V_{CE} is near 0 V. When both junctions are forward-biased, the transistor is in the saturation region of its operation. **Saturation** is the state of a BJT in which the collector current has reached a maximum and is independent of the base current.

As V_{CC} is increased, V_{CE} increases as the collector current increases. This is indicated by the portion of the characteristic curve between points A and B in Figure 4–10(b). I_C increases as V_{CC} is increased because V_{CE} remains less than 0.7 V due to the forward-biased base-collector junction.

Ideally, when V_{CE} exceeds 0.7 V, the base-collector junction becomes reverse-biased and the transistor goes into the *active,* or **linear**, *region* of its operation. Once the base-collector junction is reverse-biased, I_C levels off and remains essentially constant for a given value of I_B as V_{CE} continues to increase. Actually, I_C increases very slightly as V_{CE} increases due to widening of the base-collector depletion region. This results in fewer holes for recombination in the base region which effectively causes a slight increase in β_{DC}. This is shown by the portion of the characteristic curve between points B and C in Figure 4–10(b). For this portion of the characteristic curve, the value of I_C is determined only by the relationship expressed as $I_C = \beta_{DC}I_B$.

When V_{CE} reaches a sufficiently high voltage, the reverse-biased base-collector junction goes into breakdown; and the collector current increases rapidly as indicated by the part of the curve to the right of point C in Figure 4–10(b). A transistor should never be operated in this breakdown region.

A family of collector characteristic curves is produced when I_C versus V_{CE} is plotted for several values of I_B, as illustrated in Figure 4–10(c). When $I_B = 0$, the transistor is in the cutoff region although there is a very small collector leakage current as indicated. **Cutoff** is the nonconducting state of a transistor. The amount of collector leakage current for $I_B = 0$ is exaggerated on the graph for illustration.

EXAMPLE 4–3

Sketch an ideal family of collector curves for the circuit in Figure 4–11 for $I_B = 5\ \mu A$ to 25 μA in 5 μA increments. Assume $\beta_{DC} = 100$ and that V_{CE} does not exceed breakdown.

▶ **FIGURE 4–11**

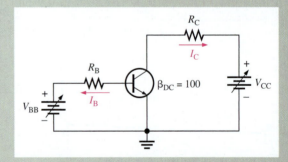

Solution   Using the relationship $I_C = \beta_{DC}I_B$, values of I_C are calculated and tabulated in Table 4–1. The resulting curves are plotted in Figure 4–12.

▶ **TABLE 4–1**

I_B	I_C
5 μA	0.5 mA
10 μA	1.0 mA
15 μA	1.5 mA
20 μA	2.0 mA
25 μA	2.5 mA

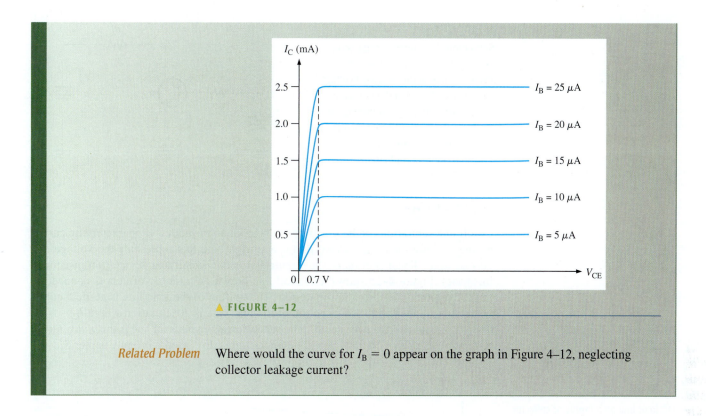

▲ FIGURE 4–12

Related Problem   Where would the curve for $I_B = 0$ appear on the graph in Figure 4–12, neglecting collector leakage current?

Cutoff

As previously mentioned, when $I_B = 0$, the transistor is in the cutoff region of its operation. This is shown in Figure 4–13 with the base lead open, resulting in a base current of zero. Under this condition, there is a very small amount of collector leakage current, I_{CEO}, due mainly to thermally produced carriers. Because I_{CEO} is extremely small, it will usually be neglected in circuit analysis so that $V_{CE} = V_{CC}$. In cutoff, neither the base-emitter nor the base-collector junctions are forward-biased. The subscript CEO represents collector-to-emitter with the base open.

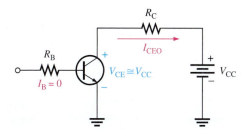

◀ FIGURE 4–13

Cutoff: Collector leakage current (I_{CEO}) is extremely small and is usually neglected. Base-emitter and base-collector junctions are reverse-biased.

Saturation

When the base-emitter junction becomes forward-biased and the base current is increased, the collector current also increases ($I_C = \beta_{DC}I_B$) and V_{CE} decreases as a result of more drop across the collector resistor ($V_{CE} = V_{CC} - I_C R_C$). This is illustrated in Figure 4–14. When V_{CE} reaches its saturation value, $V_{CE(sat)}$, the base-collector junction becomes forward-biased and I_C can increase no further even with a continued increase in I_B. At the point of saturation, the relation $I_C = \beta_{DC}I_B$ is no longer valid. $V_{CE(sat)}$ for a transistor occurs somewhere below the knee of the collector curves, and it is usually only a few tenths of a volt.

▶ FIGURE 4–14

Saturation: As I_B increases due to increasing V_{BB}, I_C also increases and V_{CE} decreases due to the increased voltage drop across R_C. When the transistor reaches saturation, I_C can increase no further regardless of further increase in I_B. Base-emitter and base-collector junctions are forward-biased.

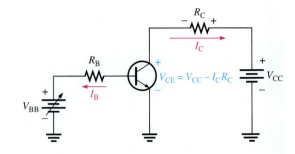

DC Load Line

Cutoff and saturation can be illustrated in relation to the collector characteristic curves by the use of a load line. A **load line** is a straight line that represents the voltage and current in the linear portion of the circuit that is connected to a device (a transistor in this case). Figure 4–15 shows a dc load line drawn on a family of curves connecting the cutoff point and the saturation point. The bottom of the load line is at ideal cutoff where $I_C = 0$ and $V_{CE} = V_{CC}$. The top of the load line is at saturation where $I_C = I_{C(sat)}$ and $V_{CE} = V_{CE(sat)}$. In between cutoff and saturation along the load line is the *active region* of the transistor's operation. Load line operation is discussed more in Chapter 5.

▶ FIGURE 4–15

DC load line on a family of collector characteristic curves illustrating the cutoff and saturation conditions.

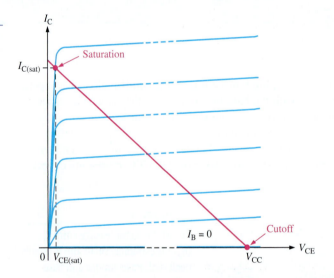

EXAMPLE 4–4

Determine whether or not the transistor in Figure 4–16 is in saturation. Assume $V_{CE(sat)} = 0.2$ V.

▶ FIGURE 4–16

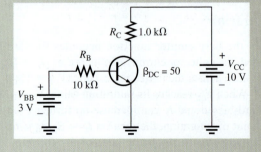

Solution First, determine $I_{C(sat)}$.

$$I_{C(sat)} = \frac{V_{CC} - V_{CE(sat)}}{R_C} = \frac{10\text{ V} - 0.2\text{ V}}{1.0\text{ k}\Omega} = \frac{9.8\text{ V}}{1.0\text{ k}\Omega} = 9.8\text{ mA}$$

Now, see if I_B is large enough to produce $I_{C(sat)}$.

$$I_B = \frac{V_{BB} - V_{BE}}{R_B} = \frac{3\text{ V} - 0.7\text{ V}}{10\text{ k}\Omega} = \frac{2.3\text{ V}}{10\text{ k}\Omega} = 0.23\text{ mA}$$

$$I_C = \beta_{DC}I_B = (50)(0.23\text{ mA}) = 11.5\text{ mA}$$

This shows that with the specified β_{DC}, this base current is capable of producing an I_C greater than $I_{C(sat)}$. Therefore, the **transistor is saturated,** and the collector current value of 11.5 mA is never reached. If you further increase I_B, the collector current remains at its saturation value of 9.8 mA.

Related Problem Determine whether or not the transistor in Figure 4–16 is saturated for the following values: $\beta_{DC} = 125$, $V_{BB} = 1.5$ V, $R_B = 6.8$ kΩ, $R_C = 180$ Ω, and $V_{CC} = 12$ V.

 Open the Multisim file EXM04-04 or LT Spice file EXS04-04 in the Examples folder on the website. Determine if the transistor is in saturation and explain how you did this.

More About β_{DC}

The β_{DC} or h_{FE} is an important BJT parameter that we need to examine further. β_{DC} is not truly constant but varies with both collector current and with temperature. Keeping the junction temperature constant and increasing I_C causes β_{DC} to increase to a maximum. A further increase in I_C beyond this maximum point causes β_{DC} to decrease. If I_C is held constant and the temperature is varied, β_{DC} changes directly with the temperature. If the temperature goes up, β_{DC} goes up and vice versa. Figure 4–17 shows the variation of β_{DC} with I_C and junction temperature (T_J) for a typical BJT.

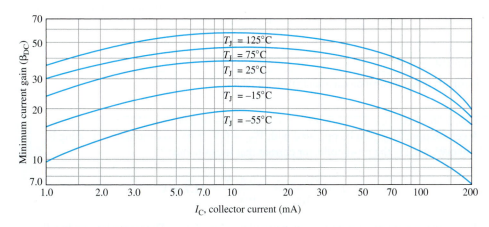

◀ **FIGURE 4–17**

Variation of β_{DC} with I_C for several temperatures.

A transistor datasheet usually specifies $\beta_{DC}(h_{FE})$ at specific I_C values. Even at fixed values of I_C and temperature, β_{DC} varies from one device to another for a given type of transistor due to inconsistencies in the manufacturing process that are unavoidable. The β_{DC} specified at a certain value of I_C is usually the minimum value, $\beta_{DC(min)}$, although the maximum and typical values are also sometimes specified.

EXAMPLE 4–7

A certain transistor has a $P_{D(max)}$ of 1 W at 25°C. The derating factor is 5 mW/°C. What is the $P_{D(max)}$ at a temperature of 70°C?

Solution The change (reduction) in $P_{D(max)}$ is

$$\Delta P_{D(max)} = (5 \text{ mW/°C})(70°C - 25°C) = (5 \text{ mW/°C})(45°C) = 225 \text{ mW}$$

Therefore, the $P_{D(max)}$ at 70°C is

$$1 \text{ W} - 225 \text{ mW} = \textbf{775 mW}$$

Related Problem A transistor has a $P_{D(max)} = 5$ W at 25°C. The derating factor is 10 mW/°C. What is the $P_{D(max)}$ at 70°C?

BJT Datasheet

A partial datasheet for the 2N3904 *npn* transistor is shown in Figure 4–20. Notice that the maximum collector-emitter voltage (V_{CEO}) is 40 V. The CEO subscript indicates that the voltage is measured from collector (C) to emitter (E) with the base open (O). In the text, we use $V_{CE(max)}$ for this parameter. Also notice that the maximum collector current is 200 mA.

The $\beta_{DC}(h_{FE})$ is specified for several values of I_C. As you can see, h_{FE} varies with I_C as we previously discussed.

The collector-emitter saturation voltage, $V_{CE(sat)}$ is 0.2 V maximum for $I_{C(sat)} = 10$ mA and increases with the current.

EXAMPLE 4–8

A 2N3904 transistor is used in the circuit of Figure 4–19 (Example 4–6). Determine the maximum value to which V_{CC} can be adjusted without exceeding a rating. Refer to the datasheet in Figure 4–20.

Solution From the datasheet,

$$P_{D(max)} = P_D = 625 \text{ mW}$$
$$V_{CE(max)} = V_{CEO} = 40 \text{ V}$$
$$I_{C(max)} = I_C = 200 \text{ mA}$$

Assume $\beta_{DC} = 100$. This is a reasonably valid assumption based on the datasheet $h_{FE} = 100$ minimum for specified conditions (β_{DC} and h_{FE} are the same parameter). As you have learned, the β_{DC} has considerable variations for a given transistor, depending on circuit conditions. Under this assumption, $I_C = 19.5$ mA and $V_{R_C} = 19.5$ V from Example 4–6.

Since I_C is much less than $I_{C(max)}$ and, ideally, will not change with V_{CC}, the maximum value to which V_{CC} can be increased before $V_{CE(max)}$ is exceeded, that is,

$$V_{CC(max)} = V_{CE(max)} + V_{R_C} = 40 \text{ V} + 19.5 \text{ V} = 59.5 \text{ V}$$

However, at the maximum value of V_{CE}, the power dissipation is

$$P_D = V_{CE(max)}I_C = (40 \text{ V})(19.5 \text{ mA}) = 780 \text{ mW}$$

Power dissipation exceeds the maximum of 625 mW specified on the datasheet. To find the maximum value of V_{CC} without exceeding $P_{D(max)}$, first find V_{CE} at a current of 19.5 mA and a P_D of 625 mW:

$$V_{CE} = 625 \text{ mW}/19.5 \text{ mA} = 32 \text{ V}$$
$$V_{CC} = 32 \text{ V} + 19.5 \text{ V} = 51.5 \text{ V}$$

Related Problem Use the datasheet in Figure 4–20 to find the maximum P_D at 50°C.

2N3904   MMBT3904   PZT3904

TO-92

SOT-23
Mark: 1A

SOT-223

NPN General Purpose Amplifier

This device is designed as a general purpose amplifier and switch.
The useful dynamic range extends to 100 mA as a switch and to
100 MHz as an amplifier.

Absolute Maximum Ratings* $T_A = 25°C$ unless otherwise noted

Symbol	Parameter	Value	Units
V_{CEO}	Collector-Emitter Voltage	40	V
V_{CBO}	Collector-Base Voltage	60	V
V_{EBO}	Emitter-Base Voltage	6.0	V
I_C	Collector Current - Continuous	200	mA
T_J, T_{stg}	Operating and Storage Junction Temperature Range	-55 to +150	°C

*These ratings are limiting values above which the serviceability of any semiconductor device may be impaired.

NOTES:
1) These ratings are based on a maximum junction temperature of 150 degrees C.
2) These are steady state limits. The factory should be consulted on applications involving pulsed or low duty cycle operations.

Thermal Characteristics $T_A = 25°C$ unless otherwise noted

Symbol	Characteristic	Max 2N3904	Max *MMBT3904	Max **PZT3904	Units
P_D	Total Device Dissipation	625	350	1,000	mW
	Derate above 25°C	5.0	2.8	8.0	mW/°C
$R_{θJC}$	Thermal Resistance, Junction to Case	83.3			°C/W
$R_{θJA}$	Thermal Resistance, Junction to Ambient	200	357	125	°C/W

*Device mounted on FR-4 PCB 1.6" X 1.6" X 0.06."
**Device mounted on FR-4 PCB 36 mm X 18 mm X 1.5 mm; mounting pad for the collector lead min. 6 cm².

Electrical Characteristics $T_A = 25°C$ unless otherwise noted

Symbol	Parameter	Test Conditions	Min	Max	Units
OFF CHARACTERISTICS					
$V_{(BR)CEO}$	Collector-Emitter Breakdown Voltage	$I_C = 1.0$ mA, $I_B = 0$	40		V
$V_{(BR)CBO}$	Collector-Base Breakdown Voltage	$I_C = 10$ μA, $I_E = 0$	60		V
$V_{(BR)EBO}$	Emitter-Base Breakdown Voltage	$I_E = 10$ μA, $I_C = 0$	6.0		V
I_{BL}	Base Cutoff Current	$V_{CE} = 30$ V, $V_{EB} = 3$V		50	nA
I_{CEX}	Collector Cutoff Current	$V_{CE} = 30$ V, $V_{EB} = 3$V		50	nA
ON CHARACTERISTICS*					
h_{FE}	DC Current Gain	$I_C = 0.1$ mA, $V_{CE} = 1.0$ V	40		
		$I_C = 1.0$ mA, $V_{CE} = 1.0$ V	70		
		$I_C = 10$ mA, $V_{CE} = 1.0$ V	100	300	
		$I_C = 50$ mA, $V_{CE} = 1.0$ V	60		
		$I_C = 100$ mA, $V_{CE} = 1.0$ V	30		
$V_{CE(sat)}$	Collector-Emitter Saturation Voltage	$I_C = 10$ mA, $I_B = 1.0$ mA		0.2	V
		$I_C = 50$ mA, $I_B = 5.0$ mA		0.3	V
$V_{BE(sat)}$	Base-Emitter Saturation Voltage	$I_C = 10$ mA, $I_B = 1.0$ mA	0.65	0.85	V
		$I_C = 50$ mA, $I_B = 5.0$ mA		0.95	V
SMALL SIGNAL CHARACTERISTICS					
f_T	Current Gain - Bandwidth Product	$I_C = 10$ mA, $V_{CE} = 20$ V, f = 100 MHz	300		MHz
C_{obo}	Output Capacitance	$V_{CB} = 5.0$ V, $I_E = 0$, f = 1.0 MHz		4.0	pF
C_{ibo}	Input Capacitance	$V_{EB} = 0.5$ V, $I_C = 0$, f = 1.0 MHz		8.0	pF
NF	Noise Figure	$I_C = 100$μA, $V_{CE} = 5.0$ V, $R_S = 1.0$kΩ,f=10 Hz to 15.7kHz		5.0	dB
SWITCHING CHARACTERISTICS					
t_d	Delay Time	$V_{CC} = 3.0$ V, $V_{BE} = 0.5$ V,		35	ns
t_r	Rise Time	$I_C = 10$ mA, $I_{B1} = 1.0$ mA		35	ns
t_s	Storage Time	$V_{CC} = 3.0$ V, $I_C = 10$mA		200	ns
t_f	Fall Time	$I_{B1} = I_{B2} = 1.0$ mA		50	ns

*Pulse Test: Pulse Width ≤300μs, Duty Cycle ≤2.0%

◀ **FIGURE 4–20**

Partial datasheet for the 2N3904. Note that $P_{D(max)}$ is listed as P_D under Thermal Characteristics but on some data sheets is listed under Absolute Maximum Ratings. For a complete 2N3904 datasheet, go to *http://www.fairchildsemi.com/ds/2N%2F2N3904.pdf*. Copyright Fairchild Semiconductor Corporation. Used by permission.

Since R_C is always considerably larger in value than r'_e, the output voltage for this configuration is greater than the input voltage. Various types of amplifiers are covered in detail in later chapters.

EXAMPLE 4–9

Determine the voltage gain and the ac output voltage in Figure 4–22 if $r'_e = 50\ \Omega$.

▶ **FIGURE 4–22**

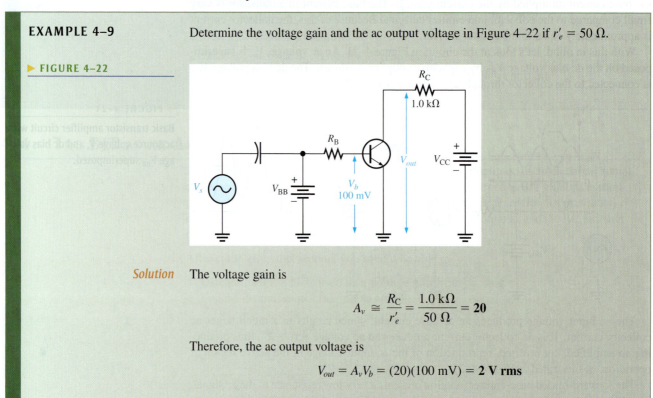

Solution   The voltage gain is

$$A_v \cong \frac{R_C}{r'_e} = \frac{1.0\ k\Omega}{50\ \Omega} = 20$$

Therefore, the ac output voltage is

$$V_{out} = A_v V_b = (20)(100\ mV) = \mathbf{2\ V\ rms}$$

Related Problem   What value of R_C in Figure 4–22 will it take to have a voltage gain of 50?

SECTION 4–4 CHECKUP

1. What is amplification?
2. How is voltage gain defined?
3. Name two factors that determine the voltage gain of an amplifier.
4. What is the voltage gain of a transistor amplifier that has an output of 5 V rms and an input of 250 mV rms?
5. A transistor connected as in Figure 4–22 has an $r'_e = 20\ \Omega$. If R_C is 1200 Ω, what is the voltage gain?

4–5   THE BJT AS A SWITCH

In the previous section, you saw how a BJT can be used as a linear amplifier. The second major application area is switching applications. When used as an electronic switch, a BJT is normally operated alternately in cutoff and saturation. Many digital circuits use the BJT as a switch.

After completing this section, you should be able to

❑ **Discuss how a BJT is used as a switch**
❑ Describe BJT switching operation
❑ Explain the conditions in cutoff
   ◆ Determine the cutoff voltage in terms of the dc supply voltage

□ Explain the conditions in saturation
  ◆ Calculate the collector current and the base current in saturation
□ Describe a simple application

Switching Operation

Figure 4–23 illustrates the basic operation of a BJT as a switching device. In part (a), the transistor is in the cutoff region because the base-emitter junction is not forward-biased. In this condition, there is, ideally, an *open* between collector and emitter, as indicated by the switch equivalent. In part (b), the transistor is in the saturation region because the base-emitter junction and the base-collector junction are forward-biased and the base current is made large enough to cause the collector current to reach its saturation value. In this condition, there is, ideally, a *short* between collector and emitter, as indicated by the switch equivalent. Actually, a small voltage drop across the transister of up to a few tenths of a volt normally occurs, which is the saturation voltage, $V_{CE(sat)}$.

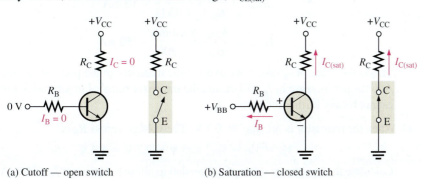

◀ FIGURE 4–23

Switching action of an ideal transistor.

(a) Cutoff — open switch   (b) Saturation — closed switch

Conditions in Cutoff As mentioned before, a transistor is in the cutoff region when the base-emitter junction is not forward-biased. Neglecting leakage current, all of the currents are zero, and V_{CE} is equal to V_{CC}.

$$V_{CE(cutoff)} = V_{CC}$$

Equation 4–8

Conditions in Saturation As you have learned, when the base-emitter junction is forward-biased and there is enough base current to produce a maximum collector current, the transistor is saturated. The formula for collector saturation current is

$$I_{C(sat)} = \frac{V_{CC} - V_{CE(sat)}}{R_C}$$

Equation 4–9

Since $V_{CE(sat)}$ is very small compared to V_{CC}, it can usually be neglected.
The minimum value of base current needed to produce saturation is

$$I_{B(min)} = \frac{I_{C(sat)}}{\beta_{DC}}$$

Equation 4–10

Normally, I_B should be significantly greater than $I_{B(min)}$ to ensure that the transistor is saturated.

EXAMPLE 4–10

(a) For the transistor circuit in Figure 4–24, what is V_{CE} when $V_{IN} = 0$ V?

(b) What minimum value of I_B is required to saturate this transistor if β_{DC} is 200? Neglect $V_{CE(sat)}$.

(c) Calculate the maximum value of R_B that will put the transistor in saturation assuming $\beta_{DC} = 200$ when $V_{IN} = 5$ V.

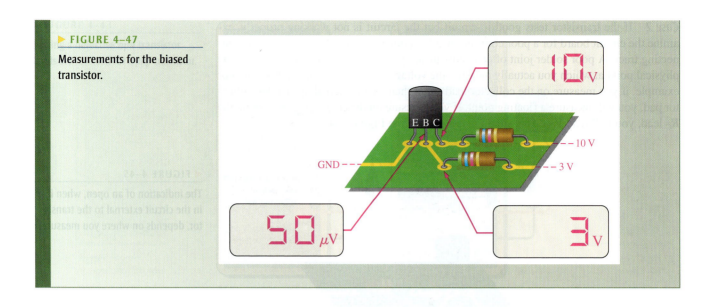

▶ FIGURE 4–47

Measurements for the biased transistor.

Leakage Measurement Very small leakage currents exist in all transistors and in most cases are small enough to neglect (usually nA). When a transistor is connected with the base open ($I_B = 0$), it is in cutoff. Ideally $I_C = 0$; but actually there is a small current from collector to emitter, as mentioned earlier, called I_{CEO} (collector-to-emitter current with base open). This leakage current is usually in the nA range. A faulty transistor will often have excessive leakage current and can be checked in a transistor tester. Another leakage current in transistors is the reverse collector-to-base current, I_{CBO}. This is measured with the emitter open. If it is excessive, a shorted collector-base junction is likely.

Gain Measurement In addition to leakage tests, the typical transistor tester also checks the β_{DC}. A known value of I_B is applied, and the resulting I_C is measured. The reading will indicate the value of the I_C/I_B ratio, although in some units only a relative indication is given.

Curve Tracers A *curve tracer* is an oscilloscope type of instrument that can display transistor characteristics such as a family of collector curves. In addition to the measurement and display of various transistor characteristics, diode curves can also be displayed.

Multisim Troubleshooting Exercises

These file circuits are in the Troubleshooting Exercises folder on the website. Open each file and determine if the circuit is working properly. If it is not working properly, determine the fault.

1. Multisim file TSM04-01

2. Multisim file TSM04-02

3. Multisim file TSM04-03

4. Multisim file TSM04-04

SECTION 4–8 CHECKUP	1. If a transistor on a circuit board is suspected of being faulty, what should you do?
	2. In a transistor bias circuit, such as the one in Figure 4–39; what happens if R_B opens?
	3. In a circuit such as the one in Figure 4–39, what are the base and collector voltages if there is an external open between the emitter and ground?

Device Application: *Security Alarm System*

A circuit using transistor switches will be developed for use in an alarm system for detecting forced entry into a building. In its simplest form, the alarm system will accommodate four zones with any number of openings. It can be expanded to cover additional zones. For the purposes of this application, a zone is one room in a house or other building. The sensor used for each opening can be either a mechanical switch, a magnetically operated switch, or an optical sensor. Detection of an intrusion can be used to initiate an audible alarm signal and/or to initiate transmission of a signal over the phone line to a monitoring service.

Designing the Circuit

A basic block diagram of the system is shown in Figure 4–48. The sensors for each zone are connected to the switching circuits, and the output of the switching circuit goes to an audible alarm circuit and/or to a telephone dialing circuit. The focus of this application is the transistor switching circuits.

▶ **FIGURE 4–48**

Block diagram of security alarm system.

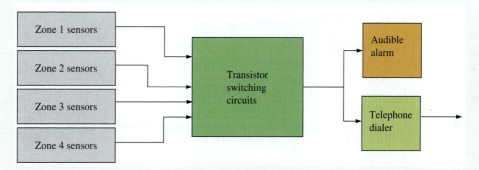

A zone sensor detects when a window or door is opened. They are normally in a closed position and are connected in series to a dc voltage source, as shown in Figure 4–49(a). When a window or door is opened, the corresponding sensor creates an open circuit, as shown in part (b). The sensors are represented by switch symbols.

▶ **FIGURE 4–49**

Zone sensor configuration.

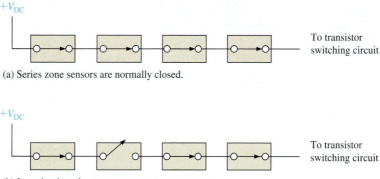

(a) Series zone sensors are normally closed.

(b) Intrusion into the zone causes a sensor to open.

A circuit for one zone is shown in Figure 4–50. It consists of two BJTs, Q_1 and Q_2. As long as the zone sensors are closed, Q_1 is in the *on* state (saturated). The very low saturation voltage at the Q_1 collector keeps Q_2 *off*. Notice that the collector of Q_2 is left open with no load connected. This allows for all four of the zone circuit outputs to be tied together and a common load connected externally to drive the alarm and/or dialing circuits. If one of the zone sensors opens, indicating a break-in, Q_1 turns *off* and its collector voltage goes to V_{CC}. This turns on Q_2, causing it to saturate. The *on* state of Q_2 will then activate the audible alarm and the telephone dialing sequence.

▶ **FIGURE 4–50**

One of the four identical transistor switching circuits.

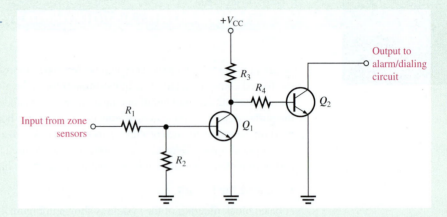

1. Refer to the partial datasheet for the 2N2222A in Figure 4–51 and determine the value of the collector resistor R_3 to limit the current to 10 mA with a +12 V dc supply voltage.

Absolute Maximum Ratings * T_a=25°C unless otherwise noted

Symbol	Parameter	Value	Units
V_{CEO}	Collector-Emitter Voltage	40	V
V_{CBO}	Collector-Base Voltage	75	V
V_{EBO}	Emitter-Base Voltage	6.0	V
I_C	Collector Current	1.0	A
T_{STG}	Operating and Storage Junction Temperature Range	- 55 ~ 150	°C

* These ratings are limiting values above which the serviceability of any semiconductor device may be impaired

NOTES:
1) These ratings are based on a maximum junction temperature of 150 degrees C.
2) These are steady state limits. The factory should be consulted on applications involving pulsed or low duty cycle operations

Electrical Characteristics T_a=25°C unless otherwise noted

Symbol	Parameter	Test Condition	Min.	Max.	Units
Off Characteristics					
$BV_{(BR)CEO}$	Collector-Emitter Breakdown Voltage *	$I_C = 10\text{mA}, I_B = 0$	40		V
$BV_{(BR)CBO}$	Collector-Base Breakdown Voltage	$I_C = 10\mu\text{A}, I_E = 0$	75		V
$BV_{(BR)EBO}$	Emitter-Base Breakdown Voltage	$I_E = 10\mu\text{A}, I_C = 0$	6.0		V
I_{CEX}	Collector Cutoff Current	$V_{CE} = 60\text{V}, V_{EB(off)} = 3.0\text{V}$		10	nA
I_{CBO}	Collector Cutoff Current	$V_{CB} = 60\text{V}, I_E = 0$		0.01	μA
		$V_{CB} = 60\text{V}, I_E = 0, T_a = 125°\text{C}$		10	μA
I_{EBO}	Emitter Cutoff Current	$V_{EB} = 3.0\text{V}, I_C = 0$		10	μA
I_{BL}	Base Cutoff Current	$V_{CE} = 60\text{V}, V_{EB(off)} = 3.0\text{V}$		20	μA
On Characteristics					
h_{FE}	DC Current Gain	$I_C = 0.1\text{mA}, V_{CE} = 10\text{V}$	35		
		$I_C = 1.0\text{mA}, V_{CE} = 10\text{V}$	50		
		$I_C = 10\text{mA}, V_{CE} = 10\text{V}$	75		
		$I_C = 10\text{mA}, V_{CE} = 10\text{V}, T_a = -55°\text{C}$	35		
		$I_C = 150\text{mA}, V_{CE} = 10\text{V} *$	100	300	
		$I_C = 150\text{mA}, V_{CE} = 10\text{V} *$	50		
		$I_C = 500\text{mA}, V_{CE} = 10\text{V} *$	40		
$V_{CE(sat)}$	Collector-Emitter Saturation Voltage *	$I_C = 150\text{mA}, V_{CE} = 10\text{V}$		0.3	V
		$I_C = 500\text{mA}, V_{CE} = 10\text{V}$		1.0	V
$V_{BE(sat)}$	Base-Emitter Saturation Voltage *	$I_C = 150\text{mA}, V_{CE} = 10\text{V}$	0.6	1.2	V
		$I_C = 500\text{mA}, V_{CE} = 10\text{V}$		2.0	V

* Pulse Test: Pulse Width ≤ 300μs, Duty Cycle ≤ 2.0%

▲ **FIGURE 4–51**

Partial datasheet for the 2N2222A transistor. Copyright Fairchild Semiconductor Corporation. Used by permission.

2. Using the minimum β_{DC} or h_{FE} from the datasheet, determine the base current required to saturate Q_1 at $I_C = 10$ mA.

3. To ensure saturation, calculate the value of R_1 necessary to provide sufficient base current to Q_1 from the +12 V sensor input. R_2 can be any arbitrarily high value to assure the base of Q_1 is near ground when there is no input voltage.

4. Calculate the value of R_4 so that a sufficient base current is supplied to Q_2 to ensure saturation for a load of 620 Ω. This simulates the actual load of the alarm and dialing circuits.

Simulation

The switching circuit is simulated with Multisim, as shown in Figure 4–52. A switch connected to a 12 V source simulates the zone input and a 620 Ω load resistor is connected to

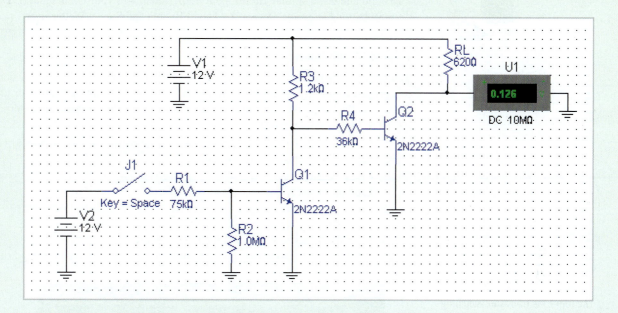

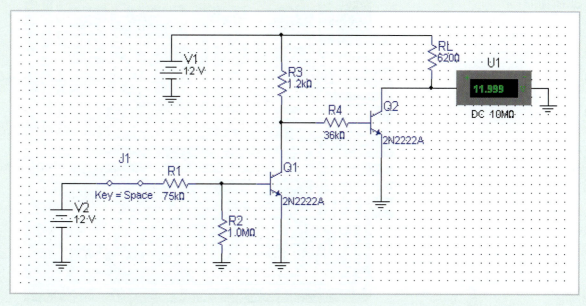

▲ **FIGURE 4–52**

Simulation of the switching circuit.

the output to represent the actual load. When the zone switch is open, Q_2 is saturated as indicated by 0.126 V at its collector. When the zone switch is closed, Q_2 is *off* as indicated by the 11.999 V at its collector.

    5. How does the Q_2 saturation voltage compare to the value specified on the datasheet?

 Simulate the circuit using your Multisim or LT Spice software. Measure and observe the operation.

Prototyping and Testing

Now that the circuit has been simulated, it is connected on a protoboard and tested for proper operation.

Lab Experiment

 To build and test a similar circuit, go to Experiment 4 in your lab manual (*Laboratory Exercises for Electronic Devices* by David Buchla and Steven Wetterling).

Printed Circuit Board

The transistor switching circuit prototype has been built and tested. It is now committed to a printed circuit layout, as shown in Figure 4–53. Notice that there are four identical circuits on the board, one for each zone to be monitored. The outputs are externally connected to form a single input.

    6. Compare the printed circuit board to the schematic in Figure 4–50 and verify that they agree. Identify each component.

    7. Compare the resistor values on the printed circuit board to those that you calculated previously. They should closely agree.

    8. Label the input and output pins on the printed circuit board according to their function.

    9. Describe how you would test the circuit board.

    10. Explain how the system can be expanded to monitor six zones instead of four.

▶ FIGURE 4–53

The four-zone transistor switching circuit board.

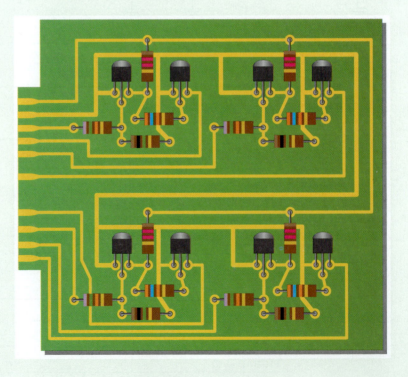

SUMMARY OF BIPOLAR JUNCTION TRANSISTORS

SYMBOLS

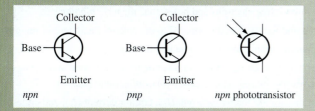

npn   *pnp*   *npn* phototransistor

CURRENTS AND VOLTAGES

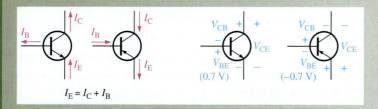

$$I_E = I_C + I_B$$

V_{BE} (0.7 V) V_{BE} (−0.7 V)

AMPLIFICATION

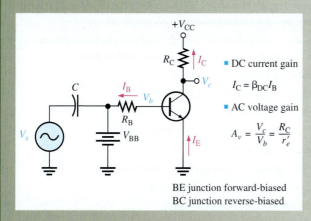

- DC current gain

$$I_C = \beta_{DC} I_B$$

- AC voltage gain

$$A_v = \frac{V_c}{V_b} = \frac{R_C}{r_e'}$$

BE junction forward-biased
BC junction reverse-biased

SWITCHING

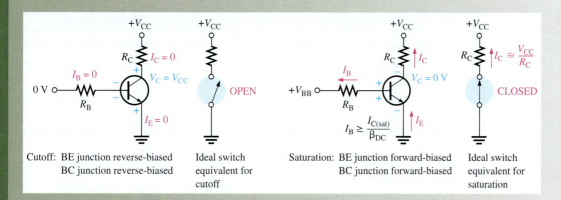

Cutoff: BE junction reverse-biased   Ideal switch   Saturation: BE junction forward-biased   Ideal switch
        BC junction reverse-biased   equivalent for   BC junction forward-biased   equivalent for
                               cutoff   saturation

SUMMARY

Section 4–1
- The BJT (bipolar junction transistor) is constructed with three regions: base, collector, and emitter.
- The BJT has two *pn* junctions, the base-emitter junction and the base-collector junction.
- Current in a BJT consists of both free electrons and holes, thus the term *bipolar*.
- The base region is very thin and lightly doped compared to the collector and emitter regions.
- The two types of bipolar junction transistor are the *npn* and the *pnp*.

Section 4–2
- To operate as an amplifier, the base-emitter junction must be forward-biased and the base-collector junction must be reverse-biased. This is called *forward-reverse bias*.
- The three currents in the transistor are the base current (I_B), emitter current (I_E), and collector current (I_C).
- I_B is very small compared to I_C and I_E.

Section 4–3
- The dc current gain of a transistor is the ratio of I_C to I_B and is designated β_{DC}. Values typically range from less than 20 to several hundred.
- β_{DC} is usually referred to as h_{FE} on transistor datasheets.
- The ratio of I_C to I_E is called α_{DC}. Values typically range from 0.95 to 0.99.
- There is a variation in β_{DC} over temperature and also from one transistor to another of the same type.

Section 4–4
- When a transistor is forward-reverse biased, the voltage gain depends on the internal emitter resistance and the external collector resistance.
- Voltage gain is the ratio of output voltage to input voltage.
- Internal transistor resistances are represented by a lowercase *r*.

Section 4–5
- A transistor can be operated as an electronic switch in cutoff and saturation.
- In cutoff, both *pn* junctions are reverse-biased and there is essentially no collector current. The transistor ideally behaves like an open switch between collector and emitter.
- In saturation, both *pn* junctions are forward-biased and the collector current is maximum. The transistor ideally behaves like a closed switch between collector and emitter.
- Transistors are used as switches in digital logic circuits.

Section 4–6
- In a phototransistor, base current is produced by incident light.
- A phototransistor can be either a two-lead or a three-lead device.
- An optocoupler consists of an LED and a photodiode or phototransistor.
- Optocouplers are used to electrically isolate circuits.

Section 4–7
- There are many types of transistor packages using plastic, metal, or ceramic.
- Two basic package types are through-hole and surface mount.

Section 4–8
- It is best to check a transistor in-circuit before removing it.
- Common faults in transistor circuits are open junctions, low β_{DC}, excessive leakage currents, and external opens and shorts on the circuit board.

KEY TERMS

Key terms and other bold terms in the chapter are defined in the end-of-book glossary.

Amplification   The process of increasing the power, voltage, or current by electronic means.

AND gate   A digital circuit in which the output is at a high level voltage when all of the inputs are at a high level voltage.

Base   One of the semiconductor regions in a BJT. The base is very thin and lightly doped compared to the other regions.

Beta (β)   The ratio of dc collector current to dc base current in a BJT; current gain from base to collector.

BJT   A bipolar junction transistor constructed with three doped semiconductor regions separated by two *pn* junctions.

Collector   The largest of the three semiconductor regions of a BJT.

Cutoff The nonconducting state of a transistor.

Emitter The most heavily doped of the three semiconductor regions of a BJT.

Gain The amount by which an electrical signal is increased or amplified.

Linear Characterized by a straight-line relationship of the transistor currents.

Load Line A load line is a straight line that represents the voltage and current in the linear portion of circuit that is connected to a device (a transistor in this case).

OR gate A digital circuit in which the output is at a high level voltage when one or more inputs are at a high level voltage.

Phototransistor A transistor in which base current is produced when light strikes the photosensitive semiconductor base region.

Saturation The state of a BJT in which the collector current has reached a maximum and is independent of the base current.

KEY FORMULAS

4–1	$I_E = I_C + I_B$	Transistor currents
4–2	$\beta_{DC} = \dfrac{I_C}{I_B}$	DC current gain
4–3	$V_{BE} \cong 0.7\ \text{V}$	Base-to-emitter voltage (silicon)
4–4	$I_B = \dfrac{V_{BB} - V_{BE}}{R_B}$	Base current
4–5	$V_{CE} = V_{CC} - I_C R_C$	Collector-to-emitter voltage (common-emitter)
4–6	$V_{CB} = V_{CE} - V_{BE}$	Collector-to-base voltage
4–7	$A_v \cong \dfrac{R_C}{r'_e}$	Approximate ac voltage gain
4–8	$V_{CE(cutoff)} = V_{CC}$	Cutoff condition
4–9	$I_{C(sat)} = \dfrac{V_{CC} - V_{CE(sat)}}{R_C}$	Collector saturation current
4–10	$I_{B(min)} = \dfrac{I_{C(sat)}}{\beta_{DC}}$	Minimum base current for saturation
4–11	$I_C = \beta_{DC} I_\lambda$	Phototransistor collector current

TRUE/FALSE QUIZ

Answers can be found at www.pearsonhighered.com/floyd.

1. A bipolar junction transistor has three terminals.
2. The three regions of a BJT are base, emitter, and cathode.
3. For operation in the linear or active region, the base-emitter junction of a transistor is forward-biased.
4. Two types of BJT are *npn* and *pnp*.
5. The base current and collector current are approximately equal.
6. The dc voltage gain of a transistor is designated β_{DC}.
7. Cutoff and saturation are the two normal states of a linear transistor amplifier.
8. When a transistor is saturated, the collector current is maximum.
9. β_{DC} and h_{FE} are two different transistor parameters.
10. Voltage gain of a transistor amplifier depends on the collector resistor and the internal ac resistance.
11. Amplification is the output voltage divided by the input current.
12. A transistor in cutoff acts as an open switch.

CIRCUIT-ACTION QUIZ
Answers can be found at www.pearsonhighered.com/floyd.

1. If a transistor with a higher β_{DC} is used in Figure 4–9, the collector current will

   (a) increase   (b) decrease   (c) not change

2. If a transistor with a higher β_{DC} is used in Figure 4–9, the emitter current will

   (a) increase   (b) decrease   (c) not change

3. If a transistor with a higher β_{DC} is used in Figure 4–9, the base current will

   (a) increase   (b) decrease   (c) not change

4. If V_{BB} is reduced in Figure 4–16, the collector current will

   (a) increase   (b) decrease   (c) not change

5. If V_{CC} in Figure 4–16 is increased, the base current will

   (a) increase   (b) decrease   (c) not change

6. If the amplitude of V_{in} in Figure 4–22 is decreased, the ac output voltage amplitude will

   (a) increase   (b) decrease   (c) not change

7. If the transistor in Figure 4–24 is saturated and the base current is increased, the collector current will

   (a) increase   (b) decrease   (c) not change

8. If R_C in Figure 4–24 is reduced in value, the value of $I_{C(sat)}$ will

   (a) increase   (b) decrease   (c) not change

9. If the transistor in Figure 4–39 is open from collector to emitter, the voltage across R_C will

   (a) increase   (b) decrease   (c) not change

10. If the transistor in Figure 4–39 is open from collector to emitter, the collector voltage will

    (a) increase   (b) decrease   (c) not change

11. If the base resistor in Figure 4–39 is open, the transistor collector voltage will

    (a) increase   (b) decrease   (c) not change

12. If the emitter in Figure 4–39 becomes disconnected from ground, the collector voltage will

    (a) increase   (b) decrease   (c) not change

SELF-TEST
Answers can be found at www.pearsonhighered.com/floyd.

Section 4–1

1. The three terminals of a bipolar junction transistor are called

   (a) p, n, p (b) n, p, n (c) input, output, ground   (d) base, emitter, collector

2. In a *pnp* transistor, the *p* regions are

   (a) base and emitter   (b) base and collector   (c) emitter and collector

Section 4–2

3. For operation as an amplifier, the base of an *npn* transistor must be

   (a) positive with respect to the emitter

   (b) negative with respect to the emitter

   (c) positive with respect to the collector

   (d) 0 V

4. The emitter current is always

   (a) greater than the base current

   (b) less than the collector current

   (c) greater than the collector current

   (d) answers (a) and (c)

Section 4–3

5. The β_{DC} of a transistor is its

   (a) current gain   (b) voltage gain

   (c) power gain   (d) internal resistance

6. If I_C is 50 times larger than I_B, then β_{DC} is

    (a) 0.02   (b) 100   (c) 50   (d) 500

7. The approximate voltage across the forward-biased base-emitter junction of a silicon BJT is

    (a) 0 V   (b) 0.7 V   (c) 0.3 V   (d) V_{BB}

8. The bias condition for a transistor to be used as a linear amplifier is called

    (a) forward-reverse

    (b) forward-forward

    (c) reverse-reverse

    (d) collector bias

Section 4–4 9. If the output of a transistor amplifier is 5 V rms and the input is 100 mV rms, the voltage gain is

    (a) 5   (b) 500   (c) 50   (d) 100

10. When a lowercase r' is used in relation to a transistor, it refers to

    (a) a low resistance   (b) a wire resistance

    (c) an internal ac resistance   (d) a source resistance

11. In a given transistor amplifier, $R_C = 2.2 \text{ k}\Omega$ and $r'_e = 20 \text{ }\Omega$, the voltage gain is

    (a) 2.2   (b) 110   (c) 20   (d) 44

Section 4–5 12. When operated in cutoff and saturation, the transistor acts like a

    (a) linear amplifier   (b) switch

    (c) variable capacitor   (d) variable resistor

13. In cutoff, V_{CE} is

    (a) 0 V   (b) minimum   (c) maximum

    (d) equal to V_{CC} (e) answers (a) and (b)   (f) answers (c) and (d)

14. In saturation, V_{CE} is

    (a) 0.7 V   (b) equal to V_{CC} (c) minimum   (d) maximum

15. To saturate a BJT,

    (a) $I_B = I_{C(sat)}$ (b) $I_B > I_{C(sat)}/\beta_{DC}$

    (c) V_{CC} must be at least 10 V   (d) the emitter must be grounded

16. Once in saturation, a further increase in base current will

    (a) cause the collector current to increase

    (b) not affect the collector current

    (c) cause the collector current to decrease

    (d) turn the transistor off

Section 4–6 17. In a phototransistor, base current is

    (a) set by a bias voltage

    (b) directly proportional to light intensity

    (c) inversely proportional to light intensity

    (d) not a factor

18. The relationship between the collector current and a light-generated base current is

    (a) $I_C = \beta_{DC}I_\lambda$ (b) $I_C = \alpha_{DC}I_\lambda$ (c) $I_C = \lambda I_\lambda$ (d) $I_C = \beta_{DC}^2 I_\lambda$

19. An optocoupler usually consists of

    (a) two LEDs   (b) an LED and a photodiode

    (c) an LED and a phototransistor   (d) both (b) and (c)

Section 4–8 20. In a transistor amplifier, if the base-emitter junction is open, the collector voltage is

    (a) V_{CC} (b) 0 V   (c) floating   (d) 0.2 V

21. A DMM measuring on open transistor junction shows

    (a) 0 V   (b) 0.7 V   (c) OL   (d) V_{CC}

PROBLEMS

Answers to all odd-numbered problems are at the end of the book.

BASIC PROBLEMS

Section 4–1 **Bipolar Junction Transistor (BJT) Structure**

1. Describe the difference in the structures of the *npn* and *pnp* transistors.
2. What does the term *bipolar* refer to?
3. What are the majority carriers in the base region of an *npn* transistor called?
4. Explain the purpose of a thin, lightly doped base region.

Section 4–2 **Basic BJT Operation**

5. Why is the base current in a transistor so much less than the collector current?
6. In a certain transistor circuit, the base current is 2% of the 30 mA emitter current. Determine the collector current.
7. For normal operation of a *pnp* transistor, the base must be (+ or −) with respect to the emitter, and (+ or −) with respect to the collector.
8. What is the value of I_C for $I_E = 5.34$ mA and $I_B = 475$ μA?

Section 4–3 **BJT Characteristics and Parameters**

9. What is the α_{DC} when $I_C = 8.23$ mA and $I_E = 8.69$ mA?
10. A certain transistor has an $I_C = 25$ mA and an $I_B = 200$ μA. Determine the β_{DC}.
11. What is the β_{DC} of a transistor if $I_C = 20.3$ mA and $I_E = 20.5$ mA?
12. What is the α_{DC} if $I_C = 5.35$ mA and $I_B = 50$ μA?
13. A certain transistor exhibits an α_{DC} of 0.96. Determine I_C when $I_E = 9.35$ mA.
14. A base current of 50 μA is applied to the transistor in Figure 4–54, and a voltage of 5 V is dropped across R_C. Determine the β_{DC} of the transistor.

▶ **FIGURE 4–54**

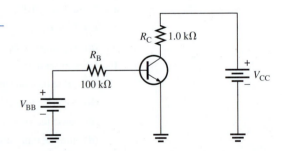

15. Calculate α_{DC} for the transistor in Problem 14.
16. Assume that the transistor in the circuit of Figure 4–54 is replaced with one having a β_{dc} of 200. Determine I_B, I_C, I_E, and V_{CE} given that $V_{CC} = 10$ V and $V_{BB} = 3$ V.
17. If V_{CC} is increased to 15 V in Figure 4–54, how much do the currents and V_{CE} change?
18. Determine each current in Figure 4–55. What is the β_{DC}?

▶ **FIGURE 4–55**

Multisim or LT Spice file circuits are identified with a logo and are in the Problems folder on the website. Filenames correspond to figure numbers (e.g., FGM04-54 and FGS04-55).

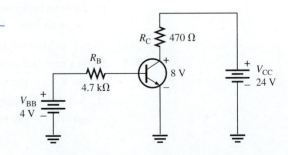

19. Find V_{CE}, V_{BE}, and V_{CB} in both circuits of Figure 4–56.

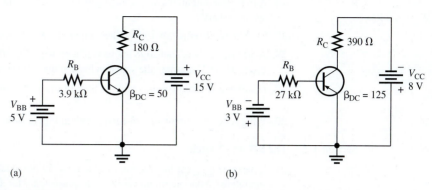

(a)   (b)

▲ FIGURE 4–56

20. Determine whether or not the transistors in Figure 4–56 are saturated.

21. Find I_B, I_E, and I_C in Figure 4–57. $\alpha_{DC} = 0.98$.

▶ FIGURE 4–57

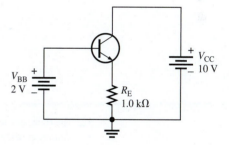

22. Determine the terminal voltages of each transistor with respect to ground for each circuit in Figure 4–58. Also determine V_{CE}, V_{BE}, and V_{CB}.

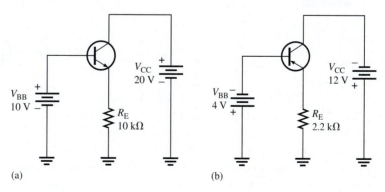

(a)   (b)

▲ FIGURE 4–58

23. If the β_{DC} in Figure 4–58(a) changes from 100 to 150 due to a temperature increase, what is the change in collector current?

24. A certain transistor is to be operated at a collector current of 50 mA. How high can V_{CE} go without exceeding a $P_{D(max)}$ of 1.2 W?

25. The power dissipation derating factor for a certain transistor is 1 mW/°C. The $P_{D(max)}$ is 0.5 W at 25°C. What is $P_{D(max)}$ at 100°C?

Section 4–4

The BJT as an Amplifier

26. A transistor amplifier has a voltage gain of 50. What is the output voltage when the input voltage is 100 mV?

27. To achieve an output of 10 V with an input of 300 mV, what voltage gain is required?

28. A 50 mV signal is applied to the base of a properly biased transistor with $r'_e = 10 \ \Omega$ and $R_C = 560 \ \Omega$. Determine the signal voltage at the collector.

29. Determine the value of the collector resistor in an *npn* transistor amplifier with $\beta_{DC} = 250$, $V_{BB} = 2.5 \ \text{V}$, $V_{CC} = 9 \ \text{V}$, $V_{CE} = 4 \ \text{V}$, and $R_B = 100 \ \text{k}\Omega$.

30. What is the dc current gain of each circuit in Figure 4–56?

Section 4–5

The BJT as a Switch

31. Determine $I_{C(sat)}$ for the transistor in Figure 4–59. What is the value of I_B necessary to produce saturation? What minimum value of V_{IN} is necessary for saturation? Assume $V_{CE(sat)} = 0 \ \text{V}$.

 ▶ **FIGURE 4–59**

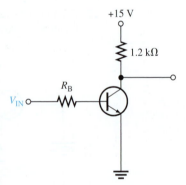

32. The transistor in Figure 4–60 has a β_{DC} of 50. Determine the value of R_B required to ensure saturation when V_{IN} is 5 V. What must V_{IN} be to cut off the transistor? Assume $V_{CE(sat)} = 0 \ \text{V}$.

 ▶ **FIGURE 4–60**

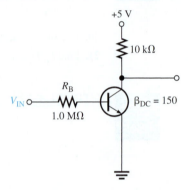

33. What input voltage is required to saturate the transistor in Figure 4–26(a)? Assume $\beta_{DC} = 100$.

34. For the circuit in Figure 4–26(b), what is the output if both inputs = 0.3 V? Assume $\beta_{DC} = 100$?

Section 4–6

The Phototransistor

35. A certain phototransistor in a circuit has a $\beta_{DC} = 200$. If $I_\lambda = 100 \ \mu\text{A}$, what is the collector current?

36. Determine the emitter current in the phototransistor circuit in Figure 4–61 if, for each lm/m^2 of light intensity, 1 μA of base current is produced in the phototransistor.

▶ FIGURE 4–61

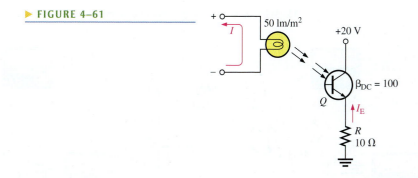

37. A particular optical coupler has a current transfer ratio of 30%. If the input current is 100 mA, what is the output current?

38. The optical coupler shown in Figure 4–62 is required to deliver at least 10 mA to the external load. If the current transfer ratio is 60%, how much current must be supplied to the input?

▶ FIGURE 4–62

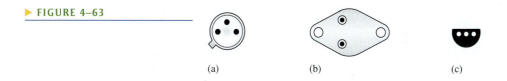

Section 4–7 Transistor Categories and Packaging

39. Identify the leads on the transistors in Figure 4–63. Bottom views are shown.

▶ FIGURE 4–63

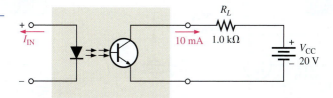

(a)   (b)   (c)

40. What is the most probable category of each transistor in Figure 4–64?

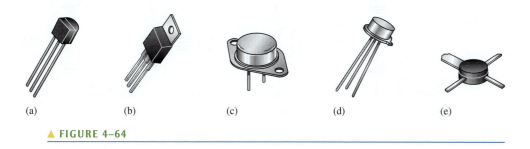

(a)   (b)   (c)   (d)   (e)

▲ FIGURE 4–64

Section 4–8 Troubleshooting

41. In an out-of-circuit test of a good *npn* transistor, what should an analog ohmmeter indicate when its positive probe is touching the emitter and the negative probe is touching the base? When its positive probe is touching the base and the negative probe is touching the collector?

42. What is the most likely problem, if any, in each circuit of Figure 4–65? Assume a β_{DC} of 75.

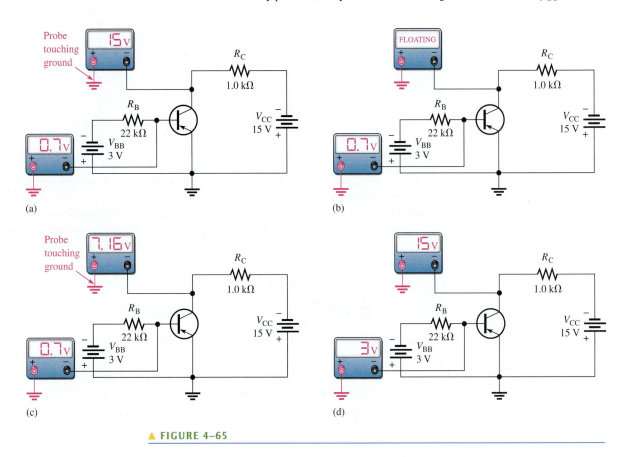

(a)   (b)

(c)   (d)

▲ FIGURE 4–65

43. What is the value of the β_{DC} of each transistor in Figure 4–66?

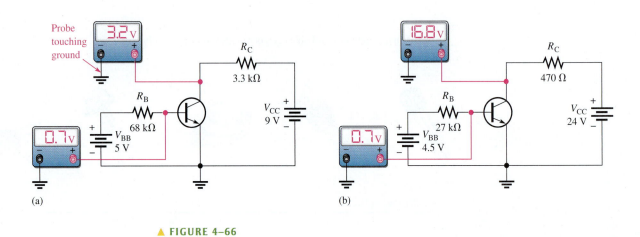

(a)   (b)

▲ FIGURE 4–66

DEVICE APPLICATION PROBLEMS

44. Calculate the power dissipation in each resistor in Figure 4–52 for both states of the circuit.

45. Determine the minimum value of load resistance that Q_2 can drive without exceeding the maximum collector current specified on the datasheet.

46. Develop a wiring diagram for the printed circuit board in Figure 4–53 for connecting it in the security alarm system. The input/output pins are numbered from 1 to 10 starting at the top.

DATASHEET PROBLEMS

47. Refer to the partial transistor datasheet in Figure 4–20.

    (a) What is the maximum collector-to-emitter voltage for a 2N3904?

    (b) How much continuous collector current can the 2N3904 handle?

    (c) How much power can a 2N3904 dissipate if the ambient temperature is 25°C?

    (d) How much power can a 2N3904 dissipate if the ambient temperature is 50°C?

    (e) What is the minimum h_{FE} of a 2N3904 if the collector current is 1 mA?

48. Refer to the transistor datasheet in Figure 4–20. An MMBT3904 is operating in an environment where the ambient temperature is 65°C. What is the most power that it can dissipate?

49. Refer to the transistor datasheet in Figure 4–20. A PZT3904 is operating with an ambient temperature of 45°C. What is the most power that it can dissipate?

50. Refer to the transistor datasheet in Figure 4–20. Determine if any rating is exceeded in each circuit of Figure 4–67 based on minimum specified values.

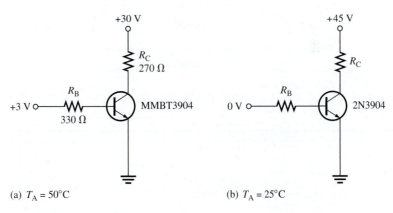

(a) $T_A = 50°C$ (b) $T_A = 25°C$

▲ **FIGURE 4–67**

51. Refer to the transistor datasheet in Figure 4–20. Determine whether or not the transistor is saturated in each circuit of Figure 4–68 based on the maximum specified value of h_{FE}.

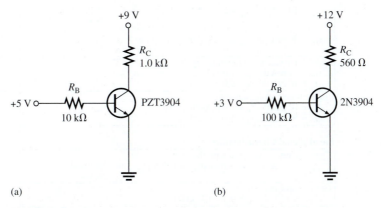

(a)   (b)

▲ **FIGURE 4–68**

52. Refer to the partial transistor datasheet in Figure 4–69. Determine the minimum and maximum base currents required to produce a collector current of 10 mA in a 2N3946. Assume that the transistor is not in saturation and $V_{CE} = 1$ V.

2N3946
2N3947

General-Purpose
Transistors

NPN Silicon

Maximum Ratings

Rating	Symbol	Value	Unit
Collector-Emitter voltage	V_{CEO}	40	V dc
Collector-Base voltage	V_{CBO}	60	V dc
Emitter-Base voltage	V_{EBO}	6.0	V dc
Collector current — continuous	I_C	200	mA dc
Total device dissipation @ $T_A = 25°C$ Derate above 25°C	P_D	0.36 2.06	Watts mW/°C
Total device dissipation @ $T_C = 25°C$ Derate above 25°C	P_D	1.2 6.9	Watts mW/°C
Operating and storage junction Temperature range	T_J, T_{stg}	−65 to +200	°C

Thermal Characteristics

Characteristic	Symbol	Max	Unit
Thermal resistance, junction to case	$R_{\theta JC}$	0.15	°C/mW
Thermal resistance, junction to ambient	$R_{\theta JA}$	0.49	°C/mW

Electrical Characteristics ($T_A = 25°C$ unless otherwise noted.)

Characteristic		Symbol	Min	Max	Unit
OFF Characteristics					
Collector-Emitter breakdown voltage ($I_C = 10$ mA dc)		$V_{(BR)CEO}$	40	–	V dc
Collector-Base breakdown voltage ($I_C = 10$ μA dc, $I_E = 0$)		$V_{(BR)CBO}$	60	–	V dc
Emitter-Base breakdown voltage ($I_E = 10$ μA dc, $I_C = 0$)		$V_{(BR)EBO}$	6.0	–	V dc
Collector cutoff current ($V_{CE} = 40$ V dc, $V_{OB} = 3.0$ V dc) ($V_{CE} = 40$ V dc, $V_{OB} = 3.0$ V dc, $T_A = 150°C$)		I_{CEX}	– –	0.010 15	μA dc
Base cutoff current ($V_{CE} = 40$ V dc, $V_{OB} = 3.0$ V dc)		I_{BL}	–	.025	μA dc
ON Characteristics					
DC current gain ($I_C = 0.1$ mA dc, $V_{CE} = 1.0$ V dc) ($I_C = 1.0$ mA dc, $V_{CE} = 1.0$ V dc) ($I_C = 10$ mA dc, $V_{CE} = 1.0$ V dc) ($I_C = 50$ mA dc, $V_{CE} = 1.0$ V dc)	2N3946 2N3947 2N3946 2N3947 2N3946 2N3947 2N3946 2N3947	h_{FE}	30 60 45 90 50 100 20 40	– – – – 150 300 – –	–
Collector-Emitter saturation voltage ($I_C = 10$ mA dc, $I_B = 1.0$ mA dc) ($I_C = 50$ mA dc, $I_B = 5.0$ mA dc)		$V_{CE(sat)}$	– –	0.2 0.3	V dc
Base-Emitter saturation voltage ($I_C = 10$ mA dc, $I_B = 1.0$ mA dc) ($I_C = 50$ mA dc, $I_B = 5.0$ mA dc)		$V_{BE(sat)}$	0.6 –	0.9 1.0	V dc
Small-Signal Characteristics					
Current gain — Bandwidth product ($I_C = 10$ mA dc, $V_{CE} = 20$ V dc, $f = 100$ MHz)	2N3946 2N3947	f_T	250 300	– –	MHz
Output capacitance ($V_{CB} = 10$ V dc, $I_E = 0, f = 100$ kHz)		C_{obo}	–	4.0	pF

▲ **FIGURE 4–69**

53. For each of the circuits in Figure 4–70, determine if there is a problem based on the datasheet information in Figure 4–69. Use the maximum specified h_{FE}.

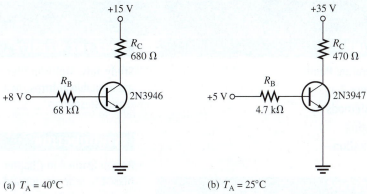

(a) $T_A = 40°C$ (b) $T_A = 25°C$

▲ **FIGURE 4–70**

ADVANCED PROBLEMS

54. Derive a formula for α_{DC} in terms of β_{DC}.

55. A certain 2N3904 dc bias circuit with the following values is in saturation. $I_B = 500\ \mu A$, $V_{CC} = 10$ V, and $R_C = 180\ \Omega$, $h_{FE} = 150$. If you increase V_{CC} to 15 V, does the transistor come out of saturation? If so, what is the collector-to-emitter voltage and the collector current?

56. Design a dc bias circuit for a 2N3904 operating from a collector supply voltage of 9 V and a base-bias voltage of 3 V that will supply 150 mA to a resistive load that acts as the collector resistor. The circuit must not be in saturation. Assume the minimum specified β_{DC} from the datasheet.

57. Modify the design in Problem 56 to use a single 9 V dc source rather than two different sources. Other requirements remain the same.

58. Design a dc bias circuit for an amplifier in which the voltage gain is to be a minimum of 50 and the output signal voltage is to be "riding" on a dc level of 5 V. The maximum input signal voltage at the base is 10 mV rms. $V_{CC} = 12$ V, and $V_{BB} = 4$ V. Assume $r'_e = 8\ \Omega$.

MULTISIM TROUBLESHOOTING PROBLEMS

These file circuits are in the Troubleshooting Problems folder on the website.

59. Open file TPM04-59 and determine the fault.

60. Open file TPM04-60 and determine the fault.

61. Open file TPM04-61 and determine the fault.

62. Open file TPM04-62 and determine the fault.

63. Open file TPM04-63 and determine the fault.

64. Open file TPM04-64 and determine the fault.

65. Open file TPM04-65 and determine the fault.

66. Open file TPM04-66 and determine the fault.

transistor are shown in Figure 5–2(b); we will use these curves to graphically illustrate the effects of dc bias.

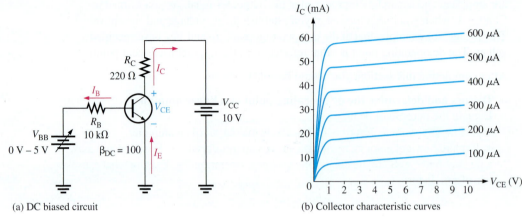

(a) DC biased circuit

(b) Collector characteristic curves

▲ **FIGURE 5–2**

A dc-biased transistor circuit with variable bias voltage (V_{BB}) for generating the collector characteristic curves shown in part (b).

In Figure 5–3, we assign three values to I_B and observe what happens to I_C and V_{CE}. First, V_{BB} is adjusted to produce an I_B of 200 μA, as shown in Figure 5–3(a). Since $I_C = \beta_{DC} I_B$, the collector current is 20 mA, as indicated, and

$$V_{CE} = V_{CC} - I_C R_C = 10 \text{ V} - (20 \text{ mA})(220 \text{ }\Omega) = 10 \text{ V} - 4.4 \text{ V} = 5.6 \text{ V}$$

This Q-point is shown on the graph of Figure 5–3(a) as Q_1.

Next, as shown in Figure 5–3(b), V_{BB} is increased to produce an I_B of 300 μA and an I_C of 30 mA.

$$V_{CE} = 10 \text{ V} - (30 \text{ mA})(220 \text{ }\Omega) = 10 \text{ V} - 6.6 \text{ V} = 3.4 \text{ V}$$

The Q-point for this condition is indicated by Q_2 on the graph.

Finally, as in Figure 5–3(c), V_{BB} is increased to give an I_B of 400 μA and an I_C of 40 mA.

$$V_{CE} = 10 \text{ V} - (40 \text{ mA})(220 \text{ }\Omega) = 10 \text{ V} - 8.8 \text{ V} = 1.2 \text{ V}$$

Q_3 is the corresponding Q-point on the graph.

DC Load Line   The dc operation of a transistor circuit can be described graphically using a **dc load line**. This is a straight line drawn on the characteristic curves from the saturation value where $I_C = I_{C(sat)}$ on the *y*-axis to the cutoff value where $V_{CE} = V_{CC}$ on the *x*-axis, as shown in Figure 5–4(a). The load line is determined by the external circuit (V_{CC} and R_C), not the transistor itself, which is described by the characteristic curves.

In Figure 5–3, the equation for I_C is

$$I_C = \frac{V_{CC} - V_{CE}}{R_C} = \frac{V_{CC}}{R_C} - \frac{V_{CE}}{R_C} = -\frac{V_{CE}}{R_C} + \frac{V_{CC}}{R_C} = -\left(\frac{1}{R_C}\right)V_{CE} + \frac{V_{CC}}{R_C}$$

This is the equation of a straight line with a slope of $-1/R_C$, an *x* intercept of $V_{CE} = V_{CC}$, and a *y* intercept of V_{CC}/R_C, which is $I_{C(sat)}$.

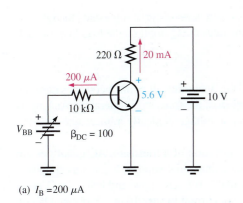
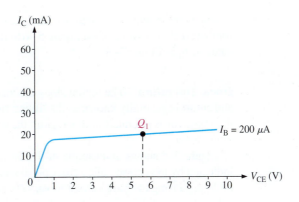

(a) $I_B = 200\ \mu A$

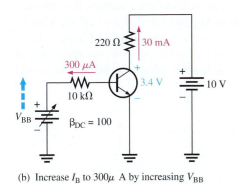
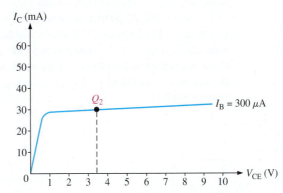

(b) Increase I_B to 300μ A by increasing V_{BB}

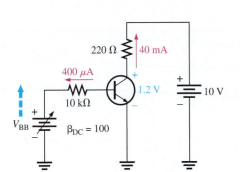
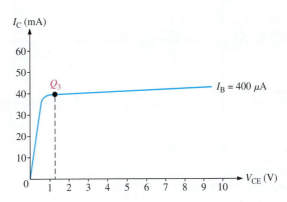

(c) Increase I_B to 400μ A by increasing V_{BB}

▲ FIGURE 5–3

Illustration of Q-point adjustment.

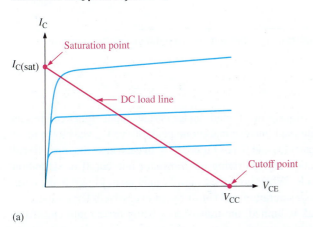

(a)

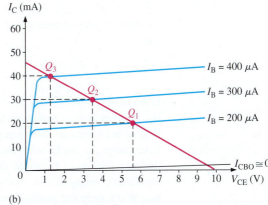

(b)

▲ FIGURE 5–4

The dc load line.

The point at which the load line intersects a characteristic curve represents the Q-point for that particular value of I_B. Figure 5–4(b) illustrates the Q-point on the load line for each value of I_B in Figure 5–3.

Linear Operation   The region along the load line including all points between saturation and cutoff is generally known as the **linear region** of the transistor's operation. As long as the transistor is operated in this region, the output voltage is ideally a linear reproduction of the input.

Figure 5–5 shows an example of the linear operation of a transistor. AC quantities are indicated by lowercase italic subscripts. Assume a sinusoidal voltage, V_{in}, is superimposed on V_{BB}, causing the base current to vary sinusoidally 100 μA above and below its Q-point value of 300 μA. This, in turn, causes the collector current to vary 10 mA above and below its Q-point value of 30 mA. As a result of the variation in collector current, the collector-to-emitter voltage varies 2.2 V above and below its Q-point value of 3.4 V. Point A on the load line in Figure 5–5 corresponds to the positive peak of the sinusoidal input voltage. Point B corresponds to the negative peak, and point Q corresponds to the zero value of the sine wave, as indicated. V_{CEQ}, I_{CQ}, and I_{BQ} are dc Q-point values with no input sinusoidal voltage applied.

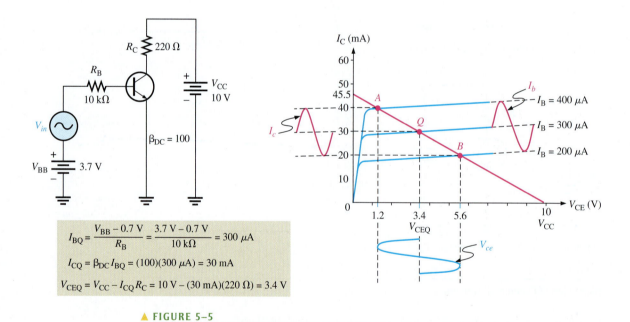

$$I_{BQ} = \frac{V_{BB} - 0.7\ V}{R_B} = \frac{3.7\ V - 0.7\ V}{10\ k\Omega} = 300\ \mu A$$

$$I_{CQ} = \beta_{DC}I_{BQ} = (100)(300\ \mu A) = 30\ mA$$

$$V_{CEQ} = V_{CC} - I_{CQ}R_C = 10\ V - (30\ mA)(220\ \Omega) = 3.4\ V$$

▲ **FIGURE 5–5**

Variations in collector current and collector-to-emitter voltage as a result of a variation in base current.

Waveform Distortion   As previously mentioned, under certain input signal conditions the location of the Q-point on the load line can cause one peak of the V_{ce} waveform to be limited or clipped, as shown in parts (a) and (b) of Figure 5–6. In each case the input signal is too large for the Q-point location and is driving the transistor into cutoff or saturation during a portion of the input cycle. When both peaks are limited as in Figure 5–6(c), the transistor is being driven into both saturation and cutoff by an excessively large input signal. When only the positive peak is limited, the transistor is being driven into cutoff but not saturation. When only the negative peak is limited, the transistor is being driven into saturation but not cutoff.

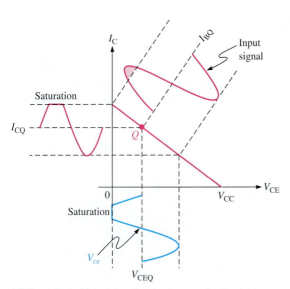

(a) Transistor is driven into saturation because the Q-point is too close to saturation for the given input signal.

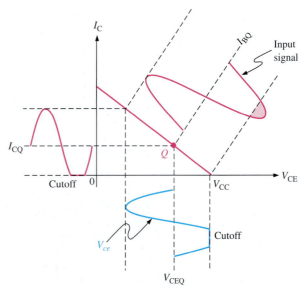

(b) Transistor is driven into cutoff because the Q-point is too close to cutoff for the given input signal.

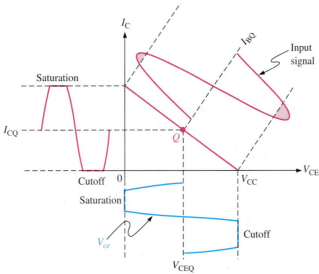

(c) Transistor is driven into both saturation and cutoff because the input signal is too large.

▲ FIGURE 5–6

Graphical load line illustration of a transistor being driven into saturation and/or cutoff.

EXAMPLE 5–1

Determine the Q-point for the circuit in Figure 5–7 and draw the dc load line. Find the maximum peak value of base current for linear operation. Assume $\beta_{DC} = 200$.

Solution   The Q-point is defined by the values of I_C and V_{CE}.

$$I_B = \frac{V_{BB} - V_{BE}}{R_B} = \frac{10\ V - 0.7\ V}{47\ k\Omega} = 198\ \mu A$$

$$I_C = \beta_{DC} I_B = (200)(198\ \mu A) = \textbf{39.6 mA}$$

$$V_{CE} = V_{CC} - I_C R_C = 20\ V - 13.07\ V = \textbf{6.93 V}$$

▶ FIGURE 5–7

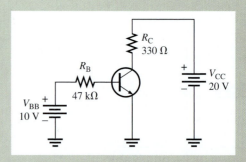

The Q-point is at $I_C = 39.6$ mA and at $V_{CE} = 6.93$ V.

Since $I_{C(cutoff)} = 0$, you need to know $I_{C(sat)}$ to determine how much variation in collector current can occur and still maintain linear operation of the transistor.

$$I_{C(sat)} = \frac{V_{CC}}{R_C} = \frac{20 \text{ V}}{330 \text{ }\Omega} = 60.6 \text{ mA}$$

The dc load line is graphically illustrated in Figure 5–8, showing that before saturation is reached, I_C can increase an amount ideally equal to

$$I_{C(sat)} - I_{CQ} = 60.6 \text{ mA} - 39.6 \text{ mA} = 21.0 \text{ mA}$$

However, I_C can decrease by 39.6 mA before cutoff ($I_C = 0$) is reached. Therefore, the limiting excursion is 21 mA because the *Q-point is closer to saturation than to cutoff.* The 21 mA is the maximum peak variation of the collector current. Actually, it would be slightly less in practice because $V_{CE(sat)}$ is not quite zero.

▶ FIGURE 5–8

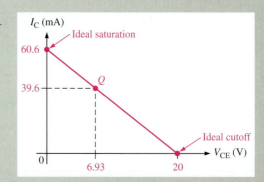

Determine the maximum peak variation of the base current as follows:

$$I_{b(peak)} = \frac{I_{c(peak)}}{\beta_{DC}} = \frac{21 \text{ mA}}{200} = \mathbf{105 \text{ }\mu A}$$

Related Problem Find the Q-point for the circuit in Figure 5–7, and determine the maximum peak value of base current for linear operation for the following circuit values: $\beta_{DC} = 100$, $R_C = 1.0$ kΩ, and $V_{CC} = 24$ V.

*Answers can be found at www.pearsonhighered.com/floyd.

Open the Multisim file EXM05-01 or LT Spice file EXS05-01 in the Examples folder on the website. Measure I_C and V_{CE} and compare with the calculated values.

1. What are the upper and lower limits on a dc load line in terms of V_{CE} and I_C?
2. Define *Q-point*.
3. At what point on the load line does saturation occur? At what point does cutoff occur?
4. For maximum V_{ce}, where should the Q-point be placed?

5–2 VOLTAGE-DIVIDER BIAS

You will now study a method of biasing a transistor for linear operation using a single-source resistive voltage divider. This is the most widely used biasing method. Four other methods are covered in Section 5–3.

After completing this section, you should be able to

❑ **Analyze a voltage-divider biased circuit**
 ◆ Define the term *stiff voltage-divider* ◆ Calculate currents and voltages in a voltage-divider biased circuit
❑ Explain the loading effects in voltage-divider bias
 ◆ Describe how dc input resistance at the transistor base affects the bias
❑ Apply Thevenin's theorem to the analysis of voltage-divider bias
 ◆ Analyze both *npn* and *pnp* circuits

Up to this point a separate dc source, V_{BB}, was used to bias the base-emitter junction because it could be varied independently of V_{CC} and it helped to illustrate transistor operation. A more practical bias method is to use V_{CC} as the single bias source, as shown in Figure 5–9. To simplify the schematic, the battery symbol is omitted and replaced by a line termination circle with a voltage indicator (V_{CC}) as shown.

A dc bias voltage at the base of the transistor can be developed by a resistive voltage-divider that consists of R_1 and R_2, as shown in Figure 5–9. V_{CC} is the dc collector supply voltage. Two current paths are between point *A* and ground: one through R_2 and the other through the base-emitter junction of the transistor and R_E.

Generally, voltage-divider bias circuits are designed so that the base current is much smaller than the current (I_2) through R_2 in Figure 5–9. In this case, the voltage-divider circuit is very straightforward to analyze because the loading effect of the base current can be ignored. A voltage divider in which the base current is small compared to the current in R_2 is said to be a **stiff voltage divider** because the base voltage is relatively independent of different transistors and temperature effects.

To analyze a voltage-divider circuit in which I_B is small compared to I_2, first calculate the voltage on the base using the unloaded voltage-divider rule:

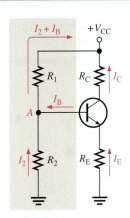

▲ FIGURE 5–9

Voltage-divider bias.

$$V_B \cong \left(\frac{R_2}{R_1 + R_2}\right)V_{CC}$$

Equation 5–1

Once you know the base voltage, you can find the voltages and currents in the circuit, as follows:

$$V_E = V_B - V_{BE}$$

Equation 5–2

and

$$I_C \cong I_E = \frac{V_E}{R_E}$$

Equation 5–3

Then,

$$V_C = V_{CC} - I_C R_C$$

Equation 5–4

Once you know V_C and V_E, you can determine V_{CE}.

$$V_{CE} = V_C - V_E$$

EXAMPLE 5–2

Determine V_{CE} and I_C in the stiff voltage-divider biased transistor circuit of Figure 5–10 if $\beta_{DC} = 100$.

▶ **FIGURE 5–10**

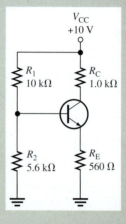

Solution The base voltage is

$$V_B \cong \left(\frac{R_2}{R_1 + R_2} \right) V_{CC} = \left(\frac{5.6 \text{ k}\Omega}{15.6 \text{ k}\Omega} \right) 10 \text{ V} = 3.59 \text{ V}$$

So,

$$V_E = V_B - V_{BE} = 3.59 \text{ V} - 0.7 \text{ V} = 2.89 \text{ V}$$

and

$$I_E = \frac{V_E}{R_E} = \frac{2.89 \text{ V}}{560 \ \Omega} = 5.16 \text{ mA}$$

Therefore,

$$I_C \cong I_E = \textbf{5.16 mA}$$

and

$$V_C = V_{CC} - I_C R_C = 10 \text{ V} - (5.16 \text{ mA})(1.0 \text{ k}\Omega) = 4.84 \text{ V}$$
$$V_{CE} = V_C - V_E = 4.84 \text{ V} - 2.89 \text{ V} = \textbf{1.95 V}$$

Related Problem If the voltage divider in Figure 5–10 was not stiff, how would V_B be affected?

 Open the Multisim file EXM05-02 or LT Spice file EXS05-02 in the Examples folder on the website. Measure I_C and V_{CE}. If these results do not agree very closely with those in the Example, what original assumption was incorrect?

The basic analysis developed in Example 5–2 is all that is needed for most voltage-divider circuits, but there may be cases where you need to analyze the circuit with more accuracy. Ideally, a voltage-divider circuit is stiff, which means that the transistor does not appear as a significant load to the voltage divider. All circuit design involves trade-offs, and one trade-off is that stiff voltage dividers require smaller resistors, which are not always desirable because of potential loading effects on the driving circuit and added power requirements. If the circuit designer wanted to raise the input resistance to avoid loading the driving stage, the divider string might not be stiff; more detailed analysis would be required to calculate circuit parameters. To determine if the divider is stiff, you need to examine the dc input resistance looking in at the base as shown in Figure 5–11.

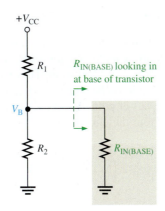

Stiff:

$$R_{IN(BASE)} \cong 10R_2$$

$$V_B \cong \left(\frac{R_2}{R_1 + R_2}\right) V_{CC}$$

Not stiff:

$$R_{IN(BASE)} < 10R_2$$

$$V_B = \left(\frac{R_2 \| R_{IN(BASE)}}{R_1 + R_2 \| R_{IN(BASE)}}\right) V_{CC}$$

Loading Effects of Voltage-Divider Bias

DC Input Resistance at the Transistor Base   The dc input resistance of the transistor is proportional to β_{DC}, so it will change for different transistors. When a transistor is operating in its linear region, the emitter current (I_E) is $\beta_{DC}I_B$. When the emitter resistor is viewed from the base circuit, the resistor appears to be larger than its actual value because of the dc current gain in the transistor. That is, $R_{IN(BASE)} = V_B/I_B = V_B/(I_E/\beta_{DC})$.

$$R_{IN(BASE)} = \frac{\beta_{DC}V_B}{I_E}$$

Equation 5–5

This is the effective load on the voltage divider illustrated in Figure 5–11.

You can quickly estimate the loading effect by comparing $R_{IN(BASE)}$ to the resistor R_2 in the voltage divider. As long as $R_{IN(BASE)}$ is at least ten times larger than R_2, the loading effect will be 10% or less and the voltage divider is stiff. If $R_{IN(BASE)}$ is less than ten times R_2, it should be combined in parallel with R_2.

EXAMPLE 5–3

Determine the dc input resistance looking in at the base of the transistor in Figure 5–12. $\beta_{DC} = 125$ and $V_B = 4$ V.

▶ **FIGURE 5–12**

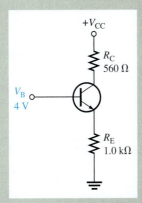

Solution

$$I_E = \frac{V_B - 0.7 \text{ V}}{R_E} = \frac{3.3 \text{ V}}{1.0 \text{ k}\Omega} = 3.3 \text{ mA}$$

$$R_{IN(BASE)} = \frac{\beta_{DC}V_B}{I_E} = \frac{125(4 \text{ V})}{3.3 \text{ mA}} = \mathbf{152 \text{ k}\Omega}$$

Related Problem   What is $R_{IN(BASE)}$ in Figure 5–12 if $\beta_{DC} = 60$ and $V_B = 2$ V?

Thevenin's Theorem Applied to Voltage-Divider Bias

To analyze a voltage-divider biased transistor circuit for base current loading effects, we will apply Thevenin's theorem to evaluate the circuit. First, let's get an equivalent base-emitter circuit for the circuit in Figure 5–13(a) using Thevenin's theorem. Looking out from the base terminal, the bias circuit can be redrawn as shown in Figure 5–13(b). Apply Thevenin's theorem to the circuit left of point A, with V_{CC} replaced by a short to ground and the transistor disconnected from the circuit. The voltage at point A with respect to ground is

$$V_{TH} = \left(\frac{R_2}{R_1 + R_2}\right)V_{CC}$$

and the resistance is

$$R_{TH} = \frac{R_1 R_2}{R_1 + R_2}$$

▶ **FIGURE 5–13**

Thevenizing the bias circuit.

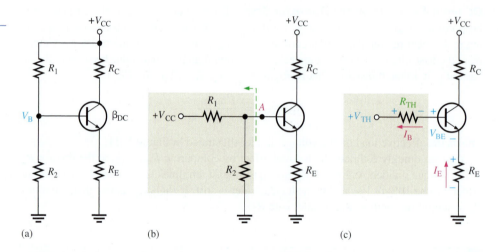

(a)　　　　　(b)　　　　　(c)

The Thevenin equivalent of the bias circuit, connected to the transistor base, is shown in the beige box in Figure 5–13(c). Applying Kirchhoff's voltage law around the equivalent base-emitter loop gives

$$V_{TH} - V_{R_{TH}} - V_{BE} - V_{R_E} = 0$$

Substituting, using Ohm's law, and solving for V_{TH},

$$V_{TH} = I_B R_{TH} + V_{BE} + I_E R_E$$

Substituting I_E/β_{DC} for I_B,

$$V_{TH} = I_E(R_E + R_{TH}/\beta_{DC}) + V_{BE}$$

Then solving for I_E,

Equation 5–6

$$I_E = \frac{V_{TH} - V_{BE}}{R_E + R_{TH}/\beta_{DC}}$$

If R_{TH}/β_{DC} is small compared to R_E, the result is the same as for an unloaded voltage divider.

Voltage-divider bias is widely used because reasonably good bias stability is achieved with a single supply voltage.

Voltage-Divider Biased PNP Transistor As you know, a *pnp* transistor requires bias polarities opposite to the *npn*. This can be accomplished with a negative collector supply voltage, as in Figure 5–14(a), or with a positive emitter supply voltage, as in Figure 5–14(b).

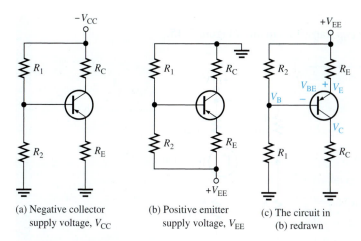

(a) Negative collector supply voltage, V_{CC}

(b) Positive emitter supply voltage, V_{EE}

(c) The circuit in (b) redrawn

▲ **FIGURE 5–14**

Voltage-divider biased *pnp* transistor.

In a schematic, the *pnp* is often drawn upside down so that the supply voltage is at the top of the schematic and ground at the bottom, as in Figure 5–14(c).

The analysis procedure is the same as for an *npn* transistor circuit using Thevenin's theorem and Kirchhoff's voltage law, as demonstrated in the following steps with reference to Figure 5–14. For Figure 5–14(a), applying Kirchhoff's voltage law around the base-emitter circuit gives

$$V_{TH} + I_B R_{TH} - V_{BE} + I_E R_E = 0$$

By Thevenin's theorem,

$$V_{TH} = \left(\frac{R_2}{R_1 + R_2}\right) V_{CC}$$

$$R_{TH} = \frac{R_1 R_2}{R_1 + R_2}$$

The base current is

$$I_B = \frac{I_E}{\beta_{DC}}$$

The equation for I_E is

$$I_E = \frac{-V_{TH} + V_{BE}}{R_E + R_{TH}/\beta_{DC}}$$

Equation 5–7

For Figure 5–14(b), the analysis is as follows:

$$-V_{TH} + I_B R_{TH} - V_{BE} + I_E R_E - V_{EE} = 0$$

$$V_{TH} = \left(\frac{R_1}{R_1 + R_2}\right) V_{EE}$$

$$R_{TH} = \frac{R_1 R_2}{R_1 + R_2}$$

$$I_B = \frac{I_E}{\beta_{DC}}$$

The equation for I_E is

$$I_E = \frac{V_{TH} + V_{BE} - V_{EE}}{R_E + R_{TH}/\beta_{DC}}$$

Equation 5–8

When a voltmeter is connected to the emitter, it provides a high-resistance current path through its internal impedance, resulting in a forward-biased base-emitter junction. Therefore, the emitter voltage is $V_E = V_B - V_{BE}$. The amount of the forward voltage drop across the BE junction depends on the current. $V_{BE} = 0.7$ V is assumed for purposes of illustration, but it may be much less. The result is an emitter voltage as follows:

$$V_E = V_B - V_{BE} = 3.2 \text{ V} - 0.7 \text{ V} = 2.5 \text{ V}$$

Fault 3: Base Internally Open   An internal transistor fault is more likely to happen than an open resistor. Again, the transistor is nonconducting so $I_C = 0$ A and $V_C = V_{CC} = 10$ V. Just as for the case of the open R_E, the voltage divider produces 3.2 V at the external base connection. The voltage at the external emitter connection is 0 V because there is no emitter current through R_E and, thus, no voltage drop.

Fault 4: Emitter Internally Open   Again, the transistor is nonconducting, so $I_C = 0$ A and $V_C = V_{CC} = 10$ V. Just as for the case of the open R_E and the internally open base, the voltage divider produces 3.2 V at the base. The voltage at the external emitter lead is 0 V because that point is open and connected to ground through R_E. Notice that Faults 3 and 4 produce identical symptoms.

Fault 5: Collector Internally Open   Since there is an internal open in the transistor collector, there is no I_C and, therefore, $V_C = V_{CC} = 10$ V. In this situation, the voltage divider is loaded by R_E through the forward-biased BE junction, as shown by the approximate equivalent circuit in Figure 5–26. The base voltage and emitter voltage are determined as follows:

$$V_B \cong \left(\frac{R_2 \| R_E}{R_1 + R_2 \| R_E} \right) V_{CC} + 0.7 \text{ V}$$

$$= \left(\frac{427 \ \Omega}{10.427 \text{ k}\Omega} \right) 10 \text{ V} + 0.7 \text{ V} = 0.41 \text{ V} + 0.7 \text{ V} = 1.11 \text{ V}$$

$$V_E = V_B - V_{BE} = 1.11 \text{ V} - 0.7 \text{ V} = 0.41 \text{ V}$$

▶ **FIGURE 5–26**

Equivalent bias circuit for an internally open collector.

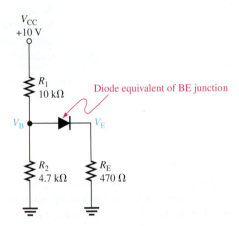

There are two possible additional faults for which the transistor is conducting or appears to be conducting, based on the collector voltage measurement. These are indicated in Figure 5–27.

Fault 6: Resistor R_C Open   For this fault, which is illustrated in Figure 5–27(a), the collector voltage may lead you to think that the transistor is in saturation, but actually it is

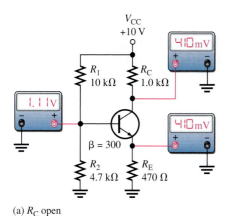

(a) R_C open

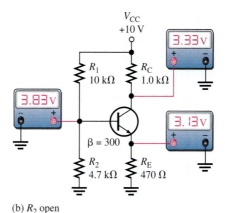

(b) R_2 open

◀ FIGURE 5–27

Faults for which the transistor is conducting or appears to be conducting.

nonconducting. Obviously, if R_C is open, there can be no collector current. In this situation, the equivalent bias circuit is the same as for Fault 5, as illustrated in Figure 5–26. Therefore, $V_B = 1.11$ V and since the BE junction is forward-biased,

$$V_E = V_B - V_{BE} = 1.11 \text{ V} - 0.7 \text{ V} = 0.41 \text{ V}$$

When a voltmeter is connected to the collector to measure V_C, a current path is provided through the internal impedance of the meter and the BC junction is forward-biased by V_B. Therefore,

$$V_C = V_B - V_{BC} = 1.11 \text{ V} - 0.7 \text{ V} = 0.41 \text{ V}$$

Again the forward drops across the internal transistor junctions depend on the current. We are using 0.7 V for illustration, but the forward drops may be much less.

Fault 7: Resistor R_2 Open   When R_2 opens as shown in Figure 5–27(b), the base voltage and base current increase from their normal values because the voltage divider is now formed by R_1 and $R_{IN(BASE)}$. The circuit is equivalent to base bias with a small value bias resistor. In this case, the base voltage is determined by the emitter voltage ($V_B = V_E + V_{BE}$).

First, verify whether the transistor is in saturation or not. The collector saturation current and the base current required to produce saturation are determined as follows (assuming $V_{CE(sat)} = 0.2$ V):

$$I_{C(sat)} = \frac{V_{CC} - V_{CE(sat)}}{R_C + R_E} = \frac{9.8 \text{ V}}{1.47 \text{ k}\Omega} = 6.67 \text{ mA}$$

$$I_{B(sat)} = \frac{I_{C(sat)}}{\beta_{DC}} = \frac{6.67 \text{ mA}}{300} = 22.2 \text{ }\mu\text{A}$$

Assuming the transistor is saturated, the maximum base current is determined.

$$I_{E(sat)} \cong 6.67 \text{ mA}$$
$$V_E = I_{E(sat)}R_E = 3.13 \text{ V}$$
$$V_B = V_E + V_{BE} = 3.83 \text{ V}$$
$$R_{IN(BASE)} = \frac{\beta_{DC}V_B}{I_E} = \frac{(300)(3.83 \text{ V})}{6.67 \text{ mA}} = 172 \text{ k}\Omega$$
$$I_B = \frac{V_{CC}}{R_1 + R_{IN(BASE)}} = \frac{10 \text{ V}}{182 \text{ k}\Omega} = 54.9 \text{ mA}$$

Since this amount of base current is more than enough to produce saturation, the transistor is definitely saturated. Therefore, V_E, V_B, and V_C are as follows:

$$V_E = 3.13 \text{ V}$$
$$V_B = 3.83 \text{ V}$$
$$V_C = V_{CC} - I_{C(sat)}R_C = 10 \text{ V} - (6.67 \text{ mA})(1.0 \text{ k}\Omega) = 3.33 \text{ V}$$

Lab Experiment

> To build and test a similar circuit, go to Experiment 5 in your lab manual (*Laboratory Exercises for Electronic Devices* by David Buchla and Steven Wetterling).

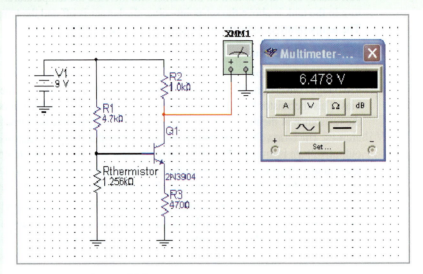

(a) Circuit output voltage at 60° C

R_{therm} = 1.481 kΩ R_{therm} = 1.753 kΩ R_{therm} = 2.084 kΩ R_{therm} = 2.490 kΩ

(b) Circuit output voltages at 65°, 70°, 75°, and 80°

▲ **FIGURE 5–30**

Operation of the temperature-to-voltage conversion circuit over temperature.

The Printed Circuit Board

A partially completed printed circuit board is shown in Figure 5–31. Indicate how you would add conductive traces to complete the circuit and show the input/output terminal functions.

▶ **FIGURE 5–31**

Partially complete temperature-to-voltage conversion circuit PC board.

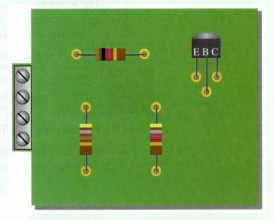

SUMMARY OF TRANSISTOR BIAS CIRCUITS

npn transistors are shown. Supply voltage polarities are reversed for *pnp* transistors.

VOLTAGE-DIVIDER BIAS

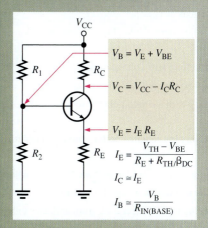

$$V_B = V_E + V_{BE}$$

$$V_C = V_{CC} - I_C R_C$$

$$V_E = I_E R_E$$

$$I_E = \frac{V_{TH} - V_{BE}}{R_E + R_{TH}/\beta_{DC}}$$

$$I_C \cong I_E$$

$$I_B \cong \frac{V_B}{R_{IN(BASE)}}$$

EMITTER BIAS

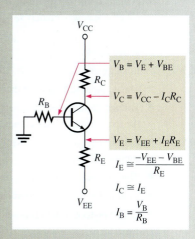

$$V_B = V_E + V_{BE}$$

$$V_C = V_{CC} - I_C R_C$$

$$V_E = V_{EE} + I_E R_E$$

$$I_E \cong \frac{-V_{EE} - V_{BE}}{R_E}$$

$$I_C \cong I_E$$

$$I_B = \frac{V_B}{R_B}$$

COLLECTOR-FEEDBACK BIAS

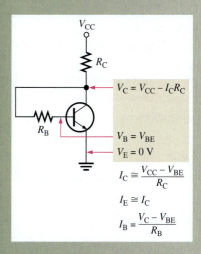

$$V_C = V_{CC} - I_C R_C$$

$$V_B = V_{BE}$$

$$V_E = 0 \text{ V}$$

$$I_C \cong \frac{V_{CC} - V_{BE}}{R_C}$$

$$I_E \cong I_C$$

$$I_B = \frac{V_C - V_{BE}}{R_B}$$

BASE BIAS

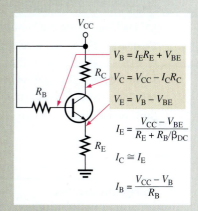

$$V_B = I_E R_E + V_{BE}$$

$$V_C = V_{CC} - I_C R_C$$

$$V_E = V_B - V_{BE}$$

$$I_E = \frac{V_{CC} - V_{BE}}{R_E + R_B/\beta_{DC}}$$

$$I_C \cong I_E$$

$$I_B = \frac{V_{CC} - V_B}{R_B}$$

EMITTER-FEEDBACK BIAS

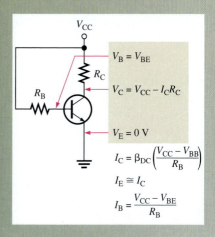

$$V_B = V_{BE}$$

$$V_C = V_{CC} - I_C R_C$$

$$V_E = 0 \text{ V}$$

$$I_C = \beta_{DC}\left(\frac{V_{CC} - V_{BB}}{R_B}\right)$$

$$I_E \cong I_C$$

$$I_B = \frac{V_{CC} - V_{BE}}{R_B}$$

SUMMARY

Section 5–1 ◆ The purpose of biasing a circuit is to establish a proper stable dc operating point (Q-point).

◆ The Q-point of a circuit is defined by specific values for I_C and V_{CE}. These values are called the coordinates of the Q-point.

◆ A dc load line passes through the Q-point on a transistor's collector curves intersecting the vertical axis at approximately $I_{C(sat)}$ and the horizontal axis at $V_{CE(off)}$.

◆ The linear (active) operating region of a transistor lies along the load line below saturation and above cutoff.

Section 5–2 ◆ Loading effects on the bias circuit can be neglected for a stiff voltage divider.

◆ The dc input resistance at the base of a BJT is approximately $\beta_{DC}R_E$.

◆ Voltage-divider bias provides good Q-point stability with a single-polarity supply voltage. It is the most common bias circuit.

Section 5–3 ◆ Emitter bias generally provides good Q-point stability but requires both positive and negative supply voltages.

◆ The base bias circuit arrangement has poor stability because its Q-point varies directly with β_{DC}.

◆ Emitter-feedback bias combines base bias with the addition of an emitter resistor.

◆ Collector-feedback bias provides good stability using negative feedback from collector to base.

KEY TERMS

Key terms and other bold terms in the chapter are defined in the end-of-book glossary.

DC load line   A straight line plot of I_C and V_{CE} for a transistor circuit.

Feedback   The process of returning a portion of a circuit's output back to the input in such a way as to oppose or aid a change in the output.

Linear region   The region of operation along the load line between saturation and cutoff.

Q-point   The dc operating (bias) point of an amplifier specified by voltage and current values.

Stiff voltage divider   A voltage divider for which loading effects can be neglected.

KEY FORMULAS

Voltage-Divider Bias

5–1 $\qquad V_B \cong \left(\dfrac{R_2}{R_1 + R_2} \right) V_{CC}$ for a stiff voltage divider

5–2 $\qquad V_E = V_B - V_{BE}$

5–3 $\qquad I_C \cong I_E = \dfrac{V_E}{R_E}$

5–4 $\qquad V_C = V_{CC} - I_C R_C$

5–5 $\qquad R_{IN(BASE)} = \dfrac{\beta_{DC} V_B}{I_E}$

5–6 $\qquad I_E = \dfrac{V_{TH} - V_{BE}}{R_E + R_{TH}/\beta_{DC}}$

5–7 $\qquad I_E = \dfrac{-V_{TH} + V_{BE}}{R_E + R_{TH}/\beta_{DC}}$

5–8 $\qquad I_E = \dfrac{V_{TH} + V_{BE} - V_{EE}}{R_E + R_{TH}/\beta_{DC}}$

Emitter Bias

5–9 $\qquad I_E = \dfrac{-V_{EE} - V_{BE}}{R_E + R_B/\beta_{DC}}$

Base Bias

5–10 $\qquad V_{CE} = V_{CC} - I_C R_C$

5–11 $\qquad I_C = \beta_{DC}\left(\dfrac{V_{CC} - V_{BE}}{R_B}\right)$

Emitter-Feedback Bias

5–12 $\qquad I_E = \dfrac{V_{CC} - V_{BE}}{R_E + R_B/\beta_{DC}}$

Collector-Feedback Bias

5–13 $\qquad I_C = \dfrac{V_{CC} - V_{BE}}{R_C + R_B/\beta_{DC}}$

5–14 $\qquad V_{CE} = V_{CC} - I_C R_C$

TRUE/FALSE QUIZ

Answers can be found at www.pearsonhighered.com/floyd.

1. DC bias establishes the dc operating point for an amplifier.
2. Q-point is the quadratic point in a bias circuit.
3. The dc load line intersects the horizontal axis of a transistor characteristic curve at $V_{CE} = V_{CC}$.
4. The dc load line intersects the vertical axis of a transistor characteristic curve at $I_C = 0$.
5. The linear region of a transistor's operation lies between saturation and cutoff.
6. Voltage-divider bias is rarely used.
7. Input resistance at the base of the transistor can affect voltage-divider bias.
8. Stiff voltage-divider bias is essentially independent of base loading.
9. Emitter bias uses one dc supply voltage.
10. Negative feedback is employed in collector-feedback bias.
11. Base bias is less stable than voltage-divider bias.
12. A *pnp* transistor requires bias voltage polarities opposite to an *npn* transistor.
13. A voltage written with a single subscript is referenced to ground.
14. When a transistor is saturated, $V_{CE} = V_{CC}$.
15. When a BJT amplifier is biased for linear operation, V_{CE} should be about 0.7 V.
16. If a transistor amplifier has $V_C = 0$ V, the fault may be the power supply.

CIRCUIT-ACTION QUIZ

Answers can be found at www.pearsonhighered.com/floyd.

1. If V_{BB} in Figure 5–7 is increased, the Q-point value of collector current will
   (a) increase (b) decrease (c) not change
2. If V_{BB} in Figure 5–7 is increased, the Q-point value of V_{CE} will
   (a) increase (b) decrease (c) not change
3. If the value of R_2 in Figure 5–10 is reduced, the base voltage will
   (a) increase (b) decrease (c) not change
4. If the value of R_1 in Figure 5–10 is increased, the emitter current will
   (a) increase (b) decrease (c) not change
5. If R_E in Figure 5–15 is decreased, the collector current will
   (a) increase (b) decrease (c) not change
6. If R_B in Figure 5–18 is reduced, the base-to-emitter voltage will
   (a) increase (b) decrease (c) not change
7. If V_{CC} in Figure 5–20 is increased, the base-to-emitter voltage will
   (a) increase (b) decrease (c) not change

Section 5–4 **Troubleshooting**

35. Determine the meter readings in Figure 5–43 if R_1 is open.

▶ FIGURE 5–43

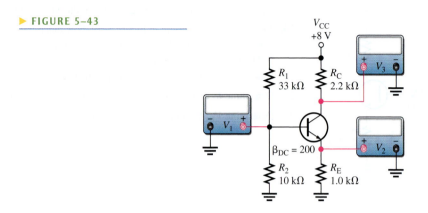

36. Assume the emitter becomes shorted to ground in Figure 5–43 by a solder splash or stray wire clipping. What do the meters read? When you correct the problem, what do the meters read?

37. Determine the most probable failures, if any, in each circuit of Figure 5–44, based on the indicated measurements.

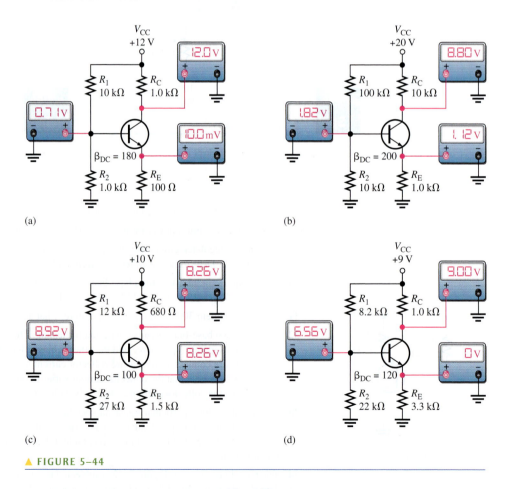

(a)

(b)

(c)

(d)

▲ FIGURE 5–44

38. Determine if the DMM readings 2 through 4 in the breadboard circuit of Figure 5–45 are correct. If they are not, isolate the problem(s). The transistor is a *pnp* device with a specified dc beta range of 35 to 100.

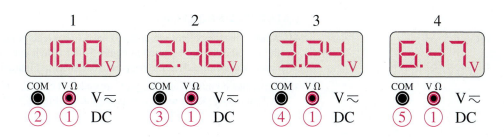

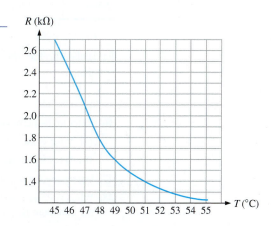

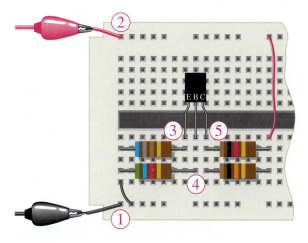

▲ FIGURE 5–45

39. Determine each meter reading in Figure 5–45 for each of the following faults:

    (a) the 680 Ω resistor open   **(b)** the 5.6 kΩ resistor open

    (c) the 10 kΩ resistor open   **(d)** the 1.0 kΩ resistor open

    (e) a short from emitter to ground   **(f)** an open base-emitter junction

DEVICE APPLICATION PROBLEMS

40. Determine V_B, V_E, and V_C in the temperature-to-voltage conversion circuit in Figure 5–29(a) if R_1 fails open.

41. What faults will cause the transistor in the temperature-to-voltage conversion circuit to go into cutoff?

42. A thermistor with the characteristic curve shown in Figure 5–46 is used in the circuit of Figure 5–29(a). Calculate the output voltage for temperatures of 45°C, 48°C, and 53°C. Assume a stiff voltage divider.

43. Explain how you would identify an open collector-base junction in the transistor in Figure 5–29(a).

▶ FIGURE 5–46

DATASHEET PROBLEMS

44. Analyze the temperature-to-voltage conversion circuit in Figure 5–47 at the temperature extremes indicated on the graph in Figure 5–46 for both minimum and maximum specified datasheet values of h_{FE}. Refer to the partial datasheet in Figure 5–48.

45. Verify that no maximum ratings are exceeded in the temperature-to-voltage conversion circuit in Figure 5–47. Refer to the partial datasheet in Figure 5–48.

▶ FIGURE 5–47

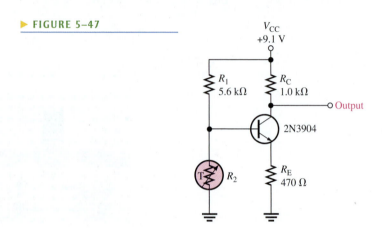

Absolute Maximum Ratings* $T_A = 25°C$ unless otherwise noted

Symbol	Parameter	Value	Units
V_{CEO}	Collector-Emitter Voltage	40	V
V_{CBO}	Collector-Base Voltage	60	V
V_{EBO}	Emitter-Base Voltage	6.0	V
I_C	Collector Current - Continuous	200	mA
T_J, T_{stg}	Operating and Storage Junction Temperature Range	-55 to +150	°C

*These ratings are limiting values above which the serviceability of any semiconductor device may be impaired.

NOTES:
1) These ratings are based on a maximum junction temperature of 150 degrees C.
2) These are steady state limits. The factory should be consulted on applications involving pulsed or low duty cycle operations.

ON CHARACTERISTICS*

h_{FE}	DC Current Gain	$I_C = 0.1$ mA, $V_{CE} = 1.0$ V	40		
		$I_C = 1.0$ mA, $V_{CE} = 1.0$ V	70		
		$I_C = 10$ mA, $V_{CE} = 1.0$ V	100	300	
		$I_C = 50$ mA, $V_{CE} = 1.0$ V	60		
		$I_C = 100$ mA, $V_{CE} = 1.0$ V	30		
$V_{CE(sat)}$	Collector-Emitter Saturation Voltage	$I_C = 10$ mA, $I_B = 1.0$ mA		0.2	V
		$I_C = 50$ mA, $I_B = 5.0$ mA		0.3	V
$V_{BE(sat)}$	Base-Emitter Saturation Voltage	$I_C = 10$ mA, $I_B = 1.0$ mA	0.65	0.85	V
		$I_C = 50$ mA, $I_B = 5.0$ mA		0.95	V

▲ FIGURE 5–48

Partial datasheet for the 2N3904 transistor. Copyright Fairchild Semiconductor Corporation. Used by permission.

46. Refer to the partial datasheet in Figure 5–49.

   (a) What is the maximum collector current for a 2N2222A?

   (b) What is the maximum reverse base-emitter voltage for a 2N2218A?

47. Determine the maximum power dissipation for a 2N2222A at 100°C.

48. When you increase the collector current in a 2N2219A from 1 mA to 500 mA, how much does the minimum β_{DC} (h_{FE}) change?

Maximum Ratings

Rating	Symbol	2N2218 2N2219 2N2221 2N2222	2N2218A 2N2219A 2N2221A 2N2222A	2N5581 2N5582	Unit
Collector-Emitter voltage	V_{CEO}	30	40	40	V dc
Collector-Base voltage	V_{CBO}	60	75	75	V dc
Emitter-Base voltage	V_{EBO}	5.0	6.0	6.0	V dc
Collector current — continuous	I_C	800	800	800	mA dc
		2N2218,A 2N2219,A	2N2221,A 2N2222,A	2N5581 2N5582	
Total device dissipation @ $T_A = 25°C$ Derate above 25°C	P_D	0.8 4.57	0.5 2.28	0.6 3.33	Watt mW/°C
Total device dissipation @ $T_C = 25°C$ Derate above 25°C	P_D	3.0 17.1	1.2 6.85	2.0 11.43	Watt mW/°C
Operating and storage junction Temperature range	T_J, T_{stg}	−65 to +200			°C

Electrical Characteristics ($T_A = 25°C$ unless otherwise noted.)

Characteristic	Symbol	Min	Max	Unit
Off Characteristics				
Collector-Emitter breakdown voltage ($I_C = 10$ mA dc, $I_B = 0$) Non-A Suffix A-Suffix, 2N5581, 2N5582	$V_{(BR)CEO}$	30 40	— —	V dc
Collector-Base breakdown voltage ($I_C = 10$ μA dc, $I_E = 0$) Non-A Suffix A-Suffix, 2N5581, 2N5582	$V_{(BR)CBO}$	60 75	— —	V dc
Emitter-Base breakdown voltage ($I_E = 10$ μA dc, $I_C = 0$) Non-A Suffix A-Suffix, 2N5581, 2N5582	$V_{(BR)EBO}$	5.0 6.0	— —	V dc
Collector cutoff current ($V_{CE} = 60$ V dc, $V_{EB(off)} = 3.0$ V dc) A-Suffix, 2N5581, 2N5582	I_{CEX}	—	10	nA dc
Collector cutoff current ($V_{CB} = 50$ V dc, $I_E = 0$) Non-A Suffix ($V_{CB} = 60$ V dc, $I_E = 0$) A-Suffix, 2N5581, 2N5582 ($V_{CB} = 50$ V dc, $I_E = 0$, $T_A = 150°C$) Non-A Suffix ($V_{CB} = 60$ V dc, $I_E = 0$, $T_A = 150°C$) A-Suffix, 2N5581, 2N5582	I_{CBO}	— — — —	0.01 0.01 10 10	μA dc
Emitter cutoff current ($V_{EB} = 3.0$ V dc, $I_C = 0$) A-Suffix, 2N5581, 2N5582	I_{EBO}	—	10	nA dc
Base cutoff current ($V_{CE} = 60$ V dc, $V_{EB(off)} = 3.0$ V dc) A-Suffix	I_{BL}	—	20	nA dc
On Characteristics				
DC current gain ($I_C = 0.1$ mA dc, $V_{CE} = 10$ V dc) 2N2218,A, 2N2221,A, 2N5581(1) 2N2219,A, 2N2222,A, 2N5582(1)	h_{FE}	20 35	— —	—
($I_C = 1.0$ mA dc, $V_{CE} = 10$ V dc) 2N2218,A, 2N2221,A, 2N5581 2N2219,A, 2N2222,A, 2N5582		25 50	— —	
($I_C = 10$ mA dc, $V_{CE} = 10$ V dc) 2N2218,A, 2N2221,A, 2N5581(1) 2N2219,A, 2N2222,A, 2N5582(1)		35 75	— —	
($I_C = 10$ mA dc, $V_{CE} = 10$ V dc, $T_A = -55°C$) 2N2218,A, 2N2221,A, 2N5581 2N2219,A, 2N2222,A, 2N5582		15 35	— —	
($I_C = 150$ mA dc, $V_{CE} = 10$ V dc) 2N2218,A, 2N2221,A, 2N5581 2N2219,A, 2N2222,A, 2N5582		40 100	120 300	
($I_C = 150$ mA dc, $V_{CE} = 1.0$ V dc) 2N2218,A, 2N2221,A, 2N5581 2N2219,A, 2N2222,A, 2N5582		20 50	— —	
($I_C = 500$ mA dc, $V_{CE} = 10$ V dc) 2N2218, 2N2221 2N2219, 2N2222 2N2218A, 2N2221A, 2N5581 2N2219A, 2N2222A, 2N5582		20 30 25 40	— — — —	
Collector-Emitter saturation voltage ($I_C = 150$ mA dc, $I_B = 15$ mA dc) Non-A Suffix A-Suffix, 2N5581, 2N5582	$V_{CE(sat)}$	— —	0.4 0.3	V dc
($I_C = 500$ mA dc, $I_B = 50$ mA dc) Non-A Suffix A-Suffix, 2N5581, 2N5582		— —	1.6 1.0	
Base-Emitter saturation voltage ($I_C = 150$ mA dc, $I_B = 15$ mA dc) Non-A Suffix A-Suffix, 2N5581, 2N5582	$V_{BE(sat)}$	0.6 0.6	1.3 1.2	V dc
($I_C = 500$ mA dc, $I_B = 50$ mA dc) Non-A Suffix A-Suffix, 2N5581, 2N5582		— —	2.6 2.0	

▲ **FIGURE 5–49**

Partial datasheet for 2N2218A–2N2222A.

ADVANCED PROBLEMS

49. Design a circuit using base bias that operates from a 15 V dc voltage and draws a maximum current from the dc source ($I_{CC(max)}$) of 10 mA. The Q-point values are to be $I_C = 5$ mA and $V_{CE} = 5$ V. The transistor is a 2N3904. Assume a midpoint value for β_{DC}.

50. Design a circuit using emitter bias that operates from dc voltages of $+12$ V and -12 V. The maximum I_{CC} is to be 20 mA and the Q-point is at 10 mA and 4 V. The transistor is a 2N3904.

51. Design a circuit using voltage-divider bias for the following specifications: $V_{CC} = 9$ V, $I_{CC(max)} = 5$ mA, $I_C = 1.5$ mA, and $V_{CE} = 3$ V. The transistor is a 2N3904.

52. Design a collector-feedback circuit using a 2N2222A with $V_{CC} = 5$ V, $I_C = 10$ mA, and $V_{CE} = 1.5$ V.

53. Can you replace the 2N3904 in Figure 5–47 with a 2N2222A and maintain the same range of output voltage over a temperature range from 45°C to 55°C?

54. Refer to the datasheet graph in Figure 5–50 and the partial datasheet in Figure 5–49. Determine the minimum dc current gain for a 2N2222A at -55°C, 25°C, and 175°C for $V_{CE} = 1$ V.

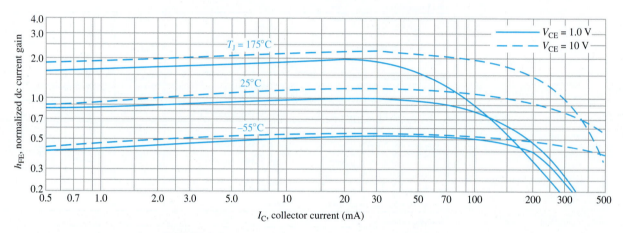

▲ **FIGURE 5–50**

55. A design change is required in the valve interface circuit of the temperature-control system shown in Figure 5–28. The new design will have a valve interface input resistance of 10 kΩ. Determine the effect this change has on the temperature-to-voltage conversion circuit.

56. Investigate the feasibility of redesigning the temperature-to-voltage conversion circuit in Figure 5–29 to operate from a dc supply voltage of 5.1 V and produce the same range of output voltages determined in the Device Application over the required thermistor temperature range from 60°C to 80°C.

MULTISIM TROUBLESHOOTING PROBLEMS

These file circuits are in the Troubleshooting Problems folder on the website.

57. Open file TPM05-57 and determine the fault.

58. Open file TPM05-58 and determine the fault.

59. Open file TPM05-59 and determine the fault.

60. Open file TPM05-60 and determine the fault.

61. Open file TPM05-61 and determine the fault.

62. Open file TPM05-62 and determine the fault.

6

BJT AMPLIFIERS

CHAPTER OBJECTIVES

◆ Describe amplifier operation
◆ Discuss transistor models
◆ Describe and analyze the operation of common-emitter amplifiers
◆ Describe and analyze the operation of common-collector amplifiers
◆ Describe and analyze the operation of common-base amplifiers
◆ Describe and analyze the operation of multistage amplifiers
◆ Discuss the differential amplifier and its operation
◆ Troubleshoot amplifier circuits

KEY TERMS

◆ *r* parameter
◆ Common-emitter
◆ ac ground
◆ Input resistance
◆ Output resistance
◆ Attenuation
◆ Bypass capacitor
◆ Common-collector

◆ Emitter-follower
◆ Common-base
◆ Decibel
◆ Differential amplifier
◆ Common mode
◆ CMRR (Common-mode rejection ratio)

VISIT THE WEBSITE

Study aids, Multisim, and LT Spice files for this chapter are available at https://www.pearsonhighered.com/careersresources/

INTRODUCTION

The things you learned about biasing a transistor in Chapter 5 are now applied in this chapter where bipolar junction transistor (BJT) circuits are used as small-signal amplifiers. The term *small-signal* refers to the use of signals that take up a relatively small percentage of an amplifier's operational range. Additionally, you will learn how to reduce an amplifier to an equivalent dc and ac circuit for easier analysis, and you will learn about multistage amplifiers. The differential amplifier is also covered.

DEVICE APPLICATION PREVIEW

The Device Application in this chapter involves a preamplifier circuit for a public address system. The complete system includes the preamplifier, a power amplifier, and a dc power supply. You will focus on the preamplifier in this chapter and then on the power amplifier in Chapter 7.

intersects the horizontal axis (V_{CE}) at the ac value of the collector-to-emitter cutoff voltage $V_{ce(cutoff)}$. These values are determined as follows:

$$I_{c(sat)} = V_{CEQ}/R_c + I_{CQ}$$
$$V_{ce(cutoff)} = V_{CEQ} + I_{CQ}R_c$$

Where R_c is the parallel combination of R_C and R_L.

Lines projected from the peaks of the base current, across to the I_C axis, and down to the V_{CE} axis, indicate the peak-to-peak variations of the collector current and collector-to-emitter voltage, as shown. The ac load line differs from the dc load line because the capacitors C_1 and C_2 effectively change the resistance seen by the ac signal. In the circuit in Figure 6–2, notice that the ac collector resistance is R_L in parallel with R_C, which is less than the dc collector resistance R_C alone. This difference between the dc and the ac load lines is covered further in Chapter 7 in relation to power amplifiers.

EXAMPLE 6–1

Given the Q-point value of $I_{CQ} = 4$ mA, $V_{CEQ} = 2$ V, $R_C = 1$ kΩ, and $R_L = 10$ kΩ for a certain amplifier, determine the ac load line values of $I_{c(sat)}$ and $V_{ce(cutoff)}$.

Solution   The ac load line values of $I_{c(sat)}$ and $V_{ce(cutoff)}$ are

$$R_c = R_C \| R_L = 1\text{ kΩ} \| 10\text{ kΩ} = 909\text{ Ω}$$
$$I_{c(sat)} = V_{CEQ}/R_c + I_{CQ} = 2\text{ V}/909\text{Ω} + 4\text{ mA} = 6.2\text{ mA}$$
$$V_{ce(cutoff)} = V_{CEQ} + I_{CQ}R_c = 2\text{ V} + 4\text{ mA }(909\text{ Ω}) = 5.64\text{ V}$$

Related Problem*   If the Q-point is changed to 3 V and 6 mA, what is the intersection values of the ac load line on the two axes?

* Answers can be found at www.pearsonhighered.com/floyd.

EXAMPLE 6–2

The ac load line operation of a certain amplifier extends 10 μA above and below the Q-point base current value of 50 μA, as shown in Figure 6–4. Determine the resulting peak-to-peak values of collector current and collector-to-emitter voltage from the graph.

▶ **FIGURE 6–4**

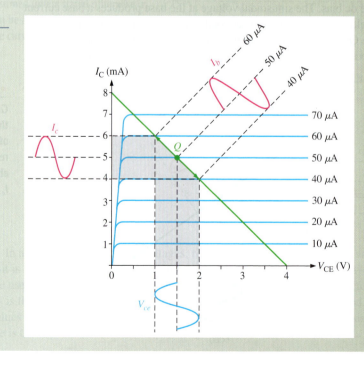

Solution Projections on the graph of Figure 6–4 show the collector current varying from 6 mA to 4 mA for a peak-to-peak value of **2 mA** and the collector-to-emitter voltage varying from 1 V to 2 V for a peak-to-peak value of **1 V**.

Related Problem What are the Q-point values of I_C and V_{CE} in Figure 6–4?

**SECTION 6–1
CHECKUP**
Answers can be found at www
.pearsonhighered.com/floyd.

1. When I_b is at its positive peak, I_c is at its _____ peak, and V_{ce} is at its _____ peak.
2. What is the difference between V_{CE} and V_{ce}?
3. What is the difference between R_e and r_e'?
4. Why is the ac resistance seen by the collector different from the dc resistance?

6–2 TRANSISTOR AC MODELS

To visualize the operation of a transistor in an amplifier circuit, it is often useful to represent the device by a model circuit. A transistor model circuit uses various internal transistor parameters to represent its operation. Transistor models are described in this section based on resistance or r parameters. Another system of parameters, called hybrid or h parameters, is briefly described.

After completing this section, you should be able to

❑ **Discuss transistor models**
❑ List and define the r parameters
❑ Describe the r-parameter transistor model
❑ Determine r_e' using a formula
❑ Compare ac beta and dc beta
❑ List and define the h parameters

r Parameters

The r parameters that are commonly used for BJTs are given in Table 6–1. Strictly speaking, α_{ac} and β_{ac} are current ratios, not r parameters, but they are used with the resistance parameters to model basic transistor circuits. The italic lowercase letter r with a prime denotes resistances internal to the transistor.

r PARAMETERS	DESCRIPTION
r_e'	ac emitter resistance
r_b'	ac base resistance
r_c'	ac collector resistance
α_{ac}	ac alpha (I_c/I_e)
β_{ac}	ac beta (I_c/I_b)

◀ **TABLE 6–1**

r parameters.

r-Parameter Transistor Model

An **r-parameter** model for a BJT is shown in Figure 6–5(a). For most general analysis work, it can be simplified as follows: The effect of the ac base resistance (r_b') is usually

Relationships of *h* Parameters and *r* Parameters

The ac current ratios, α_{ac} and β_{ac}, convert directly from *h* parameters as follows:

$$\alpha_{ac} = h_{fb}$$
$$\beta_{ac} = h_{fe}$$

Because datasheets often provide only common-emitter *h* parameters, the following formulas show how to convert them to *r* parameters. We will use *r* parameters throughout the text because they are easier to apply and more practical.

$$r'_e = \frac{h_{re}}{h_{oe}}$$

$$r'_c = \frac{h_{re} + 1}{h_{oe}}$$

$$r'_b = h_{ie} - \frac{h_{re}}{h_{oe}}(1 + h_{fe})$$

SECTION 6–2 CHECKUP

1. Define each of the parameters: α_{ac}, β_{ac}, r'_e, r'_b, r'_c.
2. Which *h* parameter is equivalent to β_{ac}?
3. If $I_E = 15$ mA, what is the approximate value of r'_e.
4. What is the difference between β_{ac} and β_{DC}?

6–3   THE COMMON-EMITTER AMPLIFIER

As you have learned, a BJT can be represented in an ac model circuit. Three amplifier configurations are the common-emitter, the common-base, and the common-collector. The common-emitter (CE) configuration has the emitter as the common terminal, or ground, to an ac signal. CE amplifiers exhibit high voltage gain and high current gain. The common-collector and common-base configurations are covered in the Sections 6–4 and 6–5.

After completing this section, you should be able to

❑ **Describe and analyze the operation of common-emitter amplifiers**
❑ Discuss a common-emitter amplifier with voltage-divider bias
  ◆ Show input and output signals   ◆ Discuss phase inversion
❑ Perform a dc analysis
  ◆ Represent the amplifier by its dc equivalent circuit
❑ Perform an ac analysis
  ◆ Represent the amplifier by its ac equivalent circuit   ◆ Define *ac ground*
  ◆ Discuss the voltage at the base   ◆ Discuss the input resistance at the base and the output resistance
❑ Analyze the amplifier for voltage gain
  ◆ Define *attenuation*   ◆ Define *bypass capacitor*   ◆ Describe the effect of an emitter bypass capacitor on voltage gain   ◆ Discuss voltage gain without a bypass capacitor   ◆ Explain the effect of a load on voltage gain
❑ Discuss the stability of the voltage gain
  ◆ Define *stability*   ◆ Explain the purpose of swamping r'_e and the effect on input resistance
❑ Determine current gain and power gain

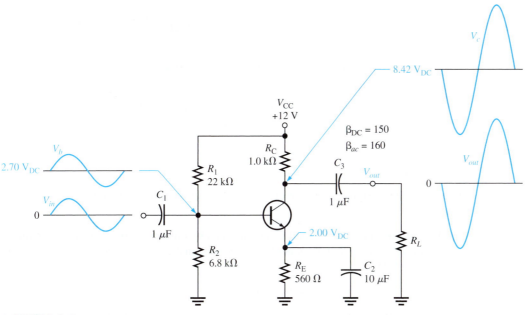

▲ FIGURE 6–8

A common-emitter amplifier.

Figure 6–8 shows a **common-emitter** amplifier with voltage-divider bias and coupling capacitors C_1 and C_3 on the input and output and a bypass capacitor, C_2, from emitter to ground. The input signal, V_{in}, is capacitively coupled to the base terminal, the output signal, V_{out}, is capacitively coupled from the collector to the load. The amplified output is 180° out of phase with the input. Because the ac signal is applied to the base terminal as the input and taken from the collector terminal as the output, the emitter is common to both the input and output signals. There is no signal at the emitter because the bypass capacitor effectively shorts the emitter to ground at the signal frequency. All amplifiers have a combination of both ac and dc operation, which must be considered, but keep in mind that the common-emitter designation refers to the ac operation.

Phase Inversion   The output signal is 180° out of phase with the input signal. As the input signal voltage changes, it causes the ac base current to change, resulting in a change in the collector current from its Q-point value. If the base current increases, the collector current increases above its Q-point value, causing an increase in the voltage drop across R_C. This increase in the voltage across R_C means that the voltage at the collector decreases from its Q-point. So, any change in input signal voltage results in an opposite change in collector signal voltage, which is a phase inversion.

DC Analysis

To analyze the amplifier in Figure 6–8, the dc bias values must first be determined. To do this, a dc equivalent circuit is developed by removing the coupling and bypass capacitors because they appear open as far as the dc bias is concerned. This also removes the load resistor and signal source. The dc equivalent circuit is shown in Figure 6–9.

Theveninizing the bias circuit and applying Kirchhoff's voltage law to the base-emitter circuit,

$$R_{TH} = \frac{R_1 R_2}{R_1 + R_2} = \frac{(6.8\ k\Omega)(22\ k\Omega)}{6.8\ k\Omega + 22\ k\Omega} = 5.19\ k\Omega$$

$$V_{TH} = \left(\frac{R_2}{R_1 + R_2}\right)V_{CC} = \left(\frac{6.8\ k\Omega}{6.8\ k\Omega + 22\ k\Omega}\right)12\ V = 2.83\ V$$

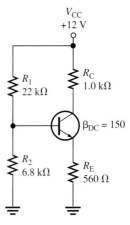

▲ FIGURE 6–9

DC equivalent circuit for the amplifier in Figure 6–8.

EXAMPLE 6–4

Determine the signal voltage at the base of the transistor in Figure 6–13. This circuit is the ac equivalent of the amplifier in Figure 6–8 with a 10 mV rms, 300 Ω signal source. I_E was previously found to be 3.58 mA.

▶ **FIGURE 6–13**

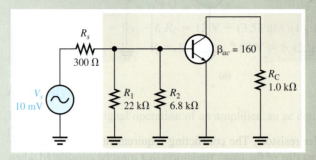

Solution First, determine the ac emitter resistance.

$$r'_e \cong \frac{25 \text{ mV}}{I_E} = \frac{25 \text{ mV}}{3.58 \text{ mA}} = 6.98 \text{ }\Omega$$

Then,

$$R_{in(base)} = \beta_{ac}r'_e = 160(6.98 \text{ }\Omega) = 1.12 \text{ k}\Omega$$

Next, determine the total input resistance viewed from the source.

$$R_{in(tot)} = R_1 \parallel R_2 \parallel R_{in(base)} = \frac{1}{\dfrac{1}{22 \text{ k}\Omega} + \dfrac{1}{6.8 \text{ k}\Omega} + \dfrac{1}{1.12 \text{ k}\Omega}} = 920 \text{ }\Omega$$

The source voltage is divided down by R_s and $R_{in(tot)}$, so the signal voltage at the base is the voltage across $R_{in(tot)}$.

$$V_b = \left(\frac{R_{in(tot)}}{R_s + R_{in(tot)}}\right)V_s = \left(\frac{920 \text{ }\Omega}{1221 \text{ }\Omega}\right)10 \text{ mV} = \textbf{7.53 mV}$$

As you can see, there is significant attenuation (reduction) of the source voltage due to the source resistance and amplifier's input resistance combining to act as a voltage divider.

Related Problem Determine the signal voltage at the base of Figure 6–13 if the source resistance is 75 Ω and another transistor with an ac beta of 200 is used.

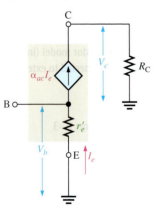

▲ **FIGURE 6–14**

Model circuit for obtaining ac voltage gain.

Voltage Gain

The ac voltage gain expression for the common-emitter amplifier is developed using the model circuit in Figure 6–14. The gain is the ratio of ac output voltage at the collector (V_c) to ac input voltage at the base (V_b).

$$A_v = \frac{V_{out}}{V_{in}} = \frac{V_c}{V_b}$$

Notice in the figure that $V_c = \alpha_{ac}I_eR_C \cong I_eR_C$ and $V_b = I_er'_e$. Therefore,

$$A_v = \frac{I_eR_C}{I_er'_e}$$

The I_e terms cancel, so

$$A_v = \frac{R_C}{r_e'}$$

Equation 6–5

Equation 6–5 is the voltage gain from base to collector. To get the overall gain of the amplifier from the source voltage to collector, the attenuation of the input circuit must be included.

Attenuation is the reduction in signal voltage as it passes through a circuit and corresponds to a gain of less than 1. For example, if the signal amplitude is reduced by half, the attenuation is 2, which can be expressed as a gain of 0.5 because gain is the reciprocal of attenuation. Suppose a source produces a 10 mV input signal and the source resistance combined with the load resistance results in a 2 mV output signal. In this case, the attenuation is 10 mV/2 mV = 5. That is, the input signal is reduced by a factor of 5. This can be expressed in terms of gain as $1/5 = 0.2$.

Assume that the amplifier in Figure 6–15 has a voltage gain from base to collector of A_v and the attenuation from the source to the base is V_s/V_b. This attenuation is produced by the source resistance and total input resistance of the amplifier acting as a voltage divider and can be expressed as

$$\text{Attenuation} = \frac{V_s}{V_b} = \frac{R_s + R_{in(tot)}}{R_{in(tot)}}$$

The overall voltage gain of the amplifier, A_v', is the voltage gain from base to collector, V_c/V_b, times the reciprocal of the attenuation, V_b/V_s.

$$A_v' = \left(\frac{V_c}{V_b}\right)\left(\frac{V_b}{V_s}\right) = \frac{V_c}{V_s}$$

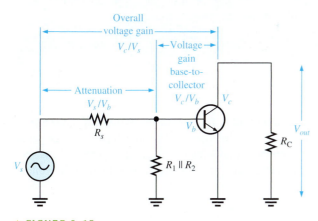

▲ **FIGURE 6–15**

Base circuit attenuation and overall voltage gain.

Effect of the Emitter Bypass Capacitor on Voltage Gain   The emitter **bypass capacitor**, which is C_2 in Figure 6–8, provides an effective short to the ac signal around the emitter resistor, thus keeping the emitter at ac ground, as you have seen. With the bypass capacitor, the gain of a given amplifier is maximum and equal to R_C/r_e'.

The value of the bypass capacitor must be large enough so that its reactance over the frequency range of the amplifier is very small (ideally 0 Ω) compared to R_E. A good rule of thumb is that the capacitive reactance, X_C, of the bypass capacitor should be at least 10 times smaller than R_E at the minimum frequency for which the amplifier must operate.

$$10X_C \leq R_E$$

EXAMPLE 6–5

Select a minimum value for the emitter bypass capacitor, C_2, in Figure 6–16 if the amplifier must operate over a frequency range from 200 Hz to 10 kHz.

▶ **FIGURE 6–16**

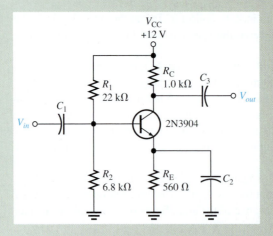

Solution

The X_C of the bypass capacitor, C_2, should be at least ten times less than R_E.

$$X_{C2} = \frac{R_E}{10} = \frac{560\ \Omega}{10} = 56\ \Omega$$

Determine the capacitance value at the minimum frequency of 200 Hz as follows:

$$C_2 = \frac{1}{2\pi f X_{C2}} = \frac{1}{2\pi(200\ \text{Hz})(56\ \Omega)} = \textbf{14.2}\ \boldsymbol{\mu}\textbf{F}$$

This is the minimum value for the bypass capacitor for this circuit. You can always use a larger value, although cost and physical size may impose limitations.

Related Problem

If the minimum frequency is reduced to 100 Hz, what value of bypass capacitor must you use?

Voltage Gain Without the Bypass Capacitor   To see how the bypass capacitor affects ac voltage gain, let's remove it from the circuit in Figure 6–16 and compare voltage gains.

Without the bypass capacitor, the emitter is no longer at ac ground. Instead, R_E is seen by the ac signal between the emitter and ground and effectively adds to r'_e in the voltage gain formula.

Equation 6–6

$$A_v = \frac{R_C}{r'_e + R_E}$$

The effect of R_E is to decrease the ac voltage gain.

EXAMPLE 6–6

Calculate the base-to-collector voltage gain of the amplifier in Figure 6–16 both without and with an emitter bypass capacitor if there is no load resistor.

Solution

From Example 6–4, $r'_e = 6.98\ \Omega$ for this same amplifier. Without C_2, the gain is

$$A_v = \frac{R_C}{r'_e + R_E} = \frac{1.0\ \text{k}\Omega}{567\ \Omega} = \textbf{1.76}$$

With C_2, the gain is

$$A_v = \frac{R_C}{r'_e} = \frac{1.0 \text{ k}\Omega}{6.98 \text{ }\Omega} = 143$$

As you can see, the bypass capacitor makes quite a difference.

Related Problem   Determine the base-to-collector voltage gain in Figure 6–16 with R_E bypassed, for the following circuit values: $R_C = 1.8 \text{ k}\Omega$, $R_E = 1.0 \text{ k}\Omega$, $R_1 = 33 \text{ k}\Omega$, and $R_2 = 6.8 \text{ k}\Omega$.

Effect of a Load on the Voltage Gain   A **load** is the amount of current drawn from the output of an amplifier or other circuit through a load resistance. When a resistor, R_L, is connected to the output through the coupling capacitor C_3, as shown in Figure 6–17(a), it creates a load on the circuit. The collector resistance at the signal frequency is effectively R_C in parallel with R_L. Remember, the upper end of R_C is effectively at ac ground. The ac equivalent circuit is shown in Figure 6–17(b). The total ac collector resistance is

$$R_c = \frac{R_C R_L}{R_C + R_L}$$

Replacing R_C with R_c in the voltage gain expression gives

$$A_v = \frac{R_c}{r'_e}$$

Equation 6–7

When $R_c < R_C$ because of R_L, the voltage gain is reduced. However, if $R_L \gg R_C$, then $R_c \cong R_C$ and the load has very little effect on the gain.

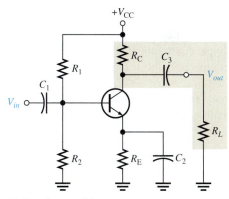

(a) Complete amplifier

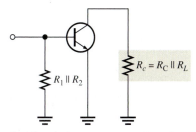

(b) AC equivalent ($X_{C1} = X_{C2} = X_{C3} = 0$)

▲ **FIGURE 6–17**

A common-emitter amplifier with an ac (capacitively) coupled load.

EXAMPLE 6–7   Calculate the base-to-collector voltage gain of the amplifier in Figure 6–16 when a load resistance of 5 kΩ is connected to the output. The emitter is effectively bypassed and $r'_e = 6.98 \text{ }\Omega$.

Solution   The ac collector resistance is

$$R_c = \frac{R_C R_L}{R_C + R_L} = \frac{(1.0 \text{ k}\Omega)(5 \text{ k}\Omega)}{6 \text{ k}\Omega} = 833 \text{ }\Omega$$

Therefore,

$$A_v = \frac{R_c}{r'_e} = \frac{833 \ \Omega}{6.98 \ \Omega} = \mathbf{119}$$

The unloaded gain was found to be 143 in Example 6–6.

Related Problem Determine the base-to-collector voltage gain in Figure 6–16 when a 10 kΩ load resistance is connected from collector to ground. Change the resistance values as follows: $R_C = 1.8 \ k\Omega$, $R_E = 1.0 \ k\Omega$, $R_1 = 33 \ k\Omega$, and $R_2 = 6.8 \ k\Omega$. The emitter resistor is effectively bypassed and $r'_e = 18.5 \ \Omega$.

Stability of the Voltage Gain

Stability is a measure of how well an amplifier maintains its design values over changes in temperature or for a transistor with a different β. Although bypassing R_E does produce the maximum voltage gain, there is a stability problem because the ac voltage gain is dependent on r'_e since $A_v = R_C/r'_e$. Also, r'_e depends on I_E and on temperature. This causes the gain to be unstable over changes in temperature because when r'_e increases, the gain decreases and vice versa.

With no bypass capacitor, the gain is decreased because R_E is now in the ac circuit ($A_v = R_C/(r'_e + R_E)$). However, with R_E unbypassed, the gain is much less dependent on r'_e. If $R_E \gg r'_e$, the gain is essentially independent of r'_e because

$$A_v \cong \frac{R_C}{R_E}$$

Swamping r'_e to Stabilize the Voltage Gain *Swamping* is a method used to minimize the effect of r'_e without reducing the voltage gain to its minimum value. This method "swamps" out the effect of r'_e on the voltage gain. Swamping is, in effect, a compromise between having a bypass capacitor across R_E and having no bypass capacitor at all. Whenever a bypass capacitor is used, its reactance should be small compared to the ac emitter resistance at the lowest frequency at which the amplifier will be used.

In a swamped amplifier, R_E is partially bypassed so that a reasonable gain can be achieved, and the effect of r'_e on the gain is greatly reduced or eliminated. The total external emitter resistance, R_E, is formed with two separate emitter resistors, R_{E1} and R_{E2}, as indicated in Figure 6–18. One of the resistors, R_{E2}, is bypassed and the other is not.

Both resistors ($R_{E1} + R_{E2}$) affect the dc bias while only R_{E1} affects the ac voltage gain.

$$A_v = \frac{R_C}{r'_e + R_{E1}}$$

▶ **FIGURE 6–18**

A swamped amplifier uses a partially bypassed emitter resistance to minimize the effect of r'_e on the gain in order to achieve gain stability.

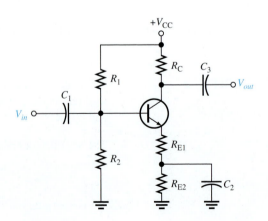

If R_{E1} is at least ten times larger than r'_e, then the effect of r'_e is minimized and the approximate voltage gain for the swamped amplifier is

$$A_v \cong \frac{R_C}{R_{E1}}$$

Equation 6–8

EXAMPLE 6–8

Determine the voltage gain of the swamped amplifier in Figure 6–19. Assume that the bypass capacitor has a negligible reactance for the frequency at which the amplifier is operated. Assume $r'_e = 15\ \Omega$.

▶ **FIGURE 6–19**

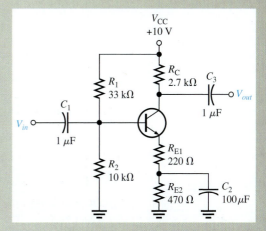

Solution R_{E2} is bypassed by C_2. R_{E1} is more than ten times r'_e so the approximate voltage gain is

$$A_v \cong \frac{R_C}{R_{E1}} = \frac{2.7\ \text{k}\Omega}{220\ \Omega} = \mathbf{12}$$

Related Problem What would be the voltage gain without C_2? What would be the approximate voltage gain if R_{E1} and R_{E2} were exchanged?

The Effect of Swamping on the Amplifier's Input Resistance The ac input resistance, looking in at the base of a common-emitter amplifier with R_E completely bypassed, is $R_{in} = \beta_{ac}r'_e$. When the emitter resistance is partially bypassed, the portion of the resistance that is unbypassed is seen by the ac signal and results in an increase in the ac input resistance by appearing in series with r'_e. The formula is

$$R_{in(base)} = \beta_{ac}(r'_e + R_{E1})$$

Equation 6–9

EXAMPLE 6–9

For the amplifier in Figure 6–20,

(a) Determine the dc collector voltage.

(b) Determine the ac collector voltage.

(c) Draw the total collector voltage waveform and the total output voltage waveform.

▶ FIGURE 6–20

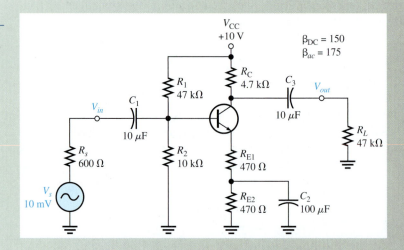

Solution   **(a)** Determine the dc bias values using the dc equivalent circuit in Figure 6–21.

▶ FIGURE 6–21

DC equivalent for the circuit in Figure 6–20.

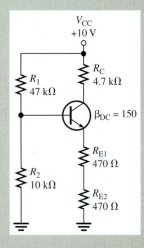

Apply Thevenin's theorem and Kirchhoff's voltage law to the base-emitter circuit in Figure 6–21.

$$R_{TH} = \frac{R_1 R_2}{R_1 + R_2} = \frac{(47 \text{ k}\Omega)(10 \text{ k}\Omega)}{47 \text{ k}\Omega + 10 \text{ k}\Omega} = 8.25 \text{ k}\Omega$$

$$V_{TH} = \left(\frac{R_2}{R_1 + R_2}\right)V_{CC} = \left(\frac{10 \text{ k}\Omega}{47 \text{ k}\Omega + 10 \text{ k}\Omega}\right)10 \text{ V} = 1.75 \text{ V}$$

$$I_E = \frac{V_{TH} - V_{BE}}{R_E + R_{TH}/\beta_{DC}} = \frac{1.75 \text{ V} - 0.7 \text{ V}}{940 \text{ }\Omega + 55 \text{ }\Omega} = 1.06 \text{ mA}$$

$$I_C \cong I_E = 1.06 \text{ mA}$$

$$V_E = I_E(R_{E1} + R_{E2}) = (1.06 \text{ mA})(940 \text{ }\Omega) = 1 \text{ V}$$

$$V_B = V_E + 0.7 \text{ V} = 1 \text{ V} + 0.7 \text{ V} = 1.7 \text{ V}$$

$$V_C = V_{CC} - I_C R_C = 10 \text{ V} - (1.06 \text{ mA})(4.7 \text{ k}\Omega) = \mathbf{5.02 \text{ V}}$$

(a) The ac analysis is based on the ac equivalent circuit in Figure 6–22.

▶ FIGURE 6–22

AC equivalent for the circuit in Figure 6–20.

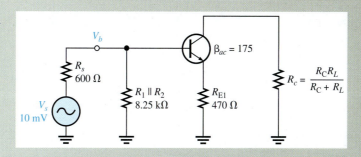

The first thing to do in the ac analysis is calculate r'_e.

$$r'_e \cong \frac{25 \text{ mV}}{I_E} = \frac{25 \text{ mV}}{1.06 \text{ mA}} = 23.6 \text{ }\Omega$$

Next, determine the attenuation in the base circuit. Looking from the 600 Ω source, the total R_{in} is

$$R_{in(tot)} = R_1 \parallel R_2 \parallel R_{in(base)}$$
$$R_{in(base)} = \beta_{ac}(r'_e + R_{E1}) = 175(494 \text{ }\Omega) = 86.5 \text{ k}\Omega$$

Therefore,

$$R_{in(tot)} = 47 \text{ k}\Omega \parallel 10 \text{ k}\Omega \parallel 86.5 \text{ k}\Omega = 7.53 \text{ k}\Omega$$

The attenuation from source to base is

$$\text{Attenuation} = \frac{V_s}{V_b} = \frac{R_s + R_{in(tot)}}{R_{in(tot)}} = \frac{600 \text{ }\Omega + 7.53 \text{ k}\Omega}{7.53 \text{ k}\Omega} = 1.08$$

Before A_v can be determined, you must know the ac collector resistance R_c.

$$R_c = \frac{R_C R_L}{R_C + R_L} = \frac{(4.7 \text{ k}\Omega)(47 \text{ k}\Omega)}{4.7 \text{ k}\Omega + 47 \text{ k}\Omega} = 4.27 \text{ k}\Omega$$

The voltage gain from base to collector is

$$A_v \cong \frac{R_c}{R_{E1}} = \frac{4.27 \text{ k}\Omega}{470 \text{ }\Omega} = 9.09$$

The overall voltage gain is the reciprocal of the attenuation times the amplifier voltage gain.

$$A'_v = \left(\frac{V_b}{V_s}\right) A_v = (0.93)(9.09) = 8.45$$

The source produces 10 mV rms, so the rms voltage at the collector is

$$V_c = A'_v V_s = (8.45)(10 \text{ mV}) = \mathbf{84.5 \text{ mV}}$$

(b) The total collector voltage is the signal voltage of 84.5 mV rms riding on a dc level of 4.74 V, as shown in Figure 6–23(a), where approximate peak values are determined as follows:

$$\text{Max } V_{c(p)} = V_C + 1.414 \text{ } V_c = 5.02 \text{ V} + (84.5 \text{ mV})(1.414) = 5.14 \text{ V}$$
$$\text{Min } V_{c(p)} = V_C - 1.414 \text{ } V_c = 5.02 \text{ V} - (84.5 \text{ mV})(1.414) = 4.90 \text{ V}$$

The coupling capacitor, C_3, keeps the dc level from getting to the output. So, V_{out} is equal to the ac component of the collector voltage ($V_{out(p)} = (84.5 \text{ mV})(1.414) = 119 \text{ mV}$),

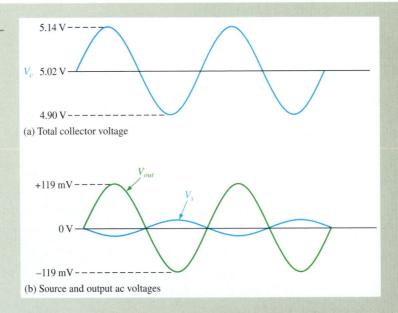

(a) Total collector voltage

(b) Source and output ac voltages

as indicated in Figure 6–23(b). The source voltage, V_s, is shown to emphasize the phase inversion.

Related Problem   What is A_v in Figure 6–20 with R_L removed?

 Open the Multisim file EXM06-09 or the LT Spice file EXS06-09 in the Examples folder on the website. Measure the dc and the ac values of the collector voltage and compare with the calculated values.

Current Gain

The focus in this section has been on voltage gain because that is the principal use for a CE amplifier. However, for completeness, we end this section with a discussion of current and power gain. The current gain from base to collector is I_c/I_b or β_{ac}. However, the overall current gain of the common-emitter amplifier is

Equation 6–10

$$A_i = \frac{I_c}{I_s}$$

I_s is the total signal input current produced by the source, part of which (I_b) is base current and part of which (I_{bias}) goes through the bias circuit ($R_1 \parallel R_2$), as shown in Figure 6–24. The source "sees" a total resistance of $R_s + R_{in(tot)}$. The total current produced by the source is

$$I_s = \frac{V_s}{R_s + R_{in(tot)}}$$

▶ FIGURE 6–24

Signal currents (directions shown are for the positive half-cycle of V_s).

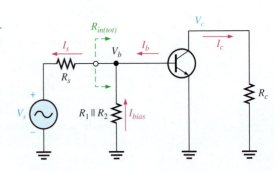

The ac current in the collector circuit is composed of current in the collector resistor, R_C and current in the load resistor R_L. If you are interested in current gain to the load, you need to apply the current divider rule to these resistors to determine ac load current.

Power Gain

As mentioned, CE amplifiers are rarely used to provide power gain. However, for completeness, the overall power gain is the product of the overall voltage gain (A'_v) and the overall current gain (A_i).

$$A_p = A'_v A_i$$

Equation 6–11

where $A'_v = V_c/V_s$. If you are interested in power gain to just the load, then the formula uses voltage gain times the current gain to the load rather than the overall current gain.

SECTION 6–3 CHECKUP

1. In the dc equivalent circuit of an amplifier, how are the capacitors treated?
2. When the emitter resistor is bypassed with a capacitor, how is the gain of the amplifier affected?
3. Explain swamping.
4. List the elements included in the total input resistance of a common-emitter amplifier.
5. What elements determine the overall voltage gain of a common-emitter amplifier?
6. When a load resistor is capacitively coupled to the collector of a CE amplifier, is the voltage gain increased or decreased?
7. What is the phase relationship of the input and output voltages of a CE amplifier?

6–4   THE COMMON-COLLECTOR AMPLIFIER

The **common-collector** (CC) amplifier is usually referred to as an emitter-follower (EF). The input is applied to the base through a coupling capacitor, and the output is at the emitter. The voltage gain of a CC amplifier is approximately 1, and its main advantages are its high input resistance and current gain.

After completing this section, you should be able to

- ❏ **Describe and analyze the operation of common-collector amplifiers**
- ❏ Discuss the emitter-follower amplifier with voltage-divider bias
- ❏ Analyze the amplifier for voltage gain
  - ◆ Explain the term *emitter-follower*
- ❏ Discuss and calculate input resistance
- ❏ Determine output resistance
- ❏ Determine current gain to the load
- ❏ Determine power gain to the load
- ❏ Describe the Darlington pair
  - ◆ Discuss an application
- ❏ Discuss the Sziklai pair

An **emitter-follower** circuit with voltage-divider bias is shown in Figure 6–25. Notice that the input signal is capacitively coupled to the base, the output signal is capacitively coupled from the emitter, and the collector is at ac ground. There is no phase inversion, and the output is approximately the same amplitude as the input.

▶ FIGURE 6–25

Emitter-follower with voltage-divider bias.

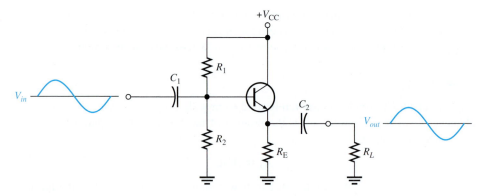

Voltage Gain

As in all amplifiers, the voltage gain is $A_v = V_{out}/V_{in}$. The capacitive reactances are assumed to be negligible at the frequency of operation. For the emitter-follower, as shown in the ac model in Figure 6–26,

$$V_{out} = I_e R_e$$

and

$$V_{in} = I_e(r'_e + R_e)$$

Therefore, the voltage gain is

$$A_v = \frac{I_e R_e}{I_e(r'_e + R_e)}$$

The I_e current terms cancel, and the base-to-emitter voltage gain expression simplifies to

$$A_v = \frac{R_e}{r'_e + R_e}$$

where R_e is the parallel combination of R_E and RL. If there is no load, then $R_e = R_E$. Notice that the voltage gain is always less than 1. If $R_e \gg r'_e$, then a good approximation is

Equation 6–12

$$A_v \cong 1$$

Since the output voltage is at the emitter, it is in phase with the base voltage, so there is no inversion from input to output. Because there is no inversion and because the voltage gain is approximately 1, the output voltage closely follows the input voltage in both phase and amplitude; thus the term *emitter-follower*.

▶ FIGURE 6–26

Emitter-follower model for voltage gain derivation.

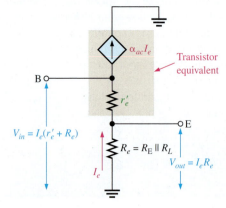

Input Resistance

The emitter-follower is characterized by a high input resistance and low output resistance; this is what makes it a useful circuit. Because of the high input resistance, it can be used as a buffer to minimize loading effects when a circuit is driving a low-resistance load. The derivation of the input resistance, looking in at the base of the common-collector amplifier, is similar to that for the common-emitter amplifier. In a common-collector circuit, however, the emitter resistor is *never* bypassed because the output is taken across R_e, which is R_E in parallel with R_L.

$$R_{in(base)} = \frac{V_{in}}{I_{in}} = \frac{V_b}{I_b} = \frac{I_e(r'_e + R_e)}{I_b}$$

Since $I_e \cong I_c = \beta_{ac}I_b$,

$$R_{in(base)} \cong \frac{\beta_{ac}I_b(r'_e + R_e)}{I_b}$$

The I_b terms cancel; therefore,

$$R_{in(base)} \cong \beta_{ac}(r'_e + R_e)$$

If $R_e \gg r'_e$, then the input resistance at the base is simplified to

$$R_{in(base)} \cong \beta_{ac}R_e \qquad \text{Equation 6–13}$$

The bias resistors in Figure 6–25 appear in parallel with $R_{in(base)}$, looking from the input source; and just as in the common-emitter circuit, the total input resistance is

$$R_{in(tot)} = R_1 \parallel R_2 \parallel R_{in(base)}$$

Output Resistance

With the load removed, the output resistance, looking into the emitter of the emitter-follower, is approximated as follows:

$$R_{out} \cong \left(\frac{R_s}{\beta_{ac}}\right) \parallel R_E \qquad \text{Equation 6–14}$$

R_s is the resistance of the input source. The derivation of Equation 6–14, found in "Derivations of Selected Equations" at www.pearsonhighered.com/floyd, is relatively involved and several assumptions have been made. The output resistance is very low, making the emitter-follower useful for driving low-resistance loads.

Current Gain

Although the voltage gain is less than 1, the current gain is not. The current gain for the emitter-follower in Figure 6–25 is

$$A_i = \frac{I_e}{I_{in}} \qquad \text{Equation 6–15}$$

where $I_{in} = V_{in}/R_{in(tot)}$.

Notice that I_e in Equation 6–15 includes both emitter and load currents. If you want only the current gain to the load, you can apply the current divider rule.

Power Gain

The common-collector power gain is the product of the voltage gain and the current gain. For the emitter-follower, the power gain is approximately equal to the current gain because the voltage gain is approximately 1.

$$A_p = A_v A_i$$

Since $A_v \cong 1$, the total power gain is

$$A_p \cong A_i \qquad \text{Equation 6–16}$$

The power gain to the load is approximately equal to the current gain to the load; use the current divider rule to determine the load current.

EXAMPLE 6–10

Determine the total input resistance of the emitter-follower in Figure 6–27. Also find the voltage gain, current gain, and power gain in terms of power delivered to the load, R_L. Assume $\beta_{ac} = 175$ and that the capacitive reactances are negligible at the frequency of operation.

▶ **FIGURE 6–27**

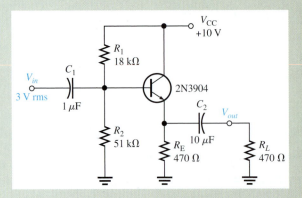

Solution   The ac emitter resistance external to the transistor is

$$R_e = R_E \parallel R_L = 470\ \Omega \parallel 470\ \Omega = 235\ \Omega$$

The approximate resistance, looking in at the base, is

$$R_{in(base)} \cong \beta_{ac}R_e = (175)(235\ \Omega) = 41.1\ \text{k}\Omega$$

The total input resistance is

$$R_{in(tot)} = R_1 \parallel R_2 \parallel R_{in(base)} = 18\ \text{k}\Omega \parallel 51\ \text{k}\Omega \parallel 41.1\ \text{k}\Omega = \mathbf{10.1\ k\Omega}$$

The voltage gain is $A_v \cong 1$. By using r'_e, you can determine a more precise value of A_v if necessary.

$$V_E = \left(\frac{R_2}{R_1 + R_2}\right)V_{CC} - V_{BE} = \left(\frac{51\ \text{k}\Omega}{18\ \text{k}\Omega + 51\ \text{k}\Omega}\right)10\ \text{V} - 0.7\ \text{V}$$
$$= (0.739)(10\ \text{V}) - 0.7\ \text{V} = 6.69\ \text{V}$$

Therefore,

$$I_E = \frac{V_E}{R_E} = \frac{6.69\ \text{V}}{470\ \Omega} = 14.2\ \text{mA}$$

and

$$r'_e \cong \frac{25\ \text{mV}}{I_E} = \frac{25\ \text{mV}}{14.2\ \text{mA}} = 1.76\ \Omega$$

So,

$$A_v = \frac{R_e}{r'_e + R_e} = \frac{235\ \Omega}{237\ \Omega} = \mathbf{0.992}$$

The small difference in A_v as a result of considering r'_e is insignificant in most cases.

The total current gain is $A_i = I_e/I_{in}$. The calculations are as follows:

$$I_e = \frac{V_e}{R_e} = \frac{A_v V_b}{R_e} \cong \frac{(0.992)(3\ \text{V})}{235\ \Omega} = \frac{2.98\ \text{V}}{235\ \Omega} = 12.7\ \text{mA}$$

$$I_{in} = \frac{V_{in}}{R_{in(tot)}} = \frac{3\ \text{V}}{10.1\ \text{k}\Omega} = 297\ \mu\text{A}$$

$$A_i = \frac{I_e}{I_{in}} = \frac{12.7\ \text{mA}}{297\ \mu\text{A}} = \mathbf{42.8}$$

The total power gain is

$$A_p \cong A_i = 42.8$$

Since $R_L = R_E$, one-half of the power is dissipated in R_E and one-half in R_L. Therefore, in terms of power to the load, the power gain is

$$A_{p(load)} = \frac{A_p}{2} = \frac{42.8}{2} = \textbf{21.4}$$

If R_E is not equal to R_L, apply the current divider rule to find the current in the load resistor.

Related Problem   If R_L in Figure 6–27 is decreased in value, does power gain to the load increase or decrease?

 Open the Multisim file EXM06-10 or the LT Spice file EXS06-10 in the Examples folder on the website. Measure the voltage gain and compare with the calculated value.

The Darlington Pair

As you have seen, β_{ac} is a major factor in determining the input resistance of an amplifier. The β_{ac} of the transistor limits the maximum achievable input resistance you can get from a given emitter-follower circuit.

One way to boost input resistance is to use a **Darlington pair,** as shown in Figure 6–28. The collectors of two transistors are connected, and the emitter of the first drives the base of the second. This configuration achieves β_{ac} multiplication as shown in the following steps. The emitter current of the first transistor is

$$I_{e1} \cong \beta_{ac1}I_{b1}$$

This emitter current becomes the base current for the second transistor, producing a second emitter current of

$$I_{e2} \cong \beta_{ac2}I_{e1} = \beta_{ac1}\beta_{ac2}I_{b1}$$

Therefore, the effective current gain of the Darlington pair is

$$\beta_{ac} = \beta_{ac1}\beta_{ac2}$$

Neglecting r'_e by assuming that it is much smaller than R_E, the input resistance is

$$R_{in} = \boldsymbol{\beta_{ac1}\beta_{ac2}R_E}$$

Equation 6–17

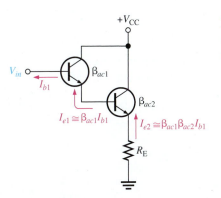

◀ FIGURE 6–28

A Darlington pair multiplies β_{ac}, thus increasing the input resistance.

An Application   The emitter-follower is often used as an interface between a circuit with a high output resistance and a low-resistance load. In such an application, the emitter-follower is called a *buffer*.

Suppose a common-emitter amplifier with a $1.0\,\text{k}\Omega$ collector resistance must drive a low-resistance load such as an $8\,\Omega$ low-power speaker. If the speaker is capacitively coupled to the output of the amplifier, the $8\,\Omega$ load appears—to the ac signal—in parallel with the $1.0\,\text{k}\Omega$ collector resistor. This results in an ac collector resistance of

$$R_c = R_C \| R_L = 1.0\,\text{k}\Omega \| 8\,\Omega = 7.94\,\Omega$$

Obviously, this is not acceptable because most of the voltage gain is lost ($A_v = R_c/r'_e$). For example, if $r'_e = 5\,\Omega$, the voltage gain is reduced from

$$A_v = \frac{R_C}{r'_e} = \frac{1.0\,\text{k}\Omega}{5\,\Omega} = 200$$

with no load to

$$A_v = \frac{R_c}{r'_e} = \frac{7.94\,\Omega}{5\,\Omega} = 1.59$$

with an $8\,\Omega$ speaker load.

An emitter-follower using a Darlington pair can be used to interface the amplifier and the speaker, as shown in Figure 6–29. This circuit is discussed further in Section 7–1 with power amplifiers, but it serves as a good example of the application of a CC amplifier.

FYI

The circuit arrangement in Figure 6–29 is useful for low-power applications ($<1\,\text{W}$ load power) but is inefficient and wasteful for higher power requirements.

▶ **FIGURE 6–29**

A Darlington emitter-follower used as a buffer between a common emitter amplifier and a low resistance load such as a speaker.

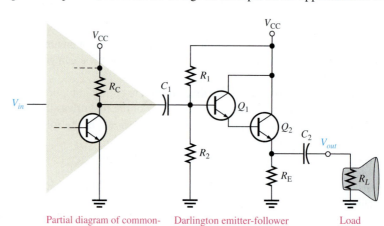

Partial diagram of common-emitter amplifier   Darlington emitter-follower   Load

---

EXAMPLE 6–11

In Figure 6–29 for the common-emitter amplifier, $V_{CC} = 10\,\text{V}$, $R_C = 1.0\,\text{k}\Omega$ and $r'_e = 5\,\Omega$. For the Darlington emitter-follower, $R_1 = 10\,\text{k}\Omega$, $R_2 = 22\,\text{k}\Omega$, $R_E = 22\,\Omega$, $R_L = 8\,\Omega$, $V_{CC} = 10\,\text{V}$, and $\beta_{DC} = \beta_{ac} = 100$ for each transistor. Neglect $R_{IN(BASE)}$ of the Darlington.

(a) Determine the voltage gain of the common-emitter amplifier.

(b) Determine the voltage gain of the Darlington emitter-follower.

(c) Determine the overall voltage gain and compare to the gain of the common-emitter amplifier driving the speaker directly without the Darlington emitter-follower.

Solution **(a)** To determine A_v for the common-emitter amplifier, first find r'_e for the Darlington emitter-follower.

$$V_B = \left(\frac{R_2}{R_1 + R_2}\right)V_{CC} = \left(\frac{22\,\text{k}\Omega}{32\,\text{k}\Omega}\right)10\,\text{V} = 6.88\,\text{V}$$

$$I_E = \frac{V_E}{R_E} = \frac{V_B - 2V_{BE}}{R_E} = \frac{6.88\,\text{V} - 1.4\,\text{V}}{22\,\Omega} = \frac{5.48\,\text{V}}{22\,\Omega} = 250\,\text{mA}$$

$$r'_e = \frac{25\,\text{mV}}{I_E} = \frac{25\,\text{mV}}{250\,\text{mA}} = 100\,\text{m}\Omega$$

Note that R_E must dissipate a power of

$$P_{R_E} = I_E^2 R_E = (250 \text{ mA})^2(22 \ \Omega) = 1.38 \text{ W}$$

and transistor Q_2 must dissipate

$$P_{Q2} = (V_{CC} - V_E)I_E = (4.52 \text{ V})(250 \text{ mA}) = 1.13 \text{ W}$$

Next, the ac emitter resistance of the Darlington emitter-follower is

$$R_e = R_E \parallel R_L = 22 \ \Omega \parallel 8 \ \Omega = 5.87 \ \Omega$$

The total input resistance of the Darlington emitter-follower is

$$R_{in(tot)} = R_1 \parallel R_2 \parallel \beta_{ac}^2(r_e' + R_e)$$
$$= 10 \text{ k}\Omega \parallel 22 \text{ k}\Omega \parallel 100^2(100 \text{ m}\Omega + 5.87 \ \Omega) = 6.16 \text{ k}\Omega$$

The effective ac collector resistance of the common-emitter amplifier is

$$R_c = R_C \parallel R_{in(tot)} = 1.0 \text{ k}\Omega \parallel 6.16 \text{ k}\Omega = 860 \ \Omega$$

The voltage gain of the common-emitter amplifier is

$$A_v = \frac{R_c}{r_e'} = \frac{860 \ \Omega}{5 \ \Omega} = \mathbf{172}$$

(b) The effective ac emitter resistance was found in part (a) to be 5.87 Ω. The voltage gain for the Darlington emitter-follower is

$$A_v = \frac{R_e}{r_e' + R_e} = \frac{5.87 \ \Omega}{100 \text{ m}\Omega + 5.87 \ \Omega} = \mathbf{0.99}$$

(c) The overall voltage gain is

$$A_v' = A_{v(EF)}A_{v(CE)} = (0.99)(172) = \mathbf{170}$$

If the common-emitter amplifier drives the speaker directly, the gain is 1.59 as we previously calculated.

Related Problem Using the same circuit values, determine the voltage gain of the common-emitter amplifier in Figure 6–29 if a single transistor is used in the emitter-follower in place of the Darlington pair. Assume $\beta_{DC} = \beta_{ac} = 100$. Explain the difference in the voltage gain without the Darlington pair.

The Sziklai Pair

The **Sziklai pair,** shown in Figure 6–30, is similar to the Darlington pair except that it consists of two types of transistors, an *npn* and a *pnp*. This configuration is sometimes known as a *complementary Darlington* or a *compound transistor*. The current gain is about the same as in the Darlington pair, as illustrated. The difference is that the Q_2 base current is the Q_1 collector current instead of emitter current, as in the Darlington arrangement.

An advantage of the Sziklai pair, compared to the Darlington, is that it takes less voltage to turn it on because only one barrier potential has to be overcome. A Sziklai pair is sometimes used in conjunction with a Darlington pair as the output stage of power amplifiers. This makes it easier to obtain exact matches of the output transistors, resulting in improved thermal stability and better sound quality in audio applications.

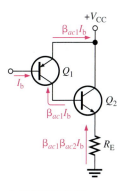

▲ **FIGURE 6–30**

The Sziklai pair.

1. What is a common-collector amplifier called?

2. What is the ideal maximum voltage gain of a common-collector amplifier?

3. What characteristic of the common-collector amplifier makes it a useful circuit?

4. What is a Darlington pair?

5. How does a Darlington pair differ from a Sziklai pair?

6–5 THE COMMON-BASE AMPLIFIER

The common-base (CB) amplifier provides high voltage gain with a maximum current gain of 1. Since it has a low input resistance, the CB amplifier is the most appropriate type for certain applications where sources tend to have very low-resistance outputs.

After completing this section, you should be able to

◻ **Describe and analyze the operation of common-base amplifiers**
◻ Determine the voltage gain
  ◆ Explain why there is no phase inversion
◻ Discuss and calculate input resistance
◻ Determine output resistance
◻ Determine current gain
◻ Determine power gain

FYI

The CB amplifier is useful at high frequencies when impedance matching is required because input impedance can be controlled and because noninverting amps have better frequency response than inverting amps.

A typical **common-base** amplifier is shown in Figure 6–31. The base is the common terminal and is at ac ground because of capacitor C_2. The input signal is capacitively coupled to the emitter. The output is capacitively coupled from the collector to a load resistor.

Voltage Gain

The voltage gain from emitter to collector is developed as follows ($V_{in} = V_e$, $V_{out} = V_c$).

$$A_v = \frac{V_{out}}{V_{in}} = \frac{V_c}{V_e} = \frac{I_c R_c}{I_e(r'_e \parallel R_E)} \cong \frac{I_e R_c}{I_e(r'_e \parallel R_E)}$$

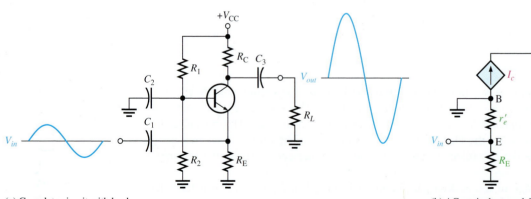

(a) Complete circuit with load   (b) AC equivalent model

▲ **FIGURE 6–31**

Common-base amplifier with voltage-divider bias.

If $R_E \gg r'_e$, then

$$A_v \cong \frac{R_c}{r'_e}$$

Equation 6–18

where $R_c = R_C \parallel R_L$. Notice that the gain expression is the same as for the common-emitter amplifier. However, there is no phase inversion from emitter to collector.

Input Resistance

The resistance, looking in at the emitter, is

$$R_{in(emitter)} = \frac{V_{in}}{I_{in}} = \frac{V_e}{I_e} = \frac{I_e(r'_e \parallel R_E)}{I_e}$$

If $R_E \gg r'_e$, then

$$R_{in(emitter)} \cong r'_e$$

Equation 6–19

R_E is typically much greater than r'_e, so the assumption that $r'_e \parallel R_E \cong r'_e$ is usually valid. The input resistance can be set to a desired value within limits by using a swamping resistor. This is useful in communication systems and other applications where you need to match a source impedance to prevent a reflected signal.

Output Resistance

Looking into the collector, the ac collector resistance, r'_c, appears in parallel with R_C. As you have previously seen in connection with the CE amplifier, r'_c is typically much larger than R_C, so a good approximation for the output resistance is

$$R_{out} \cong R_C$$

Equation 6–20

Current Gain

The current gain is the output current divided by the input current. I_c is the ac output current, and I_e is the ac input current. Since $I_c \cong I_e$, the current gain is approximately 1.

$$A_i \cong 1$$

Equation 6–21

Power Gain

The CB amplifier is primarily a voltage amplifier, so power gain is not too important. Since the current gain is approximately 1 for the common-base amplifier and $A_p = A_v A_i$, the total power gain is approximately equal to the voltage gain.

$$A_P \cong A_v$$

Equation 6–22

This power gain includes power to the collector resistor and to the load resistor. If you want the power gain only to the load, then divide V_{out}^2/R_L by the input power.

EXAMPLE 6–12

Find the input resistance, voltage gain, current gain, and power gain for the amplifier in Figure 6–32. $\beta_{DC} = 250$.

Solution First, find I_E so that you can determine r'_e. Then $R_{in} \cong r'_e$.

$$R_{TH} = \frac{R_1 R_2}{R_1 + R_2} = \frac{(56\ \text{k}\Omega)(12\ \text{k}\Omega)}{56\ \text{k}\Omega + 12\ \text{k}\Omega} = 9.88\ \text{k}\Omega$$

$$V_{TH} = \left(\frac{R_2}{R_1 + R_2}\right)V_{CC} = \left(\frac{12\ \text{k}\Omega}{56\ \text{k}\Omega + 12\ \text{k}\Omega}\right)10\ \text{V} = 1.76\ \text{V}$$

$$I_E = \frac{V_{TH} - V_{BE}}{R_E + R_{TH}/\beta_{DC}} = \frac{1.76\ \text{V} - 0.7\ \text{V}}{1.0\ \text{k}\Omega + 39.5\ \Omega} = 1.02\ \text{mA}$$

▶ FIGURE 6–32

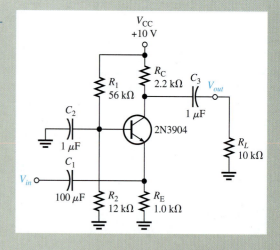

Therefore,

$$R_{in} \cong r'_e = \frac{25 \text{ mV}}{I_E} = \frac{25 \text{ mV}}{1.02 \text{ mA}} = \mathbf{24.5 \ \Omega}$$

Calculate the voltage gain as follows:

$$R_c = R_C \| R_L = 2.2 \text{ k}\Omega \| 10 \text{ k}\Omega = 1.8 \text{ k}\Omega$$

$$A_v = \frac{R_c}{r'_e} = \frac{1.8 \text{ k}\Omega}{24.5 \ \Omega} = \mathbf{73.5}$$

Also, $A_i \cong \mathbf{1}$ and $A_p \cong A_v = \mathbf{73.5}$.

Related Problem   Find A_v in Figure 6–32 if $\beta_{DC} = 50$.

  Open the Multisim file EXM06-12 or the LT Spice file EXS06-12 in the Examples folder on the website. Measure the voltage gain and compare with the calculated value.

SECTION 6–5 CHECKUP

1. Can the same voltage gain be achieved with a common-base as with a common-emitter amplifier?

2. Does the common-base amplifier have a low or a high input resistance?

3. What is the maximum current gain in a common-base amplifier?

4. Does a common-base amplifier invert the input signal?

6–6   MULTISTAGE AMPLIFIERS

Two or more amplifiers can be connected in a **cascaded** arrangement with the output of one amplifier driving the input of the next. Each amplifier in a cascaded arrangement is known as a **stage**. The basic purpose of a multistage arrangement is to increase the overall voltage gain. Although discrete multistage amplifiers are not as common as they once were, a familiarization with this area provides insight into how circuits affect each other when they are connected together.

After completing this section, you should be able to

❑ **Describe and analyze the operation of multistage amplifiers**
❑ Determine the overall voltage gain of multistage amplifiers
   ◆ Express the voltage gain in decibels (dB)
❑ Discuss and analyze capacitively-coupled multistage amplifiers
   ◆ Describe loading effects   ◆ Determine the voltage gain of each stage in
   a two-stage amplifier   ◆ Determine the overall voltage gain   ◆ Determine
   the dc voltages
❑ Describe direct-coupled multistage amplifiers

Multistage Voltage Gain

The overall voltage gain, A_v', of cascaded amplifiers, as shown in Figure 6–33, is the product of the individual voltage gains.

$$A_v' = A_{v1}A_{v2}A_{v3}\ldots A_{vn}$$

Equation 6–23

where n is the number of stages.

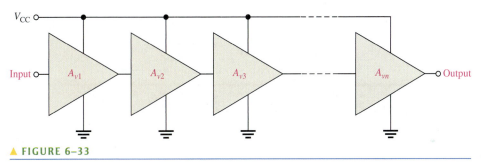

▲ FIGURE 6–33

Cascaded amplifiers. Each triangular symbol represents a separate amplifier.

Amplifier voltage gain is often expressed in **decibels** (dB) as follows:

$$A_{v(\mathbf{dB})} = 20 \log A_v$$

Equation 6–24

This is particularly useful in **multistage** systems because the overall voltage gain in dB is the *sum* of the individual voltage gains in dB.

$$A_{v(\mathrm{dB})}' = A_{v1(\mathrm{dB})} + A_{v2(\mathrm{dB})} + \cdots + A_{vn(\mathrm{dB})}$$

EXAMPLE 6–13

A certain cascaded amplifier arrangement has the following voltage gains: $A_{v1} = 10$, $A_{v2} = 15$, and $A_{v3} = 20$. What is the overall voltage gain? Also express each gain in decibels (dB) and determine the total voltage gain in dB.

Solution

$$A_v' = A_{v1}A_{v2}A_{v3} = (10)(15)(20) = \mathbf{3000}$$

$$A_{v1(\mathrm{dB})} = 20 \log 10 = \mathbf{20.0\ dB}$$

$$A_{v2(\mathrm{dB})} = 20 \log 15 = \mathbf{23.5\ dB}$$

$$A_{v3(\mathrm{dB})} = 20 \log 20 = \mathbf{26.0\ dB}$$

$$A_{v(\mathrm{dB})}' = 20.0\ \mathrm{dB} + 23.5\ \mathrm{dB} + 26.0\ \mathrm{dB} = \mathbf{69.5\ dB}$$

Related Problem   In a certain multistage amplifier, the individual stages have the following voltage gains: $A_{v1} = 25$, $A_{v2} = 5$, and $A_{v3} = 12$. What is the overall gain? Express each gain in dB and determine the total voltage gain in dB.

Capacitively Coupled Multistage Amplifier

For purposes of illustration, we will use the two-stage capacitively coupled amplifier in Figure 6–34. Notice that both stages are identical common-emitter amplifiers with the output of the first stage capacitively coupled to the input of the second stage. Capacitive coupling prevents the dc bias of one stage from affecting that of the other but allows the ac signal to pass without attenuation because $X_C \cong 0 \, \Omega$ at the frequency of operation. Notice, also, that the transistors are labeled Q_1 and Q_2.

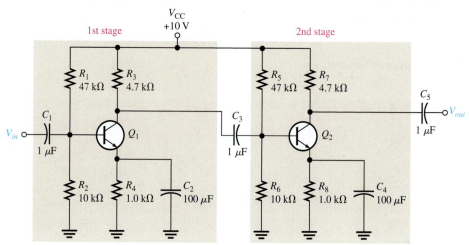

$\beta_{DC} = \beta_{ac} = 150$ for Q_1 and Q_2

▲ **FIGURE 6–34**

A two-stage common-emitter amplifier.

Loading Effects   In determining the voltage gain of the first stage, you must consider the loading effect of the second stage. Because the coupling capacitor C_3 effectively appears as a short at the signal frequency, the total input resistance of the second stage presents an ac load to the first stage.

Looking from the collector of Q_1, the two biasing resistors in the second stage, R_5 and R_6, appear in parallel with the input resistance at the base of Q_2. In other words, the signal at the collector of Q_1 "sees" R_3, R_5, R_6, and $R_{in(base2)}$ of the second stage all in parallel to ac ground. Thus, the effective ac collector resistance of Q_1 is the total of all these resistances in parallel, as Figure 6–35 illustrates. The voltage gain of the first stage is reduced by the loading of the second stage because the effective ac collector resistance of the first stage is less than the actual value of its collector resistor, R_3. Remember that $A_v = R_c/r'_e$.

Voltage Gain of the First Stage   The ac collector resistance of the first stage is

$$R_{c1} = R_3 \| R_5 \| R_6 \| R_{in(base2)}$$

Remember that lowercase italic subscripts denote ac quantities such as for R_c.

You can verify that $I_E = 1.05 \, \text{mA}$, $r'_e = 23.8 \, \Omega$, and $R_{in(base2)} = 3.57 \, \text{k}\Omega$. The effective ac collector resistance of the first stage is as follows:

$$R_{c1} = 4.7 \, \text{k}\Omega \| 47 \, \text{k}\Omega \| 10 \, \text{k}\Omega \| 3.57 \, \text{k}\Omega = 1.63 \, \text{k}\Omega$$

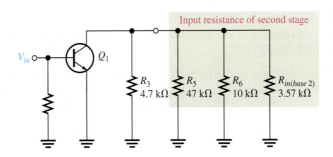

AC equivalent of first stage in Figure 6–34, showing loading from second stage input resistance.

Therefore, the base-to-collector voltage gain of the first stage is

$$A_{v1} = \frac{R_{c1}}{r_e'} = \frac{1.63 \text{ k}\Omega}{23.8 \ \Omega} = 68.5$$

This can also be expressed as $A_{v1} = 36.7$ dB

Voltage Gain of the Second Stage   The second stage has no load resistor, so the ac collector resistance is R_7, and the gain is

$$A_{v2} = \frac{R_7}{r_e'} = \frac{4.7 \text{ k}\Omega}{23.8 \ \Omega} = 197$$

In dB, this is expressed as $A_{v2} = 45.9$ dB.

Compare this to the gain of the first stage, and notice how much the loading from the second stage reduced the gain.

Overall Voltage Gain   The overall amplifier gain with no load on the output is

$$A_v' = A_{v1}A_{v2} = (68.5)(197) \cong 13,495$$

If an input signal of 100 μV, for example, is applied to the first stage and if there is no attenuation in the input base circuit due to the source resistance, an output from the second stage of $(100 \ \mu\text{V})(13,495) \cong 1.35$ V will result. The overall voltage gain can be expressed in dB as follows:

$$A_{v(dB)}' = 20 \log (13,495) = 82.6 \text{ dB}$$

Notice that the overall gain is also the sum of the two stages: 36.7 dB + 45.9 dB = 82.6 dB.

You can also view the amplifier as two identical gain stages separated by an attenuation network composed of a resistive divider. The unloaded gain of each stage (197) is used for the amplifiers and loading effects between stages are treated separately; these resistances form a voltage divider. The divider resistances are the source resistance of Q_1 which is the collector resistor (R_3) and the input resistance of Q_2, which is composed of $R_5 \parallel R_6 \parallel R_{in(base)}$. A simplified view of the amplifier is drawn in Figure 6–36.

Applying the voltage divider rule to the attenuation network:

$$Gain = \frac{2.49 \text{ k}\Omega}{2.49 \text{ k}\Omega + 4.7 \text{ k}\Omega} = 0.346$$

Expressed in dB, the attenuation network has a gain of –9.22 dB.

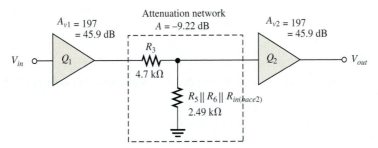

Simplifying the two-stage amplifier shown in Figure 6–34.

Thus the overall gain is $(197)(0.346)(197) = 13,428$ (difference is due to round off).

Expressed in dB, the overall gain is the sum of the three parts: $45.9 \text{ dB} - 9.22 \text{ dB} + 45.9 \text{ dB} = 82.6 \text{ dB}$.

DC Voltages in the Capacitively Coupled Multistage Amplifier Since both stages in Figure 6–34 are identical, the dc voltages for Q_1 and Q_2 are the same. Since $\beta_{DC}R_4 \gg R_2$ and $\beta_{DC}R_8 \gg R_6$, the dc base voltage for Q_1 and Q_2 is

$$V_B \cong \left(\frac{R_2}{R_1 + R_2}\right)V_{CC} = \left(\frac{10 \text{ k}\Omega}{57 \text{ k}\Omega}\right)10 \text{ V} = 1.75 \text{ V}$$

The dc emitter and collector voltages are as follows:

$$V_E = V_B - 0.7 \text{ V} = 1.05 \text{ V}$$

$$I_E = \frac{V_E}{R_4} = \frac{1.05 \text{ V}}{1.0 \text{ k}\Omega} = 1.05 \text{ mA}$$

$$I_C \cong I_E = 1.05 \text{ mA}$$

$$V_C = V_{CC} - I_C R_3 = 10 \text{ V} - (1.05 \text{ mA})(4.7 \text{ k}\Omega) = 5.07 \text{ V}$$

Direct-Coupled Multistage Amplifiers

A basic two-stage, direct-coupled amplifier is shown in Figure 6–37. Notice that there are no coupling or bypass capacitors in this circuit. The dc collector voltage of the first stage provides the base-bias voltage for the second stage. Because of the direct coupling, this type of amplifier has a better low-frequency response than the capacitively coupled type in which the reactance of coupling and bypass capacitors at very low frequencies may become excessive. The increased reactance of capacitors at lower frequencies produces gain reduction in capacitively coupled amplifiers.

Direct-coupled amplifiers can be used to amplify low frequencies all the way down to dc (0 Hz) without loss of voltage gain because there are no capacitive reactances in the circuit. The disadvantage of direct-coupled amplifiers, on the other hand, is that small changes in the dc bias voltages from temperature effects or power-supply variation are amplified by the succeeding stages, which can result in a significant drift in the dc levels throughout the circuit.

▶ **FIGURE 6–37**

A basic two-stage direct-coupled amplifier.

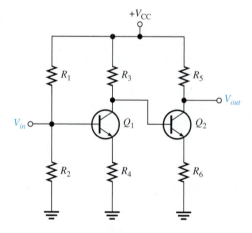

SECTION 6–6 CHECKUP	1. What does the term *stage* mean?
	2. How is the overall voltage gain of a multistage amplifier determined?
	3. Express a voltage gain of 500 in dB.
	4. Discuss a disadvantage of a capacitively coupled amplifier.

6–7 THE DIFFERENTIAL AMPLIFIER

A **differential amplifier** is an amplifier that produces outputs that are a function of the difference between two input voltages. The differential amplifier has two basic modes of operation: differential (in which the two inputs are different) and common mode (in which the two inputs are the same). The differential amplifier is important in operational amplifiers, which are covered beginning in Chapter 12.

After completing this section, you should be able to

❑ **Describe the differential amplifier and its operation**
❑ Discuss the basic operation
   ◆ Calculate dc currents and voltages
❑ Discuss the modes of signal operation
   ◆ Describe single-ended differential input operation ◆ Describe double-ended differential input operation ◆ Determine common-mode operation
❑ Define and determine the common-mode rejection ratio (CMRR)

Basic Operation

A basic differential amplifier (diff-amp) circuit is shown in Figure 6–38. Notice that the differential amplifier has two inputs and two outputs.

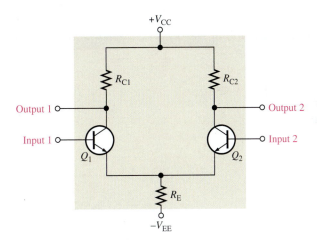

◀ **FIGURE 6–38**

Basic differential amplifier.

The following discussion is in relation to Figure 6–39 and consists of a basic dc analysis of the diff-amp's operation. First, when both inputs are grounded (0 V), the emitters are at −0.7 V, as indicated in Figure 6–39(a). It is assumed that the transistors are identically matched by careful process control during manufacturing so that their dc emitter currents are the same when there is no input signal. Thus,

$$I_{E1} = I_{E2}$$

Since both emitter currents combine through R_E,

$$I_{E1} = I_{E2} = \frac{I_{R_E}}{2}$$

where

$$I_{R_E} = \frac{V_E - V_{EE}}{R_E}$$

Based on the approximation that $I_C \cong I_E$,

$$I_{C1} = I_{C2} \cong \frac{I_{R_E}}{2}$$

Since both collector currents and both collector resistors are equal (when the input voltage is zero),

$$V_{C1} = V_{C2} = V_{CC} - I_{C1}R_{C1}$$

This condition is illustrated in Figure 6–39(a).

Next, input 2 is left grounded, and a positive bias voltage is applied to input 1, as shown in Figure 6–39(b). The positive voltage on the base of Q_1 increases I_{C1} and raises the emitter voltage to

$$V_E = V_B - 0.7 \text{ V}$$

This action reduces the forward bias (V_{BE}) of Q_2 because its base is held at 0 V (ground), thus causing I_{C2} to decrease. The net result is that the increase in I_{C1} causes a decrease in V_{C1}, and the decrease in I_{C2} causes an increase in V_{C2}, as shown.

Finally, input 1 is grounded and a positive bias voltage is applied to input 2, as shown in Figure 6–39(c). The positive bias voltage causes Q_2 to conduct more, thus increasing I_{C2}.

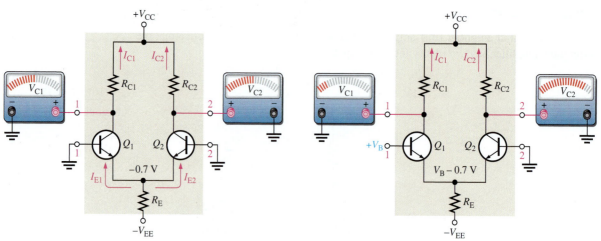

(a) Both inputs grounded

(b) Bias voltage on input 1 with input 2 grounded

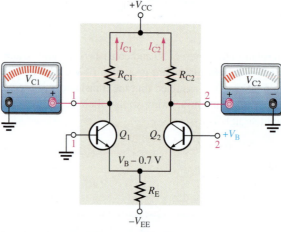

(c) Bias voltage on input 2 with input 1 grounded

▲ FIGURE 6–39

Basic operation of a differential amplifier (ground is zero volts) showing relative changes in voltages.

Also, the emitter voltage is raised. This reduces the forward bias of Q_1, since its base is held at ground, and causes I_{C1} to decrease. The result is that the increase in I_{C2} produces a decrease in V_{C2}, and the decrease in I_{C1} causes V_{C1} to increase, as shown.

Modes of Signal Operation

Single-Ended Differential Input   When a diff-amp is operated with this input configuration, one input is grounded and the signal voltage is applied only to the other input, as shown in Figure 6–40. In the case where the signal voltage is applied to input 1 as in part (a), an inverted, amplified signal voltage appears at output 1 as shown. Also, a signal voltage appears in phase at the emitter of Q_1. Since the emitters of Q_1 and Q_2 are common, the emitter signal becomes an input to Q_2, which functions as a common-base amplifier. The signal is amplified by Q_2 and appears, noninverted, at output 2. This action is illustrated in part (a).

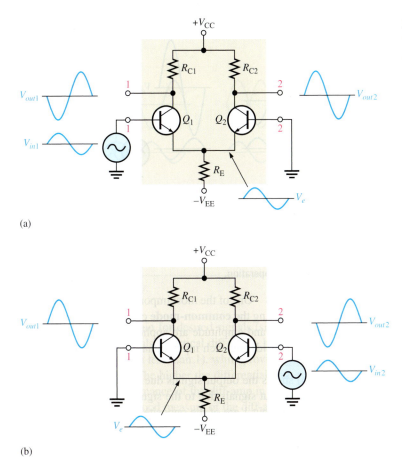

◀ FIGURE 6–40

Single-ended differential input operation.

In the case where the signal is applied to input 2 with input 1 grounded, as in Figure 6–40(b), an inverted, amplified signal voltage appears at output 2. In this situation, Q_1 acts as a common-base amplifier, and a noninverted, amplified signal appears at output 1.

Double-Ended Differential Inputs   In this input configuration, two opposite-polarity (out-of-phase) signals are applied to the inputs, as shown in Figure 6–41(a). Each input affects the outputs, as you will see in the following discussion.

Figure 6–41(b) shows the output signals due to the signal on input 1 acting alone as a single-ended input. Figure 6–41(c) on page 292 shows the output signals due to the signal on input 2 acting alone as a single-ended input. Notice in parts (b) and (c) that the signals on output 1 are of the same polarity. The same is also true for output 2. By applying the superposition theorem and summing both output 1 signals and both output 2 signals, you get the total output signals, as shown in Figure 6–41(d).

Solution $A_{v(d)} = 2000$, and $A_{cm} = 0.2$. Therefore,

$$\text{CMRR} = \frac{A_{v(d)}}{A_{cm}} = \frac{2000}{0.2} = \mathbf{10{,}000}$$

Expressed in decibels,

$$\text{CMRR} = 20 \log (10{,}000) = \mathbf{80\ dB}$$

Related Problem   Determine the CMRR and express it in decibels for an amplifier with a differential voltage gain of 8500 and a common-mode gain of 0.25.

A CMRR of 10,000 means that the desired input signal (differential) is amplified 10,000 times more than the unwanted noise (common-mode). For example, if the amplitudes of the differential input signal and the common-mode noise are equal, the desired signal will appear on the output 10,000 times greater in amplitude than the noise. Thus, the noise or interference has been essentially eliminated.

SECTION 6–7 CHECKUP

1. Distinguish between double-ended and single-ended differential inputs.
2. Define *common-mode rejection*.
3. For a given value of differential gain, does a higher CMRR result in a higher or lower common-mode gain?
4. What is the difference between a common-mode signal and a differential signal?

6–8   TROUBLESHOOTING

In working with any circuit, you must first know how it is supposed to work before you can troubleshoot it for a failure. The two-stage capacitively coupled amplifier discussed in Section 6–6 is used to illustrate a typical troubleshooting procedure.

After completing this section, you should be able to

❑ **Troubleshoot amplifier circuits**
❑ Discuss a troubleshooting procedure
  ◆ Describe the analysis phase   ◆ Describe the planning phase   ◆ Describe the measurement phase

When you are faced with having to troubleshoot a circuit, the first thing you need is a schematic with the proper dc and signal voltages labeled. You must know what the correct voltages in the circuit should be before you can identify an incorrect voltage. Schematics of some circuits are available with voltages indicated at certain points. If this is not the case, you must use your knowledge of the circuit operation to determine the correct voltages. Figure 6–43 is the schematic for the two-stage amplifier that was analyzed in Section 6–6. The correct voltages are indicated at each point.

Troubleshooting Procedure

The analysis, planning, and measurement approach to troubleshooting, discussed in Chapter 2, will be used.

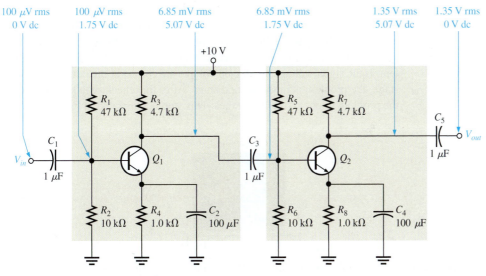

▲ FIGURE 6–43

A two-stage common-emitter amplifier with correct voltages indicated. Both transistors have dc and ac betas of 150. Different values of β will produce slightly different results.

Analysis   It has been found that there is no output voltage, V_{out}. You have also determined that the circuit did work properly and then failed. A visual check of the circuit board or assembly for obvious problems such as broken or poor connections, solder splashes, wire clippings, or burned components turns up nothing. You conclude that the problem is most likely a faulty component in the amplifier circuit or an open connection. Also, the dc supply voltage may not be correct or may be missing.

Planning   You decide to use a DMM to check the dc levels and an oscilloscope to check the ac signals at certain test points. A function generator with the signal attenuated by a voltage divider to 100 μV will be used to apply a test signal to the input. Also, you decide to apply the half-splitting method to trace the voltages in the circuit and use an in-circuit transistor tester if a transistor is suspected of being faulty.

Measurement   To determine the faulty component in a multistage amplifier, use the general five-step troubleshooting procedure which is illustrated as follows.

   Step 1: *Perform a power check.* Assume the dc supply voltage is correct as indicated in Figure 6–44.

   Step 2: *Check the input and output voltages.* Assume the measurements indicate that the input signal voltage is correct. However, there is no output signal voltage or the output signal voltage is much less than it should be, as shown by the diagram in Figure 6–44.

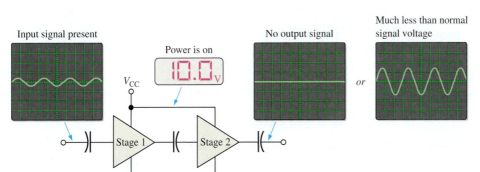

◄ FIGURE 6–44

Initial check of a faulty two-stage amplifier.

Step 3: *Apply the half-splitting method of signal tracing.* Check the voltages at the output of the first stage. No signal voltage or a much less than normal signal voltage indicates that the problem is probably in the first stage (an incorrect load could be the problem). An incorrect dc voltage also indicates a first-stage problem. If the signal voltage and the dc voltage are correct at the output of the first stage, the problem is in the second stage. After this check, you have narrowed the problem to one of the two stages. This step is illustrated in Figure 6–45.

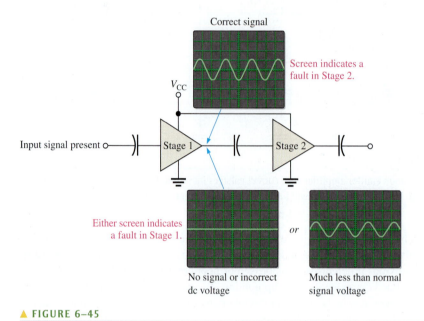

▲ **FIGURE 6–45**

Half-splitting signal tracing isolates the faulty stage.

Step 4: *Apply fault analysis.* Focus on the faulty stage and determine the component failure that can produce the incorrect output.

Symptom: DC voltages incorrect.

Likely faults: A failure of any resistor or the transistor will produce an incorrect dc bias voltage. A leaky bypass or coupling capacitor will also affect the dc bias voltages. Further measurements in the stage are necessary to isolate the faulty component.

Incorrect ac voltages and the most likely fault(s) are illustrated in Figure 6–46 as follows:

(a) *Symptom 1:* Signal voltage at output missing; dc voltage correct.
Symptom 2: Signal voltage at base missing; dc voltage correct.
Likely fault: Input coupling capacitor open. This prevents the signal from getting to the base.

(b) *Symptom:* Correct signal at base but no output signal.
Likely fault: Transistor base open.

(c) *Symptom:* Signal voltage at output much less than normal; dc voltage correct.
Likely fault: Bypass capacitor open.

Step 5: *Replace or repair.* With the power turned off, replace the defective component or repair the defective connection. Turn on the power, and check for proper operation.

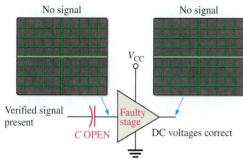

(a) Coupling capacitor open

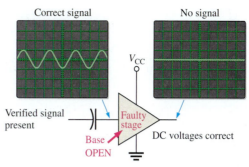

(b) Transistor base open

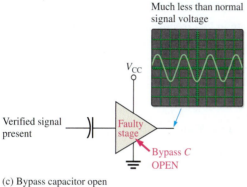

(c) Bypass capacitor open

▲ FIGURE 6–46

Troubleshooting a faulty stage.

EXAMPLE 6–15

The two-stage amplifier in Figure 6–43 has malfunctioned such that there is no output signal with a verified input. Specify the step-by-step troubleshooting procedure for an assumed fault.

Solution   Assume there are no visual or other indications of a problem such as a charred resistor, solder splash, wire clipping, broken connection, or extremely hot component. The troubleshooting procedure for a certain fault scenario is as follows:

Step 1: There is power to the circuit as indicated by a correct V_{CC} measurement.

Step 2: There is a verified input signal voltage, but no output signal voltage is measured.

Step 3: The signal voltage and the dc voltage at the collector of Q_1 are correct. This means that the problem is in the second stage or the coupling capacitor C_3 between the stages.

Step 4: The correct signal voltage and dc bias voltage are measured at the base of Q_2. This eliminates the possibility of a fault in C_3 or the second stage bias circuit.

  The collector of Q_2 is at 10 V and there is no signal voltage. This measurement, made directly on the transistor collector, indicates that either the collector is shorted to V_{CC} or the transistor is internally open. It is unlikely that the collector resistor R_7 is shorted but to verify, turn off the power and use an ohmmeter to check.

  The possibility of a short is eliminated by the ohmmeter check. The other possible faults are (a) transistor Q_2 internally open or (b) emitter resistor or connection open. Use a transistor tester and/or ohmmeter to check each of these possible faults with power off.

Step 5: Replace the faulty component or repair open connection and retest the circuit for proper operation.

Related Problem Determine the possible fault(s) if, in Step 4, you find no signal voltage at the base of Q_2 but the dc voltage is correct.

Multisim Troubleshooting Exercises

These file circuits are in the Troubleshooting Exercises folder on the website. Open each file and determine if the circuit is working properly. If it is not working properly, determine the fault.

1. Multisim file TSM06-01

2. Multisim file TSM06-02

3. Multisim file TSM06-03

4. Multisim file TSM06-04

5. Multisim file TSM06-05

SECTION 6–8 CHECKUP

1. If C_4 in Figure 6–43 were open, how would the output signal be affected? How would the dc level at the collector of Q_2 be affected?

2. If R_5 in Figure 6–43 were open, how would the output signal be affected?

3. If the coupling capacitor C_3 in Figure 6–43 shorted out, would any of the dc voltages in the amplifier be changed? If so, which ones?

Device Application: *Audio Preamplifier for PA System*

An audio preamplifier is to be developed for use in a small portable public address (PA) system. The preamplifier will have a microphone input, and its output will drive a power amplifier to be developed in Chapter 7. A block diagram of the complete PA system is shown in Figure 6–47(a), and its physical configuration is shown in part (b). The dc supply voltages are provided by a battery pack or by an electronic power supply.

The Circuit

A two-stage audio voltage preamplifier is shown in Figure 6–48. The first stage is a common-emitter *pnp* with voltage-divider bias, and the second stage is a common-emitter *npn* with voltage-divider bias. It has been decided that the amplifier should operate from 30 V dc to get a large enough signal voltage swing to provide a maximum of 6 W to the speaker. Because small IC regulators such as the 78xx and 79xx series are not available

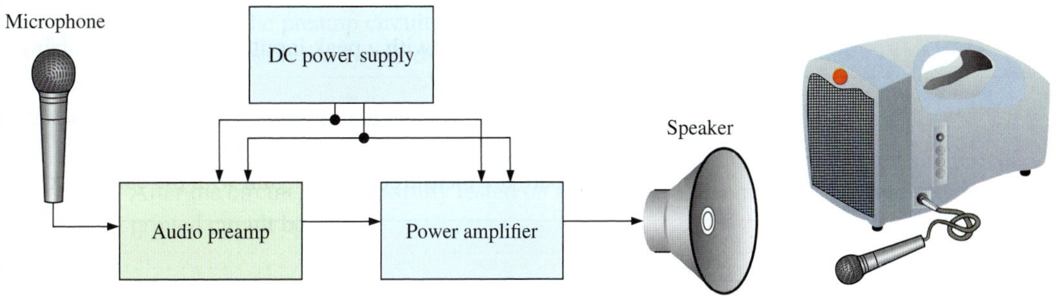

Microphone

DC power supply

Speaker

Audio preamp → Power amplifier

(a) PA system block diagram

(b) Physical configuration

▲ FIGURE 6–47

The public address system.

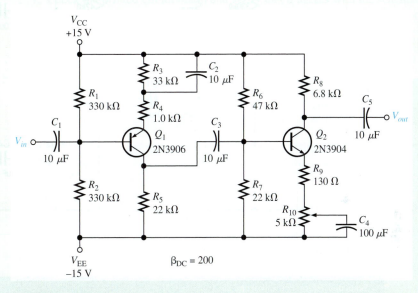

▲ FIGURE 6–48

Two-stage voltage preamplifier.

above 24 V, dual ± 15 V dc supplies are used in this particular system instead of a single supply. The operation is essentially the same as if a single $+30$ V dc source had been used. The potentiometer at the output provides gain adjustment for volume control. The input to the first stage is from the microphone, and the output of the second stage will drive a power amplifier to be developed in Chapter 7. The power amplifier will drive the speaker. The preamp is to operate with a peak input signal range of from 25 mV to 50 mV. The minimum range of voltage gain adjustment is from 90 to 170.

1. Calculate the theoretical voltage gain of the first stage when the second stage is set for maximum gain.
2. Calculate the theoretical maximum voltage gain of the second stage.
3. Determine the overall theoretical voltage gain.
4. Calculate the circuit power dissipation with no signal (quiescent).

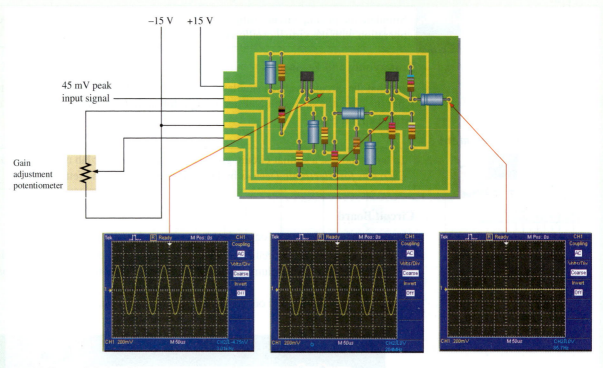

(a) Test result for board 1

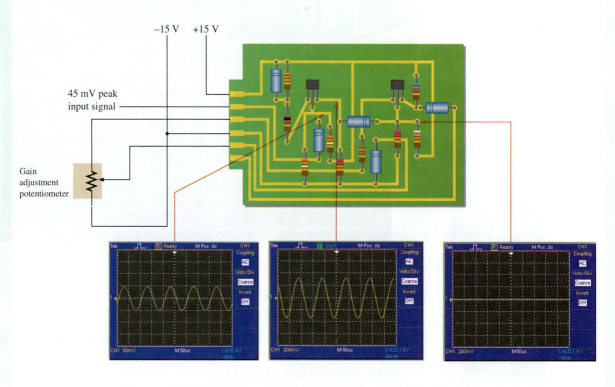

(b) Test result for board 2

▲ FIGURE 6–51

Test of two faulty preamp boards.

SUMMARY OF THE COMMON-EMITTER AMPLIFIER

CIRCUIT WITH VOLTAGE-DIVIDER BIAS

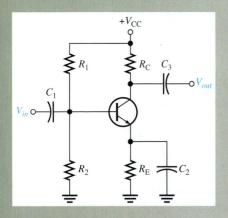

- Input is at the base. Output is at the collector.

- There is a phase inversion from input to output.

- C_1 and C_3 are coupling capacitors for the input and output signals.

- C_2 is the emitter-bypass capacitor.

- All capacitors must have a negligible reactance at the frequency of operation, so they appear as shorts.

- Emitter is at ac ground due to the bypass capacitor.

EQUIVALENT CIRCUITS AND FORMULAS

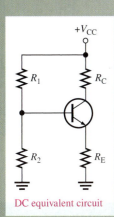

DC equivalent circuit

- DC formulas:

$$R_{TH} = \frac{R_1 R_2}{R_1 + R_2}$$

$$V_{TH} = \left(\frac{R_2}{R_1 + R_2}\right) V_{CC}$$

$$I_E = \frac{V_{TH} - V_{BE}}{R_E + R_{TH}/\beta_{DC}}$$

$$V_E = I_E R_E$$
$$V_B = V_E + V_{BE}$$
$$V_C = V_{CC} - I_C R_C$$

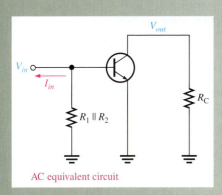

AC equivalent circuit

- AC formulas:

$$r'_e = \frac{25 \text{ mV}}{I_E}$$

$$R_{in(base)} = \beta_{ac} r'_e$$

$$R_{out} \cong R_C$$

$$A_v = \frac{R_C}{r'_e}$$

$$A_i = \frac{I_c}{I_{in}}$$

$$A_p = A'_v A_i$$

EQUIVALENT CIRCUITS AND FORMULAS

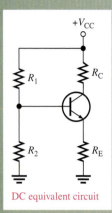

DC equivalent circuit

■ DC formulas:

$$R_{TH} = \frac{R_1 R_2}{R_1 + R_2}$$

$$V_{TH} = \left(\frac{R_2}{R_1 + R_2}\right) V_{CC}$$

$$I_E = \frac{V_{TH} - V_{BE}}{R_E + R_{TH}/\beta_{DC}}$$

$$V_E = I_E R_E$$

$$V_B = V_E + V_{BE}$$

$$V_C = V_{CC} - I_C R_C$$

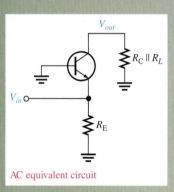

AC equivalent circuit

■ AC formulas:

$$r_e' = \frac{25\ mV}{I_E}$$

$$R_{in(emitter)} \cong r_e'$$

$$R_{out} \cong R_C$$

$$A_v \cong \frac{R_c}{r_e'}$$

$$A_i \cong 1$$

$$A_p \cong A_v$$

SUMMARY OF DIFFERENTIAL AMPLIFIER

CIRCUIT WITH DIFFERENTIAL INPUTS

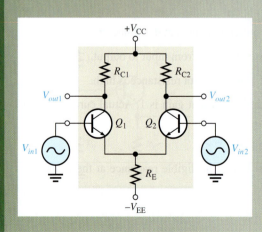

■ Double-ended differential inputs (shown)

Signal on both inputs

Input signals are out of phase

■ Single-ended differential inputs (not shown)

Signal on one input only

One input connected to ground

CIRCUIT WITH COMMON-MODE INPUTS

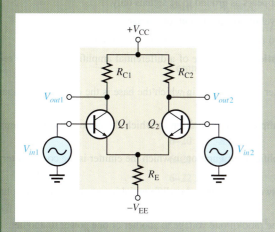

- Both input signals are the same phase, frequency, and amplitude.

- Common-mode rejection ratio:

$$\text{CMRR} = \frac{A_{v(d)}}{A_{cm}}$$

$$\text{CMRR} = 20 \log \left(\frac{A_{v(d)}}{A_{cm}} \right)$$

SUMMARY

Section 6–1
◆ A small-signal amplifier uses only a small portion of its load line under signal conditions.

◆ The ac load line differs from the dc load line because the effective ac output resistance is less than the dc output resistance.

Section 6–2
◆ *r* parameters are easily identifiable and applicable with a transistor's circuit operation.

◆ *h* parameters are important because manufacturers' datasheets specify transistors using *h* parameters.

Section 6–3
◆ A common-emitter amplifier has high voltage, current, and power gains, but a relatively low input resistance.

◆ Swamping is a method of stabilizing the voltage gain.

Section 6–4
◆ A common-collector amplifier has high input resistance and high current gain, but its voltage gain is approximately 1.

◆ A Darlington pair provides beta multiplication for increased input resistance.

◆ A common-collector amplifier is known as an emitter-follower.

Section 6–5
◆ The common-base amplifier has a high voltage gain, but it has a very low input resistance and its current gain is approximately 1.

◆ Equations for basic configurations of common-emitter, common-collector, and common-base amplifiers are given in the Key Formula list.

Section 6–6
◆ The total gain of a multistage amplifier is the product of the individual gains (sum of the individual dB gains).

◆ Single-stage amplifiers can be connected in sequence with capacitively-coupling and direct coupling methods to form multistage amplifiers.

Section 6–7
◆ A differential input voltage appears between the inverting and noninverting inputs of a differential amplifier.

◆ In the differential mode, a diff-amp can be operated with single-ended or double-ended inputs.

◆ In single-ended operation, there is a signal on one input and the other input is grounded.

◆ In double-ended operation, two signals that are 180° out of phase are on the inputs.

◆ Common-mode occurs when equal in-phase voltages are applied to both input terminals.

3. If the value of R_C in Figure 6–8 is increased, V_{out} will

(a) increase   (b) decrease   (c) not change

4. If the amplitude of V_{in} in Figure 6–8 is decreased, V_{out} will

(a) increase   (b) decrease   (c) not change

5. If C_2 in Figure 6–27 is shorted, the average value of the output voltage will

(a) increase   (b) decrease   (c) not change

6. If the value of R_E in Figure 6–27 is increased, the voltage gain will

(a) increase   (b) decrease   (c) not change

7. If the value of C_1 in Figure 6–27 is increased, V_{out} will

(a) increase   (b) decrease   (c) not change

8. If the value of R_C in Figure 6–32 is increased, the current gain will

(a) increase   (b) decrease   (c) not change

9. If C_2 and C_4 in Figure 6–34 are increased in value, V_{out} will

(a) increase   (b) decrease   (c) not change

10. If the value of R_4 in Figure 6–34 is reduced, the overall voltage gain will

(a) increase   (b) decrease   (c) not change

SELF-TEST

Answers can be found at www.pearsonhighered.com/floyd.

Section 6–1

1. A small-signal amplifier

(a) uses only a small portion of its load line

(b) always has an output signal in the mV range

(c) goes into saturation once on each input cycle

(d) is always a common-emitter amplifier

Section 6–2

2. The parameter h_{fe} corresponds to

(a) β_{DC} (b) β_{ac} (c) r'_e (d) r'_c

3. If the dc emitter current in a certain transistor amplifier is 3 mA, the approximate value of r'_e is

(a) $3\,k\Omega$ (b) $3\,\Omega$ (c) $8.33\,\Omega$ (d) $0.33\,k\Omega$

Section 6–3

4. A certain common-emitter amplifier has a voltage gain of 100. If the emitter bypass capacitor is removed,

(a) the circuit will become unstable   (b) the voltage gain will decrease

(c) the voltage gain will increase   (d) the Q-point will shift

5. For a common-emitter amplifier, $R_C = 1.0\,k\Omega$, $R_E = 390\,\Omega$, $r'_e = 15\,\Omega$, and $\beta_{ac} = 75$. Assuming that R_E is completely bypassed at the operating frequency, the voltage gain is

(a) 66.7   (b) 2.56   (c) 2.47   (d) 75

6. In the circuit of Question 5, if the frequency is reduced to the point where $X_{C(bypass)} = R_E$, the voltage gain

(a) remains the same   (b) is less   (c) is greater

7. In a common-emitter amplifier with voltage-divider bias, $R_{in(base)} = 68\,k\Omega$, $R_1 = 33\,k\Omega$, and $R_2 = 15\,k\Omega$. The total ac input resistance is

(a) $68\,k\Omega$ (b) $8.95\,k\Omega$ (c) $22.2\,k\Omega$ (d) $12.3\,k\Omega$

8. A CE amplifier is driving a $10\,k\Omega$ load. If $R_C = 2.2\,k\Omega$ and $r'_e = 10\,\Omega$, the voltage gain is approximately

(a) 220   (b) 1000   (c) 10   (d) 180

Section 6–4

9. For a common-collector amplifier, $R_E = 100\,\Omega$, $r'_e = 10\,\Omega$, and $\beta_{ac} = 150$. The ac input resistance at the base is

(a) $1500\,\Omega$ (b) $15\,k\Omega$ (c) $110\,\Omega$ (d) $16.5\,k\Omega$

10. If a 10 mV signal is applied to the base of the emitter-follower circuit in Question 9, the output signal is approximately

    (a) 100 mV   **(b)** 150 mV   **(c)** 1.5 V   **(d)** 10 mV

11. In a certain emitter-follower circuit, the current gain is 50. The power gain is approximately

    (a) $50A_v$ **(b)** 50   **(c)** 1   **(d)** answers (a) and (b)

12. In a Darlington pair configuration, each transistor has an ac beta of 125. If R_E is 560 Ω, the input resistance is

    (a) 560 Ω **(b)** 70 kΩ **(c)** 8.75 MΩ **(d)** 140 kΩ

Section 6–5   13. The input resistance of a common-base amplifier is

    (a) very low   **(b)** very high

    (c) the same as a CE   **(d)** the same as a CC

Section 6–6   14. Each stage of a four-stage amplifier has a voltage gain of 15. The overall voltage gain is

    (a) 60   **(b)** 15   **(c)** 50,625   **(d)** 3078

15. The overall gain found in Question 14 can be expressed in decibels as

    (a) 94.1 dB   **(b)** 47.0 dB   **(c)** 35.6 dB   **(d)** 69.8 dB

Section 6–7   16. A differential amplifier

    (a) is used in op-amps   **(b)** has one input and one output

    (c) has two outputs   **(d)** answers (a) and (c)

17. When a differential amplifier is operated single-ended,

    (a) the output is grounded

    (b) one input is grounded and a signal is applied to the other

    (c) both inputs are connected together

    (d) the output is not inverted

18. In the double-ended differential mode,

    (a) opposite polarity signals are applied to the inputs

    (b) the gain is 1

    (c) the outputs are different amplitudes

    (d) only one supply voltage is used

19. In the common mode,

    (a) both inputs are grounded

    (b) the outputs are connected together

    (c) an identical signal appears on both inputs

    (d) the output signals are in-phase

PROBLEMS

Answers to all odd-numbered problems are at the end of the book.

BASIC PROBLEMS

Section 6–1   **Amplifier Operation**

1. What is the lowest value of dc collector current to which a transistor having the characteristic curves in Figure 6–4 can be biased and still retain linear operation with a peak-to-peak base current swing of 20 μA?

2. What is the highest value of I_C under the conditions described in Problem 1?

3. Describe the end points on an ac load line.

Section 6–2   **Transistor AC Models**

4. Define all of the r parameters and all of the h parameters.

5. If the dc emitter current in a transistor is 3 mA, what is the value of r'_e?

6. If the h_{fe} of a transistor is specified as 200, determine β_{ac}.

27. Find the overall current gain A_i in Figure 6–56.

◄ **FIGURE 6–56**

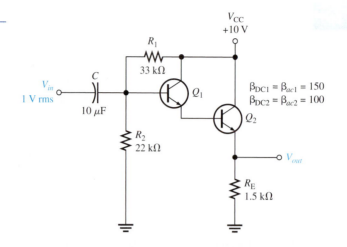

Section 6–5 **The Common-Base Amplifier**

28. What is the main disadvantage of the common-base amplifier compared to the common-emitter and the emitter-follower amplifiers?

29. Find $R_{in(emitter)}$, A_v, A_i, and A_p for the unloaded amplifier in Figure 6–57.

30. Match the following generalized characteristics with the appropriate amplifier configuration.

    (a) Unity current gain, high voltage gain, very low input resistance

    (b) High current gain, high voltage gain, low input resistance

    (c) High current gain, unity voltage gain, high input resistance

◄ **FIGURE 6–57**

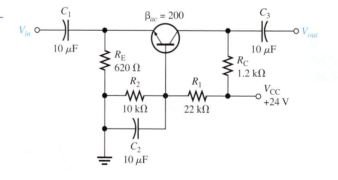

Section 6–6 **Multistage Amplifiers**

31. Each of two cascaded amplifier stages has an $A_v = 20$. What is the overall gain?

32. Each of three cascaded amplifier stages has a dB voltage gain of 10 dB. What is the overall voltage gain in dB? What is the actual overall voltage gain?

33. For the two-stage, capacitively coupled amplifier in Figure 6–58, find the following values:

    (a) voltage gain of each stage

    (b) overall voltage gain

    (c) Express the gains found in (a) and (b) in dB.

34. If the multistage amplifier in Figure 6–58 is driven by a 75 Ω, 50 μV source and the second stage is loaded with an $R_L = 18$ kΩ, determine

    (a) voltage gain of each stage

    (b) overall voltage gain

    (c) Express the gains found in (a) and (b) in dB.

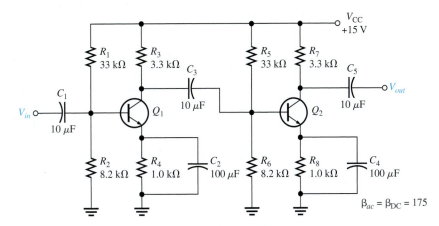

▲ FIGURE 6–58

35. Figure 6–59 shows a direct-coupled (i.e., with no coupling capacitors between stages) two-stage amplifier. The dc bias of the first stage sets the dc bias of the second. Determine all dc voltages for both stages and the overall ac voltage gain.

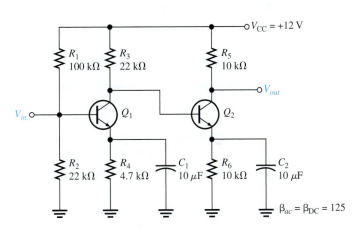

▲ FIGURE 6–59

36. Express the following voltage gains in dB:

(a) 12   (b) 50   (c) 100   (d) 2500

37. Express the following voltage gains in dB as standard voltage gains:

(a) 3 dB   (b) 6 dB   (c) 10 dB   (d) 20 dB   (e) 40 dB

Section 6–7   The Differential Amplifier

38. The dc base voltages in Figure 6–60 are zero. Using your knowledge of transistor analysis, determine the dc differential output voltage. Assume that Q_1 has an $\alpha = 0.980$ and Q_2 has an $\alpha = 0.975$.

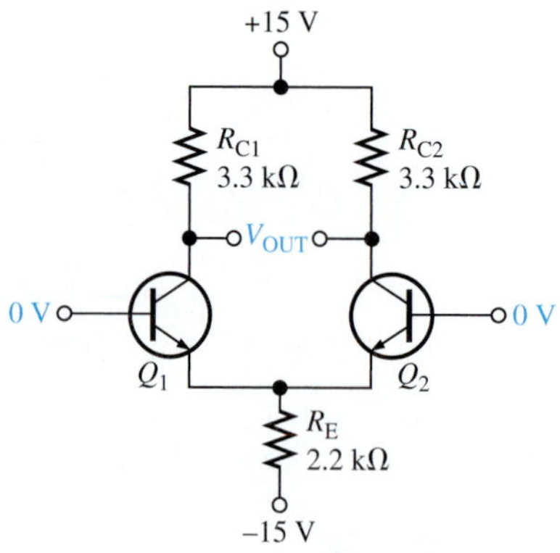

▶ FIGURE 6–60

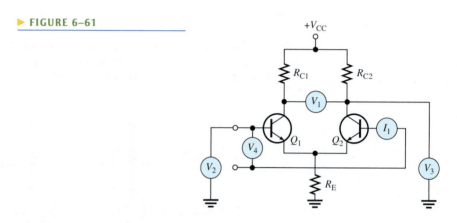

39. Identify the quantity being measured by each meter in Figure 6–61.

▶ FIGURE 6–61

40. A differential amplifier stage has collector resistors of 5.1 kΩ each. If $I_{C1} = 1.35$ mA and $I_{C2} = 1.29$ mA, what is the differential output voltage?

41. Identify the type of input and output configuration for each basic differential amplifier in Figure 6–62.

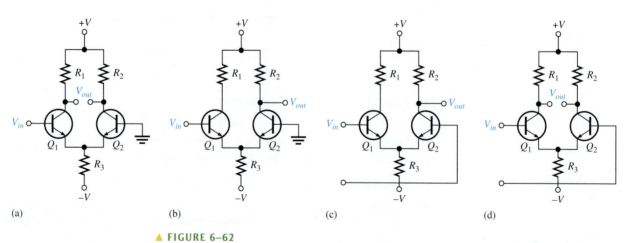

(a)   (b)   (c)   (d)

▲ FIGURE 6–62

Section 6–8   **Troubleshooting**

42. Assume that the coupling capacitor C_3 is shorted in Figure 6–34. What dc voltage will appear at the collector of Q_1?

43. Assume that R_5 opens in Figure 6–34. Will Q_2 be in cutoff or in conduction? What dc voltage will you observe at the Q_2 collector?

44. Refer to Figure 6–58 and determine the general effect of each of the following failures:

 (a) C_2 open

 (b) C_3 open

 (c) C_4 open

 (d) C_2 shorted

 (e) base-collector junction of Q_1 open

 (f) base-emitter junction of Q_2 open

45. Assume that you must troubleshoot the amplifier in Figure 6–58. Set up a table of test point values, input, output, and all transistor terminals that include both dc and rms values that you expect to observe when a 300 Ω test signal source with a 25 μV rms output is used.

APPLICATION ACTIVITY PROBLEMS

46. Refer to the public address system block diagram in Figure 6–47. You are asked to repair a system that is not working. After a preliminary check, you find that there is no output signal from the power amplifier or from the preamplifier. Based on this check and assuming that only one of the blocks is faulty, which block can you eliminate as the faulty one? What would you check next?

47. What effect would each of the following faults in the amplifier of Figure 6–63 have on the output signal?

 (a) Open C_1 (b) Open C_2 (c) Open C_3 (d) Open C_4

 (e) Q_1 collector internally open   (f) Q_2 emitter shorted to ground

▶ FIGURE 6–63

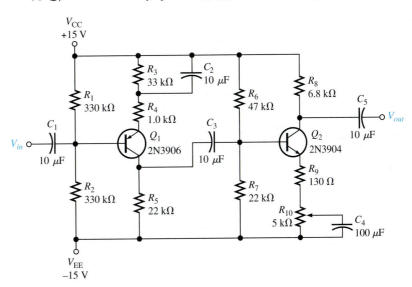

48. Suppose a 220 Ω resistor is incorrectly installed in the R_7 position of the amplifier in Figure 6–63. What effect does this have on the circuit?

49. The connection from R_1 to the supply voltage V_1 in Figure 6–63 has opened.

 (a) What happens to Q_1?

 (b) What is the dc voltage at the Q_1 collector?

 (c) What is the dc voltage at the Q_2 collector?

DATASHEET PROBLEMS

50. Refer to the 2N3946/2N3947 partial datasheet in Figure 6–64 on page 318. Determine the minimum value for each of the following r parameters:

 (a) β_{ac} (b) r'_e (c) r'_c

51. Repeat Problem 50 for maximum values.

52. Should you use a 2N3946 or a 2N3947 transistor in a certain application if the criterion is maximum current gain?

7–1 THE CLASS A POWER AMPLIFIER

When an amplifier is biased such that it always operates in the linear region where the output signal is an amplified replica of the input signal, it is a **class A** amplifier. The discussion of amplifiers in the previous chapters apply to class A operation. Power amplifiers are those amplifiers that have the objective of delivering power to a load. This means that components must be considered in terms of their ability to dissipate heat, and efficiency becomes an important consideration in the design.

After completing this section, you should be able to

❑ **Explain and analyze the operation of class A amplifiers**
❑ Discuss transistor heat dissipation
   ◆ Describe the purpose of a heat sink
❑ Discuss the importance of a centered Q-point
   ◆ Describe the relationship of the dc and ac load lines with the Q-point
   ◆ Describe the effects of a noncentered Q-point on the output waveform
❑ Determine power gain
❑ Define *dc quiescent power*
❑ Discuss and determine output signal power
❑ Define and determine the *efficiency* of a power amplifier

In a small-signal amplifier, the ac signal moves over a small percentage of the total ac load line. When the output signal is larger and approaches the limits of the ac load line, the amplifier is a **large-signal** type. Both large-signal and small-signal amplifiers are considered to be class A if they operate in the linear region at all times, as illustrated in Figure 7–1. Class A power amplifiers are large-signal amplifiers with the objective of providing power (rather than voltage) to a load. As a rule of thumb, an amplifier may be considered to be a power amplifier if it is rated for more than 1 W and it is necessary to consider the problem of heat dissipation in components.

▶ **FIGURE 7–1**

Basic class A amplifier operation. Output is shown 180° out of phase with the input (inverted).

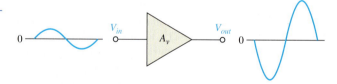

Heat Dissipation

Power transistors (and other power devices) must dissipate a large amount of internally generated heat. For BJT power transistors, the collector terminal is the critical junction; for this reason, the transistor's case is always connected to the collector terminal. The case of all power transistors is designed to provide a large contact area between it and an external heat sink. Heat from the transistor flows through the case to the heat sink and then dissipates in the surrounding air. Heat sinks vary in size, number of fins, and type of material. Their size depends on the heat dissipation requirement and the maximum ambient temperature in which the transistor is to operate. In high-power applications (a few hundred watts), a cooling fan may be necessary.

Centered Q-Point

Recall that the dc and ac load lines intersect at the Q-point. When the Q-point is at the center of the ac load line, a maximum class A signal can be obtained. You can see this concept by examining the graph of the load line for a given amplifier in Figure 7–2(a). This graph shows the ac load line with the Q-point at its center. The collector current can vary

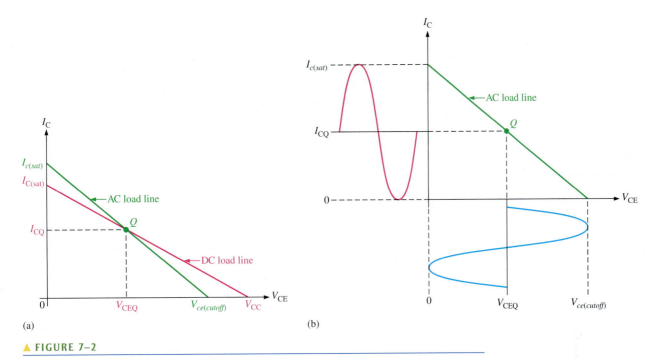

(a)   (b)

▲ **FIGURE 7–2**

Maximum class A output occurs when the Q-point is centered on the ac load line.

from its Q-point value, I_{CQ}, up to its saturation value, $I_{c(sat)}$, and down to its cutoff value of zero. Likewise, the collector-to-emitter voltage can swing from its Q-point value, V_{CEQ}, up to its cutoff value, $V_{ce(cutoff)}$, and down to its saturation value of near zero. This operation is indicated in Figure 7–2(b). The peak value of the collector current equals I_{CQ}, and the peak value of the collector-to-emitter voltage equals V_{CEQ} in this case. This signal is the maximum that can be obtained from the class A amplifier. Actually, the output cannot quite reach saturation or cutoff, so the practical maximum is slightly less.

If the Q-point is not centered on the ac load line, the output signal is limited. Figure 7–3 shows an ac load line with the Q-point moved away from center toward cutoff. The output variation is limited by cutoff in this case. The collector current can only swing down to near zero and an equal amount above I_{CQ}. The collector-to-emitter voltage can only swing up to its

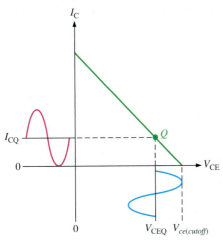

(a) Amplitude of V_{ce} and I_c limited by cutoff

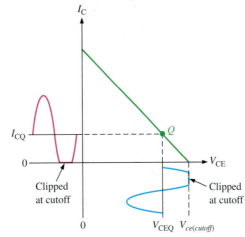

(b) Transistor driven into cutoff by a further increase in input amplitude

▲ **FIGURE 7–3**

Q-point closer to cutoff.

cutoff value and an equal amount below V_{CEQ}. This situation is illustrated in Figure 7–3(a). If the amplifier is driven any further than this, it will "clip" at cutoff, as shown in Figure 7–3(b).

Figure 7–4 shows an ac load line with the Q-point moved away from center toward saturation. In this case, the output variation is limited by saturation. The collector current can only swing up to near saturation and an equal amount below I_{CQ}. The collector-to-emitter voltage can only swing down to its saturation value and an equal amount above V_{CEQ}. This situation is illustrated in Figure 7–4(a). If the amplifier is driven any further, it will "clip" at saturation, as shown in Figure 7–4(b).

▶ FIGURE 7–4

Q-point closer to saturation.

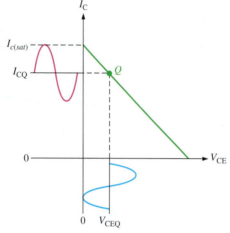 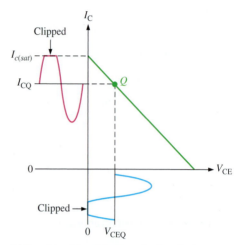

(a) Amplitude of V_{ce} and I_c limited by saturation

(b) Transistor driven into saturation by a further increase in input amplitude

Power Gain

A power amplifier delivers power to a load. The **power gain** of an amplifier is the ratio of the output power (power delivered to the load) to the input power. In general, power gain is

Equation 7–1

$$A_p = \frac{P_L}{P_{in}}$$

where A_p is the power gain, P_L is signal power delivered to the load, and P_{in} is signal power delivered to the amplifier.

The power gain can be computed by any of several formulas, depending on what is known. Frequently, the easiest way to obtain power gain is from input resistance, load resistance, and voltage gain. To see how this is done, recall that power can be expressed in terms of voltage and resistance as

$$P = \frac{V^2}{R}$$

For ac power, the voltage is expressed as rms. The output power delivered to the load is

$$P_L = \frac{V_L^2}{R_L}$$

The input power delivered to the amplifier is

$$P_{in} = \frac{V_{in}^2}{R_{in}}$$

By substituting into Equation 7–1, the following useful relationship is produced:

$$A_p = \frac{V_L^2}{V_{in}^2}\left(\frac{R_{in}}{R_L}\right)$$

Since $V_L/V_{in} = A_v$,

$$A_p = A_v^2\left(\frac{R_{in}}{R_L}\right)$$

Equation 7–2

Recall from Chapter 6 that for a voltage-divider biased amplifier,

$$R_{in(tot)} = R_1 \parallel R_2 \parallel R_{in(base)}$$

and that for a CE or CC amplifier,

$$R_{in(base)} = \beta_{ac}R_e$$

Equation 7–2 shows that the power gain of an amplifier is the voltage gain squared times the ratio of the input resistance to the output load resistance. The formula can be applied to any amplifier. For example, assume a common-collector (CC) amplifier has an input resistance of 5 kΩ and a load resistance of 100 Ω. Since a CC amplifier has a voltage gain of approximately 1, the power gain is

$$A_p = A_v^2\left(\frac{R_{in}}{R_L}\right) = 1^2\left(\frac{5\,k\Omega}{100\,\Omega}\right) = 50$$

For a CC amplifier, A_p is just the ratio of the input resistance to the output load resistance.

DC Quiescent Power

The power dissipation of a transistor with no signal input is the product of its Q-point current and voltage.

$$P_{DQ} = I_{CQ}V_{CEQ}$$

Equation 7–3

The only way a class A power amplifier can supply power to a load is to maintain a quiescent current that is at least as large as the peak current requirement for the load current. A signal will not increase the power dissipated by the transistor but actually causes less total power to be dissipated. The **dc quiescent power**, given in Equation 7–3, is the maximum power that a class A amplifier must handle. The transistor's power rating must exceed this value.

Output Power

In general, the output signal power is the product of the rms load current and the rms load voltage. The maximum unclipped ac signal occurs when the Q-point is centered on the ac load line. For a CE amplifier with a centered Q-point, the maximum peak voltage swing is

$$V_{c(max)} = I_{CQ}R_c$$

The rms value is $0.707V_{c(max)}$.
    The maximum peak current swing is

$$I_{c(max)} = \frac{V_{CEQ}}{R_c}$$

The rms value is $0.707I_{c(max)}$.
    To find the maximum signal power output, use the rms values of maximum current and voltage. The maximum power out from a class A amplifier is

$$P_{out(max)} = (0.707I_c)(0.707V_c)$$

$$P_{out(max)} = 0.5I_{CQ}V_{CEQ}$$

Equation 7–4

EXAMPLE 7–1

Determine the voltage gain and the power gain of the class A power amplifier in Figure 7–5. Assume $\beta_{ac} = 200$ for all transistors.

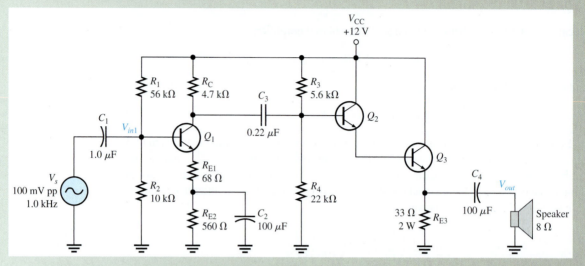

▲ **FIGURE 7–5**

Solution

Notice that the first stage (Q_1) is a voltage-divider biased common-emitter with a swamping resistor (R_{E1}). The second stage (Q_2 and Q_3) is a Darlington voltage-follower configuration. The speaker is the load.

First stage: The ac collector resistance of the first stage is R_C in parallel with the input resistance to the second stage.

Start by finding $R_{in(tot)(Q2)}$

$$R_{in(tot)(Q2)} = R_1 \| R_2 \| \beta_{ac}^2(r'_{e(Q2)} + R_e)$$

In this case, $r'_{e(Q2)}$ is small enough to ignore.

$$= 5.6 \text{ k}\Omega \| 22 \text{ k}\Omega \| 200^2(33 \ \Omega \| 8 \ \Omega)$$
$$= 4.44 \text{ k}\Omega$$
$$R_c = R_C \| R_{in(tot)(Q2)}$$
$$= 4.7 \text{ k}\Omega \| 4.44 \text{ k}\Omega$$
$$= 2.28 \text{ k}\Omega$$

The voltage gain of the first stage is the ac collector resistance, R_c, divided by the ac emitter resistance, which is the sum of $R_{E1} + r'_{e(Q1)}$. The approximate value of $r'_{e(Q1)}$ is determined by first finding I_E.

$$V_B \cong \left(\frac{R_2}{R_1 + R_2}\right)V_{CC} = \left(\frac{10 \text{ k}\Omega}{66 \text{ k}\Omega}\right)12 \text{ V} = 1.82 \text{ V}$$

$$I_E = \frac{V_B - 0.7 \text{ V}}{R_{E1} + R_{E2}} = \frac{1.82 \text{ V} - 0.7 \text{ V}}{628 \ \Omega} = 1.78 \text{ mA}$$

$$r'_{e(Q1)} = \frac{25 \text{ mV}}{I_E} = \frac{25 \text{ mV}}{1.78 \text{ mA}} = 14 \ \Omega$$

Using the value of r'_e, determine the voltage gain of the first stage with the loading of the second stage taken into account.

$$A_{v1} = -\frac{R_c}{R_{E1} + r'_{e(Q1)}} = -\frac{2.28 \text{ k}\Omega}{68 \ \Omega + 14 \ \Omega} = -27.8$$

The negative sign is for inversion.

The total input resistance of the first stage is equal to the bias resistors in parallel with the ac input resistance at the base of Q_1.

$$R_{in(tot)1} = R_1 \parallel R_2 \parallel \beta_{ac(Q1)}(R_{E1} + r'_{e(Q1)})$$
$$= 56\,\text{k}\Omega \parallel 10\,\text{k}\Omega \parallel 200(68\,\Omega + 14\,\Omega) = 8.4\,\text{k}\Omega$$

Second stage: The voltage gain of the Darlington emitter-follower is approximately equal to 1.

$$A_{v2} \cong 1$$

Overall amplifier: The overall voltage gain is the product of the first and second stage voltage gains. Since the second stage has a gain of approximately 1, the overall gain is approximately equal to the gain of the first stage.

$$A_{v(tot)} = A_{v1}A_{v2} = (-27.8)(1) = -27.8$$

Power gain: The power gain of the amplifier can be calculated using Equation 7–2 and rounded to three significant figures.

$$A_p = A_{v(tot)}^2\left(\frac{R_{in(tot)1}}{R_L}\right) = (-27.8)^2\left(\frac{8.4\,\text{k}\Omega}{8\,\Omega}\right) = \mathbf{811{,}000}$$

Related Problem* What happens to the power gain if a second 8 Ω speaker is connected in parallel with the first one?

---

* Answers can be found at www.pearsonhighered.com/floyd.

Efficiency

The **efficiency** of any amplifier is the ratio of the output signal power supplied to a load to the total power from the dc supply. The maximum output signal power that can be obtained is given by Equation 7–4. The average power supply current, I_{CC}, is equal to I_{CQ} and the supply voltage is at least $2V_{CEQ}$. Therefore, the total dc power is

$$P_{DC} = I_{CC}V_{CC} = 2I_{CQ}V_{CEQ}$$

The maximum efficiency, η_{max}, of a capacitively coupled class A amplifier is

$$\eta_{max} = \frac{P_{out}}{P_{DC}} = \frac{0.5I_{CQ}V_{CEQ}}{2I_{CQ}V_{CEQ}} = 0.25$$

The maximum efficiency of a capacitively coupled class A amplifier cannot be higher than 0.25, or 25%, and, in practice, is usually considerably less (about 10%). Although the efficiency can be made higher by transformer coupling the signal to the load, there are drawbacks to transformer coupling. These drawbacks include the size and cost of transformers as well as potential distortion problems when the transformer core begins to saturate. In general, the low efficiency of class A amplifiers limits their usefulness to small power applications that require usually less than 1 W.

EXAMPLE 7–2

Determine the efficiency of the power amplifier in Figure 7–5 (Example 7–1).

Solution The efficiency is the ratio of the signal power in the load to the power supplied by the dc source. The input voltage is 100 mV peak-to-peak which is 35.4 mV rms. The input power is, therefore,

$$P_{in} = \frac{V_{in}^2}{R_{in}} = \frac{(35.4\,\text{mV})^2}{8.4\,\text{k}\Omega} = 149\,\text{nW}$$

The output power is

$$P_{out} = P_{in}A_p = (149 \text{ nW})(811,000) = 121 \text{ mW}$$

Most of the power from the dc source is supplied to the output stage. The current in the output stage can be computed from the dc emitter voltage of Q_3.

$$V_{E(Q3)} \cong \left(\frac{22 \text{ k}\Omega}{27.6 \text{ k}\Omega}\right)12 \text{ V} - 1.4 \text{ V} = 8.2 \text{ V}$$

$$I_{E(Q3)} = \frac{V_{E(Q3)}}{R_E} = \frac{8.2 \text{ V}}{33 \text{ }\Omega} = 0.25 \text{ A}$$

Neglecting the other transistor and bias currents, which are very small, the total dc supply current is about 0.25 A. The power from the dc source is

$$P_{DC} = I_{CC}V_{CC} = (0.25 \text{ A})(12 \text{ V}) = 3 \text{ W}$$

Therefore, the efficiency of the amplifier for this input is

$$\eta = \frac{P_{out}}{P_{DC}} = \frac{121 \text{ mW}}{3 \text{ W}} \cong \mathbf{0.04}$$

This represents an efficiency of 4% and illustrates why class A is not a good choice for a power amplifier.

Related Problem Explain what happens to the efficiency if R_{E3} were replaced with the speaker. What problem does this have?

SECTION 7–1 CHECKUP
Answers can be found at www.pearsonhighered.com/floyd.

1. What is the purpose of a heat sink?
2. Which lead of a BJT is connected to the case?
3. What are the two types of clipping with a class A power amplifier?
4. What is the maximum efficiency for a class A amplifier?
5. How can the power gain of a CC amplifier be expressed in terms of a ratio of resistances?

7–2 THE CLASS B AND CLASS AB PUSH-PULL AMPLIFIERS

When an amplifier is biased at cutoff so that it operates in the linear region for 180° of the input cycle and is in cutoff for 180°, it is a **class B** amplifier. Class AB amplifiers are biased to conduct for slightly more than 180°. The primary advantage of a class B or class AB amplifier over a class A amplifier is that either one is more efficient than a class A amplifier; you can get more output power for a given amount of input power. A disadvantage of class B or class AB is that it is more difficult to implement the circuit in order to get a linear reproduction of the input waveform. The term *push-pull* refers to a common type of class B or class AB amplifier circuit in which two transistors are used on alternating half-cycles to reproduce the input waveform at the output.

After completing this section, you should be able to

❑ **Explain and analyze the operation of class B and class AB amplifiers**
❑ Describe class B operation
  ◆ Discuss Q-point location

◘ Describe class B push-pull operation
  ◆ Discuss transformer coupling ◆ Explain *complementary symmetry transistors*
  ◆ Explain crossover distortion
◘ Bias a push-pull amplifier for class AB operation
  ◆ Define *class AB* ◆ Explain class AB ac signal operation
◘ Describe a single-supply push-pull amplifier
◘ Discuss class B/AB power
  ◆ Calculate maximum output power ◆ Calculate dc input power
  ◆ Determine efficiency
◘ Determine the ac input resistance of a push-pull amplifier
◘ Discuss the Darlington class AB amplifier
  ◆ Determine ac input resistance
◘ Describe the Darlington/complementary Darlington class AB amplifier

Class B Operation

The class B operation is illustrated in Figure 7–6, where the output waveform is shown relative to the input in terms of time (t).

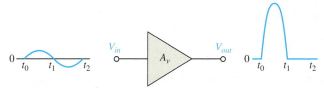

▲ **FIGURE 7–6**

Basic class B amplifier operation (noninverting).

The Q-Point Is at Cutoff The class B amplifier is biased at the cutoff point so that $I_{CQ} = 0$ and $V_{CEQ} = V_{CE(cutoff)}$. It is brought out of cutoff and operates in its linear region when the input signal drives the transistor into conduction. This is illustrated in Figure 7–7 with an emitter-follower circuit where the output is not a replica of the input.

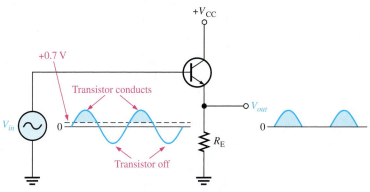

▲ **FIGURE 7–7**

Common-collector class B amplifier.

Class B Push-Pull Operation

As you can see, the circuit in Figure 7–7 only conducts for the positive half of the cycle. To amplify the entire cycle, it is necessary to add a second class B amplifier that operates on the negative half of the cycle. The combination of two class B amplifiers working together is called **push-pull** operation.

There are two common approaches for using push-pull amplifiers to reproduce the entire waveform. The first approach uses transformer coupling. The second uses two **complementary symmetry transistors;** these are a matching pair of *npn/pnp* BJTs.

Transformer Coupling Transformer coupling is illustrated in Figure 7–8. The input transformer has a center-tapped secondary that is connected to ground, producing phase inversion of one side with respect to the other. The input transformer thus converts the input signal to two out-of-phase signals for the transistors. Notice that both transistors are *npn* types. Because of the signal inversion, Q_1 will conduct on the positive part of the cycle and Q_2 will conduct on the negative part. The output transformer combines the signals by permitting current in both directions, even though one transistor is always cut off. The dc power supply voltage, VCC, is connected to the center tap of the output transformer.

◆ **FIGURE 7–8**

Transformer-coupled push-pull amplifiers. Q_1 conducts during the positive half-cycle; Q_2 conducts during the negative half-cycle. The two halves are combined by the output transformer.

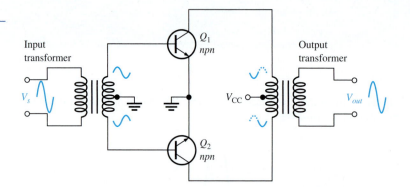

Complementary Symmetry Transistors Figure 7–9 shows one of the most popular types of push-pull class B amplifiers using two emitter-followers and both positive and negative power supplies. This is a complementary amplifier because one emitter-follower uses an *npn* transistor and the other a *pnp,* which conduct on opposite alternations of the input cycle. Notice that there is no dc base bias voltage ($V_B = 0$). Thus, only the signal voltage drives the transistors into conduction. Transistor Q_1 conducts during the positive half of the input cycle, and Q_2 conducts during the negative half.

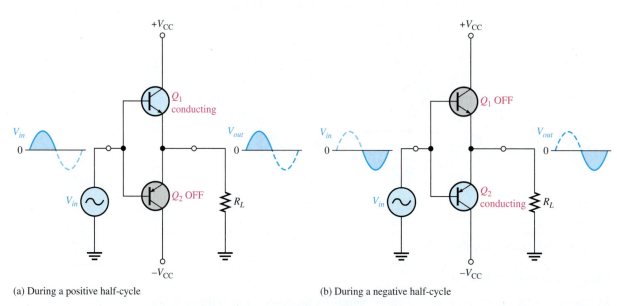

(a) During a positive half-cycle

(b) During a negative half-cycle

▲ **FIGURE 7–9**

Class B push-pull ac operation.

Crossover Distortion When the dc base voltage is zero, both transistors are off and the input signal voltage must exceed V_{BE} before a transistor conducts. Because of this, there is a time interval between the positive and negative alternations of the input when neither transistor is conducting, as shown in Figure 7–10. The resulting distortion in the output waveform is called **crossover distortion.**

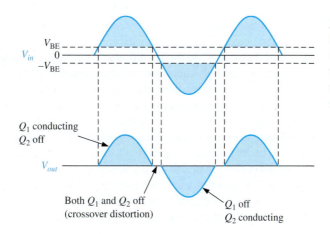

Illustration of crossover distortion in a class B push-pull amplifier. The transistors conduct only during portions of the input indicated by the shaded areas.

Biasing the Push-Pull Amplifier for Class AB Operation

To overcome crossover distortion, the biasing is adjusted to just overcome the V_{BE} of the transistors; this results in a modified form of operation called **class AB**. In class AB operation, the push-pull stages are biased into slight conduction, even when no input signal is present. This can be done with a voltage-divider and diode arrangement, as shown in Figure 7–11. When the diode characteristics of D_1 and D_2 are closely matched to the characteristics of the transistor base-emitter junctions, the current in the diodes and the current in the transistors are the same; this is called a **current mirror.** This current mirror produces the desired class AB operation and eliminates crossover distortion.

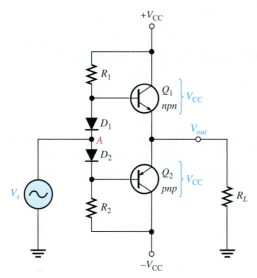

Biasing the push-pull amplifier with current-mirror diode bias to eliminate crossover distortion. The transistors form a complementary pair (one *npn* and one *pnp*).

In the bias path of the circuit in Figure 7–11, R_1 and R_2 are of equal value, as are the positive and negative supply voltages. This forces the voltage at point A (between the diodes) to equal 0 V and eliminates the need for an input coupling capacitor (provided there is no dc component to the input signal). The dc voltage on the output is also 0 V. Ideally, both diodes and both complementary transistors should be identical. In this case, the drop across D_1 equals the V_{BE} of Q_1, and the drop across D_2 equals the V_{BE} of Q_2. Since they are matched, the diode current will be the same as I_{CQ}. The diode current and I_{CQ} can be found by applying Ohm's law to either R_1 or R_2 as follows:

$$I_{CQ} = \frac{V_{CC} - 0.7 \text{ V}}{R_1}$$

This small current required of class AB operation eliminates the crossover distortion but has the potential for thermal instability if the transistor's V_{BE} drops are not matched to the diode drops or if the diodes are not in thermal equilibrium with the transistors. Heat in the power transistors decreases the base-emitter voltage and tends to increase current. If the diodes are warmed the same amount, the current is stabilized; but if the diodes are in a cooler environment, they cause I_{CQ} to increase even more. More heat is produced in an unrestrained cycle known as *thermal runaway*. To keep this from happening, the diodes should have the same thermal environment as the transistors. In some cases, a small resistor in the emitter of each transistor can alleviate thermal runaway.

Crossover distortion also occurs in transformer-coupled amplifiers like the one shown in Figure 7–8. To eliminate it in this case, 0.7 V is applied to the input transformer's secondary that just biases both transistors into conduction. The bias voltage to produce this drop can be derived from the power supply using a single diode as shown in Figure 7–12.

▶ **FIGURE 7–12**

Eliminating crossover distortion in a transformer-coupled push-pull amplifier. The biased diode compensates for the base-emitter drop of the transistors and produces class AB operation.

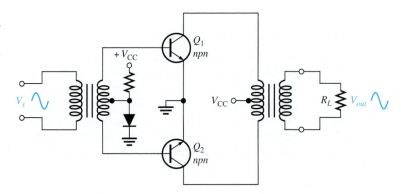

AC Operation Consider the ac load line for Q_1 of the class AB amplifier in Figure 7–11. The Q-point is slightly above cutoff. (In a true class B amplifier, the Q-point is at cutoff.) The ac cutoff voltage for a two-supply operation is at V_{CC} with an I_{CQ} as given earlier. The ac saturation current for a two-supply operation with a push-pull amplifier is

Equation 7–5

$$I_{c(sat)} = \frac{V_{CC}}{R_L}$$

The ac load line for the *npn* transistor is as shown in Figure 7–13. The dc load line can be found by drawing a line that passes through V_{CEQ} and the dc saturation current, $I_{C(sat)}$. However, the saturation current for dc is the current if the collector to emitter is shorted

▶ **FIGURE 7–13**

Load lines for a complementary symmetry push-pull amplifier. Only the load lines for the *npn* transistor are shown.

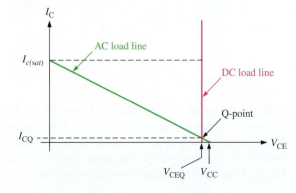

on both transistors! This assumed short across the power supplies obviously would cause maximum current from the supplies and implies the dc load line passes almost vertically through the cutoff as shown. Operation along the dc load line, such as caused by thermal runaway, could produce such a high current that the transistors are destroyed.

Figure 7–14(a) illustrates the ac load line for Q_1 of the class AB amplifier in Figure 7–14(b). In the case illustrated, a signal is applied that swings over the region of the ac load line shown in bold. At the upper end of the ac load line, the voltage across the transistor (V_{ce}) is a minimum, and the output voltage is maximum.

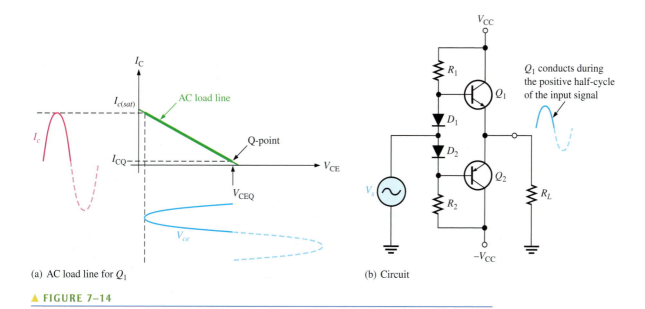

(a) AC load line for Q_1

(b) Circuit

▲ FIGURE 7–14

Under maximum conditions, transistors Q_1 and Q_2 are alternately driven from near cutoff to near saturation. During the positive alternation of the input signal, the Q_1 emitter is driven from its Q-point value of 0 to nearly V_{CC}, producing a positive peak voltage a little less than V_{CC}. Likewise, during the negative alternation of the input signal, the Q_2 emitter is driven from its Q-point value of 0 V, to near $-V_{CC}$, producing a negative peak voltage almost equal to $-V_{CC}$. Although it is possible to operate close to the saturation current, this type of operation results in clipping of the peaks of the output signal.

The ac saturation current (Equation 7–5) is also the peak output current. Each transistor can operate over nearly all of its ac load line. Recall that in class A operation, the transistor can also operate over the entire load line but with a significant difference. In class A operation, the Q-point is near the middle and there is significant current in the transistors even with no signal. In class B operation, when there is no signal, the transistors have only a very small current and therefore dissipate very little power. Thus, the efficiency of a class B amplifier can be much higher than a class A amplifier. It will be shown later that the maximum ideal efficiency of a class B amplifier is 79%. In practice, this efficiency cannot be obtained because of other losses in the circuit.

EXAMPLE 7–3	Determine the ideal maximum peak output voltage and current for the circuit shown in Figure 7–15.
Solution	The ideal maximum peak output voltage is

$$V_{out(peak)} \cong V_{CEQ} \cong V_{CC} = \textbf{20 V}$$

and

$$V_{out(rms)} = 0.707V_{out(peak)} = 0.707V_{CEQ}$$

then

$$P_{out} = 0.5I_{c(sat)}V_{CEQ}$$

Substituting $V_{CC}/2$ for V_{CEQ}, the maximum average output power is

Equation 7–6

$$P_{out} = 0.25I_{c(sat)}V_{CC}$$

DC Input Power   The dc input power comes from the V_{CC} supply and is

$$P_{DC} = I_{CC}V_{CC}$$

Since each transistor draws current for a half-cycle, the current is a half-wave signal with an average value of

$$I_{CC} = \frac{I_{c(sat)}}{\pi}$$

So,

$$P_{DC} = \frac{I_{c(sat)}V_{CC}}{\pi}$$

Efficiency   An advantage of push-pull class B and class AB amplifiers over class A is a much higher efficiency. This advantage usually overrides the difficulty of biasing the class AB push-pull amplifier to eliminate crossover distortion. Recall that efficiency, η is defined as the ratio of ac output power to dc input power.

$$\eta = \frac{P_{out}}{P_{DC}}$$

The maximum efficiency, η_{max}, for a class B amplifier (class AB is slightly less) is developed as follows, starting with Equation 7–6.

$$P_{out} = 0.25I_{c(sat)}V_{CC}$$

$$\eta_{max} = \frac{P_{out}}{P_{DC}} = \frac{0.25I_{c(sat)}V_{CC}}{I_{c(sat)}V_{CC}/\pi} = 0.25\pi$$

Equation 7–7

$$\eta_{max} = 0.79$$

or, as a percentage,

$$\eta_{max} = 79\%$$

Recall that the maximum efficiency for class A is 0.25 (25%).

EXAMPLE 7–5

Find the ideal maximum ac output power and the dc input power of the amplifier in Figure 7–18.

Solution   The ideal maximum peak output voltage is

$$V_{out(peak)} \cong V_{CEQ} = \frac{V_{CC}}{2} = \frac{20\ V}{2} = 10\ V$$

The ideal maximum peak output current is

$$I_{out(peak)} \cong I_{c(sat)} = \frac{V_{CEQ}}{R_L} = \frac{10\ V}{8\ \Omega} = 1.25\ A$$

▶ FIGURE 7–18

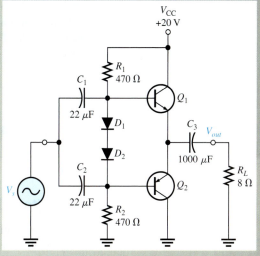

The ac output power and the dc input power are

$$P_{out} = 0.25I_{c(sat)}V_{CC} = 0.25(1.25 \text{ A})(20 \text{ V}) = \textbf{6.25 W}$$

$$P_{DC} = P_{R1} + P_{R2} + P_{Q1Q2}$$

$$= \frac{V_{R1}^2}{R_1} + \frac{V_{R2}^2}{R_2} + \frac{I_{c(sat)}V_{CC}}{\pi}$$

$$= \frac{(9.3 \text{ V})^2}{470 \text{ }\Omega} + \frac{(9.3 \text{ V})^2}{470 \text{ }\Omega} + \frac{(1.25 \text{ A})(20 \text{ V})}{\pi} = \textbf{8.33 W}$$

Related Problem   Determine the ideal maximum ac output power and the dc input power in Figure 7–18 for $V_{CC} = 15$ V and $R_L = 16$ Ω.

Input Resistance

The complementary push-pull configuration used in class B/class AB amplifiers is, in effect, two emitter-followers. The input resistance for the emitter-follower, where R_1 and R_2 are the bias resistors, is

$$R_{in} = \beta_{ac}(r_e' + R_E) \parallel R_1 \parallel R_2$$

Since $R_E = R_L$, the formula is

$$R_{in} = \beta_{ac}(r_e' + R_L) \parallel R_1 \parallel R_2$$ Equation 7–8

EXAMPLE 7–6

Assume that a preamplifier stage with an output signal voltage of 3 V rms and an output resistance of 50 Ω is driving the push-pull power amplifier in Figure 7–18 (Example 7–5). Q_1 and Q_2 in the power amplifier have a β_{ac} of 100 and an r_e' of 1.6 Ω. Determine the loading effect that the power amplifier has on the preamp stage.

Solution   Looking from the input signal source, the bias resistors appear in parallel because both go to ac ground and the ac resistance of the forward-biased diodes is very small and can be ignored. The input resistance at the emitter of either transistor is $\beta_{ac}(r_e' + R_L)$. So, the signal source sees R_1, R_2, and $\beta_{ac}(r_e' + R_L)$ all in parallel.

The ac input resistance of the power amplifier is

$$R_{in} = \beta_{ac}(r_e' + R_L) \parallel R_1 \parallel R_2 = 100(9.6 \text{ }\Omega) \parallel 470 \text{ }\Omega \parallel 470 \text{ }\Omega = 188 \text{ }\Omega$$

Tank circuit oscillations. V_r is the voltage across the tank circuit.

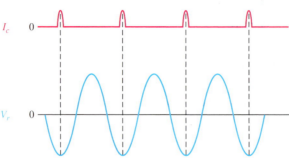

(a) An oscillation will gradually die out (decay) due to energy loss. The rate of decay depends on the efficiency of the tank circuit.

(b) Oscillation at the fundamental frequency can be sustained by short pulses of collector current.

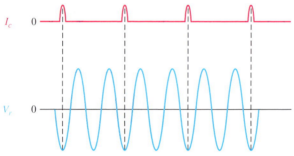

(c) Oscillation at the second harmonic frequency

Maximum Output Power

Since the voltage developed across the tank circuit has a peak-to-peak value of approximately $2V_{CC}$, the maximum output power can be expressed as

$$P_{out} = \frac{V_{rms}^2}{R_c} = \frac{(0.707V_{CC})^2}{R_c}$$

Equation 7–9

$$P_{out} = \frac{0.5V_{CC}^2}{R_c}$$

R_c is the equivalent parallel resistance of the collector tank circuit at resonance and represents the parallel combination of the coil resistance and the load resistance. It usually has a low value. The total power that must be supplied to the amplifier is

$$P_T = P_{out} + P_{D(avg)}$$

Therefore, the efficiency is

Equation 7–10

$$\eta = \frac{P_{out}}{P_{out} + P_{D(avg)}}$$

When $P_{out} \gg P_{D(avg)}$, the class C efficiency closely approaches 1 (100%).

EXAMPLE 7–8

Suppose the class C amplifier described in Example 7–7 has a V_{CC} equal to 24 V and the R_c is 100 Ω. Determine the efficiency.

Solution   From Example 7–7, $P_{D(avg)} = 4$ mW.

$$P_{out} = \frac{0.5V_{CC}^2}{R_c} = \frac{0.5(24 \text{ V})^2}{100 \text{ Ω}} = 2.88 \text{ W}$$

Therefore,

$$\eta = \frac{P_{out}}{P_{out} + P_{D(avg)}} = \frac{2.88 \text{ W}}{2.88 \text{ W} + 4 \text{ mW}} = \mathbf{0.999}$$

or, as a percentage, 99.9%.

Related Problem   What happens to the efficiency of the amplifier if R_c is increased?

Clamper Bias for a Class C Amplifier

Figure 7–26 shows a class C amplifier with a base bias clamping circuit. The base-emitter junction functions as a diode.

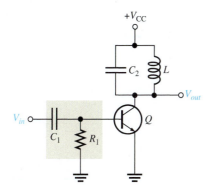

◀ **FIGURE 7–26**

Tuned class C amplifier with clamper bias.

When the input signal goes positive, capacitor C_1 is charged to the peak value with the polarity shown in Figure 7–27(a). This action produces an average voltage at the base of approximately $-V_p$. This places the transistor in cutoff except at the positive peaks, when the transistor conducts for a short interval. For good clamping action, the R_1C_1 time constant of the clamping circuit must be much greater than the period of the input signal. Parts (b) through (f) of Figure 7–27 illustrate the bias clamping action in more detail. During the time up to the positive peak of the input (t_0 to t_1), the capacitor charges to $V_p - 0.7$ V through the base-emitter diode, as shown in part (b). During the time from t_1 to t_2, as shown in part (c), the capacitor discharges very little because of the large RC time constant. The capacitor, therefore, maintains an average charge slightly less than $V_p - 0.7$ V.

Since the dc value of the input signal is zero (positive side of C_1), the dc voltage at the base (negative side of C_1) is slightly more positive than $-(V_p - 0.7$ V$)$, as indicated in Figure 7–27(d). As shown in Figure 7–27(e), the capacitor couples the ac input signal through to the base so that the voltage at the transistor's base is the ac signal riding on a dc level slightly more positive than $-(V_p - 0.7$ V$)$. Near the positive peaks of the input voltage, the base voltage goes slightly above 0.7 V and causes the transistor to conduct for a short time, as shown in Figure 7–27(f).

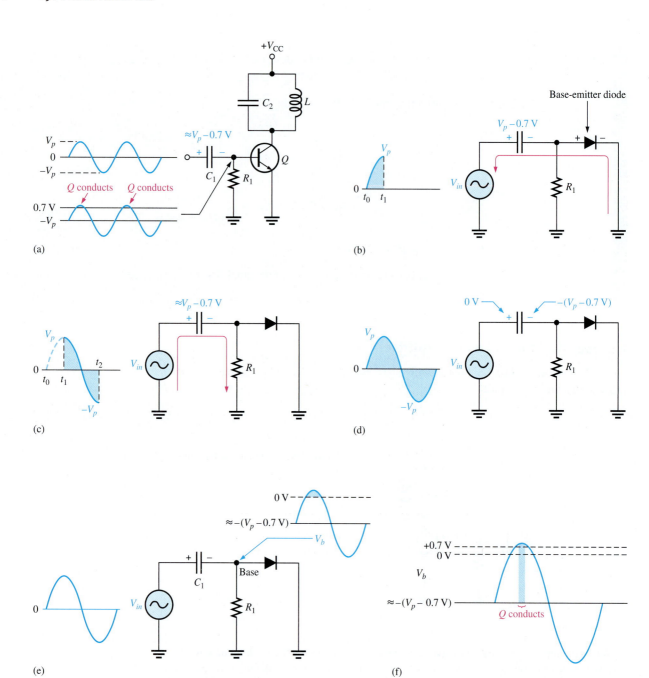

▲ FIGURE 7–27

Clamper bias action.

EXAMPLE 7–9

Determine the voltage at the base of the transistor, the resonant frequency, and the peak-to-peak value of the output signal voltage for the class C amplifier in Figure 7–28.

Solution
$$V_{s(p)} = (1.414)(1 \text{ V}) \cong 1.4 \text{ V}$$
The base is clamped at

$$-(V_{s(p)} - 0.7) = \textbf{−0.7 V dc}$$

The signal at the base has a positive peak of $+0.7$ V and a negative peak of

$$-V_{s(p)} + (-0.7 \text{ V}) = -1.4 \text{ V} - 0.7 \text{ V} = \textbf{−2.1 V}$$

▶ FIGURE 7–28

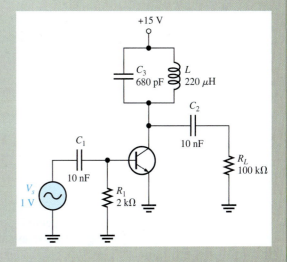

The resonant frequency is

$$f_r = \frac{1}{2\pi\sqrt{LC}} = \frac{1}{2\pi\sqrt{(220\ \mu H)(680\ pF)}} = \textbf{411 kHz}$$

The output signal has a peak-to-peak value of

$$V_{pp} = 2V_{CC} = 2(15\ V) = \textbf{30 V}$$

Related Problem   How could you make the circuit in Figure 7–28 a frequency doubler?

**SECTION 7–3
CHECKUP**

1. At what point is a class C amplifier normally biased?
2. What is the purpose of the tuned circuit in a class C amplifier?
3. A certain class C amplifier has a power dissipation of 100 mW and an output power of 1 W. What is its percent efficiency?

7–4   TROUBLESHOOTING

In this section, examples of isolating a component failure in a circuit are presented. We will use a class A amplifier and a class AB amplifier with the output voltage monitored by an oscilloscope. Several incorrect output waveforms will be examined and the most likely faults will be discussed.

After completing this section, you should be able to

❑ **Troubleshoot power amplifiers**
❑ Troubleshoot a class A amplifier for various faults
❑ Troubleshoot a class AB amplifier for various faults

Case 1: Class A

As shown in Figure 7–29, the class A power amplifier should have a normal sinusoidal output when a sinusoidal input signal is applied.

► FIGURE 7–29

Class A power amplifier with correct output voltage swing.

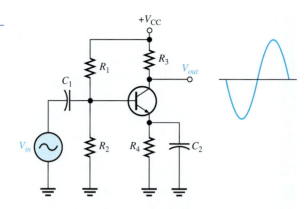

Now let's consider four incorrect output waveforms and the most likely causes in each case. In Figure 7–30(a), the scope displays a dc level equal to the dc supply voltage, indicating that the transistor is in cutoff. The two most likely causes of this condition are (1) the transistor has an open *pn* junction, or (2) R_4 is open, preventing collector and emitter current.

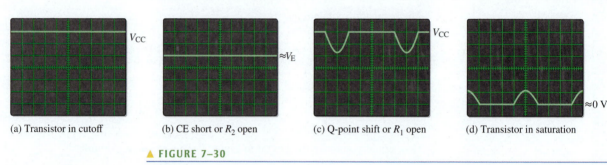

(a) Transistor in cutoff (b) CE short or R_2 open (c) Q-point shift or R_1 open (d) Transistor in saturation

▲ FIGURE 7–30

Oscilloscope displays showing output voltage for the amplifier in Figure 7–29 for several types of failures.

In Figure 7–30(b), the scope displays a dc level at the collector approximately equal to the dc emitter voltage. The two probable causes of this indication are (1) the transistor is shorted from collector to emitter, or (2) R_2 is open, causing the transistor to be biased in saturation. In the second case, a sufficiently large input signal can bring the transistor out of saturation on its negative peaks, resulting in short pulses on the output.

In Figure 7–30(c), the scope displays an output waveform that indicates the transistor is in cutoff except during a small portion of the input cycle. Possible causes of this indication are (1) the Q-point has shifted down due to a drastic out-of-tolerance change in a resistor value, or (2) R_1 is open, biasing the transistor in cutoff. The display shows that the input signal is sufficient to bring it out of cutoff for a small portion of the cycle.

In Figure 7–30(d), the scope displays an output waveform that indicates the transistor is saturated except during a small portion of the input cycle. Again, it is possible that an incorrect resistance value has caused a drastic shift in the Q-point up toward saturation, or R_2 is open, causing the transistor to be biased in saturation, and the input signal is bringing it out of saturation for a small portion of the cycle.

Case 2: Class AB

As shown in Figure 7–31, the class AB push-pull amplifier should have a sinusoidal output when a sinusoidal input signal is applied.

Two incorrect output waveforms are shown in Figure 7–32. The waveform in part (a) shows that only the positive half of the input signal is present on the output. One possible cause is that diode D_1 is open. If this is the fault, the positive half of the input signal forward-biases D_2 and causes transistor Q_2 to conduct. Another possible cause is that the base-emitter junction of Q_2 is open, so only the positive half of the input signal appears on the output because Q_1 is still working. Another possibility is an open supply voltage to $-V_{CC}$. In this case, only Q_1 is conducting normally.

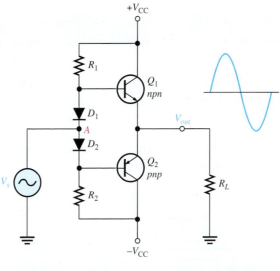

◀ FIGURE 7–31

A class AB push-pull amplifier with correct output voltage.

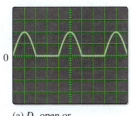
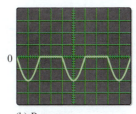

(a) D_1 open or
Q_2 base-emitter open

(b) D_2 open or
Q_1 base-emitter open

◀ FIGURE 7–32

Incorrect output waveforms for the amplifier in Figure 7–31. At low signal levels, you may see a shift in the 0 V level when a diode is open.

The waveform in Figure 7–32(b) shows that only the negative half of the input signal is present on the output. One possible cause is that diode D_2 is open. If this is the fault, the negative half of the input signal forward-biases D_1 and places the half-wave signal on the base of Q_1. Another possible cause is that the base-emitter junction of Q_1 is open, so only the negative half of the input signal appears on the output because Q_2 is still working. Another possibility is an open supply voltage to $+V_{CC}$. In this case, only Q_2 is conducting normally.

There are, of course, other faults that can occur in a push-pull amplifier, such as power line "hum" on the output. In analyzing the clues, a troubleshooter can zero in on the problem; in this case a likely cause is noise on one or more power supply lines. It could also be from an unshielded input line or from poor grounding. A good troubleshooter analyzes the clues and uses this analysis to plan where to logically test the circuit.

Multisim Troubleshooting Exercises

These file circuits are in the Troubleshooting Exercises folder on the website. Open each file and determine if the circuit is working properly. If it is not working properly, determine the fault.

1. Multisim file TSM07-01

2. Multisim file TSM07-02

3. Multisim file TSM07-03

4. Multisim file TSM07-04

SECTION 7–4 CHECKUP	1. What would you check for if you noticed clipping at both peaks of the output waveform?
	2. A significant loss of gain in the amplifier of Figure 7–29 would most likely be caused by what type of failure?
	3. What are possible problems if you notice a 60 cycle hum on the output of an audio amplifier?

Device Application: *The Complete PA System*

The class AB power amplifier follows the audio preamp and drives the speaker as shown in the PA system block diagram in Figure 7–33. In this application, the power amplifier is developed and interfaced with the preamp that was developed in Chapter 6. The maximum signal power to the speaker should be approximately 6 W for a frequency range of 70 Hz to 5 kHz. The dynamic range for the input voltage is up to 40 mV. Finally, the complete PA system is put together.

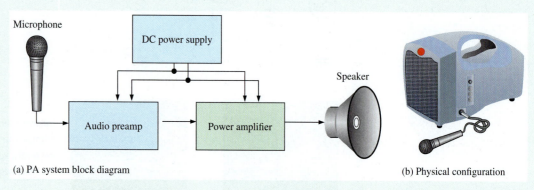

(a) PA system block diagram

(b) Physical configuration

▲ **FIGURE 7–33**

The Power Amplifier Circuit

The schematic of the push-pull power amplifier is shown in Figure 7–34. The circuit is a class AB amplifier implemented with Darlington configurations and diode current mirror bias. Both a traditional Darlington pair and a complementary Darlington (Sziklai) pair are used to provide sufficient current to an 8 Ω speaker load. The signal from the preamp is capacitively coupled to the driver stage, Q_5, which is used to prevent excessive loading

▶ **FIGURE 7–34**

Class AB power push-pull amplifier.

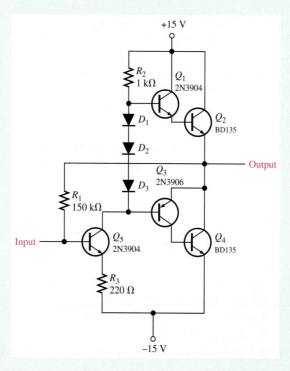

on the preamp and provide additional gain. Notice that Q_5 is biased with the dc output voltage (0 V) fed back through R_1. Also, the signal voltage fed back to the base of Q_5 is out-of-phase with the signal from the preamp and has the effect of stabilizing the gain. This is called *negative feedback*. The amplifier will deliver up to 5 W to an 8 Ω speaker.

A partial datasheet for the BD135 power transistor is shown in Figure 7–35.

1. Estimate the input resistance of the power amplifier in Figure 7–34.
2. Calculate the approximate voltage gain of the power amplifier in Figure 7–34.

► FIGURE 7–35

Partial datasheet for the BD135 power transistors. Copyright Fairchild semiconductor corporation. Used by permission.

FAIRCHILD
SEMICONDUCTOR ™

BD135/137/139

Medium Power Linear and Switching Applications
• Complement to BD136, BD138 and BD140 respectively

TO-126
1. Emitter 2.Collector 3.Base

NPN Epitaxial Silicon Transistor

Absolute Maximum Ratings T_C = 25°C unless otherwise noted

Symbol	Parameter		Value	Units
V_{CBO}	Collector-Base Voltage	: BD135	45	V
		: BD137	60	V
		: BD139	80	V
V_{CEO}	Collector-Emitter Voltage	: BD135	45	V
		: BD137	60	V
		: BD139	80	V
V_{EBO}	Emitter-Base Voltage		5	V
I_C	Collector Current (DC)		1.5	A
I_{CP}	Collector Current (Pulse)		3.0	A
I_B	Base Current		0.5	A
P_C	Collector Dissipation (T_C = 25°C)		12.5	W
P_C	Collector Dissipation (T_a = 25°C)		1.25	W
T_J	Junction Temperature		150	°C
T_{STG}	Storage Temperature		- 55 ~ 150	°C

Electrical Characteristics T_C = 25°C unless otherwise noted

Symbol	Parameter		Test Condition	Min.	Typ.	Max.	Units
V_{CEO}(sus)	Collector-Emitter Sustaining Voltage						
		: BD135	I_C = 30mA, I_B = 0	45			V
		: BD137		60			V
		: BD139		80			V
I_{CBO}	Collector Cut-off Current		V_{CB} = 30V, I_E = 0			0.1	μA
I_{EBO}	Emitter Cut-off Current		V_{EB} = 5V, I_C = 0			10	μA
h_{FE1}	DC Current Gain : ALL DEVICE		V_{CE} = 2V, I_C = 5mA	25			
h_{FE2}	: ALL DEVICE		V_{CE} = 2V, I_C = 0.5A	25			
h_{FE3}	: BD135		V_{CE} = 2V, I_C = 150mA	40		250	
	: BD137, BD139			40		160	
V_{CE}(sat)	Collector-Emitter Saturation Voltage		I_C = 500mA, I_B = 50mA			0.5	V
V_{BE}(on)	Base-Emitter ON Voltage		V_{CE} = 2V, I_C = 0.5A			1	V

h_{FE} Classification

Classification	6	10	16
h_{FE3}	40 ~ 100	63 ~ 160	100 ~ 250

Simulation

The power amplifier is simulated using Multisim with a 1 kHz input signal at near its maximum linear operation. The results are shown in Figure 7–36 where an 8.2 Ω resistor is used to closely approximate the 8 Ω speaker.

3. Calculate the power to the load in Figure 7–36.
4. What is the measured voltage gain? The input is a peak value.
5. Compare the measured gain to the calculated gain for the amplifier in Figure 7–34.

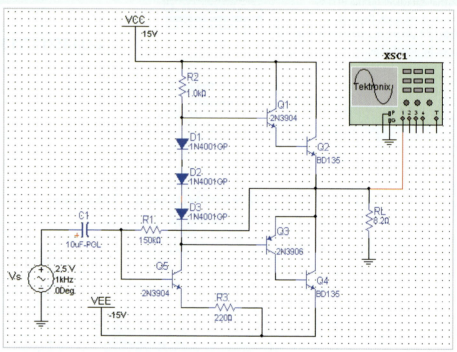

(a) Circuit screen

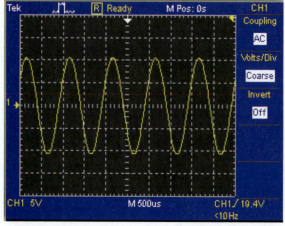

(b) Output signal

▲ **FIGURE 7–36**

Simulation of the power amplifier.

The Complete Audio Amplifier

Both the preamp and the power amp have been simulated individually. Now, they must work together to produce the required signal power to the speaker. Figure 7–37 is the simulation of the combined audio preamp and power amp. Components in the power amplifier are now numbered sequentially with the preamp components.

6. Calculate the power to the load in Figure 7–37.
7. What is the measured voltage gain of the power amplifier?
8. What is the measured overall voltage gain?

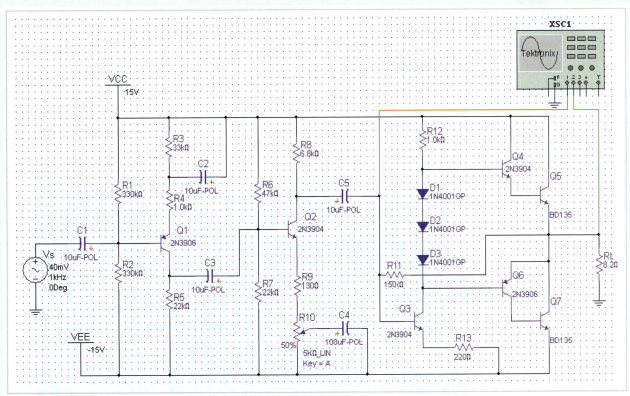

(a) Circuit screen

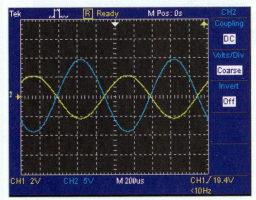

(b) Preamp output and final output

▲ FIGURE 7–37

Simulation of the complete audio amplifier.

Simulate the audio amplifier using your Multisim or LT Spice software. Observe the operation with the virtual oscilloscope.

Prototyping and Testing

Now that the circuit has been simulated, the prototype circuit is constructed and tested. After the circuit is successfully tested on a protoboard, it is ready to be finalized on a printed circuit board.

Lab Experiment

To build and test a similar circuit, go to Experiment 7 in your lab manual (*Laboratory Exercises for Electronic Devices* by David Buchla and Steven Wetterling).

Circuit Board

The power amplifier is implemented on a printed circuit board as shown in Figure 7–38. Heat sinks are used to provide additional heat dissipation from the power transistors.

9. Check the printed circuit board and verify that it agrees with the schematic in Figure 7–34. The volume control potentiometer is mounted off the PC board for easy access.

10. Label each input and output pin according to function. Locate the single backside trace.

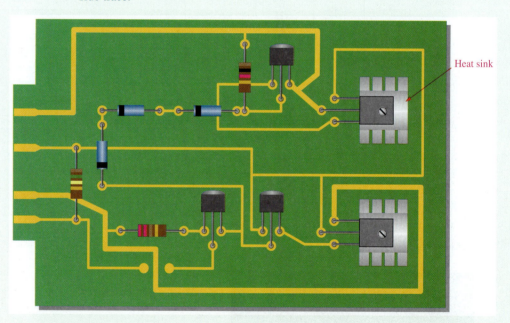

Heat sink

▲ FIGURE 7–38

Power amplifier circuit board.

Troubleshooting the Power Amplifier Board

A power amplifier circuit board has failed the production test. Test results are shown in Figure 7–39.

11. Based on the scope displays, list possible faults for the circuit board.

Putting the System Together

The preamp circuit board and the power amplifier circuit board are interconnected and the dc power supply (battery pack), microphone, speaker, and volume control potentiometer are attached, as shown in Figure 7–40.

12. Verify that the system interconnections are correct.

Test of faulty power amplifier board.

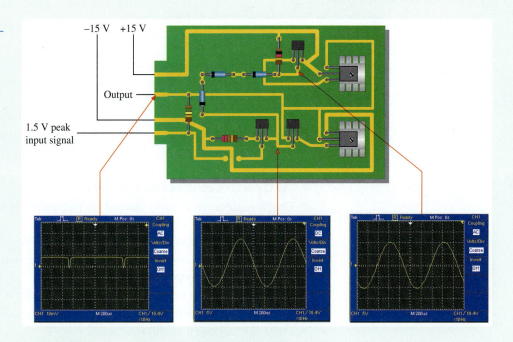

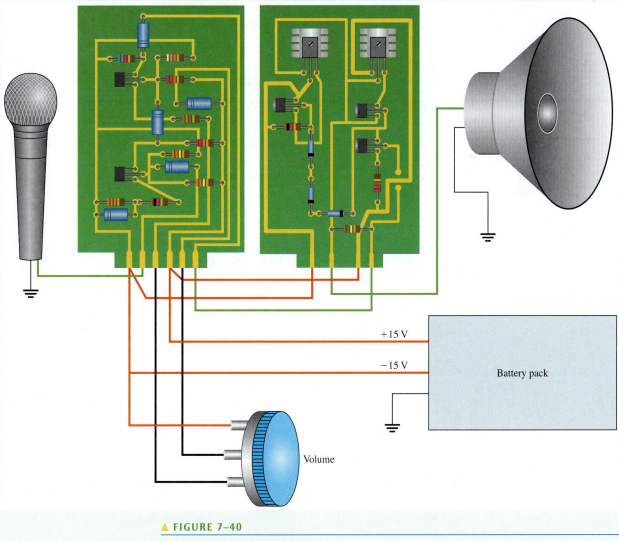

▲ FIGURE 7–40

The complete public address system.

SUMMARY

Section 7–1
- ◆ A class A power amplifier operates entirely in the linear region of the load line. The transistor conducts during the full 360° of the input cycle.
- ◆ The Q-point must be centered on the load line for maximum class A output signal swing.
- ◆ The maximum ideal efficiency of a class A power amplifier is 25%.

Section 7–2
- ◆ A class B amplifier operates in the linear region of the load line for half of the input cycle (180°), and it is in cutoff for the other half.
- ◆ The Q-point is at cutoff for class B operation.
- ◆ Class B amplifiers are normally operated in a push-pull configuration in order to produce an output that is a replica of the input.
- ◆ The maximum ideal efficiency of a class B amplifier is 79%.
- ◆ A class AB amplifier is biased slightly above cutoff and operates in the linear region for slightly more than 180° of the input cycle.
- ◆ Class AB eliminates crossover distortion found in pure class B.

Section 7–3
- ◆ A class C amplifier operates in the linear region for only a small part of the input cycle.
- ◆ The class C amplifier is biased below cutoff.
- ◆ Class C amplifiers are normally operated as tuned amplifiers to produce a sinusoidal output.
- ◆ The maximum efficiency of a class C amplifier is higher than that of either class A or class B amplifiers. Under conditions of low power dissipation and high output power, the efficiency can approach 100%.

KEY TERMS

Key terms and other bold terms in the chapter are defined in the end-of-book glossary.

Class A A type of amplifier that operates entirely in its linear (active) region.

Class AB A type of amplifier that is biased into slight conduction.

Class B A type of amplifier that operates in the linear region for 180° of the input cycle because it is biased at cutoff.

Class C A type of amplifier that operates only for a small portion of the input cycle.

Efficiency The ratio of the signal power delivered to a load to the power from the power supply of an amplifier.

Power gain The ratio of output power to input power of an amplifier.

Push-Pull A type of class B amplifier with two transistors in which one transistor conducts for one half-cycle and the other conducts for the other half-cycle.

KEY FORMULAS

The Class A Power Amplifier

7–1 $\quad A_p = \dfrac{P_L}{P_{in}}$ $\qquad$ Power gain

7–2 $\quad A_p = A_v^2\left(\dfrac{R_{in}}{R_L}\right)$ $\qquad$ Power gain in terms of voltage gain

7–3 $\quad P_{DQ} = I_{CQ}V_{CEQ}$ $\qquad$ DC quiescent power

7–4 $\quad P_{out(max)} = 0.5I_{CQ}V_{CEQ}$ $\qquad$ Maximum output power

The Class B/AB Push-Pull Amplifiers

7–5 $\quad I_{c(sat)} = \dfrac{V_{CC}}{R_L}$ $\qquad$ AC saturation current

7–6 $\quad P_{out} = 0.25I_{c(sat)}V_{CC}$ $\qquad$ Maximum average output power

7–7 $\eta_{max} = 0.79$ Maximum efficiency

7–8 $R_{in} = \beta_{ac}(r'_e + R_L) \| R_1 \| R_2$ Input resistance

The Class C Amplifier

7–9 $P_{out} = \dfrac{0.5V_{CC}^2}{R_c}$ Output power

7–10 $\eta = \dfrac{P_{out}}{P_{out} + P_{D(avg)}}$ Efficiency

TRUE/FALSE QUIZ   Answers can be found at www.pearsonhighered.com/floyd.

1. Class A power amplifiers are a type of large-signal amplifier.
2. Ideally, the Q-point should be centered on the load line in a class A amplifier.
3. The quiescent power dissipation occurs when the maximum signal is applied.
4. Efficiency is the ratio of output signal power to total power.
5. Each transistor in a class B amplifier conducts for the entire input cycle.
6. Class AB operation overcomes the problem of crossover distortion.
7. Complementary symmetry transistors must be used in a class AB amplifier.
8. A current mirror is implemented with a laser diode.
9. Darlington transistors can be used to increase the input resistance of a class AB amplifier.
10. The transistor in a class C amplifier conducts for a small portion of the input cycle.
11. The output of a class C amplifier is a replica of the input signal.
12. A class C amplifier usually employs a tuned circuit.

CIRCUIT-ACTION QUIZ   Answers can be found at www.pearsonhighered.com/floyd.

1. If the value of R_3 in Figure 7–5 is decreased, the voltage gain of the first stage will
   (a) increase   (b) decrease   (c) not change
2. If the value of R_{E2} in Figure 7–5 is increased, the voltage gain of the first stage will
   (a) increase   (b) decrease   (c) not change
3. If C_2 in Figure 7–5 opens, the dc voltage at the emitter of Q_1 will
   (a) increase   (b) decrease   (c) not change
4. If the value of R_4 in Figure 7–5 is increased, the dc voltage at the base of Q_3 will
   (a) increase   (b) decrease   (c) not change
5. If V_{CC} in Figure 7–18 is increased, the peak output voltage will
   (a) increase   (b) decrease   (c) not change
6. If the value of R_L in Figure 7–18 is increased, the ac output power will
   (a) increase   (b) decrease   (c) not change
7. If the value of R_L in Figure 7–19 is decreased, the voltage gain will
   (a) increase   (b) decrease   (c) not change
8. If the value of V_{CC} in Figure 7–19 is increased, the ac output power will
   (a) increase   (b) decrease   (c) not change
9. If the values of R_1 and R_2 in Figure 7–19 are increased, the voltage gain will
   (a) increase   (b) decrease   (c) not change
10. If the value of C_2 in Figure 7–23 is decreased, the resonant frequency will
    (a) increase   (b) decrease   (c) not change

SELF-TEST

Answers can be found at www.pearsonhighered.com/floyd.

Section 7–1   **1.** An amplifier that operates in the linear region at all times is

(a) Class A   (b) Class AB   (c) Class B   (d) Class C

2. A certain class A power amplifier delivers 5 W to a load with an input signal power of 100 mW. The power gain is

(a) 100   (b) 50   (c) 250   (d) 5

3. The peak current a class A power amplifier can deliver to a load depends on the

(a) maximum rating of the power supply   (b) quiescent current

(c) current in the bias resistors   (d) size of the heat sink

4. For maximum output, a class A power amplifier must maintain a value of quiescent current that is

(a) one-half the peak load current   (b) twice the peak load current

(c) at least as large as the peak load current   (d) just above the cutoff value

5. A certain class A power amplifier has $V_{CEQ} = 12$ V and $I_{CQ} = 1$ A. The maximum signal power output is

(a) 6 W   (b) 12 W   (c) 1 W   (b) 0.707 W

6. The efficiency of a power amplifier is the ratio of the power delivered to the load to the

(a) input signal power   (b) power dissipated in the last stage

(c) power from the dc power supply   (d) none of these answers

7. The maximum efficiency of a class A power amplifier is

(a) 25%   (b) 50%   (c) 79%   (d) 98%

Section 7–2   **8.** The transistors in a class B amplifier are biased

(a) into cutoff   (b) in saturation

(c) at midpoint of the load line   (d) right at cutoff

9. Crossover distortion is a problem for

(a) class A amplifiers   (b) class AB amplifiers

(c) class B amplifiers   (d) all of these amplifiers

10. A BJT class B push-pull amplifier with no transformer coupling uses

(a) two *npn* transistors   (b) two *pnp* transistors

(c) complementary symmetry transitors   (d) none of these

11. A current mirror in a push-pull amplifier should give an I_{CQ} that is

(a) equal to the current in the bias resistors and diodes

(b) twice the current in the bias resistors and diodes

(c) half the current in the bias resistors and diodes

(d) zero

12. The maximum efficiency of a class B push-pull amplifier is

(a) 25%   (b) 50%   (c) 79%   (d) 98%

13. The output of a certain two-supply class B push-pull amplifier has a V_{CC} of 20 V. If the load resistance is 50 Ω, the value of $I_{c(sat)}$ is

(a) 5 mA   (b) 0.4 A   (c) 4 mA   (d) 40 mA

14. The maximum efficiency of a class AB amplifier is

(a) higher than a class B   (b) the same as a class B

(c) about the same as a class A   (d) slightly less than a class B

Section 7–3   **15.** The power dissipation of a class C amplifier is normally

(a) very low   (b) very high

(c) the same as a class B   (d) the same as a class A

16. The efficiency of a class C amplifier is

(a) less than class A   (b) less than class B

(c) less than class AB   (d) greater than classes A, B, or AB

17. The transistor in a class C amplifier conducts for

    (a) more than 180° of the input cycle   **(b)** one-half of the input cycle

    (c) a very small percentage of the input cycle   **(d)** all of the input cycle

PROBLEMS

Answers to all odd-numbered problems are at the end of the book.

BASIC PROBLEMS

Section 7–1   **The Class A Power Amplifier**

1. Figure 7–41 shows a CE power amplifier in which the collector resistor serves also as the load resistor. Assume $\beta_{DC} = \beta_{ac} = 100$.

    (a) Determine the dc Q-point (I_{CQ} and V_{CEQ}).

    (b) Determine the voltage gain and the power gain.

▶ **FIGURE 7–41**

Multisim and LT Spice file circuits are identified with a logo and are in the Problems folder on the website. Filenames correspond to figure numbers (e.g., FGM07-41 and FGS07-41).

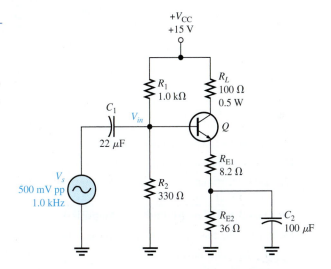

2. For the circuit in Figure 7–41, determine the following:

    (a) the power dissipated in the transistor with no load

    (b) the total power from the power supply with no load

    (c) the signal power in the load with a 500 mV input

3. Refer to the circuit in Figure 7–41. What changes would be necessary to convert the circuit to a *pnp* transistor with a positive supply? What advantage would this have?

4. Assume a CC amplifier has an input resistance of 2.2 kΩ and drives an output load of 50 Ω. What is the power gain?

5. Determine the Q-point for each amplifier in Figure 7–42.

▶ **FIGURE 7–42**

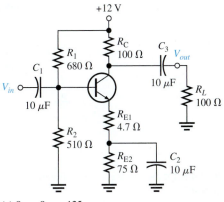

(a) $\beta_{ac} = \beta_{DC} = 125$

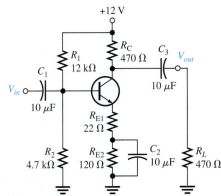

(b) $\beta_{ac} = \beta_{DC} = 120$

6. If the load resistor in Figure 7–42(a) is changed to 50 Ω, how much does the Q-point change?

7. What is the maximum peak value of collector current that can be realized in each circuit of Figure 7–42? What is the maximum peak value of output voltage in each circuit?

8. Find the power gain for each circuit in Figure 7–42. Neglect r'_e.

9. Find the Q-point for the amplifier in Figure 7–43.

▶ **FIGURE 7–43**

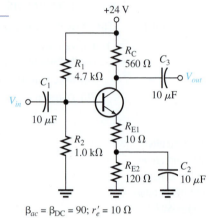

$$\beta_{ac} = \beta_{DC} = 90; \; r'_e = 10 \; \Omega$$

10. What is the voltage gain in Figure 7–43?

11. Determine the minimum power rating for the transistor in Figure 7–43.

12. Find the maximum output signal power to the load and efficiency for the amplifier in Figure 7–43 with a 500 Ω load resistor.

Section 7–2   The Class B and Class AB Push-Pull Amplifiers

13. Refer to the class AB amplifier in Figure 7–44.

   (a) Determine the dc parameters $V_{B(Q1)}$, $V_{B(Q2)}$, V_E, I_{CQ}, $V_{CEQ(Q1)}$, $V_{CEQ(Q2)}$.

   (b) For the 5 V rms input, determine the power delivered to the load resistor.

▶ **FIGURE 7–44**

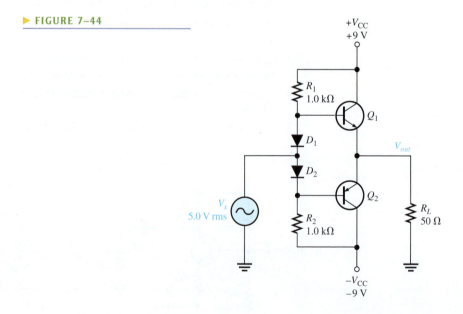

14. Draw the load line for the *npn* transistor in Figure 7–44. Label the saturation current, $I_{c(sat)}$, and show the Q-point.

15. Determine the approximate input resistance seen by the signal source for the amplifier of Figure 7–44 if $\beta_{ac} = 100$.

16. In Figure 7–44, does β_{ac} have an effect on power gain? Explain your answer.

17. Refer to the class AB amplifier in Figure 7–45 operating with a single power supply.

   (a) Determine the dc parameters $V_{B(Q1)}$, $V_{B(Q2)}$, V_E, I_{CQ}, $V_{CEQ(Q1)}$, $V_{CEQ(Q2)}$.

   (b) Assuming the input voltage is 10 V pp, determine the power delivered to the load resistor.

18. Refer to the class AB amplifier in Figure 7–45.

   (a) What is the maximum power that could be delivered to the load resistor?

   (b) Assume the power supply voltage is raised to 24 V. What is the new maximum power that could be delivered to the load resistor?

19. Refer to the class AB amplifier in Figure 7–45. What fault or faults could account for each of the following symptoms?

   (a) a positive half-wave output signal

   (b) zero volts on both bases and the emitters

   (c) no output: emitter voltage $= +15$ V

   (d) crossover distortion observed on the output waveform

20. If a 1 V rms signal source with an internal resistance of 50 Ω is connected to the amplifier in Figure 7–45, what is the actual rms signal applied to the amplifier input? Assume $\beta_{ac} = 200$.

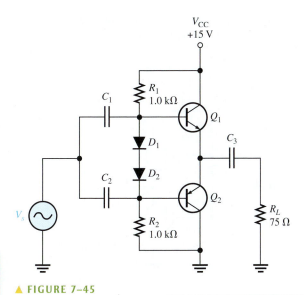

▲ FIGURE 7–45

Section 7–3 The Class C Amplifier

21. A certain class C amplifier transistor is on for 10% of the input cycle. If $V_{ce(sat)} = 0.18$ V and $I_{c(sat)} = 25$ mA, what is the average power dissipation for maximum output?

22. What is the resonant frequency of a tank circuit with $L = 10$ mH and C $= 0.001$ μF?

23. What is the maximum peak-to-peak output voltage of a tuned class C amplifier with $V_{CC} = 12$ V?

24. Determine the efficiency of the class C amplifier described in Problem 23 if $V_{CC} = 15$ V and the equivalent parallel resistance in the collector tank circuit is 50 Ω. Assume that the transistor is on for 10% of the period.

Section 7–4　Troubleshooting

25. Refer to Figure 7–46. What would you expect to observe across R_L if the base of Q_2 is opened?

▶ FIGURE 7–46

▶ FIGURE 7–46

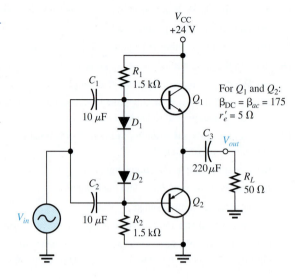

26. Assume there is no output for the circuit in Figure 7–46. List steps you would take to trouble-shoot it.

27. Determine the possible fault or faults, if any, for each circuit in Figure 7–47 based on the indicated dc voltage measurements.

▶ FIGURE 7–47

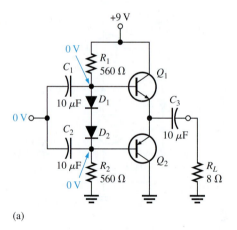

(a)

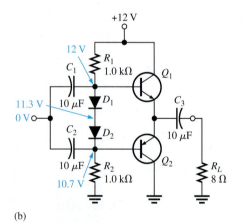

(b)

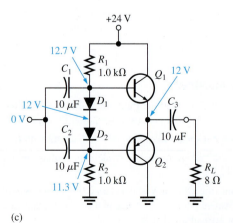

(c)

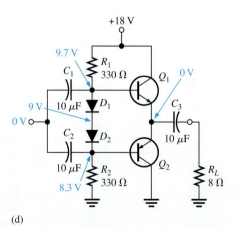

(d)

DEVICE APPLICATION PROBLEMS

28. Assume that the public address system represented by the block diagram in Figure 7–33 has quit working. You find there is no signal output from the power amplifier or the preamplifier, but you have verified that the microphone is working. Which two blocks are the most likely to be the problem? How would you narrow the choice down to one block?

29. Describe the output that would be observed in the push-pull amplifier of Figure 7–34 with a 2 V rms sinusoidal input voltage if the base-emitter junction of Q_2 opened.

30. Describe the output that would be observed in Figure 7–34 if the collector-emitter junction of Q_5 opened for the same input as in Problem 29.

31. After visually inspecting the power amplifier circuit board in Figure 7–48, describe any problems.

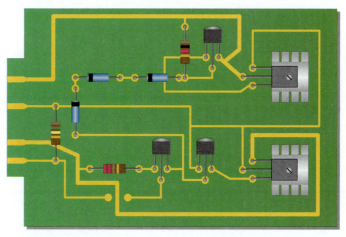

▲ **FIGURE 7–48**

DATASHEET PROBLEMS

32. Referring to the datasheet in Figure 7–49, determine the following:

    (a) minimum β_{DC} for the BD135 and the conditions

    (b) maximum collector-to-emitter voltage for the BD135

    (c) maximum power dissipation for the BD135 at a case temperature of 25°C

    (d) maximum continuous collector current for the BD135

33. Determine the maximum power dissipation for a BD135 at a case temperature of 50°C.

34. Determine the maximum power dissipation for a BD135 at an ambient temperature of 50°C.

35. Describe what happens to the dc current gain as the collector current increases.

36. Determine the approximate h_{FE} for the BD135 at I_C = 20 mA.

ADVANCED PROBLEMS

37. Explain why the specified maximum power dissipation of a power transistor at an ambient temperature of 25°C is much less than maximum power dissipation at a case temperature of 25°C.

BD135/137/139

Medium Power Linear and Switching Applications
- Complement to BD136, BD138 and BD140 respectively

TO-126
1. Emitter 2.Collector 3.Base

NPN Epitaxial Silicon Transistor

Absolute Maximum Ratings T_C = 25°C unless otherwise noted

Symbol	Parameter		Value	Units
V_{CBO}	Collector-Base Voltage	: BD135	45	V
		: BD137	60	V
		: BD139	80	V
V_{CEO}	Collector-Emitter Voltage	: BD135	45	V
		: BD137	60	V
		: BD139	80	V
V_{EBO}	Emitter-Base Voltage		5	V
I_C	Collector Current (DC)		1.5	A
I_{CP}	Collector Current (Pulse)		3.0	A
I_B	Base Current		0.5	A
P_C	Collector Dissipation (T_C = 25°C)		12.5	W
P_C	Collector Dissipation (T_a = 25°C)		1.25	W
T_J	Junction Temperature		150	°C
T_{STG}	Storage Temperature		- 55 ~ 150	°C

Electrical Characteristics T_C = 25°C unless otherwise noted

Symbol	Parameter	Test Condition	Min.	Typ.	Max.	Units
V_{CEO}(sus)	Collector-Emitter Sustaining Voltage	I_C = 30mA, I_B = 0				
	: BD135		45			V
	: BD137		60			V
	: BD139		80			V
I_{CBO}	Collector Cut-off Current	V_{CB} = 30V, I_E = 0			0.1	μA
I_{EBO}	Emitter Cut-off Current	V_{EB} = 5V, I_C = 0			10	μA
h_{FE1}	DC Current Gain : ALL DEVICE	V_{CE} = 2V, I_C = 5mA	25			
h_{FE2}	: ALL DEVICE	V_{CE} = 2V, I_C = 0.5A	25			
h_{FE3}	: BD135	V_{CE} = 2V, I_C = 150mA	40		250	
	: BD137, BD139		40		160	
V_{CE}(sat)	Collector-Emitter Saturation Voltage	I_C = 500mA, I_B = 50mA			0.5	V
V_{BE}(on)	Base-Emitter ON Voltage	V_{CE} = 2V, I_C = 0.5A			1	V

h_{FE} Classification

Classification	6	10	16
h_{FE3}	40 ~ 100	63 ~ 160	100 ~ 250

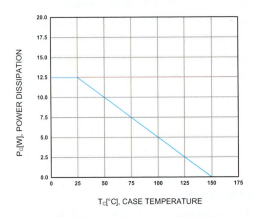

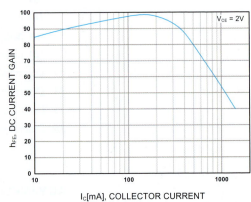

▲ **FIGURE 7–49**

38. Draw the dc and the ac load lines for the amplifier in Figure 7–50.

▶ **FIGURE 7–50**

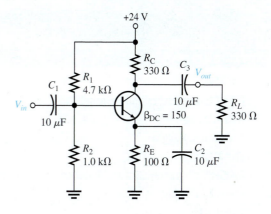

39. Design a swamped class A power amplifier that will operate from a dc supply of +15 V with an approximate voltage gain of 50. The quiescent collector current should be approximately 500 mA, and the total dc current from the supply should not exceed 750 mA. The output power must be at least 1 W.

40. The public address system in Figure 7–33 is a portable unit that is independent of ac utility power. Determine the ampere-hour rating for the +15 V and the −15 V battery supply necessary for the system to operate for 4 hours on a continuous basis.

MULTISIM TROUBLESHOOTING PROBLEMS

These file circuits are in the Troubleshooting Problems folder on the website.

41. Open file TPM07-41 and determine the fault.

42. Open file TPM07-42 and determine the fault.

43. Open file TPM07-43 and determine the fault.

44. Open file TPM07-44 and determine the fault.

45. Open file TPM07-45 and determine the fault.

8

FIELD-EFFECT TRANSISTORS (FETs)

CHAPTER OBJECTIVES

◆ Discuss the JFET and how it differs from the BJT
◆ Discuss, define, and apply JFET characteristics and parameters
◆ Discuss and analyze JFET biasing
◆ Discuss the ohmic region on a JFET characteristic curve
◆ Explain the operation of MOSFETs
◆ Discuss and apply MOSFET parameters
◆ Describe and analyze MOSFET bias circuits
◆ Discuss the IGBT
◆ Troubleshoot FET circuits

KEY TERMS

◆ JFET
◆ Drain
◆ Source
◆ Gate
◆ Pinch-off voltage
◆ Transconductance
◆ On-resistance ($R_{DS(on)}$)
◆ Power dissipation (P_D)
◆ Ohmic region
◆ MOSFET
◆ Depletion
◆ Enhancement
◆ IGBT

DEVICE APPLICATION PREVIEW

The Device Application involves the electronic control circuits for a waste water treatment system. In particular, you will focus on the application of field-effect transistors in the sensing circuits for chemical measurements.

VISIT THE WEBSITE

Study aids, Multisim files, and LT Spice files for this chapter are available at https://www.pearsonhighered.com/careersresources/

INTRODUCTION

BJTs (bipolar junction transistors) were covered in previous chapters. Now we will discuss the second major type of transistor, the FET (field-effect transistor). **FETs** are unipolar devices because, unlike BJTs that use both electron and hole current, they operate only with one type of charge carrier. The two main types of FETs are the junction field-effect transistor (JFET) and the metal oxide semiconductor field-effect transistor (MOSFET). The term *field-effect* relates to the depletion region formed in the channel of a FET as a result of a voltage applied on one of its terminals (gate).

Recall that a BJT is a current-controlled device; that is, the base current controls the amount of collector current. A FET is different. It is a voltage-controlled device, where the voltage between two of the terminals (gate and source) controls the current through the device. A major advantage of FETs is their very high input resistance. Because of their nonlinear characteristics, they are generally not as widely used in amplifiers as BJTs except where very high input impedances are required. However, FETs are the preferred device in low-voltage switching applications because they are generally faster than BJTs when turned on and off. The IGBT is generally used in high-voltage switching applications.

8–1 THE JFET

The **JFET** (junction field-effect transistor) is a type of FET that operates with a reverse-biased *pn* junction to control current in a channel. Depending on their structure, JFETs fall into either of two categories, *n* channel or *p* channel.

After completing this section, you should be able to

❑ **Discuss the JFET and how it differs from the BJT**
❑ Describe the basic structure of *n*-channel and *p*-channel JFETs
  ◆ Name the terminals ◆ Explain a channel
❑ Explain the basic operation of a JFET
❑ Identify JFET schematic symbols

Basic Structure

Figure 8–1(a) shows the basic structure of an *n*-channel JFET (junction field-effect transistor). Wire leads are connected to each end of the *n*-channel; the **drain** is at the upper end, and the **source** is at the lower end. Two *p*-type regions are diffused in the *n*-type material to form a **channel,** and both *p*-type regions are connected to the **gate** lead. The gate lead is shown connected to only one of the *p* regions, which are internally connected together. A *p*-channel JFET is shown in Figure 8–1(b).

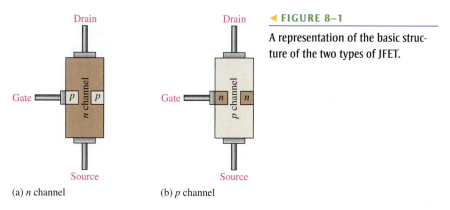

(a) *n* channel   (b) *p* channel

◄ **FIGURE 8–1**

A representation of the basic structure of the two types of JFET.

Basic Operation

To illustrate the operation of a JFET, Figure 8–2 shows dc bias voltages applied to an *n*-channel device. V_{DD} provides a drain-to-source voltage and supplies current

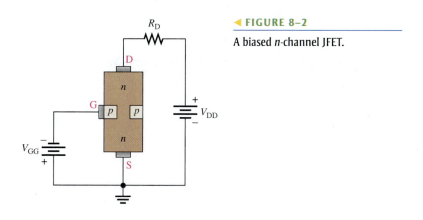

◄ **FIGURE 8–2**

A biased *n*-channel JFET.

source to drain. V_{GG} sets the reverse-bias voltage between the gate and the source, as shown.

The JFET is always operated with the gate-source pn junction reverse-biased. Reverse-biasing of the gate-source junction produces a depletion region along the *pn* junction, which extends into the channel and thus increases its resistance by restricting the channel width.

The channel width and thus the channel resistance can be controlled by varying the gate voltage, thereby controlling the amount of drain current, I_D. Figure 8–3 illustrates this concept with an *n*-channel device. The white areas represent the depletion region created by the reverse bias. It is wider toward the drain end of the channel because the reverse-bias voltage between the gate and the drain is greater than that between the gate and the source. We will discuss JFET characteristic curves and some parameters in Section 8–2.

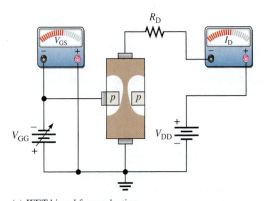

(a) JFET biased for conduction

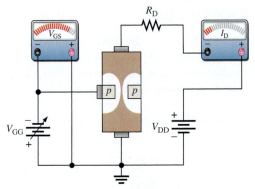

(b) Greater V_{GG} narrows the channel (between the white areas) which increases the resistance of the channel and decreases I_D.

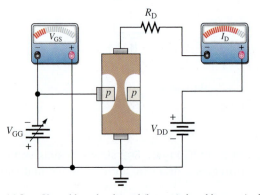

(c) Less V_{GG} widens the channel (between the white areas) which decreases the resistance of the channel and increases I_D.

▲ FIGURE 8–3

Effects of V_{GS} on channel width, resistance, and drain current ($V_{GG} = V_{GS}$).

JFET Symbols

The schematic symbols for both *n*-channel and *p*-channel JFETs are shown in Figure 8–4. Notice that the arrow on the gate points "in" for *n* channel and "out" for *p* channel.

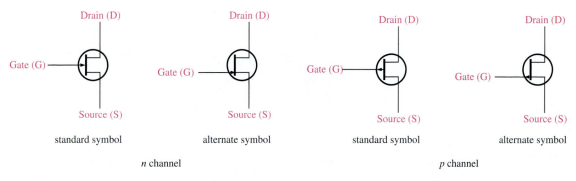

standard symbol   alternate symbol   standard symbol   alternate symbol

n channel   *p* channel

▲ FIGURE 8–4

JFET schematic symbols.

SECTION 8–1 CHECKUP Answers can be found at www .pearsonhighered.com/floyd.	1. Name the three terminals of a JFET. 2. Does an *n*-channel JFET require a positive or negative value for V_{GS}? 3. How is the drain current controlled in a JFET?

8–2 JFET Characteristics and Parameters

The JFET operates as a voltage-controlled, constant-current device. Cutoff and pinch-off as well as JFET transfer characteristics are covered in this section.

After completing this section, you should be able to

❑ **Discuss, define, and apply JFET characteristics and parameters**
❑ Discuss the drain characteristic curve
   ◆ Identify the ohmic, active, and breakdown regions of the curve
❑ Define *pinch-off voltage*
❑ Discuss breakdown
❑ Explain how gate-to-source voltage controls the drain current
❑ Discuss the cutoff voltage
❑ Compare pinch-off and cutoff
❑ Explain the JFET universal transfer characteristic
   ◆ Calculate the drain current using the transfer characteristic equation
   ◆ Interpret a JFET datasheet
❑ Discuss JFET forward transconductance
   ◆ Define *transconductance*   ◆ Calculate forward transconductance
❑ Discuss JFET input resistance and capacitance
❑ Determine the ac drain-to-source resistance

Drain Characteristic Curve

Consider the case when the gate-to-source voltage is zero ($V_{GS} = 0$ V). This is produced by shorting the gate to the source, as in Figure 8–5(a) where both are grounded. As V_{DD} (and thus V_{DS}) is increased from 0 V, I_D will increase proportionally, as shown in the graph of Figure 8–5(b) between points *A* and *B*. In this area, the channel resistance is essentially constant because the depletion region is not large enough to have significant effect. This is called the *ohmic region* because V_{DS} and I_D are related by Ohm's law. (Ohmic region is discussed further in Section 8–4.)

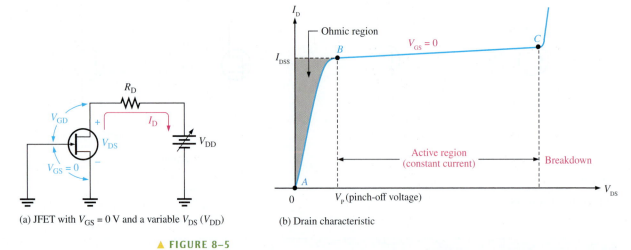

(a) JFET with $V_{GS} = 0$ V and a variable V_{DS} (V_{DD}) (b) Drain characteristic

▲ FIGURE 8–5

The drain characteristic curve of a JFET for $V_{GS} = 0$ showing pinch-off voltage.

At point B in Figure 8–5(b), the curve levels off and enters the active region where I_D becomes essentially constant. As V_{DS} increases from point B to point C, the reverse-bias voltage from gate to drain (V_{GD}) produces a depletion region large enough to offset the increase in V_{DS}, thus keeping I_D relatively constant.

Pinch-Off Voltage For $V_{GS} = 0$ V, the value of V_{DS} at which I_D becomes essentially constant (point B on the curve in Figure 8–5(b)) is the **pinch-off voltage**, V_P. For a given JFET, V_P has a fixed value. As you can see, a continued increase in V_{DS} above the pinch-off voltage produces an almost constant drain current until point C is reached. This value of drain current is I_{DSS} (*D*rain to *S*ource current with gate *S*horted) and is always specified on JFET datasheets. I_{DSS} is the *maximum* drain current that a specific JFET can produce regardless of the external circuit, and it is always specified for the condition, $V_{GS} = 0$ V.

Breakdown As shown in the graph in Figure 8–5(b), **breakdown** occurs at point C when I_D begins to increase very rapidly with any further increase in V_{DS}. Breakdown can result in irreversible damage to the device, so JFETs are always operated below breakdown and within the active region (constant current) (between points B and C on the graph). The JFET action that produces the drain characteristic curve to the point of breakdown for $V_{GS} = 0$ V is illustrated in Figure 8–6.

V_{GS} Controls I_D

Let's connect a reverse-bias voltage, V_{GG}, from gate to source as shown in Figure 8–7(a). As V_{GS} is set to increasingly more negative values by adjusting V_{GG}, a family of drain characteristic curves is produced, as shown in Figure 8–7(b). Notice that I_D decreases as the magnitude of V_{GS} is increased to larger negative values because of the narrowing of the channel. Also notice that, for each increase in V_{GS}, the JFET reaches pinch-off (where constant current begins) at values of V_{DS} less than V_P. The term *pinch-off* is not the same as pinch-off voltage, V_p. Therefore, the amount of drain current is controlled by V_{GS}, as illustrated in Figure 8–8.

Cutoff Voltage

The value of V_{GS} that makes I_D approximately zero is the **cutoff voltage, $V_{GS(off)}$**, as shown in Figure 8–8(d). The JFET must be operated between $V_{GS} = 0$ V and $V_{GS(off)}$. For this range of gate-to-source voltages, I_D will vary from a maximum of I_{DSS} to a minimum of almost zero.

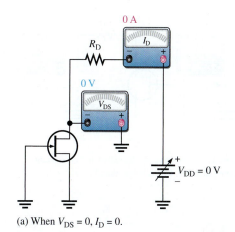

(a) When $V_{DS} = 0$, $I_D = 0$.

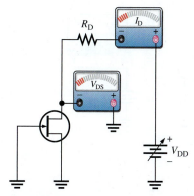

(b) I_D increases proportionally with V_{DS} in the ohmic region.

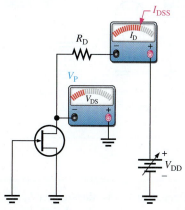

(c) When $V_{DS} = V_P$, I_D is constant and equal to I_{DSS}.

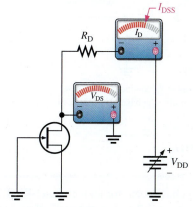

(d) As V_{DS} increases further, I_D remains at I_{DSS} until breakdown occurs.

▲ **FIGURE 8–6**

JFET action that produces the characteristic curve for $V_{GS} = 0$ V.

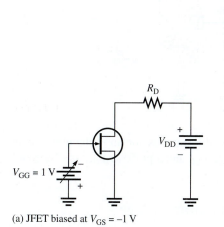

(a) JFET biased at $V_{GS} = -1$ V

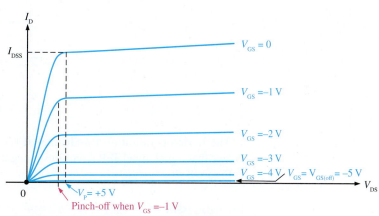

(b) Drain characteristic

▲ **FIGURE 8–7**

Pinch-off occurs at a lower V_{DS} as V_{GS} is increased to more negative values.

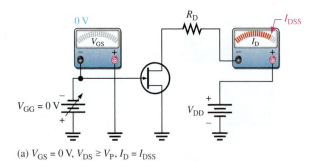

(a) $V_{GS} = 0$ V, $V_{DS} \geq V_P$, $I_D = I_{DSS}$

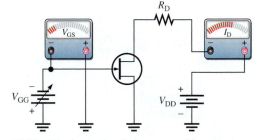

(b) When V_{GS} is negative, I_D decreases and is constant above pinch-off, which is less than V_P.

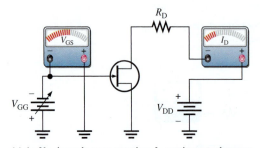

(c) As V_{GS} is made more negative, I_D continues to decrease but is constant above pinch-off, which has also decreased.

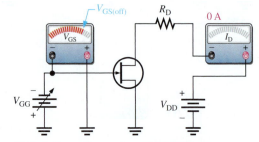

(d) Until $V_{GS} = -V_{GS(off)}$, I_D continues to decrease. When $V_{GS} \geq -V_{GS(off)}$, $I_D \cong 0$.

▲ **FIGURE 8–8**

V_{GS} controls I_D.

As you have seen, for an *n*-channel JFET, the more negative V_{GS} is, the smaller I_D becomes in the active region. When V_{GS} has a sufficiently large negative value, I_D is reduced to zero. This cutoff effect is caused by the widening of the depletion region to a point where it completely closes the channel, as shown in Figure 8–9.

▶ **FIGURE 8–9**

JFET at cutoff.

The basic operation of a *p*-channel JFET is the same as for an *n*-channel device except that a *p*-channel JFET requires a negative V_{DD} and a positive V_{GS}, as illustrated in Figure 8–10.

Comparison of Pinch-Off Voltage and Cutoff Voltage

As you have seen, there is a difference between pinch-off and cutoff voltages. There is also a connection. The pinch-off voltage V_P is the value of V_{DS} at which the drain current becomes constant and equal to I_{DSS} and is always measured at $V_{GS} = 0$ V. However,

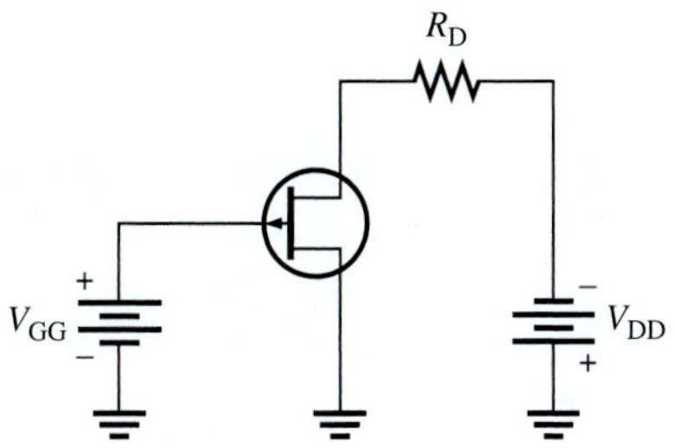

◀ **FIGURE 8–10**

A biased *p*-channel JFET.

pinch-off occurs for V_{DS} values less than V_P when V_{GS} is nonzero. So, although V_P is a constant, the minimum value of V_{DS} at which I_D becomes constant varies with V_{GS}.

$V_{GS(off)}$ and V_P are always equal in magnitude but opposite in sign. A datasheet usually will give either $V_{GS(off)}$ or V_P, but not both. However, when you know one, you have the other. For example, if $V_{GS(off)} = -5$ V, then $V_P = +5$ V, as shown in Figure 8–7(b).

EXAMPLE 8–1

For the JFET in Figure 8–11, $V_{GS(off)} = -4$ V and $I_{DSS} = 12$ mA. Determine the *minimum* value of V_{DD} required to put the device in the constant-current region of operation when $V_{GS} = 0$ V.

▶ **FIGURE 8–11**

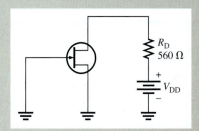

Solution   Since $V_{GS(off)} = -4$ V, $V_P = 4$ V. The minimum value of V_{DS} for the JFET to be in its constant-current region is

$$V_{DS} = V_P = 4 \text{ V}$$

In the constant-current region with $V_{GS} = 0$ V,

$$I_D = I_{DSS} = 12 \text{ mA}$$

The drop across the drain resistor is

$$V_{R_D} = I_D R_D = (12 \text{ mA})(560 \text{ } \Omega) = 6.72 \text{ V}$$

Apply Kirchhoff's law around the drain circuit.

$$V_{DD} = V_{DS} + V_{R_D} = 4 \text{ V} + 6.72 \text{ V} = \mathbf{10.7 \text{ V}}$$

This is the value of V_{DD} to make $V_{DS} = V_P$ and put the device in the constant-current region.

*Related Problem**   If V_{DD} is increased to 15 V, what is the drain current?

*Answers can be found at www.pearsonhighered.com/floyd.

EXAMPLE 8–2

A particular *p*-channel JFET has a $V_{GS(off)} = +4$ V. What is I_D when $V_{GS} = +6$ V?

Solution

The *p*-channel JFET requires a positive gate-to-source voltage. The more positive the voltage, the less the drain current. When $V_{GS} = 4$ V, $I_D = 0$. Any further increase in V_{GS} keeps the JFET cut off, so I_D remains **0**.

Related Problem

What is V_P for the JFET described in this example?

JFET Universal Transfer Characteristic

You have learned that a range of V_{GS} values from zero to $V_{GS(off)}$ controls the amount of drain current. For an *n*-channel JFET, $V_{GS(off)}$ is negative, and for a *p*-channel JFET, $V_{GS(off)}$ is positive. Because V_{GS} does control I_D, the relationship between these two quantities is very important. Figure 8–12 is a general transfer characteristic curve that illustrates graphically the relationship between V_{GS} and I_D.

▶ **FIGURE 8–12**

JFET universal transfer characteristic curve (*n*-channel).

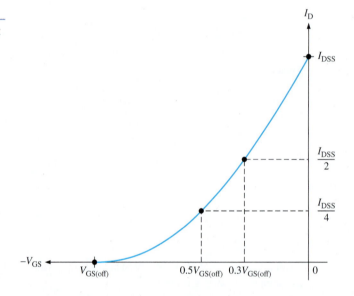

Notice that the bottom end of the curve is at a point on the V_{GS} axis equal to $V_{GS(off)}$, and the top end of the curve is at a point on the I_D axis equal to I_{DSS}. This curve shows that

$$I_D = 0 \qquad \text{when } V_{GS} = V_{GS(off)}$$

$$I_D = \frac{I_{DSS}}{4} \qquad \text{when } V_{GS} = 0.5V_{GS(off)}$$

$$I_D = \frac{I_{DSS}}{2} \qquad \text{when } V_{GS} = 0.3V_{GS(off)}$$

and

$$I_D = I_{DSS} \qquad \text{when } V_{GS} = 0$$

The transfer characteristic curve can also be developed from the drain characteristic curves by plotting values of I_D for the values of V_{GS} taken from the family of drain curves at pinch-off, as illustrated in Figure 8–13 for a specific set of curves. Each point on the transfer characteristic curve corresponds to specific values of V_{GS} and I_D on the drain curves. For example, when $V_{GS} = -2$ V, $I_D = 4.32$ mA. Also, for this specific JFET, $V_{GS(off)} = -5$ V and $I_{DSS} = 12$ mA.

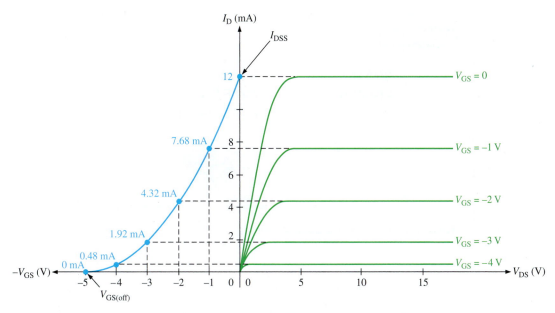

▲ FIGURE 8–13

Example of the development of an *n*-channel JFET transfer characteristic curve (blue) from the JFET drain characteristic curves (green).

A JFET transfer characteristic curve is expressed approximately as

$$I_D \cong I_{DSS}\left(1 - \frac{V_{GS}}{V_{GS(off)}}\right)^2$$

Equation 8–1

With Equation 8–1, I_D can be determined for any V_{GS} if $V_{GS(off)}$ and I_{DSS} are known. These quantities are usually available from the datasheet for a given JFET. Notice the squared term in the equation. Because of its form, a parabolic relationship is known as a *square law,* and therefore, JFETs and MOSFETs are often referred to as *square-law devices.*

The datasheet for a typical JFET series is shown in Figure 8–14.

EXAMPLE 8–3

The partial datasheet in Figure 8–14 for a 2N5459 JFET indicates that typically $I_{DSS} = 9$ mA and $V_{GS(off)} = -8$ V (maximum). Using these values, determine the drain current for $V_{GS} = 0$ V, -1 V, and -4 V.

Solution   For $V_{GS} = 0$ V,

$$I_D = I_{DSS} = \mathbf{9\ mA}$$

For $V_{GS} = -1$ V, use Equation 8–1.

$$I_D \cong I_{DSS}\left(1 - \frac{V_{GS}}{V_{GS(off)}}\right)^2 = (9\text{ mA})\left(1 - \frac{-1\text{ V}}{-8\text{ V}}\right)^2$$
$$= (9\text{ mA})(1 - 0.125)^2 = (9\text{ mA})(0.766) = \mathbf{6.89\ mA}$$

For $V_{GS} = -4$ V,

$$I_D \cong (9\text{ mA})\left(1 - \frac{-4\text{ V}}{-8\text{ V}}\right)^2 = (9\text{ mA})(1 - 0.5)^2 = (9\text{ mA})(0.25) = \mathbf{2.25\ mA}$$

Related Problem   Determine I_D for $V_{GS} = -3$ V for the 2N5459 JFET.

▶ **FIGURE 8–14**

JFET partial datasheet. Copyright
Fairchild Semiconductor
Corporation. Used by permission.

2N5457   MMBF5457
2N5458   MMBF5458
2N5459   MMBF5459

TO-92

SOT-23
Mark: 6D / 61S / 6L

NOTE: Source & Drain
are interchangeable

N-Channel General Purpose Amplifier

This device is a low level audio amplifier and switching transistors,
and can be used for analog switching applications. Sourced from
Process 55.

Absolute Maximum Ratings* TA = 25°C unless otherwise noted

Symbol	Parameter	Value	Units
V_{DG}	Drain-Gate Voltage	25	V
V_{GS}	Gate-Source Voltage	- 25	V
I_{GF}	Forward Gate Current	10	mA
T_J, T_{stg}	Operating and Storage Junction Temperature Range	-55 to +150	°C

*These ratings are limiting values above which the serviceability of any semiconductor device may be impaired.

NOTES:
1) These ratings are based on a maximum junction temperature of 150 degrees C.
2) These are steady state limits. The factory should be consulted on applications involving pulsed or low duty cycle operations.

Thermal Characteristics TA = 25°C unless otherwise noted

Symbol	Characteristic	Max		Units
		2N5457-5459	*MMBF5457-5459	
P_D	Total Device Dissipation	625	350	mW
	Derate above 25°C	5.0	2.8	mW/°C
$R_{\theta JC}$	Thermal Resistance, Junction to Case	125		°C/W
$R_{\theta JA}$	Thermal Resistance, Junction to Ambient	357	556	°C/W

*Device mounted on FR-4 PCB 1.6" X 1.6" X 0.06."

Electrical Characteristics TA = 25°C unless otherwise noted

Symbol	Parameter	Test Conditions		Min	Typ	Max	Units
OFF CHARACTERISTICS							
$V_{(BR)GSS}$	Gate-Source Breakdown Voltage	$I_G = 10\,\mu A$, $V_{DS} = 0$		- 25			V
I_{GSS}	Gate Reverse Current	$V_{GS} = -15\,V$, $V_{DS} = 0$				- 1.0	nA
		$V_{GS} = -15\,V$, $V_{DS} = 0$, $T_A = 100°C$				- 200	nA
$V_{GS(off)}$	Gate-Source Cutoff Voltage	$V_{DS} = 15\,V$, $I_D = 10\,nA$	5457	- 0.5		- 6.0	V
			5458	- 1.0		- 7.0	V
			5459	- 2.0		- 8.0	V
V_{GS}	Gate-Source Voltage	$V_{DS} = 15\,V$, $I_D = 100\,\mu A$	5457		- 2.5		V
		$V_{DS} = 15\,V$, $I_D = 200\,\mu A$	5458		- 3.5		V
		$V_{DS} = 15\,V$, $I_D = 400\,\mu A$	5459		- 4.5		V
ON CHARACTERISTICS							
I_{DSS}	Zero-Gate Voltage Drain Current*	$V_{DS} = 15\,V$, $V_{GS} = 0$	5457	1.0	3.0	5.0	mA
			5458	2.0	6.0	9.0	mA
			5459	4.0	9.0	16	mA
SMALL SIGNAL CHARACTERISTICS							
g_{fs}	Forward Transfer Conductance*	$V_{DS} = 15\,V$, $V_{GS} = 0$, f = 1.0 kHz					
			5457	1000		5000	μmhos
			5458	1500		5500	μmhos
			5459	2000		6000	μmhos
g_{os}	Output Conductance*	$V_{DS} = 15\,V$, $V_{GS} = 0$, f = 1.0 kHz			10	50	μmhos
C_{iss}	Input Capacitance	$V_{DS} = 15\,V$, $V_{GS} = 0$, f = 1.0 MHz			4.5	7.0	pF
C_{rss}	Reverse Transfer Capacitance	$V_{DS} = 15\,V$, $V_{GS} = 0$, f = 1.0 MHz			1.5	3.0	pF
NF	Noise Figure	$V_{DS} = 15\,V$, $V_{GS} = 0$, f = 1.0 kHz, $R_G = 1.0$ megohm, BW = 1.0 Hz				3.0	dB

*Pulse Test: Pulse Width≦300 ms, Duty Cycle≦2%

JFET Forward Transconductance

The forward **transconductance** (transfer conductance), g_m, is an ac quantity that is defined as a change in drain current (ΔI_D) divided by a corresponding change in gate-to-source voltage (ΔV_{GS}) with the drain-to-source voltage constant. It is expressed as a ratio and has the unit of siemens (S).

$$g_m = \frac{\Delta I_D}{\Delta V_{GS}}$$

Other common designations for this parameter are g_{fs} and y_{fs} (forward transfer admittance). As you will see in Chapter 9, g_m is an important factor in determining the voltage gain of a FET amplifier.

Because the transfer characteristic curve for a JFET is nonlinear, g_m varies in value depending on the location on the curve as set by V_{GS}. The value for g_m is greater near the top of the transfer characteristic curve (near $V_{GS} = 0$) than it is near the bottom (near $V_{GS(off)}$), as illustrated in Figure 8–15(a). Simply stated, the transconductance curve is the slope of the transfer curve. A plot of the transconductance as a function of V_{GS} is a straight line as shown in Figure 8–15(b). Notice that at the y-axis crossing point, ($V_{GS} = 0$) it has the special designation g_{m0} and that the transconductance is a linear function of V_{GS}.

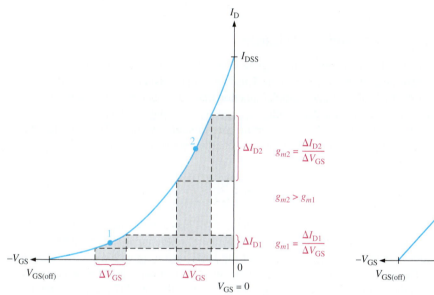

(a) g_m varies depending on the bias point (V_{GS})

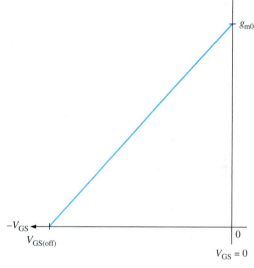

(b) Plot of g_m as a function of V_{GS}

▲ **FIGURE 8–15**

g_m varies depending on the bias point (V_{GS}).

A datasheet normally gives the value of g_m measured at $V_{GS} = 0$ V (g_{m0}). For example, the datasheet for the 2N5457 JFET specifies a minimum g_{m0} (g_{fs}) of 1000 μmhos (the mho is the same unit as the siemens (S)) with $V_{DS} = 15$ V.

Given g_{m0}, you can calculate an approximate value for g_m at any point on the transfer characteristic curve using the following formula:

$$g_m = g_{m0}\left(1 - \frac{V_{GS}}{V_{GS(off)}}\right)$$

Equation 8–2

When a value of g_{m0} is not available, you can calculate it using values of I_{DSS} and $V_{GS(off)}$. The vertical lines indicate an absolute value (no sign).

$$g_{m0} = \frac{2I_{DSS}}{|V_{GS(off)}|}$$

Equation 8–3

EXAMPLE 8–4

The following information is included on the datasheet in Figure 8–14 for a 2N5457 JFET: typically, $I_{DSS} = 3.0$ mA, $V_{GS(off)} = -6$ V maximum, and $g_{fs(max)} = 5000$ μS. Using these values, determine the forward transconductance for $V_{GS} = -4$ V, and find I_D at this point.

Solution

$g_{m0} = g_{fs} = 5000$ μS. Use Equation 8–2 to calculate g_m.

$$g_m = g_{m0}\left(1 - \frac{V_{GS}}{V_{GS(off)}}\right) = (5000 \ \mu S)\left(1 - \frac{-4 \ V}{-6 \ V}\right) = \mathbf{1667 \ \mu S}$$

Next, use Equation 8–1 to calculate I_D at $V_{GS} = -4$ V.

$$I_D \cong I_{DSS}\left(1 - \frac{V_{GS}}{V_{GS(off)}}\right)^2 = (3.0 \ mA)\left(1 - \frac{-4 \ V}{-6 \ V}\right)^2 = \mathbf{333 \ \mu A}$$

Related Problem

A given JFET has the following characteristics: $I_{DSS} = 12$ mA, $V_{GS(off)} = -5$ V, and $g_{m0} = g_{fs} = 3000$ μS. Find g_m and I_D when $V_{GS} = -2$ V.

Input Resistance and Capacitance

As you know, a JFET operates with its gate-source junction reverse-biased, which makes the input resistance at the gate very high. This high input resistance is one advantage of the JFET over the BJT. (Recall that a bipolar junction transistor operates with a forward-biased base-emitter junction.) JFET datasheets often specify the input resistance by giving a value for the gate reverse current, I_{GSS}, at a certain gate-to-source voltage. The input resistance can then be determined using the following equation, where the vertical lines indicate an absolute value (no sign):

$$R_{IN} = \left|\frac{V_{GS}}{I_{GSS}}\right|$$

For example, the 2N5457 datasheet in Figure 8–14 lists a maximum I_{GSS} of -1.0 nA for $V_{GS} = -15$ V at 25°C. I_{GSS} increases with temperature, so the input resistance decreases.

The input capacitance, C_{iss}, is a result of the JFET operating with a reverse-biased *pn* junction. Recall that a reverse-biased *pn* junction acts as a capacitor whose capacitance depends on the amount of reverse voltage. For example, the 2N5457 has a maximum C_{iss} of 7 pF for $V_{GS} = 0$.

Many discrete JFETs have lower input capacitance when compared to integrated circuit amplifiers. A low capacitance is particularly useful when working with high frequencies. By contrast, a low-capacitance integrated circuit may have input capacitances of about 20 pF, whereas many discrete JFETs have input capacitances of less than 5 pF combined with very low noise specifications. This is useful in cases where a very low-level signal is being amplified in the presence of noise or in cases where a high-impedance source is used.

EXAMPLE 8–5

A certain JFET has an I_{GSS} of -2 nA for $V_{GS} = -20$ V. Determine the input resistance.

Solution

$$R_{IN} = \left|\frac{V_{GS}}{I_{GSS}}\right| = \frac{20 \ V}{2 \ nA} = \mathbf{10,000 \ M\Omega}$$

Related Problem

Determine the input resistance for the 2N5458 from the datasheet in Figure 8–14.

AC Drain-to-Source Resistance

You learned from the drain characteristic curve that, above pinch-off, the drain current is relatively constant over a range of drain-to-source voltages. Therefore, a large change in V_{DS} produces only a very small change in I_D. The ratio of these changes is the ac drain-to-source resistance of the device, r'_{ds}.

$$r'_{ds} = \frac{\Delta V_{DS}}{\Delta I_D}$$

Datasheets often specify this parameter in terms of the output conductance, g_{os}, or output admittance, y_{os}, for $V_{GS} = 0$ V.

EXAMPLE 8–6

From the data sheet for a 2N5458 (Figure 8–14), determine the typical drain-source resistance.

Solution   The typical value of g_{os} for the 2N5458 = 10 mmhos (10 mS).

$$r'_{ds} = \frac{1}{g_{os}} = \frac{1}{10\ \mu S} = \mathbf{100\ k\Omega}$$

The high channel resistance is a disadvantage of JFETs over BJTs.

Related Problem   From the data sheet, what is the *minimum* value of r'_{ds}?

JFET Limiting Parameters   Several parameters limit operation of JFETs to prevent operating in a manner that could destroy the device. On the data sheet, they are typically listed under the heading of Absolute Maximum Ratings. Most electrical characteristics of all transistors are temperature dependent so when there is current, many specifications need to be reduced to take into account the working temperature due to internal heating. The following descriptions summarize some of the limiting parameters.

Gate Source Breakdown Voltage, $V_{(BR)GSS}$, is the voltage that will do irreparable damage if it is exceeded between the specified terminals (gate source)

On-resistance, $R_{DS(on)}$, is the ratio of drain voltage to drain current. It determines the power loss and heating loss within the transistor.

Continuous drain current, I_D, is the maximum current that can safely be carried by a FET continuously. If pulsed, this current can be exceeded depending on the width and duty cycle of pulses. I_D is derated for increasing case temperature.

Power dissipation, P_D, is the maximum power allowed for safe operation and is based in junction to case temperature.

Safe operating area, SOA, is a set of curves that define the maximum value of drain to source voltage as a function of drain current which guarantees safe operation when the device is forward biased.

**SECTION 8–2
CHECKUP**

1. The drain-to-source voltage at the pinch-off point of a particular JFET is 7 V. If the gate-to-source voltage is zero, what is V_P?
2. The V_{GS} of a certain *n*-channel JFET is increased negatively. Does the drain current increase or decrease?
3. What value must V_{GS} have to produce cutoff in a *p*-channel JFET with a $V_P = -3$ V?
4. Given an output conductance for a JFET, how do you find the drain-source resistance?

8–3   JFET BIASING

Using some of the JFET parameters discussed previously, you will now see how to dc-bias JFETs. Just as with the BJT, the purpose of biasing is to select the proper dc gate-to-source voltage to establish a desired value of drain current and, thus, a proper Q-point. Three types of bias are self-bias, voltage-divider bias, and current-source bias.

After completing this section, you should be able to

❑ **Discuss and analyze JFET biasing**
❑ Describe self-bias
   ◆ Calculate JFET currents and voltages
❑ Describe how to set the Q-point of a self-biased JFET
   ◆ Determine midpoint bias
❑ Graphically analyze a self-biased JFET
❑ Discuss voltage-divider bias
   ◆ Calculate JFET currents and voltages
❑ Graphically analyze a voltage-divider biased JFET
❑ Discuss Q-point stability
❑ Describe current-source bias

Self-Bias

Self-bias is the most common type of JFET bias. Recall that a JFET must be operated such that the gate-source junction is always reverse-biased. This condition requires a negative V_{GS} for an *n*-channel JFET and a positive V_{GS} for a *p*-channel JFET. This can be achieved using the self-bias arrangements shown in Figure 8–16. The gate resistor, R_G, does not affect the bias because it has essentially no voltage drop across it and therefore the gate remains at 0 V. R_G is necessary only to force the gate to be at 0 V and to isolate an ac signal from ground in amplifier applications, as you will see later.

▶ **FIGURE 8–16**

Self-biased JFETs ($I_S = I_D$ in all FETs).

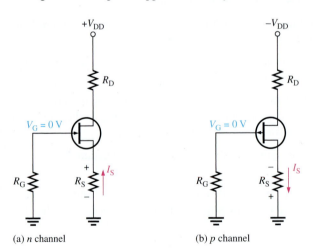

(a) *n* channel   (b) *p* channel

For the *n*-channel JFET in Figure 8–16(a), I_S produces a voltage drop across R_S and makes the source positive with respect to ground. Since $I_S = I_D$ and $V_G = 0$, then $V_S = I_D R_S$. The gate-to-source voltage is

$$V_{GS} = V_G - V_S = 0 - I_D R_S = -I_D R_S$$

Thus,

$$V_{GS} = -I_D R_S$$

For the *p*-channel JFET shown in Figure 8–16(b), the current through R_S produces a negative voltage at the source, making the gate positive with respect to the source. Therefore, since $I_S = I_D$,

$$V_{GS} = +I_D R_S$$

In the following example, the *n*-channel JFET in Figure 8–16(a) is used for illustration. Keep in mind that analysis of the *p*-channel JFET is the same except for opposite-polarity voltages. The drain voltage with respect to ground is determined as follows:

$$V_D = V_{DD} - I_D R_D$$

Since $V_S = I_D R_S$, the drain-to-source voltage is

$$V_{DS} = V_D - V_S = V_{DD} - I_D(R_D + R_S)$$

EXAMPLE 8–7

Find V_{DS} and V_{GS} in Figure 8–17. For the particular JFET in this circuit, the parameter values such as g_m, $V_{GS(off)}$, and I_{DSS} are such that a drain current (I_D) of approximately 5 mA is produced. Another JFET, even of the same type, may not produce the same results when connected in this circuit due to the variations in parameter values.

▶ **FIGURE 8–17**

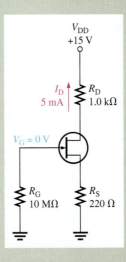

Solution

$$V_S = I_D R_S = (5 \text{ mA})(220 \ \Omega) = 1.1 \text{ V}$$

$$V_D = V_{DD} - I_D R_D = 15 \text{ V} - (5 \text{ mA})(1.0 \text{ k}\Omega) = 15 \text{ V} - 5 \text{ V} = 10 \text{ V}$$

Therefore,

$$V_{DS} = V_D - V_S = 10 \text{ V} - 1.1 \text{ V} = \textbf{8.9 V}$$

Since $V_G = 0$ V,

$$V_{GS} = V_G - V_S = 0 \text{ V} - 1.1 \text{ V} = \textbf{−1.1 V}$$

Related Problem Determine V_{DS} and V_{GS} in Figure 8–17 when $I_D = 8$ mA. Assume that $R_D = 860 \ \Omega$, $R_S = 390 \ \Omega$, and $V_{DD} = 12$ V.

Open the Multisim file EXM08-07 or the LT Spice file EXS08-07 in the Examples folder on the website. Measure I_D, V_{GS}, and V_{DS} and compare to the calculated values from the Related Problem.

Setting the Q-Point of a Self-Biased JFET

The basic approach to establishing a JFET bias point is to determine I_D for a desired value of V_{GS} or vice versa. Then calculate the required value of R_S using the following relationship. The vertical lines indicate an absolute value.

$$R_S = \left| \frac{V_{GS}}{I_D} \right|$$

For a desired value of V_{GS}, I_D can be determined in either of two ways: from the transfer characteristic curve for the particular JFET or, more practically, from Equation 8–1 using I_{DSS} and $V_{GS(off)}$ from the JFET datasheet. The next two examples illustrate these procedures.

EXAMPLE 8–8

Determine the value of R_S required to self-bias an n-channel JFET that has the transfer characteristic curve shown in Figure 8–18 at $V_{GS} = -5$ V.

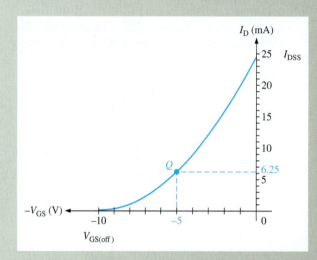

▲ **FIGURE 8–18**

Solution   From the graph, $I_D = 6.25$ mA when $V_{GS} = -5$ V. Calculate R_S.

$$R_S = \left| \frac{V_{GS}}{I_D} \right| = \frac{5\ \text{V}}{6.25\ \text{mA}} = \mathbf{800\ \Omega}$$

Related Problem   Find R_S for $V_{GS} = -3$ V.

EXAMPLE 8–9

Determine the value of R_S required to self-bias a p-channel JFET with datasheet values of $I_{DSS} = 25$ mA and $V_{GS(off)} = 15$ V. V_{GS} is to be 5 V.

Solution   Use Equation 8–1 to calculate I_D.

$$I_D \cong I_{DSS}\left(1 - \frac{V_{GS}}{V_{GS(off)}}\right)^2 = (25\ \text{mA})\left(1 - \frac{5\ \text{V}}{15\ \text{V}}\right)^2$$

$$= (25\ \text{mA})(1 - 0.333)^2 = 11.1\ \text{mA}$$

Now, determine R_S.

$$R_S = \left| \frac{V_{GS}}{I_D} \right| = \frac{5 \text{ V}}{11.1 \text{ mA}} = \mathbf{450 \ \Omega}$$

Related Problem Find the value of R_S required to self-bias a p-channel JFET with $I_{DSS} = 18$ mA and $V_{GS(off)} = 8$ V. $V_{GS} = 4$ V.

Midpoint Bias It is usually desirable to bias a JFET near the midpoint of its transfer characteristic curve where $I_D = I_{DSS}/2$. Under signal conditions, midpoint bias allows the maximum amount of drain current swing between I_{DSS} and 0. For Equation 8–1, it can be shown that I_D is approximately one-half of I_{DSS} when $V_{GS} = V_{GS(off)}/3.4$. The derivation is given in "Derivations of Selected Equations" at www.pearsonhighered.com/floyd.

$$I_D \cong I_{DSS}\left(1 - \frac{V_{GS}}{V_{GS(off)}}\right)^2 = I_{DSS}\left(1 - \frac{V_{GS(off)}/3.4}{V_{GS(off)}}\right)^2 = 0.5 I_{DSS}$$

So, by selecting $V_{GS} = V_{GS(off)}/3.4$, you should get a midpoint bias in terms of I_D.

To set the drain voltage at midpoint ($V_D = V_{DD}/2$), select a value of R_D to produce the desired voltage drop. Choose R_G arbitrarily large to prevent loading on the driving stage in a cascaded amplifier arrangement. Example 8–10 illustrates these concepts.

EXAMPLE 8–10 Looking at the datasheet in Figure 8–14, select resistor values for R_D and R_S in Figure 8–19 to set up an approximate midpoint bias. Use minimum datasheet values when given; otherwise, V_D should be approximately 6 V (one-half of V_{DD}).

▶ **FIGURE 8–19**

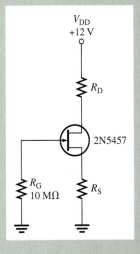

Solution For midpoint bias,

$$I_D \cong \frac{I_{DSS}}{2} = \frac{1.0 \text{ mA}}{2} = 0.5 \text{ mA}$$

and

$$V_{GS} \cong \frac{V_{GS(off)}}{3.4} = \frac{-0.5 \text{ V}}{3.4} = -147 \text{ mV}$$

Then

$$R_S = \left| \frac{V_{GS}}{I_D} \right| = \frac{147 \text{ mV}}{0.5 \text{ mA}} = \mathbf{294 \ \Omega}$$

$$V_D = V_{DD} - I_D R_D$$

$$I_D R_D = V_{DD} - V_D$$

$$R_D = \frac{V_{DD} - V_D}{I_D} = \frac{12 \text{ V} - 6 \text{ V}}{0.5 \text{ mA}} = \mathbf{12 \ k\Omega}$$

Related Problem   Select resistor values in Figure 8–19 to set up an approximate midpoint bias using a 2N5459.

 Open the Multisim file EXM08-10 or the LT Spice file EXS08-10 in the Examples folder on the website. The circuit has the calculated values for R_D and R_S from the Related Problem. Verify that an approximate midpoint bias is established by measuring V_D and I_D.

Graphical Analysis of a Self-Biased JFET

You can use the transfer characteristic curve of a JFET and certain parameters to determine the Q-point (I_D and V_{GS}) of a self-biased circuit. A circuit is shown in Figure 8–20(a), and a transfer characteristic curve is shown in Figure 8–20(b). If a curve is not available from a datasheet, you can plot it from Equation 8–1 using datasheet values for I_{DSS} and $V_{GS(off)}$.

To determine the Q-point of the circuit in Figure 8–20(a), a self-bias dc load line is established on the graph in part (b) as follows. First, calculate V_{GS} when I_D is zero.

$$V_{GS} = -I_D R_S = (0)(470 \ \Omega) = 0 \text{ V}$$

▶ **FIGURE 8–20**

A self-biased JFET and its transfer characteristic curve.

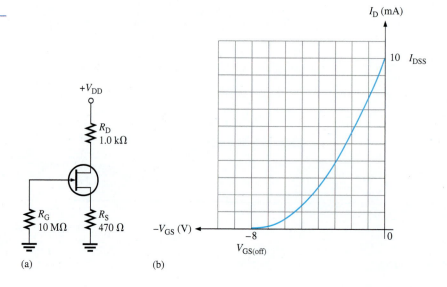

(a)   (b)

This establishes a point at the origin on the graph ($I_D = 0$, $V_{GS} = 0$). Next, calculate V_{GS} when $I_D = I_{DSS}$. From the curve in Figure 8–20(b), $I_{DSS} = 10$ mA.

$$V_{GS} = -I_D R_S = -(10 \text{ mA})(470 \text{ }\Omega) = -4.7 \text{ V}$$

This establishes a second point on the graph ($I_D = 10$ mA, $V_{GS} = -4.7$ V). Now, with two points, the load line can be drawn on the transfer characteristic curve as shown in Figure 8–21. The point where the load line intersects the transfer characteristic curve is the Q-point of the circuit as shown, where $I_D = 5.07$ mA and $V_{GS} = -2.3$ V.

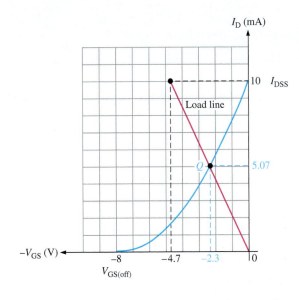

◀ **FIGURE 8–21**

The intersection of the self-bias dc load line and the transfer characteristic curve is the Q-point.

EXAMPLE 8–11

Determine the Q-point for the JFET circuit in Figure 8–22(a). The transfer characteristic curve is given in Figure 8–22(b).

▶ **FIGURE 8–22**

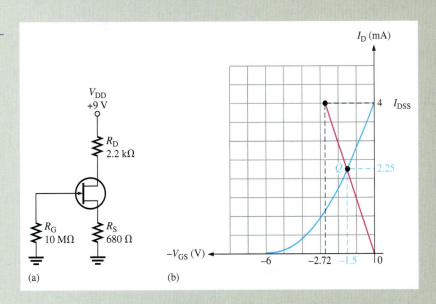

(a)

(b)

Solution   For $I_D = 0$,

$$V_{GS} = -I_D R_S = (0)(680 \text{ }\Omega) = 0 \text{ V}$$

This gives a point at the origin. From the curve, $I_{DSS} = 4$ mA; so $I_D = I_{DSS} = 4$ mA.

$$V_{GS} = -I_D R_S = -(4 \text{ mA})(680 \text{ }\Omega) = -2.72 \text{ V}$$

This gives a second point at 4 mA and -2.72 V. A line is now drawn between the two points, and the values of I_D and V_{GS} at the intersection of the line and the curve are taken from the graph, as illustrated in Figure 8–22(b). The Q-point values from the graph are

$$I_D = \textbf{2.25 mA}$$
$$V_{GS} = \textbf{−1.5 V}$$

Related Problem   If R_S is increased to 1.0 kΩ in Figure 8–22(a), what is the new Q-point?

For increased Q-point stability, the value of R_S in the self-bias circuit is increased and connected to a negative supply voltage. This is sometimes called *dual-supply bias.*

Voltage-Divider Bias

An *n*-channel JFET with voltage-divider bias is shown in Figure 8–23. The voltage at the source of the JFET must be more positive than the voltage at the gate in order to keep the gate-source junction reverse-biased.

▶ FIGURE 8–23

An *n*-channel JFET with voltage-divider bias ($I_S = I_D$).

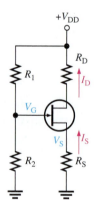

The source voltage is

$$V_S = I_D R_S$$

The gate voltage is set by resistors R_1 and R_2 as expressed by the following equation using the voltage-divider formula:

$$V_G = \left(\frac{R_2}{R_1 + R_2} \right) V_{DD}$$

The gate-to-source voltage is

$$V_{GS} = V_G - V_S$$

and the source voltage is

$$V_S = V_G - V_{GS}$$

The drain current can be expressed as

$$I_D = \frac{V_S}{R_S}$$

Substituting for V_S,

$$I_D = \frac{V_G - V_{GS}}{R_S}$$

EXAMPLE 8–12

Determine I_D and V_{GS} for the 2N4341 JFET with voltage-divider bias in Figure 8–24, given that for this particular JFET the parameter values are such that $V_D \cong 7.5$ V.

▶ **FIGURE 8–24**

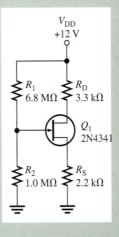

Solution

$$I_D = \frac{V_{DD} - V_D}{R_D} = \frac{12 \text{ V} - 7.5 \text{ V}}{3.3 \text{ k}\Omega} = \frac{4.4 \text{ V}}{3.3 \text{ k}\Omega} = \mathbf{1.36 \text{ mA}}$$

Calculate the gate-to-source voltage as follows:

$$V_S = I_D R_S = (1.36 \text{ mA})(2.2 \text{ k}\Omega) = 3.00 \text{ V}$$

$$V_G = \left(\frac{R_2}{R_1 + R_2}\right)V_{DD} = \left(\frac{1.0 \text{ M}\Omega}{7.8 \text{ M}\Omega}\right)12 \text{ V} = 1.54 \text{ V}$$

$$V_{GS} = V_G - V_S = 1.54 \text{ V} - 3.00 \text{ V} = \mathbf{-1.46 \text{ V}}$$

If V_D had not been given in this example, the Q-point values could have been found with the transfer characteristic curve and the intercept of the R_S line, which is started at V_G.

Related Problem

Given that $V_D = 6$ V when another JFET is inserted in the circuit of Figure 8–24, determine the Q-point.

Open the Multisim file EXM08-12 or the LT Spice file EXS08-12 in the Examples folder on the website. The circuit has the calculated values for R_D and R_S from the Related Problem. Verify that an approximate midpoint bias is established by measuring V_D and I_D.

Graphical Analysis of a JFET with Voltage-Divider Bias

An approach similar to the one used for self-bias can be used with voltage-divider bias to graphically determine the Q-point of a circuit on the transfer characteristic curve.

In a JFET with voltage-divider bias when $I_D = 0$, V_{GS} is not zero, as in the self-biased case, because the voltage divider produces a voltage at the gate independent of the drain current. The voltage-divider dc load line is determined as follows.

For $I_D = 0$,

$$V_S = I_D R_S = (0)R_S = 0 \text{ V}$$

$$V_{GS} = V_G - V_S = V_G - 0 \text{ V} = V_G$$

Therefore, one point on the line is at $I_D = 0$ and $V_{GS} = V_G$.

For $V_{GS} = 0$,

$$I_D = \frac{V_G - V_{GS}}{R_S} = \frac{V_G}{R_S}$$

A second point on the line is at $I_D = V_G/R_S$ and $V_{GS} = 0$. The generalized dc load line is shown in Figure 8–25. The point at which the load line intersects the transfer characteristic curve is the Q-point.

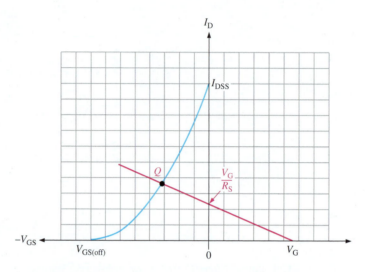

▲ FIGURE 8–25

Generalized dc load line (red) for a JFET with voltage-divider bias.

EXAMPLE 8–13

Determine the approximate Q-point for the JFET with voltage-divider bias in Figure 8–26(a), given that this particular device has a transfer characteristic curve as shown in Figure 8–26(b).

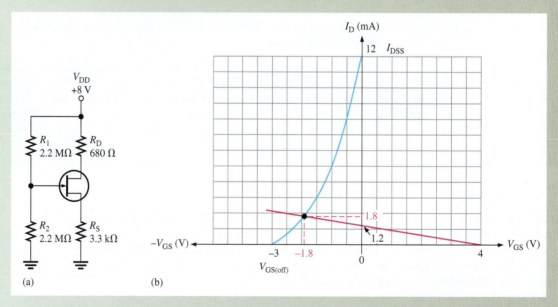

▲ FIGURE 8–26

Solution   First, establish the two points for the load line. For $I_D = 0$,

$$V_{GS} = V_G = \left(\frac{R_2}{R_1 + R_2}\right)V_{DD} = \left(\frac{2.2\ \text{M}\Omega}{4.4\ \text{M}\Omega}\right)8\ \text{V} = 4\ \text{V}$$

The first point is at $I_D = 0$ and $V_{GS} = 4$ V. For $V_{GS} = 0$,

$$I_D = \frac{V_G - V_{GS}}{R_S} = \frac{V_G}{R_S} = \frac{4\ \text{V}}{3.3\ \text{k}\Omega} = 1.2\ \text{mA}$$

The second point is at $I_D = 1.2$ mA and $V_{GS} = 0$.

The load line is drawn in Figure 8–26(b), and the approximate Q-point values of $I_D \cong \mathbf{1.8\ mA}$ and $V_{GS} \cong \mathbf{-1.8\ V}$ are picked off the graph, as indicated.

Related Problem   Change R_S to 4.7 kΩ and determine the Q-point for the circuit in Figure 8–26(a).

 Open the Multisim file EXM08-13 or the LT Spice file EXS08-13 in the Examples folder on the website. Measure the Q-point values of I_D and V_{GS} and see how they compare to the graphically determined values from the Related Problem.

Q-Point Stability

Unfortunately, the transfer characteristic of a JFET can differ considerably from one device to another of the same type. If, for example, a 2N5459 JFET is replaced in a given bias circuit with another 2N5459, the transfer characteristic curve can vary greatly, as illustrated in Figure 8–27(a). In this case, the maximum I_{DSS} is 16 mA and the minimum I_{DSS} is 4 mA. Likewise, the maximum $V_{GS(off)}$ is −8 V and the minimum $V_{GS(off)}$ is −2 V. This means that if you have a selection of 2N5459s and you randomly pick one out, it can have values anywhere within these ranges.

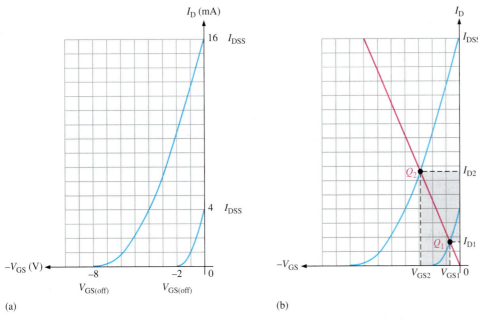

(a)   (b)

▲ **FIGURE 8–27**

Variation in the transfer characteristic of 2N5459 JFETs and the effect on the Q-point.

After completing this section, you should be able to

- **Discuss the ohmic region on a JFET characteristic curve**
  - Calculate slope and drain-to-source resistance
- Explain how a JFET can be used as a variable resistance
- Discuss JFET operation with the Q-point at the origin
  - Calculate transconductance

The ohmic region extends from the origin of the characteristic curves to the break point (where the active region begins) of the $V_{GS} = 0$ curve in a roughly parabolic shape, as shown on a typical set of curves in Figure 8–30. The characteristic curves in this region have a relatively constant slope for small values of I_D. The slope of the characteristic curve in the ohmic region is the dc drain-to-source conductance G_{DS} of the JFET.

$$\text{Slope} = G_{DS} \cong \frac{I_D}{V_{DS}}$$

Recall from your basic circuits course that resistance is the reciprocal of the conductance. Thus, the dc drain-to-source resistance is given by

$$R_{DS} = \frac{1}{G_{DS}} \cong \frac{V_{DS}}{I_D}$$

▶ FIGURE 8–30

The ohmic region is the shaded area.

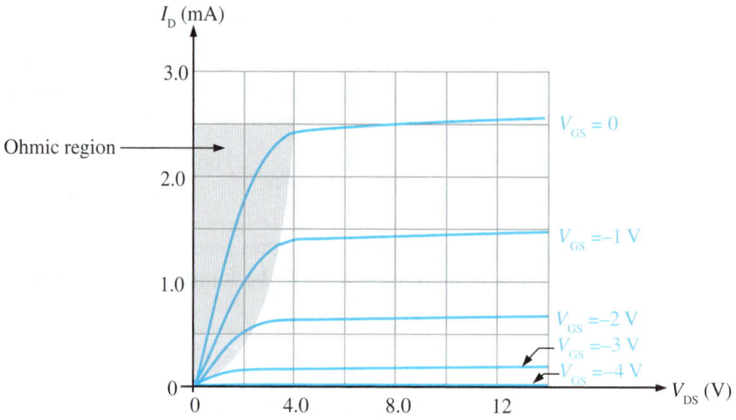

The JFET as a Variable Resistance A JFET can be biased in either the active region or the ohmic region. JFETs are often biased in the ohmic region for use as a voltage-controlled variable resistor. The control voltage is V_{GS}, and it determines the resistance by varying the Q-point. To bias a JFET in the ohmic region, the dc load line must intersect the characteristic curve in the ohmic region, as illustrated in Figure 8–31. To do this in a way that allows V_{GS} to control R_{DS}, the dc saturation current is set for a value much less than I_{DSS} so that the load line intersects most of the characteristic curves in the ohmic region, as illustrated. In this case,

$$I_{D(sat)} = \frac{V_{DD}}{R_D} = \frac{12\text{ V}}{24\text{ k}\Omega} = 0.50\text{ mA}$$

Figure 8–31 shows the operating region expanded with three Q-points shown (Q_0, Q_1, and Q_2), depending on V_{GS}.

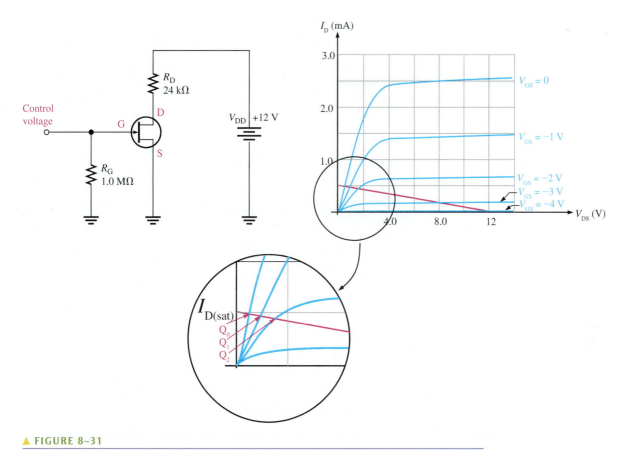

▲ **FIGURE 8–31**

The load line intersects the curves inside the ohmic region.

As you move along the load line in the ohmic region of Figure 8–31, the value of R_{DS} varies as the Q-point falls successively on curves with different slopes. The Q-point is moved along the load line by varying $V_{GS} = 0$ to $V_{GS} = -2$ V, in this case. As this happens, the slope of each successive curve is less than the previous one. A decrease in slope corresponds to less I_D and more V_{DS}, which implies an increase in R_{DS}. This change in resistance can be exploited in a number of applications where voltage control of a resistance is useful.

EXAMPLE 8–15   An *n*-channel JFET is biased in the ohmic region as shown in Figure 8–32. The graph shows an expanded section of the load line in the ohmic region. As V_{GS} is

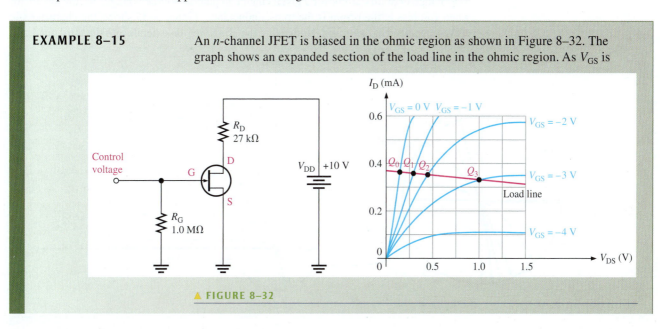

▲ **FIGURE 8–32**

varied from 0 V to -3 V as indicated, assume that the graph shows the following Q-point values:

Q_0: $I_D = 0.360$ mA, $V_{DS} = 0.13$ V

Q_1: $I_D = 0.355$ mA, $V_{DS} = 0.27$ V

Q_2: $I_D = 0.350$ mA, $V_{DS} = 0.42$ V

Q_3: $I_D = 0.33$ mA, $V_{DS} = 0.97$ V

Determine the range of R_{DS} as V_{GS} is varied from 0 V to -3 V.

Solution

$$Q_0: R_{DS} = \frac{V_{DS}}{I_D} = \frac{0.13 \text{ V}}{0.360 \text{ mA}} = 361 \text{ } \Omega$$

$$Q_1: R_{DS} = \frac{V_{DS}}{I_D} = \frac{0.27 \text{ V}}{0.355 \text{ mA}} = 760 \text{ } \Omega$$

$$Q_2: R_{DS} = \frac{V_{DS}}{I_D} = \frac{0.42 \text{ V}}{0.27 \text{ mA}} = 1.2 \text{ k}\Omega$$

$$Q_3: R_{DS} = \frac{V_{DS}}{I_D} = \frac{0.6 \text{ V}}{0.26 \text{ mA}} = 2.9 \text{ k}\Omega$$

When V_{GS} is varied from 0 V to -3 V, R_{DS} **changes from 361 Ω to 2.9 kΩ.**

Related Problem   If $I_{D(sat)}$ is reduced, what happens to the range of R_{DS} values?

Q-point at the Origin   In certain amplifiers, you may want to change the resistance seen by the ac signal without affecting the dc bias in order to control the gain. Sometimes you will see a JFET used as a variable resistance in a circuit where both I_D and V_{DS} are set at 0, which means that the Q-point is at the origin. A Q-point at the origin is achieved by using a capacitor in the drain circuit of the JFET. This makes the dc quantities $V_{DS} = 0$ V and $I_D = 0$ mA, so the only variables are V_{GS} and I_d, the ac drain current. At the origin you have the ac drain current controlled by V_{GS}. As you learned earlier, transconductance is defined as a change in drain current for a given change in gate-to-source voltage. So, the key factor when you bias at the origin is the transconductance. Figure 8–33 shows the characteristic curves expanded at the origin. Notice that the ohmic region extends into the third quadrant.

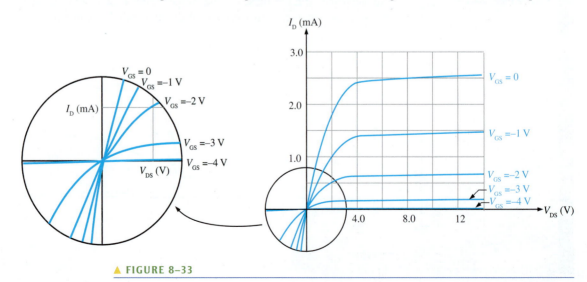

▲ **FIGURE 8–33**

At the origin, where $V_{DS} = 0$ V and $I_D = 0$ mA, the formula for transconductance, introduced earlier in this chapter, is

$$g_m = g_{m0}\left(1 - \frac{V_{GS}}{V_{GS(off)}}\right)$$

where g_m is transconductance and g_{m0} is transconductance for $V_{GS} = 0$ V. g_{m0} can be calculated from the following equation, which was also given earlier:

$$g_{m0} = \frac{2I_{DSS}}{|V_{GS(off)}|}$$

EXAMPLE 8–16

For the characteristic curve in Figure 8–33, calculate the ac drain-to-source resistance for a JFET biased at the origin if $V_{GS} = -2$ V. Assume $I_{DSS} = 2.5$ mA and $V_{GS(off)} = -4$ V.

Solution First, find the transconductance for $V_{GS} = 0$ V.

$$g_{m0} = \frac{2I_{DSS}}{|V_{GS(off)}|} = \frac{2(2.5 \text{ mA})}{4.0 \text{ V}} = 1.25 \text{ mS}$$

Next, calculate g_m at $V_{DS} = -2$ V.

$$g_m = g_{m0}\left(1 - \frac{V_{GS}}{V_{GS(off)}}\right) = 1.25 \text{ mS}\left(1 - \frac{-2 \text{ V}}{-4 \text{ V}}\right) = 0.625 \text{ mS}$$

The ac drain-to-source resistance of the JFET is the reciprocal of the transconductance.

$$r'_{ds} = \frac{1}{g_m} = \frac{1}{0.625 \text{ mS}} = 1.6 \text{ k}\Omega$$

Related Problem What is the ac drain-to-source resistance if $V_{GS} = -1$ V?

SECTION 8–4 CHECKUP

1. For a certain Q-point in the ohmic region, $I_D = 0.3$ mA and $V_{DS} = 0.6$ V. What is the resistance of the JFET when it is biased at this Q-point?

2. How does the drain-to-source resistance change as V_{GS} becomes more negative?

3. For a JFET biased at the origin, $g_m = 0.850$ mS. Determine the corresponding ac resistance.

8–5 THE MOSFET

The **MOSFET** (metal oxide semiconductor field-effect transistor) is another category of field-effect transistor. Unlike the JFET, the MOSFET has no *pn* junction structure; instead, the gate of the MOSFET is insulated from the channel by a silicon dioxide (SiO$_2$) layer. The two basic types of MOSFETs are enhancement (E) and depletion (D). Of the two types, the enhancement MOSFET is more widely used. Because polycrystalline silicon is now used for the gate material instead of metal, these devices are sometimes called IGFETs (insulated-gate FETs).

After completing this section, you should be able to

❑ **Explain the operation of MOSFETs**
❑ Discuss the enhancement MOSFET (E-MOSFET)
   ◆ Describe the structure   ◆ Identify the symbols for E-MOSFET *n*-channel and *p*-channel devices
❑ Discuss the depletion MOSFET (D-MOSFET)
   ◆ Describe the structure   ◆ Discuss the depletion and enhancement modes
   ◆ Identify the symbols for D-MOSFET *n*-channel and *p*-channel devices
❑ Discuss power MOSFETs
   ◆ Describe LDMOSFET structure   ◆ Describe VMOSFET structure
   ◆ Describe TMOSFET structure
❑ Describe the dual-gate MOSFET
   Identify the symbols for dual-gate D-MOSFETs and E-MOSFETs

MOSFET Operation

Like the JFETs we have discussed previously, MOSFETs use a relatively low gate voltage to control drain current. The main difference between JFETs and MOSFETs is the insulated gate (mentioned in the introduction). Most MOSFETs fall in the general category of either planar devices or trench devices. The first MOSFETs were all planar devices, meaning they were made on a plane with the source and drain wells diffused into the body. An insulating oxide layer is grown on the surface and metallization added for terminals. Planar devices are still widely used in linear applications. Trench devices have a "trench" that is cut into the surface with the insulating layer and metal layer then deposited in the trench. These devices have the advantage of lower on resistance and higher current specifications making them more suited for power applications.

Enhancement MOSFET (E-MOSFET)

The E-MOSFET has no structural conduction channel; hence it is a normally off device. Notice in Figure 8–34(a) that the lightly doped *p*-layer substrate extends completely to the SiO_2 layer. For an *n*-channel device, a positive gate voltage above a threshold value *induces* a channel by creating a thin layer of negative charges in the substrate region adjacent to the SiO_2 layer, as shown in Figure 8–34(b). This action is similar to what happens when a capacitor is charged. In this case the conductivity of the channel is enhanced by increasing the gate-to-source voltage and thus pulling more electrons into the channel area. For

▶ **FIGURE 8–34**

Representation of the basic E-MOSFET construction and operation (*n*-channel).

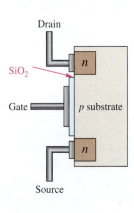

(a) Basic construction

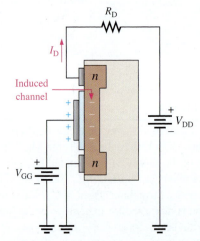

(b) Induced channel ($V_{GS} > V_{GS(th)}$)

any gate voltage below the threshold value, there is no conduction channel. For this reason, the E-MOSFET operates *only* in the enhancement mode and has no depletion mode.

The schematic symbols for the *n*-channel and *p*-channel E-MOSFETs are shown in Figure 8–35. The broken lines symbolize the absence of a physical channel. An inward-pointing substrate arrow is for *n* channel, and an outward-pointing arrow is for *p* channel. Nearly all discrete E-MOSFET devices have the substrate internally connected to the source as shown in the schematic symbol (sometimes it is shown as the body (B) terminal). The substrate is rarely needed, and it plays no role in circuit operation, hence the internal connection. In integrated circuits, the substrate is usually connected to the most negative power supply for *n*-channel devices (most positive supply for *p*-channel devices). An exception is the MIC94030, in which the substrate is brought out, making it a four-terminal device.

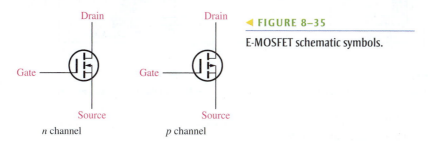

◀ **FIGURE 8–35**

E-MOSFET schematic symbols.

n channel   *p* channel

Depletion MOSFET (D-MOSFET)

Another type of MOSFET is the depletion MOSFET (D-MOSFET), and Figure 8–36 illustrates its basic structure. The drain and source are diffused into the substrate material and then connected by a narrow channel adjacent to the insulated gate. Both *n*-channel and *p*-channel devices are shown in the figure. We will use the *n*-channel device to describe the basic operation. The *p*-channel operation is the same, except the voltage polarities are opposite those of the *n*-channel.

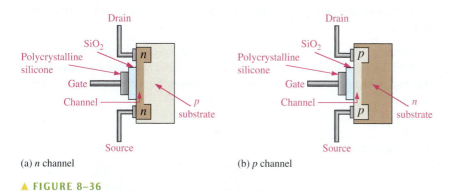

(a) *n* channel   (b) *p* channel

▲ **FIGURE 8–36**

Representation of the basic structure of D-MOSFETs.

The D-MOSFET can be operated in either of two modes—the depletion mode or the enhancement mode—and is sometimes called a *depletion/enhancement MOSFET*. Since the gate is insulated from the channel, either a positive or a negative gate voltage can be applied. The *n*-channel MOSFET operates in the depletion mode when a negative gate-to-source voltage is applied and in the enhancement mode when a positive gate-to-source voltage is applied. These devices are generally operated in the depletion mode.

Depletion Mode   Visualize the gate as one plate of a parallel-plate capacitor and the channel as the other plate. The silicon dioxide insulating layer is the dielectric. With a negative gate voltage, the negative charges on the gate repel conduction electrons from the channel, leaving positive ions in their place. Thereby, the *n* channel is depleted of some of

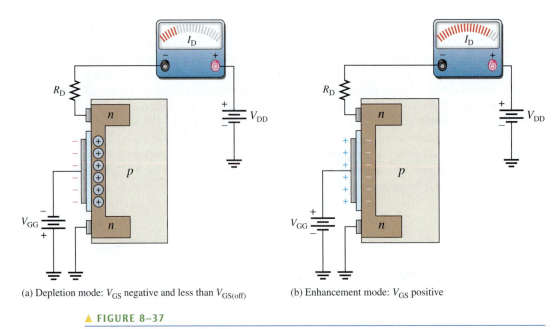

(a) Depletion mode: V_{GS} negative and less than $V_{GS(off)}$

(b) Enhancement mode: V_{GS} positive

▲ FIGURE 8–37

Operation of n-channel D-MOSFET.

its electrons, thus decreasing the channel conductivity. The greater the negative voltage on the gate, the greater the depletion of n-channel electrons. At a sufficiently negative gate-to-source voltage, $V_{GS(off)}$, the channel is totally depleted and the drain current is zero. This depletion mode is illustrated in Figure 8–37(a). Like the n-channel JFET, the n-channel D-MOSFET conducts drain current for gate-to-source voltages between $V_{GS(off)}$ and zero. In addition, the D-MOSFET conducts for values of V_{GS} above zero.

Enhancement Mode   With a positive gate voltage, more conduction electrons are attracted into the channel, thus increasing (enhancing) the channel conductivity, as illustrated in Figure 8–37(b).

D-MOSFET Symbols   The schematic symbols for both the n-channel and the p-channel depletion MOSFETs are shown in Figure 8–38. The substrate, indicated by the arrow, is normally (but not always) connected internally to the source. Sometimes, there is a separate substrate pin.

▶ FIGURE 8–38

D-MOSFET schematic symbols.

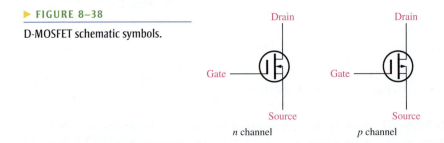

Power MOSFET Structures

The conventional enhancement MOSFETs have a long thin lateral channel as shown in the structural view in Figure 8–39. For power applications, this is a disadvantage, and manufacturers have devised various alternative structures. One structure is a large array of low-power lateral MOSFETs that are all connected in parallel and act as one. It is possible for manufacturers to use MOSFETs in this configuration because they can use matched MOSFETs with a negative temperature coefficient above a certain minimum drain current that causes the drain

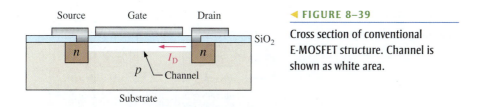

◄ FIGURE 8–39

Cross section of conventional E-MOSFET structure. Channel is shown as white area.

current to decrease with increasing temperature. If one of the parallel MOSFETs has more current than the others, it will have more heat and more channel resistance, which tends to reduce the excessive current. In general, the technique of paralleling FETs requires that the on-resistance increases with drain current to avoid thermal runaway.

Laterally Diffused MOSFET (LDMOSFET) The LDMOSFET has a lateral channel structure and is a type of enhancement MOSFET designed for power applications. This device has a shorter channel between drain and source than does the conventional E-MOSFET. The shorter channel results in lower resistance, which allows higher current and voltage. It also results in a low capacitance, making it the preferred device in high-power RF amplifiers used in communication and radar systems.

Figure 8–40 shows the basic structure of an LDMOSFET. In the substrate, n^+ means a region with higher doping level than n^-. When the gate is positive, a very short n channel is induced in the p layer between the lightly doped source and the n^- region. There is current between the drain and source through the n regions and the induced channel as indicated.

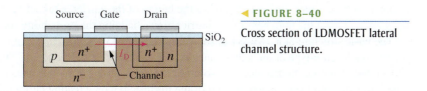

◄ FIGURE 8–40

Cross section of LDMOSFET lateral channel structure.

VMOSFET and UMOSFET The V-groove MOSFET and the U-groove MOSFET are modifications of the conventional E-MOSFET designed to achieve higher power capability by creating a shorter and wider channel with less resistance between the drain and source using a vertical channel structure. The shorter, wider channels allow for higher currents and, thus, greater power dissipation. Frequency response is also improved. The vertical structure of a VMOSFET is shown in Figure 8–41. The drain is connected to the n^+ substrate where n^+ means a higher doping level than n^-.

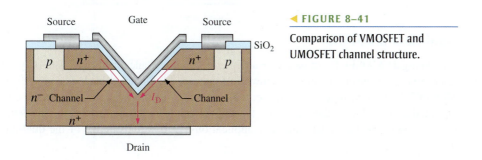

◄ FIGURE 8–41

Comparison of VMOSFET and UMOSFET channel structure.

The UMOSFET is a similar device to the VMOSFET in which the vertical channel is U-shaped as shown in Figure 8–42. The structure of the UMOSFET groove does not have the sharp point at the bottom. This reduces the electric field at the sharp corner and allows for higher voltage operation. It also generally provides a faster device with low ON state resistance, which is useful for fast switching operations and for high frequency rf amplifiers.

► **FIGURE 8–42**

Cross section of UMOSFET structure.

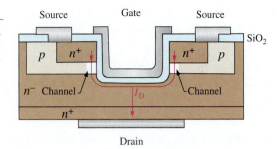

Both the VMOSFET and the UMOSFET have two source connections, a gate connection on top, and a drain connection on the bottom. In both cases, the channel is induced vertically along both sides of the groove between the drain and the source connections. The channel length is set by the thickness of the layers, which is controlled by doping densities and diffusion time rather than by mask dimensions.

Tunneling MOSFET The structure of the Tunneling MOSFET is similar to a standard MOSFET except for the way in which it switches. This distinction, called quantum tunneling, allows for low power switching of electrons. In a normal MOSFET, the gate voltage raises or lowers the *p-n* junction barrier to control current. Tunneling is a quantum effect whereby electrons do not need to cross over the barrier – instead they pass right through it and suddenly appear on the other side. The thinner the barrier, the higher the probability that tunneling can occur. This is normally a limitation for transistor design (limiting how thin barriers can be), but the tunneling MOSFET takes advantages of this peculiar property of quantum mechanics. Instead of varying the height of the barrier, a tunnel MOSFET uses the gate to control the thickness of the barrier, thus changing the probability for tunneling.

The basic structure of a TMOST is shown in Figure 8-43. As you can see it consists of a *p*-type source, an *n*-type drain, and an intrinsic area creating a P-I-N junction. The transistor is operated by increasing the gate bias voltage causing electrons from the valence band of *p*-region to flow into the conduction band of the intrinsic region, creating current through the device. Electrons tunnel between the conduction and valence bands as they move through the intrinsic region.

► **FIGURE 8–43**

Basic structure of a tunneling MOSFET.

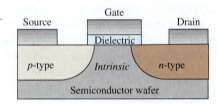

Dual-Gate MOSFETs

The dual-gate MOSFET can be either a depletion or an enhancement type. The only difference is that it has two gates, as shown in Figure 8–44. One drawback of a single-gate MOSFET is its high input capacitance, which restricts its use at higher frequencies. By using a dual-gate device, the drain-gate capacitance can be reduced, thus making the device useful in high-frequency RF amplifier applications. Another advantage of the dual-gate

► **FIGURE 8–44**

Dual-gate *n*-channel MOSFET symbols.

(a) D-MOSFET   (b) E-MOSFET

arrangement is that it allows for an automatic gain control (AGC) input in RF amplifiers. For AVG, the second gate has a gain control feedback signal applied to it that changes the overall gain of the amplifier depending on the signal strength. Another application is demonstrated in the Device Application, where the bias on the second gate is used to adjust the transconductance curve.

FINFET   The FINFET is a type of multi-gate MOSFET that offers smaller geometries and some improved operating characteristics compared to planar geometries. The smaller geometry in higher densities in integrated circuits (more devices per chip area). Figure 8-45 shows a three dimensional view of a basic FINFET. As transistors are made smaller, there are effects that make it more difficult for the gate to deplete the channel underneath (turning the transistor off). In a FINFET, designers raised the channel above the surface of the wafer (like a fin). The gate wraps around the channel, giving the gate more control as it surrounds it on three sides. The fins are made to be extremely thin (20 nm or less) so it is only capable of conducting very small currents. To increase current limits, multiple fins can be used. Basic FINFETs have several advantages including high speed but also have several disadvantages such as limited power capability. They are still a research area, particularly as smaller and faster devices are required.

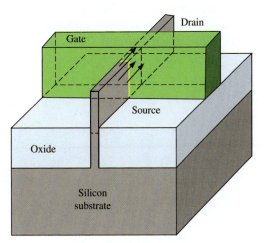

◀ FIGURE 8–45

Basic FINFET structure

SECTION 8–5 CHECKUP	1. How does the channel differ between an E-MOSFET and a D-MOSFET?
	2. If the gate-to-source voltage in an *n*-channel E-MOSFET is made more positive, does the drain current increase or decrease?
	3. If the gate-to-source voltage in an *n*-channel depletion MOSFET is made more negative, does the drain current increase or decrease?
	4. What is the advantage of a trench type of MOSFET?

8–6   MOSFET CHARACTERISTICS AND PARAMETERS

Much of the discussion concerning JFET characteristics and parameters applies equally to MOSFETs. In this section, MOSFET parameters are discussed.

After completing this section, you should be able to

❑ **Discuss and apply MOSFET parameters**
❑ Describe an E-MOSFET transfer characteristic curve
   ◆ Calculate drain current using an equation for the curve   ◆ Use an E-MOSFET datasheet

E-MOSFET Transfer Characteristic

The E-MOSFET uses only channel enhancement. Therefore, an *n*-channel device requires a positive gate-to-source voltage, and a *p*-channel device requires a negative gate-to-source voltage. Figure 8–46 shows the general transfer characteristic curves for both types of E-MOSFETs. As you can see, there is no drain current when $V_{GS} = 0$. Therefore, the E-MOSFET does not have a significant I_{DSS} parameter, as do the JFET and the D-MOSFET. Notice also that there is ideally no drain current until V_{GS} reaches a certain nonzero value called the *threshold voltage, $V_{GS(th)}$*.

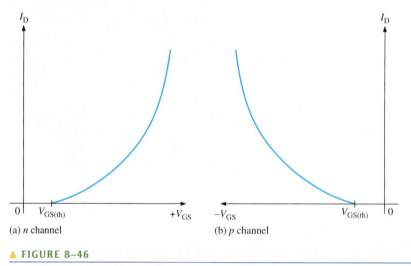

(a) *n* channel   (b) *p* channel

▲ **FIGURE 8–46**

E-MOSFET general transfer characteristic curves.

The equation for the parabolic transfer characteristic curve of the E-MOSFET differs from that of the JFET and the D-MOSFET because the curve starts at $V_{GS(th)}$ rather than $V_{GS(off)}$ on the horizontal axis and never intersects the vertical axis. The equation for the E-MOSFET transfer characteristic curve is

Equation 8–4

$$I_D = K(V_{GS} - V_{GS(th)})^2$$

The constant K depends on the particular MOSFET and can be determined from the datasheet by taking the specified value of I_D, called $I_{D(on)}$, at the given value of V_{GS} and substituting the values into Equation 8–4 as illustrated in Example 8–17.

EXAMPLE 8–17

The datasheet (see www.fairchildsemi.com) for a 2N7002 E-MOSFET gives $I_{D(on)} = 500$ mA (minimum) at $V_{GS} = 10$ V and $V_{GS(th)} = 1$ V. Determine the drain current for $V_{GS} = 5$ V.

Solution   First, solve for K using Equation 8–4.

$$K = \frac{I_{D(on)}}{(V_{GS} - V_{GS(th)})^2} = \frac{500 \text{ mA}}{(10 \text{ V} - 1 \text{ V})^2} = \frac{500 \text{ mA}}{81 \text{ V}^2} = 6.17 \text{ mA/V}^2$$

Next, using the value of K, calculate I_D for $V_{GS} = 5$ V.

$$I_D = K(V_{GS} - V_{GS(th)})^2 = (6.17 \text{ mA/V}^2)(5 \text{ V} - 1 \text{ V})^2 = \textbf{98.7 mA}$$

Related Problem   The datasheet for an E-MOSFET gives $I_{D(on)} = 100$ mA at $V_{GS} = 8$ V and $V_{GS(th)} = 4$ V. Find I_D when $V_{GS} = 6$ V.

D-MOSFET Transfer Characteristic

As previously discussed, the D-MOSFET can operate with either positive or negative gate voltages. This is indicated on the general transfer characteristic curves in Figure 8–47 for both *n*-channel and *p*-channel MOSFETs. The point on the curves where $V_{GS} = 0$ corresponds to I_{DSS}. The point where $I_D = 0$ corresponds to $V_{GS(off)}$. As with the JFET, $V_{GS(off)} = -V_P$.

The square-law expression in Equation 8–1 for the JFET curve also applies to the D-MOSFET curve, as Example 8–18 demonstrates.

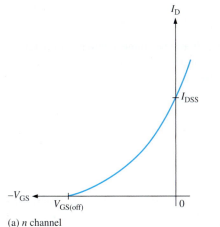

(a) *n* channel   (b) *p* channel

◀ **FIGURE 8–47**

D-MOSFET general transfer characteristic curves.

EXAMPLE 8–18

For a certain D-MOSFET, $I_{DSS} = 10$ mA and $V_{GS(off)} = -8$ V.

(a) Is this an *n*-channel or a *p*-channel?

(b) Calculate I_D at $V_{GS} = -3$ V.

(c) Calculate I_D at $V_{GS} = +3$ V.

Solution   **(a)** The device has a negative $V_{GS(off)}$; therefore, it is an *n*-**channel** MOSFET.

(b) $I_D \cong I_{DSS}\left(1 - \dfrac{V_{GS}}{V_{GS(off)}}\right)^2 = (10 \text{ mA})\left(1 - \dfrac{-3 \text{ V}}{-8 \text{ V}}\right)^2 = \textbf{3.91 mA}$

(c) $I_D \cong (10 \text{ mA})\left(1 - \dfrac{+3 \text{ V}}{-8 \text{ V}}\right)^2 = \textbf{18.9 mA}$

Related Problem For a certain D-MOSFET, $I_{DSS} = 18$ mA and $V_{GS(off)} = +10$ V.

(a) Is this an *n*-channel or a *p*-channel?

(b) Determine I_D at $V_{GS} = +4$ V.

(c) Determine I_D at $V_{GS} = -4$ V.

Handling Precautions

All MOS devices are subject to damage from electrostatic discharge (ESD). Because the gate of a MOSFET is insulated from the channel, the input resistance is extremely high (ideally infinite). The gate leakage current, I_{GSS}, for a typical MOSFET is in the pA range, whereas the gate reverse current for a typical JFET is in the nA range. The input capacitance results from the insulated gate structure. Excess static charge can be accumulated because the input capacitance combines with the very high input resistance and can result in damage to the device. To avoid damage from ESD, certain precautions should be taken when handling MOSFETs:

1. Carefully remove MOSFET devices from their packaging. They are shipped in conductive foam or special foil conductive bags. Usually they are shipped with a wire ring around the leads, which is removed just prior to installing the MOSFET in a circuit.

2. All instruments and metal benches used in assembly or test should be connected to earth ground (round or third prong of 110 V wall outlets).

3. The assembler's or handler's wrist should be connected to a commercial grounding strap, which has a high-value series resistor for safety. The resistor prevents accidental contact with voltage from becoming lethal.

4. Never remove a MOS device (or any other device, for that matter) from the circuit while the power is on.

5. Do not apply signals to a MOS device while the dc power supply is off.

MOSFET Limiting Parameters

Several parameters limit operation of FETs in general to certain absolute maximum values. Most of these parameters were presented in Section 8-2, but are worth repeating for MOSFETs. The following descriptions summarize several limiting parameters that are found on data sheets.

Drain Source Breakdown Voltage, $V_{(BR)DSS}$, is the voltage that will do irreparable damage if it is exceeded between the specified terminals (drain-source). It varies directly with temperature and is normally specified at 25° C.

Blocking voltage, BV_{DSS}, is maximum drain to source voltage that can be applied to the MOSFET.

On-resistance, $R_{DS(on)}$, is the ratio of drain voltage to drain current. It determines the power loss and heating loss within the transistor. Low on-resistance reduces heat-sinking requirements with power MOSFETs. In MOSFETs, $R_{DS(on)}$ tends to increase with temperature.

Continuous drain current, I_D, is the maximum current that can safely be carried by a FET continuously. If pulsed, this current can be exceeded depending on the width and duty cycle of pulses. I_D is derated for increasing case temperature.

Power dissipation, P_D, is the maximum power allowed for safe operation and is based in junction to case temperature.

Safe operating area, SOA, is a set of curves drawn on a log-log plot that define the maximum value of drain-source voltage as a function of drain current which guarantees safe operation when the device is forward biased.

SECTION 8–6 CHECKUP

1. What is the major difference in construction of the D-MOSFET and the E-MOSFET?
2. Name two parameters of an E-MOSFET that are not specified for D-MOSFETs.
3. What is ESD?

8–7 MOSFET BIASING

Three ways to bias a MOSFET are zero-bias, voltage-divider bias, and drain-feedback bias. Biasing is important in FET amplifiers, which you will study in the next chapter.

After completing this section, you should be able to

❑ **Describe and analyze MOSFET bias circuits**
❑ Analyze E-MOSFET bias
  ◆ Discuss and analyze voltage-divider bias ◆ Discuss and analyze drain-feedback bias
❑ Analyze D-MOSFET bias
  ◆ Discuss and analyze zero bias

E-MOSFET Bias

Because E-MOSFETs must have a V_{GS} greater than the threshold value, $V_{GS(th)}$, zero bias cannot be used. Figure 8–48 shows two ways to bias an E-MOSFET (D-MOSFETs can also be biased using these methods). An *n*-channel device is used for purposes of illustration. In either the voltage-divider or drain-feedback bias arrangement, the purpose is to make the gate voltage more positive than the source by an amount exceeding $V_{GS(th)}$. Equations for the analysis of the voltage-divider bias in Figure 8–48(a) are as follows:

$$V_{GS} = \left(\frac{R_2}{R_1 + R_2}\right)V_{DD}$$

$$V_{DS} = V_{DD} - I_D R_D$$

where $I_D = K(V_{GS} - V_{GS(th)})^2$ from Equation 8–4.

In the drain-feedback bias circuit in Figure 8–48(b), there is negligible gate current and, therefore, no voltage drop across R_G. This makes $V_{GS} = V_{DS}$.

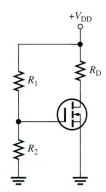

(a) Voltage-divider bias

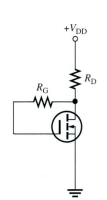

(b) Drain-feedback bias

◀ **FIGURE 8–48**

Common E-MOSFET biasing arrangements.

EXAMPLE 8–19

Determine V_{GS} and V_{DS} for the E-MOSFET circuit in Figure 8–49. Assume this particular MOSFET has minimum values of $I_{D(on)} = 200$ mA at $V_{GS} = 4$ V and $V_{GS(th)} = 2$ V.

▶ **FIGURE 8–49**

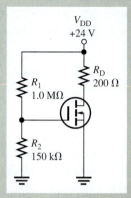

Solution   For the E-MOSFET in Figure 8–49, the gate-to-source voltage is

$$V_{GS} = \left(\frac{R_2}{R_1 + R_2} \right) V_{DD} = \left(\frac{150 \text{ k}\Omega}{1.15 \text{ M}\Omega} \right) 24 \text{ V} = \textbf{3.13 V}$$

To determine V_{DS}, first find K using the minimum value of $I_{D(on)}$ and the specified voltage values.

$$K = \frac{I_{D(on)}}{(V_{GS} - V_{GS(th)})^2} = \frac{200 \text{ mA}}{(4 \text{ V} - 2 \text{ V})^2} = \frac{200 \text{ mA}}{4 \text{ V}^2} = 50 \text{ mA/V}^2$$

Now calculate I_D for $V_{GS} = 3.13$ V.

$$I_D = K(V_{GS} - V_{GS(th)})^2 = (50 \text{ mA/V}^2)(3.13 \text{ V} - 2 \text{ V})^2$$
$$= (50 \text{ mA/V}^2)(1.13 \text{ V})^2 = 63.8 \text{ mA}$$

Finally, calculate V_{DS}.

$$V_{DS} = V_{DD} - I_D R_D = 24 \text{ V} - (63.8 \text{ mA})(200 \text{ }\Omega) = \textbf{11.2 V}$$

Related Problem   Determine V_{GS} and V_{DS} for the circuit in Figure 8–49 given $I_{D(on)} = 100$ mA at $V_{GS} = 4$ V and $V_{GS(th)} = 3$ V.

EXAMPLE 8–20

Determine the amount of drain current in Figure 8–50. The MOSFET has a $V_{GS(th)} = 3$ V.

▶ **FIGURE 8–50**

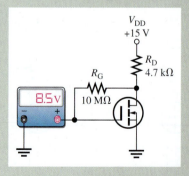

Solution The meter indicates $V_{GS} = 8.5$ V. Since this is a drain-feedback configuration,
$V_{DS} = V_{GS} = 8.5$ V.

$$I_D = \frac{V_{DD} - V_{DS}}{R_D} = \frac{15 \text{ V} - 8.5 \text{ V}}{4.7 \text{ k}\Omega} = \textbf{1.38 mA}$$

Related Problem Determine I_D if the meter in Figure 8–50 reads 5 V.

D-MOSFET Bias

The D-MOSFET can be biased with either of the bias methods described previously for the
E-MOSFET (voltage-divider bias or drain-feedback bias) as well as another simple bias
method that does not work for E-MOSFETS. Recall that D-MOSFETs can be operated
with either positive or negative values of V_{GS}. A simple bias method is to set $V_{GS} = 0$ so
that an ac signal at the gate varies the gate-to-source voltage above and below this 0 V bias
point. A MOSFET with zero bias is shown in Figure 8–51(a). Since $V_{GS} = 0$, $I_D = I_{DSS}$ as
indicated. The drain-to-source voltage is expressed as follows:

$$V_{DS} = V_{DD} - I_{DSS}R_D$$

The purpose of R_G is to accommodate an ac signal input by isolating it from ground,
as shown in Figure 8–51(b). Since there is no dc gate current, R_G does not affect the zero
gate-to-source bias.

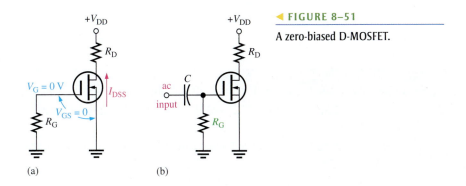

◀ **FIGURE 8–51**

A zero-biased D-MOSFET.

(a)

(b)

EXAMPLE 8–21 Determine the drain-to-source voltage in the circuit of Figure 8–52. The MOSFET
datasheet gives $V_{GS(off)} = -8$ V and $I_{DSS} = 12$ mA.

▶ **FIGURE 8–52**

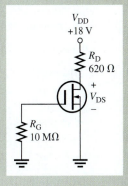

Solution   Since $I_D = I_{DSS} = 12$ mA, the drain-to-source voltage is

$$V_{DS} = V_{DD} - I_{DSS}R_D = 18 \text{ V} - (12 \text{ mA})(620 \text{ }\Omega) = \textbf{10.6 V}$$

Related Problem   Find V_{DS} in Figure 8–52 when $V_{GS(off)} = -10$ V and $I_{DSS} = 20$ mA.

Current Source Biasing

In cases where there are positive and negative supplies available, a simple addition of a current source can give even more stable biasing. Either a BJT or an FET can be used for the current source. Figure 8-53 shows the basic idea of current source biasing.

▶ **FIGURE 8–53**

A D-MOSFET with current source bias.

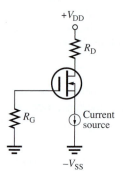

SECTION 8–7 CHECKUP	1. For a D-MOSFET biased at $V_{GS} = 0$, is the drain current equal to zero, I_{GSS}, or I_{DSS}?
	2. For an *n*-channel E-MOSFET with $V_{GS(th)} = 2$ V, V_{GS} must be in excess of what value in order to conduct?
	3. For the current-source-biased D-MOSFET in Figure 8-53, what is the gate voltage?

8–8   THE IGBT

The IGBT (insulated-gate bipolar transistor) combines features from both the MOSFET and the BJT that make it useful in high-voltage and high-current switching applications. The IGBT has largely replaced the MOSFET and the BJT in many of these applications.

After completing this section, you should be able to

❑ **Discuss the IGBT**
  ♦ Compare the IGBT to the MOSFET and the BJT   ♦ Identify the IGBT symbol
❑ Describe IGBT operation
  ♦ Explain how an IGBT is turned on and off   ♦ Discuss and analyze drain-feedback bias   ♦ Describe the IGBT equivalent circuit

The **IGBT** is a device that has the output conduction characteristics of a BJT but is voltage controlled like a MOSFET; it is an excellent choice for many high-voltage switching applications. The IGBT has three terminals: gate, collector, and emitter. One common circuit symbol is shown in Figure 8–54. As you can see, it is similar to the BJT symbol except there is an extra bar representing the gate structure of a MOSFET rather than a base.

The IGBT has MOSFET input characteristics and BJT output characteristics. BJTs are capable of higher currents than FETs, but MOSFETs have no gate current because of the insulated gate structure. IGBTs exhibit a lower saturation voltage than MOSFETs and have about the same saturation voltage as BJTs. IGBTs are superior to MOSFETs in some applications because they can handle high collector-to-emitter voltages exceeding 200 V and exhibit less saturation voltage when they are in the *on* state. IGBTs are superior to BJTs in some applications because they can switch faster. In terms of switching speed, MOSFETs switch fastest, then IGBTs, followed by BJTs, which are slowest. A general comparison of IGBTs, MOSFETs, and BJTs is given in Table 8–1.

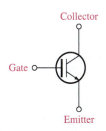

▲ **FIGURE 8–54**

A symbol for the IGBT (insulated-gate bipolar transistor).

◀ **TABLE 8–1**

Comparison of several device features for switching applications.

FEATURES	IGBT	MOSFET	BJT
Type of input drive	Voltage	Voltage	Current
Input resistance	High	High	Low
Operating frequency	Medium	High	Low
Switching speed	Medium	Fast (ns)	Slow (μs)
Saturation voltage	Low	High	Low

Operation

The IGBT is controlled by the gate voltage just like a MOSFET. Essentially, an IGBT can be thought of as a voltage-controlled BJT, but with faster switching speeds. Because it is controlled by voltage on the insulated gate, the IGBT has essentially no input current and does not load the driving source. A simplified equivalent circuit for an IGBT is shown in Figure 8–55. The input element is a MOSFET, and the output element is a bipolar transistor. When the gate voltage with respect to the emitter is less than a threshold voltage, V_{thresh}, the device is turned off. The device is turned on by increasing the gate voltage to a value exceeding the threshold voltage.

The *npnp* structure of the IGBT forms a parasitic transistor and an inherent parasitic resistance within the device, as shown in red in Figure 8–56. These parasitic components

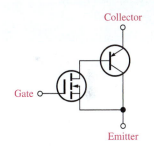

▲ **FIGURE 8–55**

Simplified equivalent circuit for an IGBT.

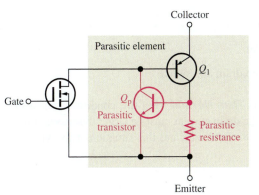

◀ **FIGURE 8–56**

Parasitic components of an IGBT that can cause latch-up.

have no effect during normal operation. However, if the maximum collector current is exceeded under certain conditions, the parasitic transistor, Q_p can turn on. If Q_p turns on, it effectively combines with Q_1 to form a parasitic element, as shown in Figure 8–56, in which a latchup condition can occur. In latch-up, the device will stay on and cannot be controlled by the gate voltage. Latch-up can be avoided by always operating within the specified limits of the device.

SECTION 8–8 CHECKUP

1. What does IGBT stand for?
2. What is a major application area for IGBTs?
3. Name an advantage of an IGBT over a power MOSFET.
4. Name an advantage of an IGBT over a power BJT.
5. What is latch-up?

8–9 TROUBLESHOOTING

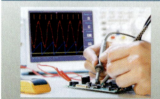

In this section, some common faults that may be encountered in FET circuits and the probable causes for each fault are discussed.

After completing this section, you should be able to

◻ **Troubleshoot FET circuits**
◻ Determine faults in self-biased JFET circuits
◻ Determine faults in MOSFET circuits
  ◆ Troubleshoot a D-MOSFET with zero bias ◆ Troubleshoot an E-MOSFET with voltage-divider bias

Faults in Self-Biased JFET Circuits

Symptom 1: $V_D = V_{DD}$ For this condition, the drain current must be zero because there is no voltage drop across R_D, as illustrated in Figure 8–57(a). As in any circuit, it is good troubleshooting practice to first check for obvious problems such as open or poor connections, as well as charred resistors. Next, disconnect power and measure suspected resistors for opens. If these are okay, the JFET is probably bad. Any of the following faults can produce this symptom:

1. No ground connection at R_S
2. R_S open
3. Open drain lead connection
4. Open source lead connection
5. FET internally open between drain and source

Symptom 2: V_D Significantly Less Than Normal For this condition, unless the supply voltage is lower than it should be, the drain current must be larger than normal because the drop across R_D is too much. Figure 8–57(b) indicates this situation. This symptom can be caused by any of the following:

1. Open R_G
2. Open gate lead
3. FET internally open at gate

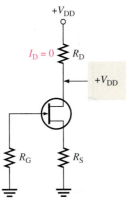 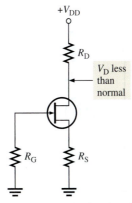

◀ FIGURE 8–57

Two symptoms in a self-biased JFET circuit.

(a) *Symptom 1*: Drain voltage equal to supply voltage

(b) *Symptom 2*: Drain voltage less than normal

Any of these three faults will cause the depletion region in the JFET to disappear and the channel to widen so that the drain current is limited only by R_D, R_S, and the small channel resistance.

Faults in D-MOSFET and E-MOSFET Circuits

One fault that is difficult to detect is when the gate opens in a zero-biased D-MOSFET. In a zero-biased D-MOSFET, the gate-to-source voltage ideally remains zero when an open occurs in the gate circuit; thus, the drain current doesn't change, and the bias appears normal, as indicated in Figure 8–58. However, static charge as a result of the open may cause I_D to behave irratically.

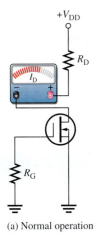 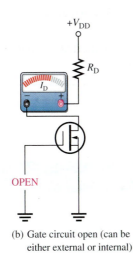

◀ FIGURE 8–58

An open fault in the gate circuit of a D-MOSFET causes no change in I_D.

(a) Normal operation

(b) Gate circuit open (can be either external or internal)

In an E-MOSFET circuit with voltage-divider bias, an open R_1 makes the gate voltage zero. This causes the transistor to be off and act like an open switch because a gate-to-source threshold voltage greater than zero is required to turn the device on. This condition is illustrated in Figure 8–59(a). If R_2 opens, the gate is at $+V_{DD}$ and the channel resistance is very low so the device approximates a closed switch. The drain current is limited only by R_D. This condition is illustrated in Figure 8–59(b).

with both positive and negative gate voltages, making it ideal for this particular application where the input voltage can have either polarity. The graph in Figure 8–63 shows that the transconductance curve depends on the value of the voltage on the second gate which, in this particular design, is set at 6 V by the R_1-R_2 voltage divider. The input from the sensor is applied to the first gate.

   3. What is the specified typical transconductance (transadmittance) for the BF998?
   4. If the drain-to-source voltage is 10 V, determine the maximum allowable drain current.
   5. If one gate is biased to 1 V, what is I_D when the other gate is 0 V?

Simulation

The pH sensor circuit is simulated in Multisim, and the results for a series of sensor input voltages are shown in Figure 8–64. The sensor is modeled as a dc source in series with an internal resistance. Notice that the output of the circuit increases as the sensor input decreases. Rheostat R_3 is used to calibrate each of the three sensor circuits so that they have an identical output voltage for a given sensor input voltage.

   6. If the output of the sensor circuit is 7 V, is the solution acidic, neutral, or basic (caustic)?
   7. Plot a graph of V_{OUT} vs. pH for each measurement in Figure 8–64.

 Simulate the pH sensor circuit using your Multisim or LT Spice software. Measure the output voltage for $V_{sensor} = 50$ mV, $V_{sensor} = 150$ mV, and $V_{sensor} = -25$ mV.

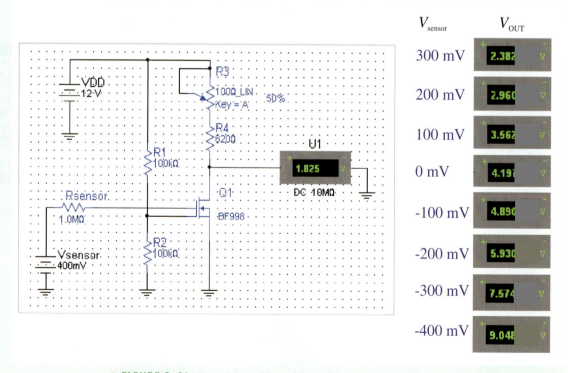

▲ FIGURE 8–64

Simulation results for the pH sensor circuit.

Prototyping and Testing

Now that the circuit has been simulated, the prototype circuit is constructed and tested. A dc voltage source can be used to provide the sensor input voltages. After the circuit is successfully tested on a protoboard, it is ready to be finalized on a printed circuit board.

Lab Experiment

To build and test a similar circuit, go to Experiment 8 in your lab manual (*Laboratory Exercises for Electronic Devices* by David Buchla and Steven Wetterling).

Circuit Board

The pH sensor circuits are implemented on a printed circuit board as shown in Figure 8–65. Each circuit monitors one of the three pH sensors in the system. Note that a single voltage divider provides +6 V to the second gate of each transistor.

8. Check the printed circuit board for correctness by comparing with the schematic in Figure 8–62.
9. Identify the connections on the back side of the board.
10. Label each input and output pin according to function.

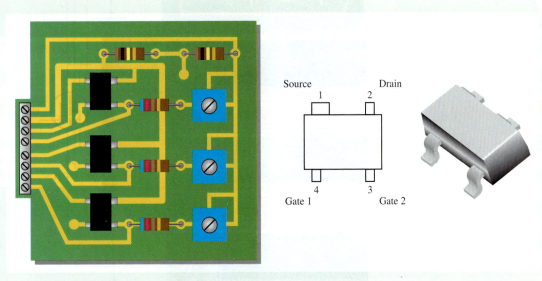

▲ FIGURE 8–65

pH sensor circuit board.

Calibration and Testing

The first step is to calibrate each of the three circuits for a pH of 7. Using a known neutral test solution in a container into which the sensors are placed, the rheostat is adjusted (if necessary) to produce the same output voltage for each circuit. In this case it is 4.197 V, as shown in Figure 8–66.

The next step is to replace the neutral solution with one that has an acidity with a known pH. All the circuits should produce the same voltage within a specified tolerance. Finally, using a basic solution with a known pH, measure the output voltages. Again, they should all agree.

▶ FIGURE 8–80

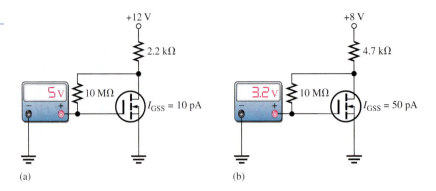

(a)   (b)

46. Determine the actual gate-to-source voltage in Figure 8–81 by taking into account the gate leakage current, I_{GSS}. Assume that I_{GSS} is 50 pA and I_D is 1 mA under the existing bias conditions.

▶ FIGURE 8–81

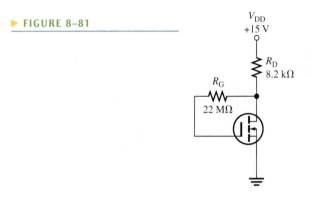

Section 8–8 **The IGBT**

47. Explain why the IGBT has a very high input resistance.

48. Explain how an excessive collector current can produce a latch-up condition in an IGBT.

Section 8–9 **Troubleshooting**

49. The current reading in Figure 8–69(a) suddenly goes to zero. What are the possible faults?

50. The current reading in Figure 8–69(b) suddenly jumps to approximately 16 mA. What are the possible faults?

51. If the supply voltage in Figure 8–69(c) is changed to −20 V, what would you see on the ammeter?

52. You measure +10 V at the drain of the MOSFET in Figure 8–77(a). The transistor checks good and the ground connections are okay. What can be the problem?

53. You measure approximately 0 V at the drain of the MOSFET in Figure 8–77(b). You can find no shorts and the transistor checks good. What is the most likely problem?

DEVICE APPLICATION PROBLEMS

54. Refer to Figure 8–61 and determine the sensor voltage for each of the following pH values.

   (a) 2   (b) 5

   (c) 7   (d) 11

55. Referring to the transconductance curves for the BF998 in Figure 8–82, determine the change in I_D when the bias on the second gate is changed from 6 V to 1 V and V_{G1S} is 0.0 V. Each curve represents a different V_{G2S} value.

56. Refer to Figure 8–64 and plot the transconductance curve (I_D vs. V_{G1S}).

Transconductance curves for BF998.

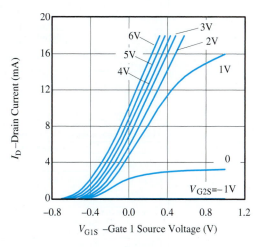

57. Refer to Figure 8–82. Determine the output voltage of the circuit in Figure 8–64 if $V_{G1S} = V_{sensor} = 0$ V and R_2 is changed to 50 kΩ.

DATASHEET PROBLEMS

58. What type of FET is the 2N5457?

59. Referring to the datasheet in Figure 8–14, determine the following:

   (a) Minimum $V_{GS(off)}$ for the 2N5457.

   (b) Maximum drain-to-source voltage for the 2N5457.

   (c) Maximum power dissipation for the 2N5458 at an ambient temperature of 25°C.

   (d) Maximum reverse gate-to-source voltage for the 2N5459.

60. Referring to Figure 8–14, determine the maximum power dissipation for a 2N5457 at an ambient temperature of 65°C.

61. Referring to Figure 8–14, determine the minimum g_{m0} for the 2N5459 at a frequency of 1 kHz.

62. Referring to Figure 8–14, what is the typical drain current in a 2N5459 for $V_{GS} = 0$ V?

63. Referring to the 2N3796 datasheet in Figure 8–83, determine the drain current for $V_{GS} = 0$ V.

64. Referring to Figure 8–83, what is the drain current for a 2N3796 when $V_{GS} = 6$ V?

65. Referring to the datasheet in Figure 8–83, determine I_D in a 2N3797 when $V_{GS} = +3$ V. Determine I_D when $V_{GS} = -2$ V.

66. Referring to Figure 8–83, how much does the maximum forward transconductance of a 2N3796 change over a range of signal frequencies from 1 kHz to 1 MHz?

67. Referring to Figure 8–83, determine the typical value of gate-to-source voltage at which the 2N3796 will go into cutoff.

Maximum Ratings

Rating	Symbol	Value	Unit
Drain-Source voltage	V_{DS}		V dc
2N3796		25	
2N3797		20	
Gate-Source voltage	V_{GS}	±10	V dc
Drain current	I_D	20	mA dc
Total device dissipation @ T_A = 25°C	P_D	200	mW
Derate above 25°C		1.14	mW/°C
Junction temperature range	T_J	+175	°C
Storage channel temperature range	T_{stg}	−65 to +200	°C

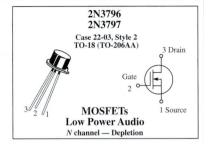

2N3796
2N3797

Case 22-03, Style 2
TO-18 (TO-206AA)

3 Drain

Gate
2

1 Source

MOSFETs
Low Power Audio
N channel — Depletion

Electrical Characteristics (T_A = 25°C unless otherwise noted.)

Characteristic		Symbol	Min	Typ	Max	Unit
OFF Characteristics						
Drain-Source breakdown voltage		$V_{(BR)DSX}$				V dc
(V_{GS} = −4.0 V, I_D = 5.0 μA)	2N3796		25	30	–	
(V_{GS} = −7.0 V, I_D = 5.0 μA)	2N3797		20	25	–	
Gate reverse current		I_{GSS}				pA dc
(V_{GS} = −10 V, V_{DS} = 0)			–	–	1.0	
(V_{GS} = −10 V, V_{DS} = 0, T_A = 150°C)			–	–	200	
Gate-Source cutoff voltage		$V_{GS(off)}$				V dc
(I_D = 0.5 μA, V_{DS} = 10 V)	2N3796		–	−3.0	−4.0	
(I_D = 2.0 μA, V_{DS} = 10 V)	2N3797		–	−5.0	−7.0	
Drain-Gate reverse current		I_{DGO}	–		1.0	pA dc
(V_{DG} = 10 V, I_S = 0)						
ON Characteristics						
Zero-Gate-Voltage drain current	2N3796	I_{DSS}				mA dc
(V_{DS} = 10 V, V_{GS} = 0)	2N3797					
			0.5	1.5	3.0	
			2.0	2.9	6.0	
On-State drain current	2N3796	$I_{D(on)}$				mA dc
(V_{DS} = 10 V, V_{GS} = +3.5 V)	2N3797		7.0	8.3	14	
			9.0	14	18	
Small-Signal Characteristics						
Forward-transfer admittance		$|y_{fs}|$				μmhos
(V_{DS} = 10 V, V_{GS} = 0, f = 1.0 kHz)	2N3796		900	1200	1800	or
	2N3797		1500	2300	3000	μS
(V_{DS} = 10 V, V_{GS} = 0, f = 1.0 MHz)	2N3796		900	–	–	
	2N3797		1500	–	–	
Output admittance		$|y_{os}|$				μmhos
(V_{DS} = 10 V, V_{GS} = 0, f = 1.0 kHz)	2N3796		–	12	25	or
	2N3797		–	27	60	μS
Input capacitance		C_{iss}				pF
(V_{DS} = 10 V, V_{GS} = 0, f = 1.0 MHz)	2N3796		–	5.0	7.0	
	2N3797		–	6.0	8.0	
Reverse transfer capacitance		C_{rss}				pF
(V_{DS} = 10 V, V_{GS} = 0, f = 1.0 MHz)			–	0.5	0.8	
Functional Characteristics						
Noise figure		NF				dB
(V_{DS} = 10 V, V_{GS} = 0, f = 1.0 kHz, R_S = 3 megohms)			–	3.8	–	

2N3796

2N3797

▲ **FIGURE 8–83**

Partial datasheet for the 2N3797 D-MOSFET.

ADVANCED PROBLEMS

68. Find V_{DS} and V_{GS} in Figure 8–84 using minimum datasheet values.

▶ **FIGURE 8–84**

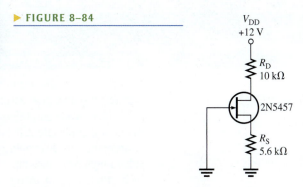

69. Determine the maximum I_D and V_{GS} for the circuit in Figure 8–85.

▶ **FIGURE8–85**

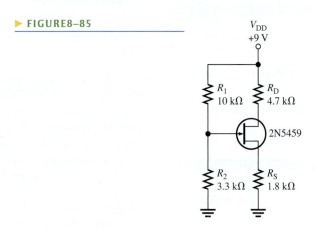

70. Determine the range of possible Q-point values from minimum to maximum for the circuit in Figure 8–84.

71. Find the drain-to-source voltage for the pH sensor circuit in Figure 8–62 when a pH of 5 is measured. Assume the rheostat is set to produce 4 V at the drain when a pH of 7 is measured.

72. Design a MOSFET circuit with zero bias using a 2N3797 that operates from a +9 V dc supply and produces a V_{DS} of 4.5 V. The maximum current drawn from the source is to be 1 mA.

73. Design a circuit using an *n*-channel E-MOSFET with the following datasheet specifications: $I_{D(on)}$ = 500 mA@ V_{GS} = 10 V and $V_{GS(th)}$ = 1 V. Use a +12 V dc supply voltage with voltage-divider bias. The voltage at the drain with respect to ground is to be +8 V. The maximum current from the supply is to be 20 mA.

MULTISIM TROUBLESHOOTING PROBLEMS

These file circuits are in the Troubleshooting Problems folder on the website.

74. Open file TPM08-74 and determine the fault.

75. Open file TPM08-75 and determine the fault.

76. Open file TPM08-76 and determine the fault.

77. Open file TPM08-77 and determine the fault.

78. Open file TPM08-78 and determine the fault.

79. Open file TPM08-79 and determine the fault.

80. Open file TPM08-80 and determine the fault.

81. Open file TPM08-81 and determine the fault.

82. Open file TPM08-82 and determine the fault.

9

FET AMPLIFIERS AND SWITCHING CIRCUITS

CHAPTER OBJECTIVES

◆ Explain and analyze the operation of common-source FET amplifiers

◆ Explain and analyze the operation of common-drain FET amplifiers

◆ Explain and analyze the operation of common-gate FET amplifiers

◆ Discuss the operation of a class D amplifier

◆ Describe how MOSFETs can be used in analog switching applications

◆ Describe how MOSFETs are used in digital switching applications

◆ Troubleshoot FET amplifiers

KEY TERMS

◆ Common-source
◆ Common-drain
◆ Source-follower
◆ Common-gate
◆ Cascode amplifier
◆ Class D amplifier
◆ Pulse-width modulation
◆ Analog switch
◆ CMOS

DEVICE APPLICATION PREVIEW

A JFET common-source amplifier and a common-gate amplifier are combined in a cascode arrangement for an active antenna. Cascode amplifiers are often used for RF (radio frequency) applications to achieve improved high-frequency performance. In this application, the cascode amplifier provides a high resistance input for a whip antenna, as well as high gain to amplify extremely small antenna signals.

VISIT THE WEBSITE

Study aids, Multisim files, and LT Spice files for this chapter are available at https://www.pearsonhighered.com/careersresources/

INTRODUCTION

Because of their extremely high input resistance and low noise, FET amplifiers are a good choice for certain applications, such as amplifying low-level signals in the first stage of a communication receiver. FETs also have the advantage in certain power amplifiers and in switching circuits because biasing is simple and more efficient. The standard amplifier configurations are common-source (CS), common-drain (CD) and common-base (CB), which are analogous to CE, CC, and CB configurations of BJTs.

FETs can be used in any of the amplifier types introduced earlier (class A, class B, and class C). In some cases, the FET circuit will perform better; in other cases, the BJT circuit is superior because BJTs have higher gain and better linearity. Another type of amplifier (class D) is introduced in this chapter because FETs are always superior to BJTs in class D and you will rarely see BJTs used in class D. The class D amplifier is a switching amplifier that is normally either in cutoff or saturation. It is used in analog power amplifiers with a circuit called a pulse-width modulator, introduced in Section 9–4.

FETs are superior to BJTs in nearly all switching applications. Various switching circuits—analog switches, analog multiplexers, and switched capacitors—are discussed. In addition, common digital switching circuits are introduced using CMOS (complementary MOS).

9–1 THE COMMON-SOURCE AMPLIFIER

When used in amplifier applications, the FET has an important advantage compared to the BJT due to the FET's extremely high input impedance. Disadvantages, however, include higher distortion and lower gain. The particular application will usually determine which type of transistor is best suited. The common-source (CS) amplifier is comparable to the common-emitter BJT amplifier that you studied in Chapter 6.

After completing this section, you should be able to

❑ **Explain and analyze the operation of common-source FET amplifiers**
❑ Discuss and analyze the FET ac model
❑ Describe and analyze common-source JFET amplifier operation
❑ Perform dc analysis of a JFET amplifier
   ◆ Use the graphical approach   ◆ Use the mathematical approach
❑ Discuss and analyze the ac equivalent circuit of a JFET amplifier
   ◆ Determine the signal voltage at the gate   ◆ Determine the voltage gain
❑ Explain the effect of an ac load on the voltage gain
❑ Discuss phase inversion
❑ Determine amplifier input resistance
❑ Describe and analyze D-MOSFET amplifier operation
❑ Describe and analyze E-MOSFET amplifier operation
   ◆ Determine input resistance

FET AC Model

An equivalent FET model is shown in Figure 9–1 for the constant current region of the characteristic curve. In part (a), the internal resistance, r'_{gs}, appears between the gate and source, and a current source equal to $g_m V_{gs}$ appears between the drain and source. Also, the internal drain-to-source resistance, r'_{ds}, is included. This is just the slope of the characteristic curve in the constant current region. In part (b), a simplified ideal model is shown. The resistance, r'_{gs}, is assumed to be extremely large so that an open circuit between the gate and source can be assumed. Also, r'_{ds} is assumed large enough to neglect. This approximation is equivalent to assuming a constant current (horizontal line) for a given drain curve.

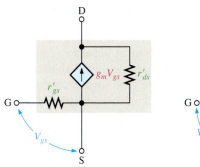

(a) Complete   (b) Simplified

◀ FIGURE 9–1

Internal FET equivalent circuits.

An ideal FET circuit model with an external ac drain resistance is shown in Figure 9–2. The ac voltage gain of this circuit is V_{out}/V_{in}, where $V_{in} = V_{gs}$ and $V_{out} = V_{ds}$. The voltage gain expression is, therefore,

$$A_v = \frac{V_{ds}}{V_{gs}}$$

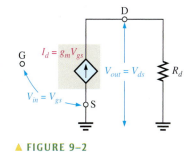

▲ FIGURE 9–2

Simplified FET equivalent circuit with an external ac drain resistance.

From the equivalent circuit in Figure 9–2,

$$V_{ds} = I_d R_d$$

and from the definition of transconductance, $g_m = I_d / V_{gs}$,

$$V_{gs} = \frac{I_d}{g_m}$$

Substituting the two preceding expressions into the equation for voltage gain yields

$$A_v = \frac{I_d R_d}{I_d / g_m} = \frac{g_m I_d R_d}{I_d}$$

Equation 9–1
$$A_v = g_m R_d$$

EXAMPLE 9–1

From the datasheet, a certain JFET has a typical g_{m0} of 6 mS and a $V_{GS(off)}$ of -5 V. Assume it is biased with $V_{GS} = -1.67$ V. With an external ac drain resistance of 1.5 kΩ, what is the ideal voltage gain?

Solution

Start by finding g_m:

$$g_m = g_{m0}\left(1 - \frac{V_{GS}}{V_{GS(off)}}\right) = 6\ \text{mS}\left(1 - \frac{-1.67\ \text{V}}{-5.0\ \text{V}}\right) = 4.0\ \text{mS}$$

$$A_v = g_m R_d = (4.0\ \text{mS})(1.5\ \text{k}\Omega) = \textbf{6.0}$$

Related Problem

What is the ideal voltage gain when $g_{m0} = 8.0$ mS, $V_{GS} = -1$ V, $V_{GS(off)} = -4$ V, and $R_d = 2.2$ kΩ?

_____

*Answers can be found at www.pearsonhighered.com/floyd.

JFET Amplifier Operation

A **common-source** JFET amplifier is one in which the ac input signal is applied to the gate and the ac output signal is taken from the drain. The source terminal is common to both the input and output signal. A common-source amplifier either has no source resistor or has a bypassed source resistor, so the source is connected to ac ground. A self-biased common-source n-channel JFET amplifier with an ac source capacitively coupled to the gate is shown in Figure 9–3(a). The resistor, R_G, serves two purposes: It keeps the gate at approximately 0 V dc (because I_{GSS} is extremely small), and its large value (usually several megohms) prevents loading of the ac signal source. A bias voltage is produced by the drop across R_S. The bypass capacitor, C_2, keeps the source of the JFET at ac ground.

A common-source amplifier has much lower gain than its BJT counterpart, the common-emitter amplifier. Its big advantage is the very high input impedance, which is particularly useful in instrumentation and measurement because low-level signals from high-impedance sources are common in these cases. JFETs are often used in combination with BJTs and operational amplifiers to take advantage of the best characteristics of each.

▶ **FIGURE 9–3**

JFET common-source amplifier.

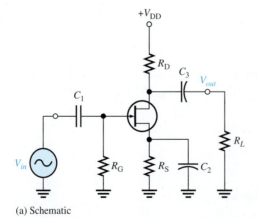

(a) Schematic

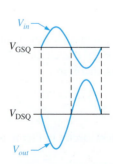

(b) Voltage waveform relationship

The input signal voltage causes the gate-to-source voltage to swing above and below its Q-point value (V_{GSQ}), causing a corresponding swing in drain current. As the drain current increases, the voltage drop across R_D also increases, causing the drain voltage to decrease. The drain current swings above and below its Q-point value in phase with the gate-to-source voltage. The drain-to-source voltage swings above and below its Q-point value (V_{DSQ}) and is 180° out of phase with the gate-to-source voltage, as illustrated in Figure 9–3(b).

A Graphical Picture The operation just described for an *n*-channel JFET is illustrated graphically on both the transfer characteristic curve and the drain characteristic curve in Figure 9–4. Part (a) shows how a sinusoidal variation, V_{gs}, produces a corresponding sinusoidal variation in I_d. As V_{gs} swings from its Q-point value to a more negative value, I_d decreases from its Q-point value. As V_{gs} swings to a less negative value, I_d increases. Figure 9–4(b) shows a view of the same operation using the drain curves. The signal at the gate drives the drain current above and below the Q-point on the load line, as indicated by the arrows. Lines projected from the peaks of the gate voltage across to the I_D axis and down to the V_{DS} axis indicate the peak-to-peak variations of the drain current and drain-to-source voltage, as shown. Because the transfer characteristic curve is nonlinear, the output will have some distortion. This can be minimized if the signal swings over a limited portion of the load line; this occurs naturally if it is only the input stage of a multistage amplifier.

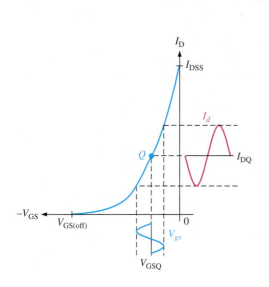

(a) JFET (*n*-channel) transfer characteristic curve showing signal operation

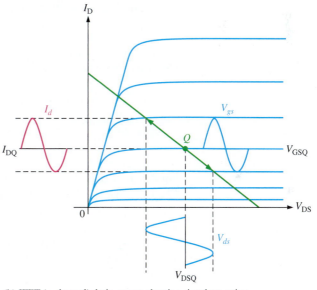

(b) JFET (*n*-channel) drain curves showing signal operation

▲ **FIGURE 9–4**

JFET characteristic curves.

DC Analysis

The first step in analyzing a JFET amplifier is to determine the dc conditions including I_D and V_S. I_D determines the Q-point for an amplifier and enables you to calculate V_D, so it is useful to determine its value. It can be found either graphically or mathematically. The graphical approach, introduced in Chapter 8 using the transfer characteristic curve, will be applied to an amplifier here. The same result can be obtained by expanding Equation 8–1, which is the mathematical description of the transfer characteristic curve. The amplifier shown in Figure 9–5 will be used to illustrate both approaches. To simplify the dc analysis, the equivalent circuit is shown in Figure 9–6; capacitors appear open to dc, so they are removed.

Graphical Approach Recall from Section 8–2 that the JFET universal transfer characteristic illustrates the relationship between the output current and the input voltage. The end points of the transfer curve are at I_{DSS} and $V_{GS(off)}$. A dc graphical solution is done by plotting the load line (for the self-biased case shown) on the same plot and reading the values of V_{GS} and I_D at the intersection of these plots (Q-point).

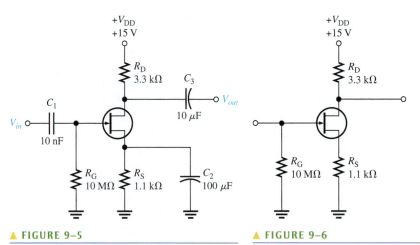

▲ **FIGURE 9–5**

JFET common-source amplifier.

▲ **FIGURE 9–6**

DC equivalent for the amplifier in Figure 9–5.

EXAMPLE 9–2

Determine I_D and V_{GS} at the Q-point for the JFET amplifier in Figure 9–6. The typical I_{DSS} for this particular JFET is 4.3 mA and $V_{GS(off)}$ is -7.7 V.

Solution

Plot the transfer characteristic curve. The end points are at I_{DSS} and $V_{GS(off)}$. You can plot two additional points quickly by noting from the universal curve in Figure 8–12 that

$$V_{GS} = 0.3V_{GS(off)} = -2.31 \text{ V} \qquad \text{when } I_D = \frac{I_{DSS}}{2} = 2.15 \text{ mA}$$

and

$$V_{GS} = 0.5V_{GS(off)} = -3.85 \text{ V} \qquad \text{when } I_D = \frac{I_{DSS}}{4} = 1.075 \text{ mA}$$

For this particular JFET, the points are plotted as shown in Figure 9–7(a). Recall from Chapter 8 that the load line starts at the origin and goes to a point where $I_D = I_{DSS}$ and $V_{GS} = I_{DSS}R_S = (-4.3 \text{ mA})(1.1 \text{ k}\Omega) = -4.73$ V as shown in Figure 9–7(b). Connect a load line from the origin to this point and read the I_D and V_{GS} values from the intersection (Q-point), as shown in Figure 9–7(b). For the graph shown, $I_D = \textbf{2.2 mA}$ and $V_{GS} = \textbf{-2.4 V}$.

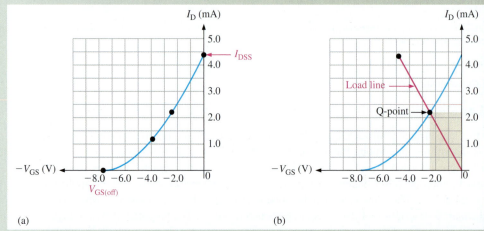

(a) (b)

▲ **FIGURE 9–7**

Related Problem

Show the Q-point if the transistor is replaced with one with an $I_{DSS} = 5.0$ mA and a $V_{GS(off)} = -8$ V.

Mathematical Approach The mathematical approach is more tedious than the graphical approach but can be simplified with tools on line or using a graphing calculator. Recall that Equation 8–1 is the formula that relates I_D to other quantities:

$$I_D = I_{DSS}\left(1 - \frac{V_{GS}}{V_{GS(off)}}\right)^2$$

For *n*-channel JFETs, both V_{GS} and $V_{GS(off)}$ are negative quantities; for *p*-channel JFETs, they are both positive. For this reason, the term represented by the fraction can be expressed as an absolute (unsigned) value without affecting the result. That is,

$$I_D = I_{DSS}\left(1 - \left|\frac{V_{GS}}{V_{GS(off)}}\right|\right)^2$$

The absolute value of V_{GS} is just $I_D R_S$ and the absolute value of $V_{GS(off)}$ is V_P. (Recall that $V_P = |V_{GS(off)}|$.) By substitution, we can express I_D in terms of known quantities:

$$I_D = I_{DSS}\left(1 - \frac{I_D R_S}{V_p}\right)^2$$

Equation 9–2

The result in Equation 9–2 has I_D on both sides. Isolating I_D requires the solution of the quadratic form, which is given in "Derivations of Selected Equations" at www.pearsonhighered .com/floyd. An easier approach is to enter Equation 9–2 into a graphing calculator such as the TI-89. The steps for determining I_D using the TI-89 are given in Example 9–3.

EXAMPLE 9–3

Determine I_D and V_{GS} at the Q-point for the JFET amplifier in Figure 9–6 using the mathematical/calculator approach. The I_{DSS} for this particular JFET is 4.3 mA and $V_{GS(off)}$ is −7.7 V.

Solution To calculate I_D using the TI-89, follow these six steps.[*]

Step 1: On the Applications screen select the Numeric Solver logo.

f(x) = 0

Numeric So …

Step 2: Press ENTER to display the Numeric Solver screen.

Enter Equation

eqn:

Step 3: Enter the equation. Each letter in the variables must be preceded by ALPHA.

Enter Equation

eqn: id=idss*(1-id*rs/vp)^2

Step 4: Press ENTER to display the variables.

Enter Equation

eqn: id=idss*(1-id*rs/vp)^2

id=

idss=

rs=

vp=

Step 5: Enter the value of each variable except id.

> **Enter Equation**
>
> eqn: id=idss*(1-id*rs/vp)^2
>
> id=
>
> idss=.0043
>
> rs=1100
>
> vp=7.7

Step 6: Move the cursor to id and Press F2 to solve. The answer is **.0021037......(2.10 mA)**.

Calculate V_{GS}.

$$V_{GS} = -I_D R_S = -(2.10\text{ mA})(1.1\text{ k}\Omega) = -2.31\text{ V}$$

Related Problem Calculate the solution for the Related Problem in Example 9–2.

The following website is a tutorial for the TI-89 calculator: http://www.math.lsu.edu/~neal/TI_89/index.html

Another approach to solving for I_D is to put Equation 9–2 into quadratic form. Recall from algebra that the standard quadratic form is $ax^2 + bx + c = 0$ and that the solution to the quadratic equation has two roots given by the general formula:

$$x = \frac{-b \pm \sqrt{b^2 - 4ac}}{2a}$$

By expanding Equation 9–2, it can be expressed in quadratic form as:

Equation 9–3

$$\left(I_{DSS}\left(\frac{R_S}{V_P}\right)^2\right)I_D^2 + \left(-2\frac{I_{DSS}R_S}{V_P} - 1\right)I_D + I_{DSS} = 0$$

$\left(I_{DSS}\left(\frac{R_S}{V_P}\right)^2\right)$ is the a coefficient

$\left(-2\frac{I_{DSS}R_S}{V_P} - 1\right)$ is the b coefficient

I_{DSS} is the c coefficient

I_D is the unknown, which is represented by x in the quadratic formula.

There are simple on-line tools that will enable you to enter the coefficients and solve for the unknown. Alternatively, you can solve for I_D by substituting into the general solution for the quadratic equation as shown in the following example.

EXAMPLE 9–4

Determine I_D and V_{GS} at the Q-point for the JFET amplifier in Figure 9–6 by solving the quadratic equation. I_{DSS} was given in Example 9–3 as 4.3 mA and $V_{GS(off)}$ was given as −7.7 V.

Solution Notice that $V_P = |V_{GS(off)}| = +7.7$ V.

Find the values for the coefficients:

$$a = \left(I_{DSS}\left(\frac{R_S}{V_P}\right)^2\right) = \left(0.0043\left(\frac{1100}{7.7}\right)^2\right) = 87.76$$

$$b = \left(-2\frac{I_{DSS}R_S}{V_P} - 1\right) = \left(-2\frac{(0.0043)(1100)}{7.7} - 1\right) = -2.228$$

$$c = 0.0043$$

Substitute these into the general solution for the quadratic equation:

$$I_D = \frac{-b \pm \sqrt{b^2 - 4ac}}{2a} = \frac{2.228 \pm \sqrt{(-2.228)^2 - 4(87.76)(0.0043)}}{2(87.76)}$$

The two roots are 23.3 mA and 2.10 mA. Since 23.3 mA is not possible, it is rejected.

$$I_D = \textbf{2.10 mA}$$

$$V_{GS} = -I_D R_S = -(2.1 \text{ mA})(1.1 \text{ k}\Omega) = \textbf{-2.31 V}$$

Related Problem   Use the quadratic formula to calculate I_D if I_{DSS} is changed to 6.0 mA and other quantities remain the same.

AC Equivalent Circuit

To analyze the signal operation of the amplifier in Figure 9–5, develop an ac equivalent circuit as follows. Replace the capacitors by effective shorts, based on the simplifying assumption that $X_C \cong 0$ at the signal frequency. Replace the dc source by a ground, based on the assumption that the voltage source has a zero internal resistance. The V_{DD} terminal is at a zero-volt ac potential and therefore acts as an ac ground.

The ac equivalent circuit is shown in Figure 9–8(a). Notice that the $+V_{DD}$ end of R_d and the source terminal are both effectively at ac ground. Recall that in ac analysis, the ac ground and the actual circuit ground are treated as the same point.

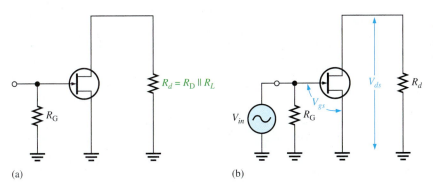

◀ **FIGURE 9–8**

AC equivalent for the amplifier in Figure 9–5.

Signal Voltage at the Gate   An ac voltage source is shown connected to the input in Figure 9–8(b). Since the input resistance to a JFET is extremely high, practically all of the input voltage from the signal source appears at the gate with very little voltage dropped across the internal source resistance.

$$V_{gs} = V_{in}$$

Voltage Gain The expression for JFET voltage gain that was given in Equation 9–1 applies to the common-source amplifier with the source terminal at ac ground.

Equation 9–4

$$A_v = g_m R_d$$

The output signal voltage V_{ds} at the drain is

$$V_{out} = V_{ds} = A_v V_{gs}$$

or

$$V_{out} = g_m R_d V_{in}$$

where $R_d = R_D \parallel R_L$ and $V_{in} = V_{gs}$.

EXAMPLE 9–5

What is the total output voltage for the unloaded amplifier in Figure 9–9? I_{DSS} is 4.3 mA; $V_{GS(off)}$ is −2.7 V.

▶ **FIGURE 9–9**

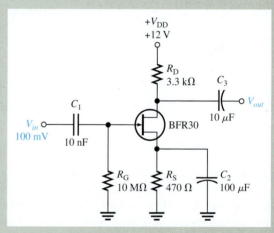

Solution Use either a graphical approach, as shown in Example 9–2, or a mathematical approach with a graphing calculator, as shown in Example 9–3, to determine I_D. The calculator solution gives

$$I_D = 1.91 \text{ mA}$$

Using this value, calculate V_D.

$$V_D = V_{DD} - I_D R_D = 12 \text{ V} - (1.91 \text{ mA})(3.3 \text{ k}\Omega) = 5.70 \text{ V}$$

Next calculate g_m as follows:

$$V_{GS} = -I_D R_S = -(1.91 \text{ mA})(470 \text{ }\Omega) = -0.90 \text{ V}$$

$$g_{m0} = \frac{2 I_{DSS}}{|V_{GS(off)}|} = \frac{2(4.3 \text{ mA})}{2.7 \text{ V}} = 3.18 \text{ mS}$$

$$g_m = g_{m0}\left(1 - \frac{V_{GS}}{V_{GS(off)}}\right) = 3.18 \text{ mS}\left(1 - \frac{-0.90 \text{ V}}{-2.7 \text{ V}}\right) = 2.12 \text{ mS}$$

Finally, find the ac output voltage.

$$V_{out} = A_v V_{in} = g_m R_D V_{in} = (2.12 \text{ mS})(3.3 \text{ k}\Omega)(100 \text{ mV}) = \textbf{700 mV}$$

Related Problem Confirm the calculator solution for I_D is correct by using the graphical method.

Effect of an AC Load on Voltage Gain

When a load is connected to an amplifier's output through a coupling capacitor, as shown in Figure 9–10(a), the ac drain resistance is effectively R_D in parallel with R_L because the upper end of R_D is at ac ground. The ac equivalent circuit is shown in Figure 9–10(b). The total ac drain resistance is

$$R_d = \frac{R_D R_L}{R_D + R_L}$$

The effect of R_L is to reduce the unloaded voltage gain, as Example 9–6 illustrates.

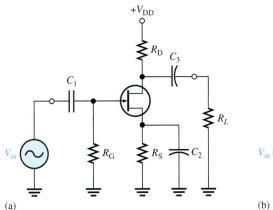

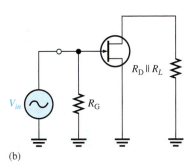

◀ FIGURE 9–10

JFET amplifier and its ac equivalent.
(a) JFET amplifier (b) ac equivalent of JFET amplifier

EXAMPLE 9–6

If a 4.7 kΩ load resistor is ac coupled to the output of the amplifier in Example 9–5, what is the resulting rms output voltage?

Solution   The ac drain resistance is

$$R_d = \frac{R_D R_L}{R_D + R_L} = \frac{(3.3\text{ k}\Omega)(4.7\text{ k}\Omega)}{8\text{ k}\Omega} = 1.94\text{ k}\Omega$$

Calculation of V_{out} yields

$$V_{out} = A_v V_{in} = g_m R_d V_{in} = (2.12\text{ mS})(1.94\text{ k}\Omega)(100\text{ mV}) = \textbf{411 mV rms}$$

The unloaded ac output voltage was 700 mV in Example 9–5.

Related Problem   If a 3.3 kΩ load resistor is ac coupled to the output of the amplifier in Example 9–5, what is the resulting rms output voltage?

Phase Inversion

The output voltage (at the drain) is 180° out of phase with the input voltage (at the gate). The phase inversion can be designated by a negative voltage gain, $-A_v$. Recall that the common-emitter BJT amplifier also exhibited a phase inversion.

Input Resistance

Because the input to a common-source amplifier is at the gate, the input resistance is extremely high. Ideally, it approaches infinity and can generally be neglected. As you know, the high input resistance is produced by the reverse-biased *pn* junction in a JFET and by the insulated gate structure in a MOSFET. The actual input resistance seen by the signal source is the gate-to-ground resistor, R_G, in parallel with the FET's input resistance, V_{GS}/I_{GSS}. The reverse leakage current, I_{GSS}, is typically given on the datasheet for a specific value of V_{GS} so that the input resistance of the device can be calculated.

Equation 9–5
$$R_{in} = R_G \left\| \left(\frac{V_{GS}}{I_{GSS}} \right) \right.$$

Since the term V_{GS}/I_{GSS} is typically much larger than R_G, the input resistance is very close to the value of R_G, as Example 9–7 shows.

EXAMPLE 9–7

What input resistance is seen by the signal source in Figure 9–11? $I_{GSS} = 30$ nA at $V_{GS} = 10$ V.

▶ **FIGURE 9–11**

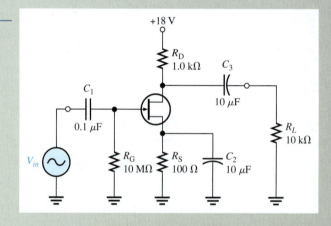

Solution   The input resistance at the gate of the JFET is

$$R_{IN(gate)} = \frac{V_{GS}}{I_{GSS}} = \frac{10 \text{ V}}{30 \text{ nA}} = 333 \text{ M}\Omega$$

The input resistance seen by the signal source is

$$R_{in} = R_G \| R_{IN(gate)} = 10 \text{ M}\Omega \| 333 \text{ M}\Omega = \mathbf{9.7 \text{ M}\Omega}$$

For all practical purposes, R_{IN} can be assumed equal to R_G.

Related Problem   How much is the total input resistance if $I_{GSS} = 1$ nA at $V_{GS} = 10$ V?

D-MOSFET Amplifier Operation

A zero-biased common-source n-channel D-MOSFET with an ac source capacitively coupled to the gate is shown in Figure 9–12. The gate is at approximately 0 V dc and the source terminal is at ground, thus making $V_{GS} = 0$ V.

▶ **FIGURE 9–12**

Zero-biased D-MOSFET common-source amplifier.

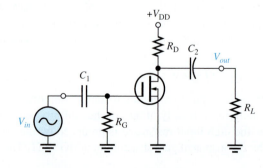

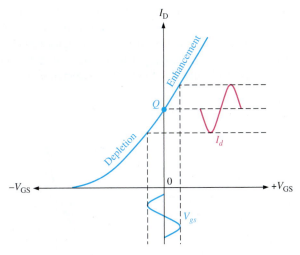

Depletion-enhancement operation of D-MOSFET shown on transfer characteristic curve.

The signal voltage causes V_{gs} to swing above and below its zero value, producing a swing in I_d, as shown in Figure 9–13. The negative swing in V_{gs} produces the depletion mode, and I_d decreases. The positive swing in V_{gs} produces the enhancement mode, and I_d increases. Note that the enhancement mode is to the right of the vertical axis ($V_{GS} = 0$), and the depletion mode is to the left. The dc analysis of this amplifier is somewhat easier than for a JFET because $I_D = I_{DSS}$ at $V_{GS} = 0$. Once I_D is known, the analysis involves calculating only V_D.

$$V_D = V_{DD} - I_D R_D$$

The ac analysis is the same as for the JFET amplifier.

E-MOSFET Amplifier Operation

A common-source n-channel E-MOSFET with voltage-divider bias with an ac source capacitively coupled to the gate is shown in Figure 9–14(a). The gate is biased with a positive voltage such that $V_{GS} > V_{GS(th)}$.

A variation of this amplifier is shown in Figure 9–14(b). Both circuits use voltage-divider bias, but in cases where a high resistance source is the driver, it is possible to increase the input resistance by adding a series resistor (R_3) and connecting the signal directly to the gate through a capacitor; the extremely high input resistance of the FET enables this configuration without affecting the dc bias. Instead of capacitively coupling the load, another option is to use it in place of the drain resistor. This has the advantage of higher efficiency for the ac signal but the disadvantage is that the load will have a dc voltage ($+V_{DD}$) on it, which increases power in the load. In (b), a speaker is shown as the load. The amplifier in both (a) and (b) is connected as a common-source class-A amplifier; for this reason it is

E-MOSFET Amplifier

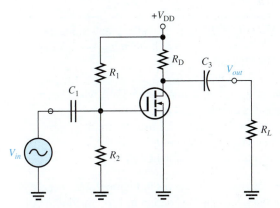

(a) A common- source class A amplifier using an E-MOSFET

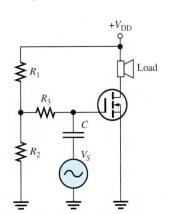

(b) A variation of the amplifier in part (a)

generally used with power levels of 1 W or less. The circuit can have excellent linearity with a VMOS transistor such as the VN66AFD E-MOSFET.

As with the JFET and D-MOSFET, the signal voltage produces a swing in V_{gs} above and below its Q-point value, V_{GSQ}. This, in turn, causes a swing in I_d above and below its Q-point value, I_{DQ}, as illustrated in Figure 9–15. Operation is entirely in the enhancement mode.

▶ **FIGURE 9–15**

E-MOSFET (*n*-channel) operation shown on transfer characteristic curve. The *p*-channel E-MOSFET is the mirror image of this, plotted in the second quadrant.

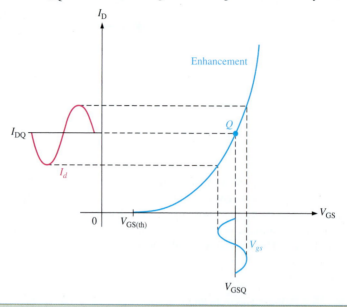

EXAMPLE 9–8

Transfer characteristic curves for a particular *n*-channel JFET, D-MOSFET, and E-MOSFET are shown in Figure 9–16. Determine the peak-to-peak variation in I_d when V_{gs} is varied ± 1 V about its Q-point value for each curve.

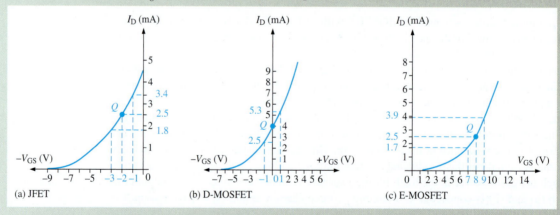

(a) JFET (b) D-MOSFET (c) E-MOSFET

▲ **FIGURE 9–16**

Solution (a) The JFET Q-point is at $V_{GS} = -2$ V and $I_D = 2.5$ mA. From the graph in Figure 9–16(a), $I_D = 3.4$ mA when $V_{GS} = -1$ V, and $I_D = 1.8$ mA when $V_{GS} = -3$ V. The peak-to-peak drain current is therefore **1.6 mA.**

(b) The D-MOSFET Q-point is at $V_{GS} = 0$ V and $I_D = I_{DSS} = 4$ mA. From the graph in Figure 9–16(b), $I_D = 2.5$ mA when $V_{GS} = -1$ V, and $I_D = 5.3$ mA when $V_{GS} = +1$ V. The peak-to-peak drain current is therefore **2.8 mA.**

(c) The E-MOSFET Q-point is at $V_{GS} = +8$ V and $I_D = 2.5$ mA. From the graph in Figure 9–16(c), $I_D = 3.9$ mA when $V_{GS} = +9$ V, and $I_D = 1.7$ mA when $V_{GS} = +7$ V. The peak-to-peak drain current is therefore **2.2 mA.**

Related Problem As the Q-point is moved toward the bottom end of the curves in Figure 9–16, does the variation in I_D increase or decrease for the same ± 1 V variation in V_{GS}? In addition to the change in the amount that I_D varies, what else will happen?

Both circuits in Figure 9–14 used voltage-divider bias to achieve a V_{GS} above threshold. The general dc analysis proceeds as follows using the E-MOSFET characteristic equation (Equation 8–4) to solve for I_D.

$$V_{GS} = \left(\frac{R_2}{R_1 + R_2}\right)V_{DD}$$

$$I_D = K(V_{GS} - V_{GS(th)})^2$$

$$V_{DS} = V_{DD} - I_D R_D$$

The voltage gain expression is the same as for the JFET and D-MOSFET circuits that have standard voltage-divider bias. The ac input resistance for the circuit in Figure 9–14(a) is

$$R_{in} = R_1 \parallel R_2 \parallel R_{IN(gate)}$$

Equation 9–6

where $R_{IN(gate)} = V_{GS}/I_{GSS}$.

EXAMPLE 9–9

A common-source amplifier using an E-MOSFET is shown in Figure 9–17. Find V_{GS}, I_D, V_{DS}, and the ac output voltage. Assume that for this particular device, $I_{D(on)} = 200$ mA at $V_{GS} = 4$ V, $V_{GS(th)} = 2$ V, and $g_m = 23$ mS. $V_{in} = 25$ mV.

▶ **FIGURE 9–17**

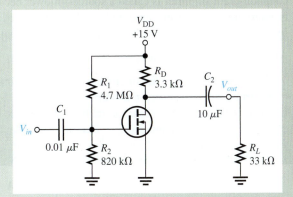

Solution

$$V_{GS} = \left(\frac{R_2}{R_1 + R_2}\right)V_{DD} = \left(\frac{820 \text{ k}\Omega}{5.52 \text{ M}\Omega}\right)15 \text{ V} = \mathbf{2.23 \text{ V}}$$

For $V_{GS} = 4$ V,

$$K = \frac{I_{D(on)}}{(V_{GS} - V_{GS(th)})^2} = \frac{200 \text{ mA}}{(4 \text{ V} - 2 \text{ V})^2} = 50 \text{ mA/V}^2$$

Therefore,

$$I_D = K(V_{GS} - V_{GS(th)})^2 = (50 \text{ mA/V}^2)(2.23 \text{ V} - 2 \text{ V})^2 = \mathbf{2.65 \text{ mA}}$$

$$V_{DS} = V_{DD} - I_D R_D = 15 \text{ V} - (2.65 \text{ mA})(3.3 \text{ k}\Omega) = \mathbf{6.26 \text{ V}}$$

$$R_d = R_D \parallel R_L = 3.3 \text{ k}\Omega \parallel 33 \text{ k}\Omega = 3 \text{ k}\Omega$$

The ac output voltage is

$$V_{out} = A_v V_{in} = g_m R_d V_{in} = (23 \text{ mS})(3 \text{ k}\Omega)(25 \text{ mV}) = \mathbf{1.73 \text{ V}}$$

Related Problem

For the E-MOSFET in Figure 9–17, $I_{D(on)} = 25$ mA at $V_{GS} = 5$ V, $V_{GS(th)} = 1.5$ V, and $g_m = 10$ mS. Find V_{GS}, I_D, V_{DS}, and the ac output voltage. $V_{in} = 25$ mV.

Open the Multisim file EXM09-09 or the LT Spice file EXS09-09 in the Examples folder on the website. Determine I_D, V_{DS}, and V_{out} using the specified value of V_{in}. Compare with the calculated values.

Electrical Characteristics ($T_A = 25°C$ unless otherwise noted.)

Characteristic		Symbol	Min	Typ	Max	Unit
OFF Characteristics						
Gate-Source breakdown voltage		$V_{(BR)GSS}$		–	–	V dc
($I_G = 10\ \mu A$ dc, $V_{DS} = 0$) 2N5460, 2N5461, 2N5462			40	–	–	
2N5463, 2N5464, 2N5465			60	–	–	
Gate reverse current		I_{GSS}				
($V_{GS} = 20$ V dc, $V_{DS} = 0$) 2N5460, 2N5461, 2N5462			–	–	5.0	nA dc
($V_{GS} = 30$ V dc, $V_{DS} = 0$) 2N5463, 2N5464, 2N5465			–	–	5.0	
($V_{GS} = 20$ V dc, $V_{DS} = 0$, $T_A = 100°C$) 2N5460, 2N5461, 2N5462			–	–	1.0	μA dc
($V_{GS} = 30$ V dc, $V_{DS} = 0$, $T_A = 100°C$) 2N5463, 2N5464, 2N5465			–	–	1.0	
Gate-Source cutoff voltage		$V_{GS(off)}$				V dc
($V_{DS} = 15$ V dc, $I_D = 1.0\ \mu A$ dc) 2N5460, 2N5463			0.75	–	6.0	
2N5461, 2N5464			1.0	–	7.5	
2N5462, 2N5465			1.8	–	9.0	
Gate-Source voltage		V_{GS}				
($V_{DS} = 15$ V dc, $I_D = 0.1$ mA dc) 2N5460, 2N5463			0.5	–	4.0	V dc
($V_{DS} = 15$ V dc, $I_D = 0.2$ mA dc) 2N5461, 2N5464			0.8	–	4.5	
($V_{DS} = 15$ V dc, $I_D = 0.4$ mA dc) 2N5462, 2N5465			1.5	–	6.0	
ON Characteristics						
Zero-gate-voltage drain current		I_{DSS}				mA dc
($V_{DS} = 15$ V dc, $V_{GS} = 0$, 2N5460, 2N5463			– 1.0	–	– 5.0	
$f = 1.0$ kHz) 2N5461, 2N5464			– 2.0	–	– 9.0	
2N5462, 2N5465			– 4.0	–	– 16	
Small-Signal Characteristics						
Forward transfer admittance		$|Y_{fs}|$				μmhos
($V_{DS} = 15$ V dc, $V_{GS} = 0, f = 1.0$ kHz) 2N5460, 2N5463			1000	–	4000	or
2N5461, 2N5464			1500	–	5000	μS
2N5462, 2N5465			2000	–	6000	
Output admittance		$|Y_{os}|$	–	–	75	μmhos or
($V_{DS} = 15$ V dc, $V_{GS} = 0, f = 1.0$ kHz)						μS
Input capacitance		C_{iss}	–	5.0	7.0	pF
($V_{DS} = 15$ V dc, $V_{GS} = 0, f = 1.0$ MHz)						
Reverse transfer capacitance		C_{rss}	–	1.0	2.0	pF
($V_{DS} = 15$ V dc, $V_{GS} = 0, f = 1.0$ MHz)						

▲ **FIGURE 9–21**

Partial datasheet for the 2N5460–2N5465 *p*-channel JFETs.

Solution   Since $R_L \gg R_S, R_s \cong R_S$. From the partial datasheet in Figure 9–21, $g_m = y_{fs} = 1000\ \mu S$ (minimum). The voltage gain is

$$A_v = \frac{g_m R_S}{1 + g_m R_S} = \frac{(1000\ \mu S)(10\ k\Omega)}{1 + (1000\ \mu S)(10\ k\Omega)} = \textbf{0.909}$$

From the datasheet, $I_{GSS} = 5$ nA (maximum) at $V_{GS} = 20$ V. Therefore,

$$R_{IN(gate)} = \frac{V_{GS}}{I_{GSS}} = \frac{20\ V}{5\ nA} = 4000\ M\Omega$$

$$R_{IN} = R_G \parallel R_{IN(gate)} = 10\ M\Omega \parallel 4000\ M\Omega \cong \textbf{10 M}\Omega$$

Related Problem   If the maximum value of g_m of the 2N5460 JFET in the source-follower of Figure 9–20 is used, what is the voltage gain?

   Open the Multisim file EXM09-10 or the LT Spice file EXS09-10 in the Examples folder on the website. Measure the voltage gain using an input voltage of 10 mV rms to see how it compares with the calculated value.

A small modification of the common-drain amplifier in Figure 9–18 can improve it by using matched FETs that are on a single chip and by using matched resistors (R_1 and R_2). The normal source resistor has been replaced by a current source consisting of Q_2 and R_2 that sources a current that is less than I_{DSS} for improved linearity. Recall from basic electronics, that an ideal current source has infinite resistance. The circuit forms a voltage follower with virtually no offset voltage (gain = 1.00); the output does not change for variations in temperature because the transistors are formed on one chip and have the same temperature profile.

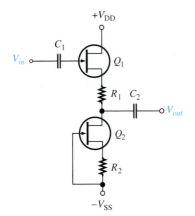

◀ **FIGURE 9–22**

A common source amplifier with a current source load.

SECTION 9–2 CHECKUP	1. What is the ideal maximum voltage gain of a common-drain amplifier?
	2. What factors influence the voltage gain of a common-drain amplifier?
	3. How does the circuit in Figure 9–22 approach the ideal gain?

9–3 THE COMMON-GATE AMPLIFIER

The common-gate FET amplifier configuration is comparable to the common-base BJT amplifier. Like the CB, the common-gate (CG) amplifier has a low input resistance. This is different from the CS and CD configurations, which have very high input resistances.

After completing this section, you should be able to

❑ **Explain and analyze the operation of common-gate FET amplifiers**
❑ Analyze common-gate JFET amplifier operation
  ◆ Determine the voltage gain ◆ Determine the input resistance
❑ Describe and analyze the cascode amplifier
  ◆ Determine the voltage gain ◆ Determine the input resistance

Common-Gate Amplifier Operation

A self-biased **common-gate** amplifier is shown in Figure 9–23. The gate is connected directly to ground. The input signal is applied at the source terminal through C_1. The output is coupled through C_2 from the drain terminal.

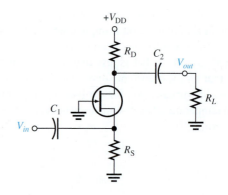

► FIGURE 9–23

JFET common-gate amplifier.

Voltage Gain   The voltage gain from source to drain is developed as follows:

$$A_v = \frac{V_{out}}{V_{in}} = \frac{V_d}{V_{gs}} = \frac{I_d R_d}{V_{gs}} = \frac{g_m V_{gs} R_d}{V_{gs}}$$

Equation 9–9

$$A_v = g_m R_d$$

where $R_d = R_D \parallel R_L$. Notice that the gain expression is the same as for the common-source JFET amplifier.

Input Resistance   As you have seen, both the common-source and common-drain configurations have extremely high input resistances because the gate is the input terminal. In contrast, the common-gate configuration where the source is the input terminal has a low input resistance. This is shown as follows. First, the input current is equal to the drain current.

$$I_{in} = I_s = I_d = g_m V_{gs}$$

Second, the input voltage equals V_{gs}.

$$V_{in} = V_{gs}$$

Therefore, the input resistance at the source terminal is

$$R_{in(source)} = \frac{V_{in}}{I_{in}} = \frac{V_{gs}}{g_m V_{gs}}$$

Equation 9–10

$$R_{in(source)} = \frac{1}{g_m}$$

If, for example, g_m has a value of 4000 μS, then

$$R_{in(source)} = \frac{1}{4000\ \mu S} = 250\ \Omega$$

EXAMPLE 9–11

Determine the minimum voltage gain and input resistance of the amplifier in Figure 9–24. V_{DD} is negative because it is a *p*-channel device.

▶ **FIGURE 9–24**

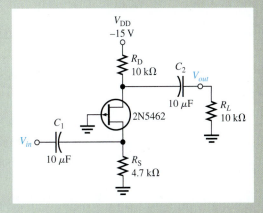

Solution   From the datasheet in Figure 9–21, $g_m = 2000\ \mu S$ minimum. This common-gate amplifier has a load resistor, so the effective drain resistance is $R_D \parallel R_L$ and the minimum voltage gain is

$$A_v = g_m(R_D \parallel R_L) = (2000\ \mu S)(10\ k\Omega \parallel 10\ k\Omega) = \mathbf{10}$$

The input resistance at the source terminal is

$$R_{in(source)} = \frac{1}{g_m} = \frac{1}{2000\ \mu S} = 500\ \Omega$$

The signal source actually sees R_S in parallel with $R_{in(source)}$, so the total input resistance is

$$R_{in} = R_{in(source)} \parallel R_S = 500\ \Omega \parallel 4.7\ k\Omega = \mathbf{452\ \Omega}$$

Related Problem   What is the input resistance in Figure 9–24 if R_S is changed to $10\ k\Omega$?

 Open the Multisim file EXM09-11 or the LT Spice file EXS09-11 in the Examples folder on the website. Measure the voltage using a 10 mV rms input voltage.

The Cascode Amplifier

The **cascode amplifier** is a two-transistor series arrangement that can be constructed from either FETs or BJTs. As a FET circuit, the input transistor is connected in a common-drain (CD) or common-source (CS) configuration; the output part of the circuit is connected as a common-gate (CG) circuit. This result is a very stable amplifier with advantages of both types including high gain, high input impedance, and excellent bandwidth. Frequently, the circuit is constructed with matched transistors formed on the same chip, ensuring that electrical characteristics are nearly identical and the temperature environment is the same for both.

A JFET cascode amplifier.

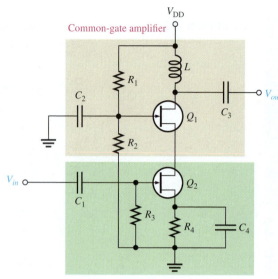

There are a number of variations and applications for cascode amplifiers. One application is found in high-frequency and RF (radio frequency) amplifiers such as the one shown in Figure 9–25. The input stage is a common-source amplifier, and its load is a common-gate amplifier connected in the drain circuit. In the circuit shown, an inductor is used in the drain circuit. This offers low resistance to dc and high reactance to the ac signal.

The cascode amplifier using JFETs provides a very high input resistance and significantly reduces capacitive effects to allow for operation at much higher frequencies than a common-source amplifier alone. Internal capacitances, which exist in every type of transistor, become significant at higher frequencies and reduce the gain of inverting amplifiers as described by the Miller effect, covered in Chapter 10. The first stage is a CS amplifier that inverts the signal. However, the gain is very low because of the low input resistance of the CB amplifier that it is driving. As a result, the effect of internal capacitances on the high-frequency response is very small. The second stage is a CG amplifier that does not invert the signal, so it can have high gain without degrading the high-frequency response. The combination of the two amplifiers provides the best of both circuits, resulting in high gain, high input resistance, and an excellent high-frequency response.

The voltage gain of the cascode amplifier in Figure 9–25 is a product of the gains of both the CS and the CG stages. However, as mentioned, the gain is primarily provided by the CG amplifier.

$$A_v = A_{v(CS)}A_{v(CG)} = (g_{m(CS)}R_d)(g_{m(CG)}X_L)$$

Since R_d of the CS amplifier stage is the input resistance of the CG stage and X_L is the reactance of the inductor in the drain of the CG stage, the voltage gain is

$$A_v = \left(g_{m(CS)}\left(\frac{1}{g_{m(CG)}}\right) \right)(g_{m(CG)}X_L) \cong g_{m(CG)}X_L$$

assuming the transconductances of both transistors are approximately the same. From the equation you can see that the voltage gain increases with frequency because X_L increases. As the frequency continues to increase, eventually capacitance effects become significant enough to begin reducing the gain.

The input resistance to the cascade amplifier is the input resistance to the CS stage.

$$R_{in} = R_3 \parallel \left(\frac{V_{GS}}{I_{GSS}}\right)$$

EXAMPLE 9–12

For the cascode amplifier in Figure 9–25, the transistors are 2N5485s and have a minimum g_m (g_{fs}) of 3500 μS. Also, $I_{GSS} = -1$ nA at $V_{GS} = 20$ V. If $R_3 = 10$ MΩ and $L = 1.0$ mH, determine the voltage gain and the input resistance at a frequency of 100 MHz.

Solution

$$A_v \cong g_{m(CG)}X_L = g_{m(CG)}(2\pi fL) = (3500\ \mu S)\ 2\pi(100\ MHz)(1.0\ mH) = \textbf{2199}$$

$$R_{in} = R_3 \left\| \left(\frac{V_{GS}}{I_{GSS}}\right)\right. = 10\ M\Omega \left\| \left(\frac{20\ V}{1\ nA}\right)\right. = \textbf{9.995 M}\Omega$$

Related Problem

What happens to the voltage gain in the cascode amplifier if the inductance value is increased?

Another popular option for implementing a cascode amplifier is to use a dual gate MOSFET, which enables the cascode amplifier to be constructed with a single transistor. (Dual gate MOSFETs are discussed in Section 8–5.) Because the cascode configuration reduces the internal capacitances, the dual gate MOSFET is useful in radio frequency applications such as an rf amplifier at the front end of a receiver. As in the case of two JFETs connected in cascode, the dual gate MOSFET has the signal connected to the lower gate and the upper gate is capacitively coupled to ac ground. Figure 9–26 shows a typical dual gate MOSFET connected as a cascode amplifier. Compare it to the JFET cascode amplifier shown in Figure 9–25.

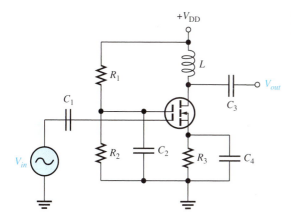

◀ **FIGURE 9–26**

A dual gate MOSFET cascode amplifier.

SECTION 9–3 CHECKUP

1. What is a major difference between a common-gate amplifier and the other two configurations?

2. What common factor determines the voltage gain and the input resistance of a common-gate amplifier?

3. Name the advantages of a cascode amplifier.

4. How does a decrease in frequency affect the gain of the cascode amplifier in Figure 9–26?

▶ FIGURE 9–30

Frequency spectrum of a PWM signal.

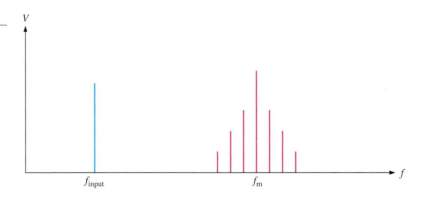

The Complementary MOSFET Stage

The MOSFETs are arranged in a common-source complementary configuration to provide power gain. Each transistor switches between the *on* state and the *off* state and when one transistor is *on,* the other one is *off,* as shown in Figure 9–31. When a transistor is *on,* there is very little voltage across it and, therefore, there is very little power dissipated even though it may have a high current through it. When a transistor is *off,* there is no current through it and, therefore, there is no power dissipated. The only time power is dissipated in the transistors is during the short switching time. Power delivered to a load can be very high because a load will have a voltage across it nearly equal to the supply voltages and a high current through it.

▶ FIGURE 9–31

Complementary MOSFETs operating as switches to amplify power.

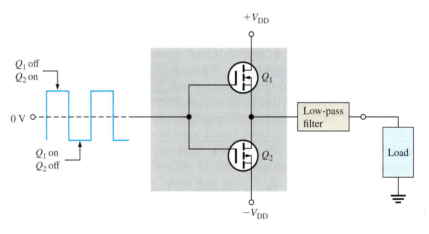

Efficiency When Q_1 is *on,* it is providing current in the load. However, ideally the voltage across it is zero so the internal power dissipated by Q_1 is

$$P_{DQ} = V_{Q1}I_L = (0 \text{ V})I_L = 0 \text{ W}$$

At the same time, Q_2 is *off* and the current through it is zero, so the internal power is

$$P_{DQ} = V_{Q2}I_L = V_{Q2}(0 \text{ A}) = 0 \text{ W}$$

Ideally, the output power to the load is $2V_QI_L$. The maximum ideal efficiency is, therefore,

$$\eta_{max} = \frac{P_{out}}{P_{tot}} = \frac{P_{out}}{P_{out} + P_{DQ}} = \frac{2V_QI_L}{2V_QI_L + 0 \text{ W}} = 1$$

As a percentage, $\eta_{max} = 100\%$.

In a practical case, each MOSFET would have a few tenths of a volt across it in the *on* state. There is also a small internal power dissipation in the low-pass filter's resistance because it is in series with the power output and a small amount of power dissipated in other components of the circuit. Also, power is dissipated during the finite switching time, distortions in the switched waveform, on-state resistance, timing errors due to finite switching time, and other subtle errors, so the ideal efficiency of 100% can never be reached in practice.

EXAMPLE 9–13

A certain class D amplifier dissipates an internal power of 100 mW in the comparator, triangular-wave generator, and filter combined. Each MOSFET in the complementary stage has a voltage of 0.4 V across it in the *on* state. The amplifier operates from ± 15 V dc sources and provides 0.5 A to the load. Neglecting any voltage dropped across the filter, determine the output power and the overall efficiency.

Solution   The output power to the load is

$$P_{out} = (V_{DD} - V_Q)I_L = (15\ V - 0.4\ V)(0.5\ A) = \mathbf{7.3\ W}$$

The total internal power dissipation ($P_{tot(int)}$) is the power in the complementary stage in the *on* state (P_{DQ}) plus the internal power in the comparator, triangular-wave generator, and filter (P_{int}).

$$P_{tot(int)} = P_{DQ} + P_{int} = (400\ mV)(0.5\ A) + 100\ mW$$
$$= 200\ mW + 100\ mW = 300\ mW$$

The efficiency is

$$\eta = \frac{P_{out}}{P_{out} + P_{tot(int)}} = \frac{7.3\ W}{7.6\ W} = \mathbf{0.961}$$

Related Problem   There is 0.5 V across each MOSFET when it is *on* and the class D amplifier operates with ± 12 V dc supply voltages. Assuming all other circuits in the amplifier dissipate 75 mW and 0.8 A is supplied to the load, determine the efficiency.

Low-Pass Filter

Ideally, the low-pass filter removes the modulating frequency and harmonics and passes only the original signal to the output. The filter has a cutoff frequency that is above the input signal frequency and below the modulating frequency and harmonics, as illustrated in Figure 9–32.

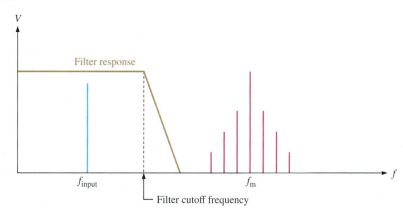

◀ **FIGURE 9–32**

The low-pass filter removes all but the input signal frequency from the PWM signal.

Signal Flow

Figure 9–33 shows the signals at each point in a class D amplifier. A small audio signal is applied and pulse-width modulated to produce a PWM signal at the output of the modulator where voltage gain is achieved. The PWM drives the complementary MOSFET stage to achieve power amplification. The PWM signal is filtered and the amplified audio signal appears on the output with sufficient power to drive a speaker.

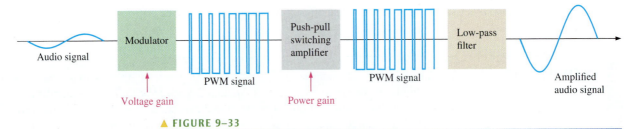

▲ **FIGURE 9–33**

Representation of signal flow in a class D amplifier

SECTION 9–4 CHECKUP	1. Name the three stages of a class D amplifier.
	2. In pulse-width modulation, to what is the pulse width proportional?
	3. How is the PWM signal changed to an audio signal?

9–5 MOSFET Analog Switching

MOSFETs are widely used in analog and digital switching applications. In the preceding section, you saw how MOSFETs are used in the switching mode in Class D amplifiers. Generally, they exhibit very low *on*-resistance, very high *off*-resistance, and fast switching times.

After completing this section, you should be able to

❑ **Describe how MOSFETs can be used in analog switching applications**
❑ Explain how a MOSFET operates as a switch
  ◆ Discuss load line operation   ◆ Discuss the ideal switch
❑ Describe a MOSFET analog switch
❑ Discuss analog switch applications
  ◆ Explain a sampling circuit   ◆ Explain an analog multiplexer   ◆ Explain a switched-capacitor circuit

MOSFET Switching Operation

E-MOSFETs are generally used for switching applications because of their threshold characteristic, $V_{GS(th)}$. When the gate-to-source voltage is less than the threshold value, the MOSFET is *off*. When the gate-to-source voltage is greater than the threshold value, the MOSFET is *on*. When V_{GS} is varied between $V_{GS(th)}$ and $V_{GS(on)}$, the MOSFET is being operated as a switch, as illustrated in Figure 9–34. In the *off* state, when $V_{GS} < V_{GS(th)}$, the device is operating at the lower end of the load line and acts like an open switch (very high R_{DS}). When V_{GS} is sufficiently greater than $V_{GS(th)}$, the device is operating at the upper end of the load line in the ohmic region and acts like a closed switch (very low R_{DS}).

The Ideal Switch   Refer to Figure 9–35(a). When the gate voltage of the *n*-channel MOSFET is $+V$, the gate is more positive than the source by an amount exceeding $V_{GS(th)}$. The MOSFET is *on* and appears as a closed switch between the drain and source. When the gate voltage is zero, the gate-to-source voltage is 0 V. The MOSFET is *off* and appears as an open switch between the drain and source.

Refer to Figure 9–35(b). When the gate voltage of the *p*-channel MOSFET is 0 V, the gate is less positive than the source by an amount exceeding $V_{GS(th)}$. The MOSFET is *on* and appears as a closed switch between the drain and source. When the gate voltage is $+V$, the gate-to-source voltage is 0 V. The MOSFET is *off* and appears as an open switch between the drain and source.

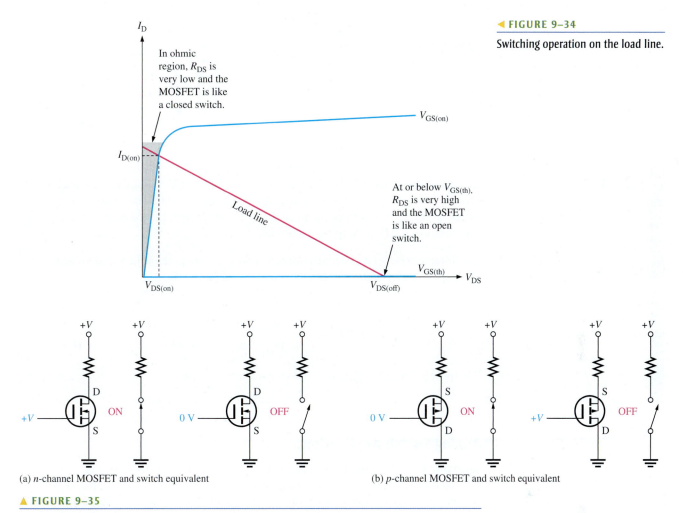

▲ FIGURE 9–35

The MOSFET as a switch.

(a) *n*-channel MOSFET and switch equivalent

(b) *p*-channel MOSFET and switch equivalent

The Analog Switch

MOSFETs are commonly used for switching analog signals. Basically, a signal applied to the drain can be switched through to the source by a voltage on the gate. A major restriction is that the signal level at the source must not cause the gate-to-source voltage to drop below $V_{GS(th)}$.

A basic *n*-channel MOSFET **analog switch** is shown in Figure 9–36. The signal at the drain is connected to the source when the MOSFET is turned on by a positive V_{GS} and is disconnected when V_{GS} is 0, as indicated.

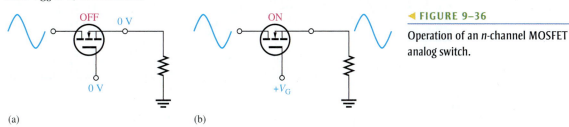

(a)   (b)

◀ FIGURE 9–36

Operation of an *n*-channel MOSFET analog switch.

When the analog switch is *on,* as illustrated in Figure 9–37, the minimum gate-to-source voltage occurs at the negative peak of the signal. The difference in V_G and $-V_{p(out)}$ is the gate-to-source voltage at the instant of the negative peak and must be equal to or greater than $V_{GS(th)}$ to keep the MOSFET in conduction.

$$V_{GS} = V_G - V_{p(out)} \geq V_{GS(th)}$$

▶ FIGURE 9–37

Signal amplitude is limited by $V_{\text{GS(th)}}$.

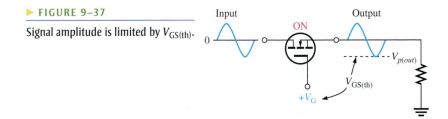

EXAMPLE 9–14

A certain analog switch similar to the one shown in Figure 9–37 uses an *n*-channel MOSFET with $V_{\text{GS(th)}} = 2$ V. A voltage of +5 V is applied at the gate to turn the switch *on*. Determine the maximum peak-to-peak input signal that can be applied, assuming no voltage drop across the switch.

Solution The difference between the gate voltage and the negative peak of the signal voltage must equal or exceed the threshold voltage. For maximum $V_{p(out)}$,

$$V_G - V_{p(out)} = V_{\text{GS(th)}}$$

$$V_{p(out)} = V_G - V_{\text{GS(th)}} = 5\text{ V} - 2\text{ V} = 3\text{ V}$$

$$V_{pp(in)} = 2V_{p(out)} = 2(3\text{ V}) = \textbf{6 V}$$

Related Problem What would happen if $V_{p(in)}$ exceeded the maximum value?

Analog Switch Applications

Sampling Circuit One application of analog switches is in analog-to-digital conversion. The analog switch is used in a *sample-and-hold* circuit to sample the input signal at a certain rate. Each sampled signal value is then temporarily stored on a capacitor until it can be converted to a digital code by an analog-to-digital converter (ADC). To accomplish this, the MOSFET is turned *on* for short intervals during one cycle of the input signal by pulses applied to the gate. The basic operation, showing only a few samples for clarity, is illustrated in Figure 9–38.

▶ FIGURE 9–38

The analog switch operating as a sampling circuit.

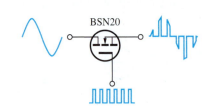

(a) Circuit action

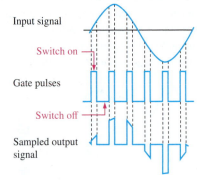

(b) Waveform diagram

The minimum rate at which a signal can be sampled and reconstructed from the samples must be *more* than twice the maximum frequency contained in the signal. The minimum sampling frequency is called the *Nyquist frequency*.

$$f_{\text{sample (min)}} > 2f_{\text{signal (max)}}$$

When a gate pulse is at its high level, the switch is turned *on* and the small portion of the input waveform occurring during that pulse appears on the output. When the pulse waveform is at its 0 V level, the switch is turned *off* and the output is also at 0 V.

EXAMPLE 9–15

An analog switch is used to sample an audio signal with a maximum frequency of 8 kHz. Determine the minimum frequency of the pulses applied to the MOSFET gate.

Solution

$$f_{sample(min)} > 2f_{signal(max)} = 2(8 \text{ kHz}) = \textbf{16 kHz}$$

The sampling frequency must be greater than 16 kHz.

Related Problem

What is the minimum sampling frequency if the highest frequency in the audio signal is 12 kHz?

Analog Multiplexer Analog multiplexers are used where two or more signals are to be routed to the same destination. For example, a two-channel analog sampling multiplexer is shown in Figure 9–39. The MOSFETs are alternately turned *on* and *off* so that first one signal sample is connected to the output and then the other. The pulses are applied to the gate of switch A, and the inverted pulses are applied to the gate of switch B. A digital circuit known as an *inverter* is used for this. When the pulses are high, switch A is *on* and switch B is *off*. When the pulses are low, switch B is *on* and switch A is *off*. This is called *time-division multiplexing* because signal A appears on the output during time intervals when the pulse is high and signal B appears during the time intervals when the pulse is low. That is, they are interleaved on a time basis for transmission on a single line.

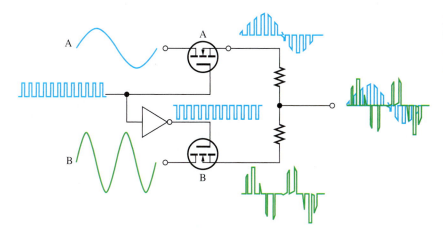

◄ FIGURE 9–39

The analog multiplexer is alternately sampling two signals and interleaving them on a single output line.

Switched-Capacitor Circuit Another application of MOSFETs is in **switched-capacitor circuits** commonly used in integrated circuit programmable analog devices known as *analog signal processors*. Because capacitors can be implemented in ICs more easily than a resistor, they are used to emulate resistors. Capacitors also take up less space on a chip than an IC resistor and dissipate almost no power, so there is much less resistive heating. Many types of analog circuits use resistors to determine voltage gain and other characteristics and by using switched capacitors to emulate resistors, dynamic programming of analog circuits can be achieved.

For example, in a certain type of IC amplifier circuit that you will study later, two external resistors are required as shown in Figure 9–40. The values of these resistors establish the voltage gain of the amplifier as $A_v = R_2/R_1$.

Inverter Notice that the circuit in Figure 9–43 actually inverts the input because when the input is 0 V or low, the output is V_{DD} or high. When the input is V_{DD} or high, the output is 0 V or low. For this reason, this circuit is called an *inverter* in digital electronics.

NAND Gate In Figure 9–44(a), two additional MOSFETs and a second input are added to the CMOS pair to create a digital circuit known as a NAND gate (in this case a two-input NAND gate). Q_4 is connected in parallel with Q_1, and Q_3 is connected in series with Q_2. When both inputs, V_A and V_B, are 0, Q_1 and Q_4 are *on* while Q_2 and Q_3 are *off*, making $V_{out} = V_{DD}$. When both inputs are equal to V_{DD}, Q_1 and Q_4 are *off* while Q_2 and Q_3 are *on*, making $V_{out} = 0$. You can verify that when the inputs are different, one at V_{DD} and the other at 0, the output is equal to V_{DD}. The operation is summarized in the table of Figure 9–44(b) and can be stated:

When V_A *AND* V_B are high, the output is low; otherwise, the output is high.

▶ **FIGURE 9–44**

CMOS NAND gate operation.

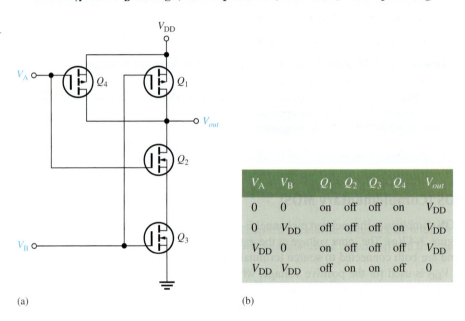

V_A	V_B	Q_1	Q_2	Q_3	Q_4	V_{out}
0	0	on	off	off	on	V_{DD}
0	V_{DD}	off	off	off	on	V_{DD}
V_{DD}	0	on	off	off	off	V_{DD}
V_{DD}	V_{DD}	off	on	on	off	0

(a)   (b)

NOR Gate In Figure 9–45(a), two additional MOSFETs and a second input are added to the CMOS pair to create a digital circuit known as a NOR gate. Q_4 is connected in parallel with Q_2, and Q_3 is connected in series with Q_1. When both inputs, V_A and V_B, are 0, Q_1 and

▶ **FIGURE 9–45**

CMOS NOR gate operation.

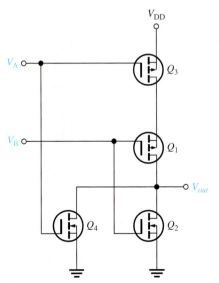

V_A	V_B	Q_1	Q_2	Q_3	Q_4	V_{out}
0	0	on	off	on	off	V_{DD}
0	V_{DD}	off	on	on	off	0
V_{DD}	0	on	off	on	off	0
V_{DD}	V_{DD}	off	on	off	on	0

(a)   (b)

Q_3 are *on* while Q_2 and Q_4 are *off*, making $V_{out} = V_{DD}$. When both inputs are equal to V_{DD}, Q_1 and Q_3 are *off* while Q_2 and Q_4 are *on*, making $V_{out} = 0$. You can verify that when the inputs are different, one at V_{DD} and the other at 0, the output is equal to 0. The operation is summarized in the table of Figure 9–45(b) and can be stated:

When V_A *OR* V_B *OR* both are high, the output is low; otherwise, the output is high.

EXAMPLE 9–16

A pulse waveform is applied to a CMOS inverter as shown in Figure 9–46. Determine the output waveform and explain the operation.

▶ FIGURE 9–46

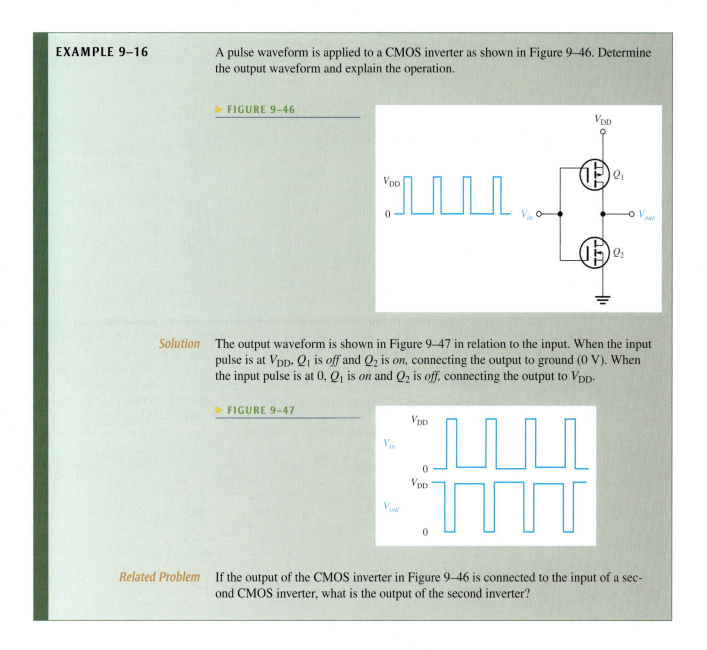

Solution   The output waveform is shown in Figure 9–47 in relation to the input. When the input pulse is at V_{DD}, Q_1 is *off* and Q_2 is *on*, connecting the output to ground (0 V). When the input pulse is at 0, Q_1 is *on* and Q_2 is *off*, connecting the output to V_{DD}.

▶ FIGURE 9–47

Related Problem   If the output of the CMOS inverter in Figure 9–46 is connected to the input of a second CMOS inverter, what is the output of the second inverter?

MOSFETs in Power Switching

The BJT was the only power transistor until the MOSFET was introduced. The BJT requires a base current to turn on, has relatively slow turn-off characteristics, and is susceptible to thermal runaway due to a negative temperature coefficient. The MOSFET, however, is voltage controlled and has a positive temperature coefficient, which prevents thermal runaway. The MOSFET can turn *off* faster than the BJT, and the low *on*-state-resistance results in conduction power losses lower than with BJTs. Power MOSFETs are used for motor control, dc-to-ac conversion, dc-to-dc conversion, load switching, and other applications that require high current and precise digital control.

Interfacing with Logic Circuits Certain E-MOSFETs are designed to interface with logic circuits. For example, the NX7002BKXB is a dual N-channel device that can directly interface one or two logic circuits with loads that require up to 330 mA of drive current. A simple interfacing circuit for a CMOS gate is shown in Figure 9–48 using one-half of the NX7002BKXB. The circuit can be used as a traditional relay driver to power high current loads. In that case, protection diodes should wired as shown in the circuit diagram. The NX7002BKXB can also interface to TTL (transistor-transistor logic) by adding a pull-up resistor to the logic output.

▶ **FIGURE 9–48**

An E-MOSFET driver for small loads.

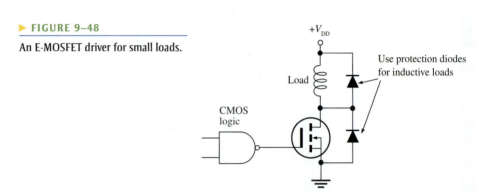

The Solid State Relay Another power switching device is the **solid state relay (SSR)**. A solid state relay is a device that uses a control voltage to open or close one or more electronic switches to a load. In a solid state relay, the input is internally coupled through a sensor that provides isolation between the input and output. The sensor can be an optical coupler or other isolation device. The input typically controls a LED, so electrically it is completely isolated from the load. The output is often an E-MOSFET but can be a thyristor (studied in Chapter 11). Figure 9–49 shows a typical MOSFET relay in which the source terminals of two internal MOSFETs are connected together, allowing it to have ac loads.

▶ **FIGURE 9–49**

A solid-state relay with an LED input and using two E-MOSFETs on the output side.

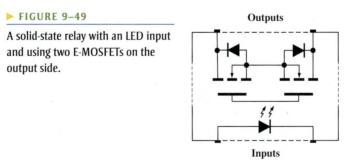

The major advantage to a solid state relay over a mechanical relay is that there are no moving parts to wear out; it is much faster, more reliable with a longer lifetime and does not have contact problems such as sparking that is common in mechanical relays. They are also insensitive to magnetic fields. The disadvantage is that, unlike its mechanical counterpart, it cannot withstand momentary overloads and has a higher on state resistance. Solid-state relays that use MOSFETs require the usual handling precautions for static sensitive devices and inputs should protected from any surge voltage. There are numerous applications for SSRs involving interfacing logic or computers to control of devices.

9–7 TROUBLESHOOTING

A technician who understands the basics of circuit operation and who can, if necessary, perform basic analysis on a given circuit is much more valuable than one who is limited to carrying out routine test procedures. In this section, you will see how to test a circuit board that has only a schematic with no specified test procedure or voltage levels. In this case, basic knowledge of how the circuit operates and the ability to do a quick circuit analysis are useful.

After completing this section, you should be able to

❑ **Troubleshoot FET amplifiers**
❑ Troubleshoot a two-stage common-source amplifier
  ◆ Explain each step in the troubleshooting procedure   ◆ Use a datasheet
  ◆ Relate the circuit board to the schematic

A Two-Stage Common-Source Amplifier

Assume that you are given a circuit board containing an audio amplifier and told simply that it is not working properly. The circuit is a two-stage CS JFET amplifier, as shown in Figure 9–50.

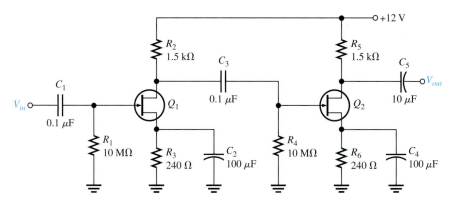

◀ **FIGURE 9–50**

A two-stage CS JFET amplifier circuit.

The problem is approached in the following sequence.

Step 1: Determine what the voltage levels in the circuit should be so that you know what to look for. First, pull a datasheet on the particular transistor (assume both Q_1 and Q_2 are found to be the same type of transistor) and determine the g_m so that you can calculate the typical voltage gain. Assume that for this particular device, a typical g_m of 5000 μS is specified. Calculate the expected typical voltage gain of each stage (notice they are identical) based on the typical value of g_m. The g_m of actual devices may be any value between the specified minimum and maximum values. Because the input resistance is very high, the second stage does not significantly load the first stage, as in a BJT amplifier. So, the unloaded voltage gain for each stage is

$$A_v = g_m R_2 = (5000 \ \mu\text{S})(1.5 \ \text{k}\Omega) = 7.5$$

Since the stages are identical, the typical overall gain should be

$$A_v' = (7.5)(7.5) = 56.3$$

Assume the dc levels have been checked and verified. You are now ready to move to ac signal checks.

Step 2: Arrange a test setup to permit connection of an input test signal, a dc supply voltage, and ground to the circuit. The schematic shows that the dc supply voltage must be $+12$ V. Choose 10 mV rms as an input test signal. This value is arbitrary (although the capability of your signal source is a factor), but small enough that the expected output signal voltage is well below the absolute peak-to-peak limit of 12 V set by the supply voltage and ground (you know that the output voltage swing cannot go higher than 12 V or lower than 0 V). Set the frequency of the sinusoidal signal source to an arbitrary value in the audio range (say 10 kHz) because you know this is an audio amplifier. The audio frequency range is generally accepted as 20 Hz to 20 kHz.

Step 3: Check the input signal at the gate of Q_1 and the output signal at the drain of Q_2 with an oscilloscope. The results are shown in Figure 9–51. The measured output voltage has a peak value of 226 mV. The expected typical peak output voltage is

$$V_{out} = V_{in}A_v' = (14.14 \ \text{mV})(56.3) = 796 \ \text{mV peak}$$

The output is much less than it should be.

Step 4: Trace the signal from the output toward the input to determine the fault. Figure 9–51 shows the oscilloscope displays of the measured signal voltages. The voltage at the gate of Q_2 is 106 mV peak, as expected (14.14 mV × 7.5 = 106 mV). This signal is properly coupled from the drain of Q_1. Therefore, the problem lies in the second stage. From the oscilloscope displays, the gain of Q_2 is much lower than it should be (213 mV/100 mV = 2.13 instead of 7.5).

Step 5: Analyze the possible causes of the observed malfunction. There are three possible reasons the gain is low:

1. Q_2 has a lower transconductance (g_m) than the specified typical value. Check the datasheet to see if the minimum g_m accounts for the lower measured gain.

2. R_5 has a lower value than shown on the schematic. An incorrect value should show up with dc voltage checks, particularly if the value is much different than specified, so this is not the likely cause in this case.

3. The bypass capacitor C_4 is open.

The best way to check the g_m is by replacing Q_2 with a new transistor of the same type and rechecking the output signal. You can make certain that R_5 is the proper value by removing one end of the resistor from the circuit board and measuring the resistance with an ohmmeter. To avoid having to unsolder a component, the best way to start isolating the fault is by checking the signal voltage at the source of Q_2. If the capacitor is working properly, there will be only a dc voltage at the source. The presence of a signal voltage at the source indicates that C_4 is open. With R_6 unbypassed, the gain expression is $g_m R_d/(1 + g_m R_s)$ rather than simply $g_m R_d$, thus resulting in less gain.

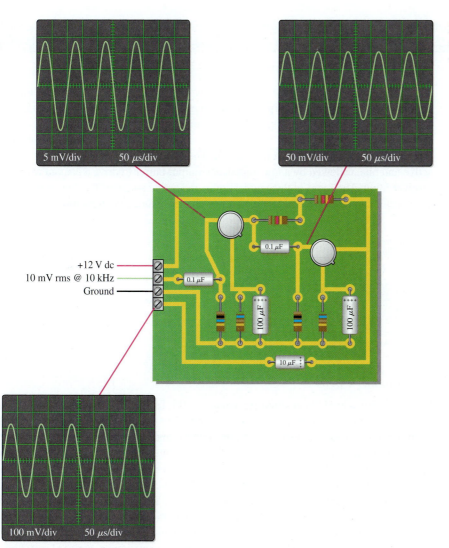

◀ FIGURE 9–51

Oscilloscope displays of signals in the two-stage JFET amplifier.

Multisim Troubleshooting Exercises

These file circuits are in the Troubleshooting Exercises folder on the website. Open each file and determine if the circuit is working properly. If it is not working properly, determine the fault.

1. Multisim file TSM09-01

2. Multisim file TSM09-02

3. Multisim file TSM09-03

4. Multisim file TSM09-04

5. Multisim file TSM09-05

SECTION 9–7 CHECKUP

1. What is the prerequisite to effective troubleshooting?

2. Assume that C_2 in the amplifier of Figure 9–50 opened. What symptoms would indicate this failure?

3. If C_3 opened in the amplifier of Figure 9–50, would the voltage gain of the first stage be affected?

Device Application: *Active Antenna*

In this application, a broadband JFET amplifier is used to provide a high input impedance and voltage gain for a whip antenna. When an antenna is connected to the input of a receiver or a coaxial cable, signal deterioration may be unacceptable due to a distant station, noisy conditions, or an impedance mismatch. An active antenna can alleviate this problem by providing a stronger signal. The block diagram in Figure 9–52 shows an active antenna, followed by a low impedance output buffer to drive a coaxial cable or a receiver input. The focus in this application is the active antenna. The low output impedance buffer can be a BJT emitter-follower or an impedance-matching transformer.

▶ **FIGURE 9–52**

An active antenna driving a receiver or a coax through a buffer.

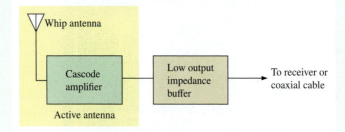

The Amplifier Circuit

Figure 9–53 is a broadband amplifier using two JFETs in a cascode arrangement commonly used in RF (radio frequency) applications. The advantage of using a JFET is that its high input impedance does not load the antenna and cause a reduction in signal voltage, resulting in poor signal reception. It also is a low-noise device and can be located close to the antenna before additional noise is picked up by the system. Generally, an antenna produces signal voltages in the microvolt range, and any signal loss because of loading or noise can significantly degrade the signal. The active antenna also provides a large voltage gain that results in a stronger signal to the receiver with improved signal-to-noise ratio. The active antenna is powered by a separate 9 V battery, which also provides isolation from noise pickup in the signal lines and is located in an enclosed metal box to provide additional isolation.

▶ **FIGURE 9–53**

Cascode amplifier for active antenna.

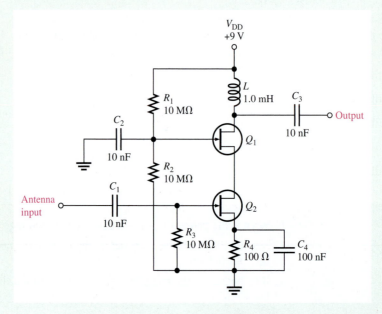

This active antenna has a voltage gain of approximately 2000 at 88 MHz and a gain of approximately 10,000 at 1 GHz which makes it applicable for the FM broadcast band, some TV channel bands, some amateur radio (HAM) bands, cell phone bands, and many others. Also below the FM band, the gain may be adequate for other radio and TV bands as well, depending on receiver requirements. The coil can be changed to optimize gain within a specified band or to adjust the band downward.

1. Research the Internet to determine the frequency band for TV channels 7–13.
2. Research the Internet to find the frequency bands allocated for cellular telephones.
3. What is the purpose of C_2 in Figure 9–53?

The transistors used in the active antenna are 2N5484 n-channel JFETs. The partial datasheet is shown in Figure 9–54.

4. Using the datasheet, determine $R_{IN(gate)}$ of the JFET (Q_2).
5. What input resistance is presented to the antenna in Figure 9–53?
6. From the datasheet, what is the minimum forward transconductance?

Simulation

The active antenna circuit is simulated in Multisim with the antenna input represented by a 10 μV peak source. The output signal is shown for 88 MHz and 1 GHz inputs in Figure 9–55 on page 476.

7. What is the significance of the 88 MHz frequency?
8. Determine the rms output voltage in Figure 9–55(b) and (c), and calculate the gain for both frequencies.

Simulate the active antenna circuit using your Multisim or LT Spice software. Measure the output voltage at 10 MHz, 100 MHz, and 500 MHz.

Prototyping and Testing

Now that the circuit has been simulated, the prototype circuit is constructed and tested. After the circuit is successfully tested, it is ready to be finalized. Because you are working at high frequencies where stray capacitances can cause unwanted resonant conditions, the circuit layout is very critical.

Lab Experiment

To build and test a similar circuit, go to Experiment 9 in your lab manual (*Laboratory Exercises for Electronic Devices* by David Buchla and Steven Wetterling).

Circuit Board

Certain considerations should be observed when laying out a printed circuit board for RF circuits. EMI (electromagnetic interference), line inductance, and stray capacitance all become important at high frequencies. A few basic features that should be incorporated on an RF circuit board are

◆ Keep traces as short and wide as possible.
◆ Do not run parallel signal lines that are in close proximity.
◆ Capacitively decouple supply voltages.
◆ Provide a large ground plane for shielding and to minimize noise.

2N5484   MMBF5484
2N5485   MMBF5485
2N5486   MMBF5486

TO-92

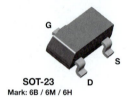

SOT-23
Mark: 6B / 6M / 6H

NOTE: Source & Drain
are interchangeable

N-Channel RF Amplifier

This device is designed primarily for electronic switching applications such as low On Resistance analog switching. Sourced from Process 50.

Absolute Maximum Ratings*   TA = 25°C unless otherwise noted

Symbol	Parameter	Value	Units
V_{DG}	Drain-Gate Voltage	25	V
V_{GS}	Gate-Source Voltage	- 25	V
I_{GF}	Forward Gate Current	10	mA
T_J, T_{stg}	Operating and Storage Junction Temperature Range	-55 to +150	°C

*These ratings are limiting values above which the serviceability of any semiconductor device may be impaired.

NOTES:
1) These ratings are based on a maximum junction temperature of 150 degrees C.
2) These are steady state limits. The factory should be consulted on applications involving pulsed or low duty cycle operations.

Thermal Characteristics   TA = 25°C unless otherwise noted

Symbol	Characteristic	Max		Units
		2N5484-5486	*MMBF5484-5486	
P_D	Total Device Dissipation	350	225	mW
	Derate above 25°C	2.8	1.8	mW/°C
$R_{\theta JC}$	Thermal Resistance, Junction to Case	125		°C/W
$R_{\theta JA}$	Thermal Resistance, Junction to Ambient	357	556	°C/W

*Device mounted on FR-4 PCB 1.6" X 1.6" X 0.06."

▲ FIGURE 9–54

Partial datasheet for the 2N5484 RF *n*-channel JFET. Copyright Fairchild Semiconductor Corporation. Used by permission.

N-Channel RF Amplifier

(continued)

Electrical Characteristics TA = 25°C unless otherwise noted

Symbol	Parameter	Test Conditions		Min	Typ	Max	Units
OFF CHARACTERISTICS							
$V_{(BR)GSS}$	Gate-Source Breakdown Voltage	I_G = - 1.0 μA, V_{DS} = 0		- 25			V
I_{GSS}	Gate Reverse Current	V_{GS} = - 20 V, V_{DS} = 0				- 1.0	nA
		V_{GS}= - 20 V, V_{DS}= 0, T_A= 100°C				- 0.2	μA
$V_{GS(off)}$	Gate-Source Cutoff Voltage	V_{DS} = 15 V, I_D = 10 nA	5484	- 0.3		- 3.0	V
			5485	- 0.5		- 4.0	V
			5486	- 2.0		- 6.0	V
ON CHARACTERISTICS							
I_{DSS}	Zero-Gate Voltage Drain Current*	V_{DS} = 15 V, V_{GS} = 0	5484	1.0		5.0	mA
			5485	4.0		10	mA
			5486	8.0		20	mA
SMALL SIGNAL CHARACTERISTICS							
g_{fs}	Forward Transfer Conductance	V_{DS} = 15 V, V_{GS} = 0, f = 1.0 kHz					
			5484	3000		6000	μmhos
			5485	3500		7000	μmhos
			5486	4000		8000	μmhos
$Re_{(yis)}$	Input Conductance	V_{DS} = 15 V, V_{GS} = 0, f = 100 MHz					
			5484			100	μmhos
		V_{DS} = 15 V, V_{GS} = 0, f = 400 MHz					
			5485 / 5486			1000	μmhos
g_{os}	Output Conductance	V_{DS} = 15 V, V_{GS} = 0, f = 1.0 kHz					
			5484			50	μmhos
			5485			60	μmhos
			5486			75	μmhos
$Re_{(yos)}$	Output Conductance	V_{DS} = 15 V, V_{GS} = 0, f = 100 MHz					
			5484			75	μmhos
		V_{DS} = 15 V, V_{GS} = 0, f = 400 MHz					
			5485 / 5486			100	μmhos
$Re_{(yfs)}$	Forward Transconductance	V_{DS} = 15 V, V_{GS} = 0, f = 100 MHz					
			5484	2500			μmhos
		V_{DS} = 15 V, V_{GS} = 0, f = 400 MHz					
			5485	3000			μmhos
			5486	3500			μmhos
C_{iss}	Input Capacitance	V_{DS} = 15 V, V_{GS} = 0, f = 1.0 MHz				5.0	pF
C_{rss}	Reverse Transfer Capacitance	V_{DS} = 15 V, V_{GS} = 0, f = 1.0 MHz				1.0	pF
C_{oss}	Output Capacitance	V_{DS} = 15 V, V_{GS} = 0, f = 1.0 MHz				2.0	pF
NF	Noise Figure	V_{DS}= 15 V, R_G= 1.0 kΩ, f = 100 MHz	5484			3.0	dB
		V_{DS}= 15 V, R_G= 1.0 kΩ, f = 400 MHz	5484		4.0		dB
		V_{DS}= 15 V , R_G= 1.0 kΩ, f = 100 MHz	5485 / 5486			2.0	dB
		V_{DS}= 15 V, R_G= 1.0 kΩ, f = 400 MHz	5485 / 5486			4.0	dB

*Pulse Test: Pulse Width ≤ 300 ms, Duty Cycle ≤ ?2%

▲ FIGURE 9–54

(continued)

▶ FIGURE 9–57

A configuration of the active antenna circuit board in a metal housing (cover removed) with a 9 V battery and an impedancematching transformer.

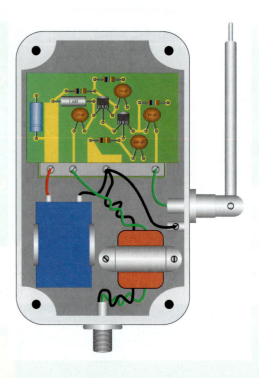

SUMMARY OF FET AMPLIFIERS

N channels are shown. V_{DD} is negative for *p* channel.

COMMON-SOURCE AMPLIFIERS

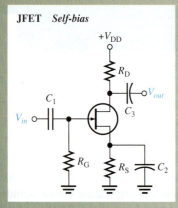

JFET *Self-bias*

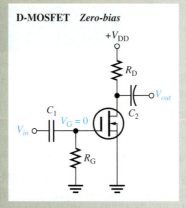

D-MOSFET *Zero-bias*

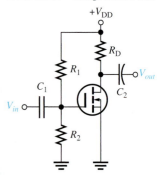

E-MOSFET *Voltage-divider bias*

- $I_D = I_{DSS}\left(1 - \dfrac{I_D R_S}{V_{GS(off)}}\right)^2$

- $A_v = g_m R_d$

- $R_{in} = R_G \parallel \left(\dfrac{V_{GS}}{I_{GSS}}\right)$

- $I_D = I_{DSS}$

- $A_v = g_m R_d$

- $R_{in} = R_G \parallel \left(\dfrac{V_{GS}}{I_{GSS}}\right)$

- $I_D = K(V_{GS} - V_{GS(th)})^2$

- $A_v = g_m R_d$

- $R_{in} = R_1 \parallel R_2 \left(\dfrac{V_{GS}}{I_{GSS}}\right)$

COMMON-DRAIN AMPLIFIER

JFET   *Self-bias*

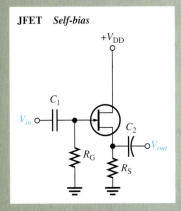

- $I_D = I_{DSS}\left(1 - \dfrac{I_D R_S}{V_{GS(off)}}\right)^2$

- $A_v = \dfrac{g_m R_s}{1 + g_m R_s}$

- $R_{in} = R_G \parallel \left(\dfrac{V_{GS}}{I_{GSS}}\right)$

COMMON-GATE AMPLIFIER

JFET   *Self-bias*

- $I_D = I_{DSS}\left(1 - \dfrac{I_D R_S}{V_{GS(off)}}\right)^2$

- $A_v = g_m R_d$

- $R_{in} = \left(\dfrac{1}{g_m}\right) \parallel R_S$

CASCODE AMPLIFIER

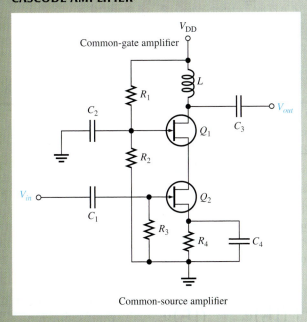

- $A_v \cong g_{m(CG)} X_L$

9–3　　$\left(I_{DSS}\left(\dfrac{R_S}{V_p}\right)^2\right)I_D^2 + \left(-2\dfrac{I_{DSS}R_S}{V_p} - 1\right)I_D + I_{DSS} = 0$　　　Self-biased JFET drain current

9–4　　$A_v = g_m R_d$　　　　　　　　　　　Voltage gain

9–5　　$R_{in} = R_G \parallel \left(\dfrac{V_{GS}}{I_{GSS}}\right)$　　　　　Input resistance, self-bias and zero-bias

9–6　　$R_{in} = R_1 \parallel R_2 \parallel R_{IN(gate)}$　　　　Input resistance, voltage-divider bias

Common-Drain Amplifier

9–7　　$A_v = \dfrac{g_m R_s}{1 + g_m R_s}$　　　　　　Voltage gain

9–8　　$R_{in} = R_G \parallel R_{IN(gate)}$　　　　　Input resistance

Common-Gate Amplifier

9–9　　$A_v = g_m R_d$　　　　　　　　　　Voltage gain

9–10　　$R_{in(source)} = \dfrac{1}{g_m}$　　　　　Input resistance

MOSFET Analog Switching

9–11　　$R = \dfrac{1}{fC}$　　　　　　　　　Emulated resistance

TRUE/FALSE QUIZ

Answers can be found at www.pearsonhighered.com/floyd.

1. A common-source (CS) amplifier has a very high input resistance.
2. The drain current in a CS amplifier can be calculated using a quadratic formula.
3. The voltage gain of a CS amplifier is the transconductance times the source resistance.
4. There is no phase inversion in a CS amplifier.
5. A CS amplifier using a D-MOSFET can operate with both positive and negative input voltages.
6. A common-drain (CD) amplifier is called a *drain-follower.*
7. The input resistance of a CD amplifier is very low.
8. The input resistance of a common-gate (CG) amplifier is very low.
9. A cascode amplifier uses both a CS and a CG amplifier.
10. The class D amplifier always operates in the linear region.
11. The class D amplifier uses pulse-width modulation.
12. An analog switch is controlled by a digital input.
13. The purpose of a switched-capacitor circuit is to emulate resistance.
14. CMOS is a device used in linear amplifiers.
15. CMOS utilizes a *pnp* MOSFET and an *npn* MOSFET connected together.

CIRCUIT-ACTION QUIZ

Answers can be found at www.pearsonhighered.com/floyd.

1. If the drain current is increased in Figure 9–9, V_{GS} will

　　(a) increase　　(b) decrease　　(c) not change

2. If the JFET in Figure 9–9 is substituted with one having a lower value of I_{DSS}, the voltage gain will

   (a) increase   (b) decrease   (c) not change

3. If the JFET in Figure 9–9 is substituted with one having a lower value of $V_{GS(off)}$, the voltage gain will

   (a) increase   (b) decrease   (c) not change

4. If the value of R_G in Figure 9–9 is increased, V_{GS} will

   (a) increase   (b) decrease   (c) not change

5. If the value of R_G in Figure 9–11 is increased, the input resistance seen by the signal source will

   (a) increase   (b) decrease   (c) not change

6. If the value of R_1 in Figure 9–17 is increased, V_{GS} will

   (a) increase   (b) decrease   (c) not change

7. If the value of R_L in Figure 9–17 is decreased, the voltage gain will

   (a) increase   (b) decrease   (c) not change

8. If the value of R_S in Figure 9–20 is increased, the voltage gain will

   (a) increase   (b) decrease   (c) not change

9. If C_4 in Figure 9–48 opens, the output signal voltage will

   (a) increase   (b) decrease   (c) not change

SELF-TEST

Answers can be found at www.pearsonhighered.com/floyd.

Section 9–1

1. In a common-source amplifier, the output voltage is

   (a) 180° out of phase with the input   (b) in phase with the input

   (c) taken at the source   (d) taken at the drain

   (e) answers (a) and (c)   (f) answers (a) and (d)

2. In a certain common-source (CS) amplifier, $V_{ds} = 3.2$ V rms and $V_{gs} = 280$ mV rms. The voltage gain is

   (a) 1   (b) 11.4   (c) 8.75   (d) 3.2

3. In a certain CS amplifier, $R_D = 1.0$ kΩ, $R_S = 560$ Ω, $V_{DD} = 10$ V, and $g_m = 4500$ μS. If the source resistor is completely bypassed, the voltage gain is

   (a) 450   (b) 45   (c) 4.5   (d) 2.52

4. Ideally, the equivalent circuit of a FET contains

   (a) a current source in series with a resistance

   (b) a resistance between drain and source terminals

   (c) a current source between gate and source terminals

   (d) a current source between drain and source terminals

5. The value of the current source in Question 4 is dependent on the

   (a) transconductance and gate-to-source voltage

   (b) dc supply voltage

   (c) external drain resistance

   (d) answers (b) and (c)

6. A certain common-source amplifier with a bypassed source resistance has a voltage gain of 10. If the source bypass capacitor is removed,

   (a) the voltage gain will increase

   (b) the transconductance will increase

   (c) the voltage gain will decrease

   (d) the Q-point will shift

7. A CS amplifier has a load resistance of 10 kΩ and $R_D = 820$ Ω. If $g_m = 5$ mS and $V_{in} = 500$ mV, the output signal voltage is

(a) 1.89 V   (b) 2.05 V   (c) 25 V   (d) 0.5 V

8. If the load resistance in Question 7 is removed, the output voltage will

(a) stay the same   (b) decrease   (c) increase   (d) be zero

Section 9–2   9. A certain common-drain (CD) amplifier with $R_S = 1.0$ kΩ has a transconductance of 6000 μS. The voltage gain is

(a) 1   (b) 0.86   (c) 0.98   (d) 6

10. The datasheet for the transistor used in a CD amplifier specifies $I_{GSS} = 5$ nA at $V_{GS} = +0$ V. If the resistor from gate to ground, R_G, is 50 MΩ, the total input resistance is approximately

(a) 50 MΩ   (b) 200 MΩ   (c) 40 MΩ   (d) 20.5 MΩ

Section 9–3   11. The common-gate (CG) amplifier differs from both the CS and CD configurations in that it has a

(a) much higher voltage gain   (b) much lower voltage gain

(c) much higher input resistance   (d) much lower input resistance

12. If you are looking for both good voltage gain and high input resistance, you must use a

(a) CS amplifier   (b) CD amplifier   (c) CG amplifier

13. A cascode amplifier can consist of

(a) a CD and a CS amplifier   (b) a CS and a CG amplifier

(c) two CD amplifiers   (d) two CG amplifiers

Section 9–4   14. The class D amplifier is similar to

(a) class C   (b) class B   (c) class A   (d) none of these

15. The class D amplifier uses

(a) frequency modulation   (b) amplitude modulation

(c) pulse-width modulation   (d) phase modulation

Section 9–5   16. E-MOSFETs are generally used for switching applications because of their

(a) threshold characteristic   (b) high input resistance

(c) linearity   (d) high gain

17. A sampling circuit must sample a signal at a minimum of

(a) one time per cycle   (b) the signal frequency

(c) more than twice the signal frequency   (b) half the signal frequency

18. The value of resistance emulated by a switched-capacitor circuit is a function of

(a) voltage and capacitance   (b) frequency and capacitance

(c) gain and transconductance   (b) frequency and transconductance

Section 9–6   19. A basic CMOS circuit uses a combination of

(a) *n*-channel MOSFETs   (b) *p*-channel MOSFETs

(c) *pnp* and *npn* BJTs   (d) an *n*-channel and a *p*-channel MOSFET

20. CMOS is commonly used in

(a) digital circuits   (b) linear circuits

(c) RF circuits   (d) power circuits

Section 9–7   21. If there is an internal open between the drain and source in a CS amplifier, the drain voltage is equal to

(a) 0 V   (b) V_{DD} (c) V_{GS} (d) V_{GD}

PROBLEMS

Answers to all odd-numbered problems are at the end of the book.

BASIC PROBLEMS

Section 9–1   **The Common-Source Amplifier**

1. Name two general approaches to the analysis of a JFET circuit.

2. A JFET has a g_m of 5 mS and an external ac drain resistance of 2.2 kΩ. What is the ideal voltage gain?

3. A FET has a $g_m = 6000\ \mu$S. Determine the rms drain current for each of the following rms values of V_{gs}.

    (a) 10 mV   **(b)** 150 mV   **(c)** 0.6 V   **(d)** 1 V

4. The gain of a certain JFET amplifier with a source resistance of zero is 20. Determine the drain resistance if the g_m is 3500 μS.

5. A certain FET amplifier has a g_m of 4.2 mS, $r'_{ds} = 12$ kΩ, and $R_D = 4.7$ kΩ. What is the voltage gain? Assume the source resistance is 0 Ω.

6. What is the gain for the amplifier in Problem 5 if the source resistance is 1.0 kΩ?

7. Identify the type of FET and its bias arrangement in Figure 9–58. Ideally, what is V_{GS}?

8. Calculate the dc voltages from each terminal to ground for the FETs in Figure 9–58.

▶ FIGURE 9–58

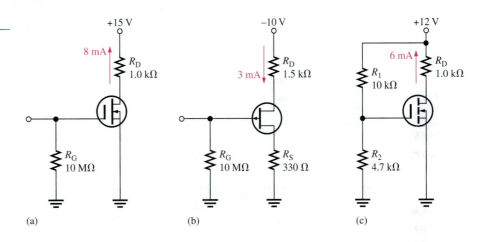

(a)   (b)   (c)

9. Identify each characteristic curve in Figure 9–59 by the type of FET that it represents.

▶ FIGURE 9–59

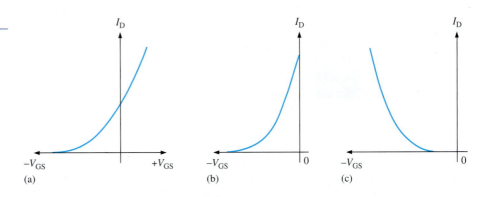

(a)   (b)   (c)

10. Refer to the JFET transfer characteristic curve in Figure 9–16(a) and determine the peak-to-peak value of I_d when V_{gs} is varied ± 1.5 V about its Q-point value.

11. Repeat Problem 10 for the curves in Figure 9–16(b) and Figure 9–16(c).

12. Given that $I_D = 2.83$ mA in Figure 9–60, find V_{DS} and V_{GS}. $V_{GS(off)} = -7$ V and $I_{DSS} = 8$ mA.

13. If a 50 mV rms input signal is applied to the amplifier in Figure 9–60, what is the peak-to-peak output voltage? $g_m = 5000\ \mu$S.

14. If a 1500 Ω load is ac coupled to the output in Figure 9–60, what is the resulting output voltage (rms) when a 50 mV rms input is applied? $g_m = 5000 \ \mu S$.

▶ FIGURE 9–60

Multisim and LT Spice file circuits are identified with a logo and are in the Problems folder on the website. Filenames correspond to figure numbers (e.g., FGM09–60 or FGS09–60).

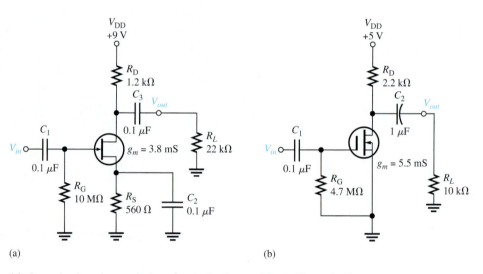

15. Determine the voltage gain of each common-source amplifier in Figure 9–61.

▶ FIGURE 9–61

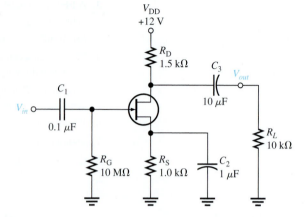

(a)   (b)

16. Draw the dc and ac equivalent circuits for the amplifier in Figure 9–62.

▶ FIGURE 9–62

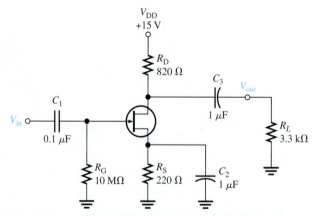

17. Determine the drain current in Figure 9–62 given that $I_{DSS} = 15$ mA and $V_{GS(off)} = -4$ V. The Q-point is centered.

18. What is the gain of the amplifier in Figure 9–62 if C_2 is removed?

19. A 4.7 kΩ resistor is connected in parallel with R_L in Figure 9–62. What is the voltage gain?

20. For the common-source amplifier in Figure 9–63, determine I_D, V_{GS}, and V_{DS} for a centered Q-point. $I_{DSS} = 9$ mA, and $V_{GS(off)} = -3$ V.

▶ FIGURE 9–63

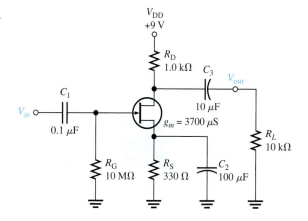

21. If a 10 mV rms signal is applied to the input of the amplifier in Figure 9–63, what is the rms value of the output signal?

22. Determine V_{GS}, I_D, and V_{DS} for the amplifier in Figure 9–64. $I_{D(on)} = 18$ mA at $V_{GS} = 10$ V, $V_{GS(th)} = 2.5$ V, and $g_m = 3000$ μS.

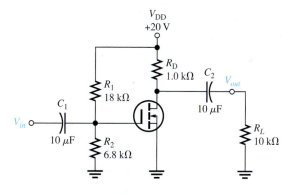

▲ FIGURE 9–64

23. Determine R_{in} seen by the signal source in Figure 9–65. $I_{GSS} = 25$ nA at $V_{GS} = -15$ V.

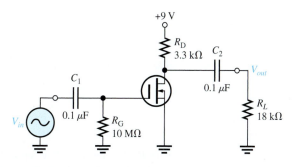

▲ FIGURE 9–65

24. Determine the total drain voltage waveform (dc and ac) and the V_{out} waveform in Figure 9–66. $g_m = 4.8$ mS and $I_{DSS} = 15$ mA. Observe that $V_{GS} = 0$.

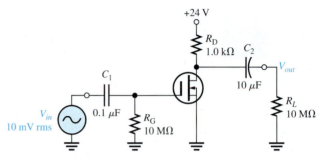

▲ **FIGURE 9–66**

25. For the unloaded amplifier in Figure 9–67, find V_{GS}, I_D, V_{DS}, and the rms output voltage V_{ds}. $I_{D(on)} = 8$ mA at $V_{GS} = 12$ V, $V_{GS(th)} = 4$ V, and $g_m = 4500$ S.

▶ **FIGURE 9–67**

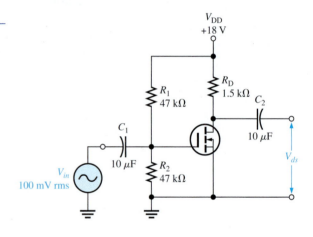

Section 9–2   **The Common-Drain Amplifier**

26. For the source-follower in Figure 9–68, determine the voltage gain and input resistance. $I_{GSS} = 50$ pA at $V_{GS} = -15$ V and $g_m = 5500$ μS.

▶ **FIGURE 9–68**

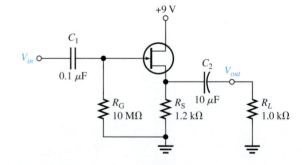

27. If the JFET in Figure 9–68 is replaced with one having a g_m of 3000 μS, what are the gain and the input resistance with all other conditions the same?

28. Find the gain of each amplifier in Figure 9–69.

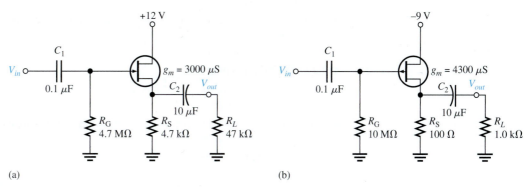

(a) (b)

▲ **FIGURE 9–69**

29. Determine the voltage gain of each amplifier in Figure 9–69 when the capacitively coupled load is changed to 10 kΩ.

30. Does R_3 in the circuit shown in Figure 9–14(b) affect the gain? Explain your answer.

31. Write an expression for the input resistance of the amplifier shown in Figure 9–14(b).

Section 9–3 **The Common-Gate Amplifier**

32. A common-gate amplifier has a $g_m = 4000\ \mu S$ and $R_d = 1.5\ k\Omega$. What is its gain?

33. What is the $R_{in(source)}$ of the amplifier in Problem 32?

34. Determine the voltage gain and input resistance of the common-gate amplifier in Figure 9–70.

▶ **FIGURE 9–70**

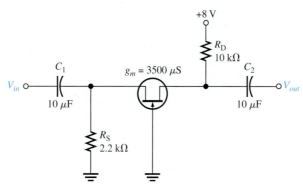

35. For a cascode amplifier like shown in Figure 9–25, $g_m = 2800\ \mu S$, $I_{GSS} = 2$ nA at $V_{GS} = 15$ V. If $R_3 = 15\ M\Omega$ and $L = 1.5$ mH, determine the voltage gain and the input resistance at $f = 100$ MHz.

Section 9–4 **The Class D Amplifier**

36. A class D amplifier has an output of ± 9 V. If the input signal is 5 mV peak-to-peak, what is the voltage gain?

37. A certain class D amplifier dissipates an internal power of 140 mW in the output filter, the comparator, and the triangular wave generator. Each complementary MOSFET has 0.25 V drop in the *on* state. The amplifier operates from ± 12 Vdc sources and provides 0.35 A to a load. Determine the efficiency.

Section 9–5 **MOSFET Analog Switching**

38. An analog switch uses an *n*-channel MOSFET with $V_{GS(th)} = 4$ V. A voltage of $+8$ V is applied to the gate. Determine the maximum peak-to-peak input signal that can be applied if the drain-to-source voltage drop is neglected.

39. An analog switch is used to sample a signal with a maximum frequency of 15 kHz. Determine the minimum frequency of the pulses applied to the MOSFET gate.

40. A switched-capacitor circuit uses a 10 pF capacitor. Determine the frequency required to emulate a 10 kΩ resistor.

41. For a frequency of 25 kHz, what is the emulated resistance in a switched-capacitor circuit if $C = 0.001\ \mu F$?

Section 9–6   MOSFET Digital Switching

42. What is the output voltage of a CMOS inverter that operates with $V_{DD} = +5$ V, when the input is 0 V? When the input is +5 V?

43. For each of the following input combinations, determine the output of a CMOS NAND gate that operates with $V_{DD} = +3.3$ V.

   (a) $V_A = 0$ V, $V_B = 0$ V

   (b) $V_A = +3.3$ V, $V_B = 0$ V

   (c) $V_A = 0$ V, $V_B = +3.3$ V

   (d) $V_A = +3.3$ V, $V_B = +3.3$ V

44. Repeat Problem 43 for a CMOS NOR gate.

45. List two advantages of the MOSFET over the BJT in power switching.

Section 9–7   Troubleshooting

46. What symptom(s) would indicate each of the following failures when a signal voltage is applied to the input in Figure 9–71?

   (a) Q_1 open from drain to source

   (b) R_3 open

   (c) C_2 shorted

   (d) C_3 open

   (d) Q_2 open from drain to source

47. If $V_{in} = 10$ mV rms in Figure 9–71, what is V_{out} for each of the following faults?

   (a) C_1 open

   (b) C_4 open

   (c) a short from the source of Q_2 to ground

   (d) Q_2 has an open gate

▶ **FIGURE 9–71**

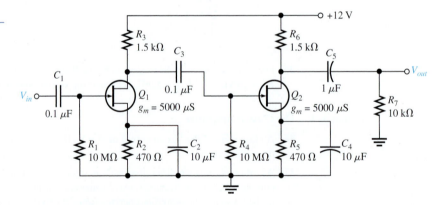

DATASHEET PROBLEMS

48. What type of FET is the 2N3796?

49. Referring to the datasheet in Figure 9–72, determine the following:

   (a) typical $V_{GS(off)}$ for the 2N3796

   (b) maximum drain-to-source voltage for the 2N3797

   (c) maximum power dissipation for the 2N3797 at an ambient temperature of 25°C

   (d) maximum gate-to-source voltage for the 2N3797

50. Referring to Figure 9–72, determine the maximum power dissipation for a 2N3796 at an ambient temperature of 55°C.

51. Referring to Figure 9–72, determine the minimum g_{m0} for the 2N3796 at a frequency of 1 kHz.

52. What is the drain current when $V_{GS} = +3.5$ V for the 2N3797?

53. Typically, what is the drain current for a zero-biased 2N3796?

54. What is the maximum possible voltage gain for a 2N3796 common-source amplifier with $R_d = 2.2$ kΩ?

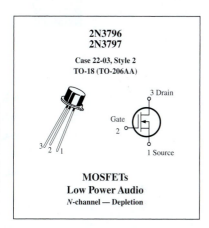

2N3796
2N3797

Case 22-03, Style 2
TO-18 (TO-206AA)

MOSFETs
Low Power Audio
N-channel — Depletion

Maximim Ratings

Rating	Symbol	Value	Unit
Drain-Source voltage	V_{DS}		V dc
2N3796		25	
2N3797		20	
Gate-Source voltage	V_{GS}	±10	V dc
Drain current	I_D	20	mA dc
Total device dissipation @ $T_A = 25°C$	P_D	200	mW
Derate above 25°C		1.14	mW/°C
Junction temperature	T_J	+175	°C
Storage channel temperature range	T_{stg}	− 65 to + 200	°C

Electrical Characteristics ($T_A = 25°C$ unless otherwise noted.)

Characteristic		Symbol	Min	Typ	Max	Unit
OFF Characteristics						
Drain-Source breakdown voltage		$V_{(BR)DSX}$				V dc
($V_{GS} = -4.0$ V, $I_D = 5.0$ μA)	2N3796		25	30	–	
($V_{GS} = -7.0$ V, $I_D = 5.0$ μA)	2N3797		20	25	–	
Gate reverse current		I_{GSS}				pA dc
($V_{GS} = -10$ V, $V_{DS} = 0$)			–	–	1.0	
($V_{GS} = -10$ V, $V_{DS} = 0$, $T_A = 150°C$)			–	–	200	
Gate-Source cutoff voltage		$V_{GS(off)}$				V dc
($I_D = 0.5$ μA, $V_{DS} = 10$ V)	2N3796		–	− 3.0	− 4.0	
($I_D = 2.0$ μA, $V_{DS} = 10$ V)	2N3797		–	− 5.0	− 7.0	
Drain-Gate reverse current		I_{DGO}				pA dc
($V_{DG} = 10$ V, $I_S = 0$)			–	–	1.0	
ON Characteristics						
Zero-gate-voltage drain current		I_{DSS}				mA dc
($V_{DS} = 10$ V, $V_{GS} = 0$)	2N3796		0.5	1.5	3.0	
	2N3797		2.0	2.9	6.0	
On-State drain current		$I_{D(on)}$				mA dc
($V_{DS} = 10$ V, $V_{GS} = +3.5$ V)	2N3796		7.0	8.3	14	
	2N3797		9.0	14	18	
Small-Signal Characteristics						
Forward transfer admittance		$|Y_{fs}|$				μmhos
($V_{DS} = 10$ V, $V_{GS} = 0$, $f = 1.0$ kHz)	2N3796		900	1200	1800	or
	2N3797		1500	2300	3000	μS
($V_{DS} = 10$ V, $V_{GS} = 0$, $f = 1.0$ MHz)	2N3796		900	–	–	
	2N3797		1500	–	–	
Output admittance		$|Y_{os}|$				μmhos or
($V_{DS} = 10$ V, $V_{GS} = 0$, $f = 1.0$ kHz)	2N3796		–	12	25	μS
	2N3797		–	27	60	
Input capacitance		C_{iss}				pF
($V_{DS} = 10$ V, $V_{GS} = 0$, $f = 1.0$ MHz)	2N3796		–	5.0	7.0	
	2N3797		–	6.0	8.0	
Reverse transfer capacitance		C_{rss}				pF
($V_{DS} = 10$ V, $V_{GS} = 0$, $f = 1.0$ MHz)			–	0.5	0.8	

▲ **FIGURE 9–72**

Partial datasheet for the 2N3796 and 2N3797 D-MOSFETs.

ADVANCED PROBLEMS

55. The MOSFET in a certain single-stage common-source amplifier has a range of forward transconductance values from 2.5 mS to 7.5 mS. If the amplifier is capacitively coupled to a variable load that ranges from 4 kΩ to 10 kΩ and the dc drain resistance is 1.0 kΩ, determine the minimum and maximum voltage gains.

56. Design an amplifier using a 2N3797 that operates from a 24 V supply voltage. The typical dc drain-to-source voltage should be approximately 12 V and the typical voltage gain should be approximately 9.

57. Modify the amplifier you designed in Problem 56 so that the voltage gain can be set at 9 for any randomly selected 2N3797.

MULTISIM TROUBLESHOOTING PROBLEMS

These file circuits are in the Troubleshooting Problems folder on the website.

58. Open file TPM09-58 and determine the fault.

59. Open file TPM09-59 and determine the fault.

60. Open file TPM09-60 and determine the fault.

61. Open file TPM09-61 and determine the fault.

62. Open file TPM09-62 and determine the fault.

63. Open file TPM09-63 and determine the fault.

64. Open file TPM09-64 and determine the fault.

65. Open file TPM09-65 and determine the fault.

66. Open file TPM09-66 and determine the fault.

AMPLIFIER FREQUENCY RESPONSE

10

CHAPTER OBJECTIVES

◆ Explain how circuit capacitances affect the frequency response of an amplifier
◆ Use the decibel (dB) to express amplifier gain
◆ Analyze the low-frequency response of an amplifier
◆ Analyze the high-frequency response of an amplifier
◆ Analyze an amplifier for total frequency response
◆ Analyze multistage amplifiers for frequency response
◆ Explain how to measure the frequency response of an amplifier

KEY TERMS

◆ Midrange gain
◆ Critical frequency
◆ Roll-off
◆ Decade
◆ Bode Plot
◆ Bandwidth
◆ Normalized
◆ Parasitic capacitance
◆ Gain-bandwidth product

DEVICE APPLICATION PREVIEW

In the Device Application, you will modify the PA system preamp from Chapter 6 to increase the low-frequency response to reduce the effects of 60 Hz interference.

VISIT THE COMPANION WEBSITE

Study aids, Multisim files, and LT Spice files for this chapter are available at https://www.pearsonhighered.com /careersresources/

INTRODUCTION

In the previous chapters on amplifiers, the effects of the input frequency on an amplifier's operation due to capacitive elements in the circuit were neglected in order to focus on other concepts. The coupling and bypass capacitors were considered to be ideal shorts and the internal transistor capacitances were considered to be ideal opens. This treatment is valid when the frequency is in an amplifier's midrange.

As you know, capacitive reactance decreases with increasing frequency and vice versa. When the frequency is low enough, the coupling and bypass capacitors can no longer be considered as shorts because their reactances are large enough to have a significant effect. Also, when the frequency is high enough, the internal transistor capacitances can no longer be considered as opens because their reactances become small enough to have a significant effect on the amplifier operation. A complete picture of an amplifier's response must take into account the full range of frequencies over which the amplifier can operate.

In this chapter, you will study the frequency effects on amplifier gain and phase shift. The coverage applies to both BJT and FET amplifiers, and a mix of both are included to illustrate the concepts.

10–1 BASIC CONCEPTS

In amplifiers, the coupling and bypass capacitors appear to be shorts to ac at the mid-band frequencies. At low frequencies the capacitive reactance of these capacitors affect the gain and phase shift of signals, so they must be taken into account. The **frequency response** of an amplifier is the change in gain or phase shift over a specified range of input signal frequencies.

After completing this section, you should be able to

❑ **Explain how circuit capacitances affect the frequency response of an amplifier**
  ◆ Define frequency response
❑ Discuss the effect of coupling capacitors
  ◆ Recall the formula for capacitive reactance
❑ Discuss the effect of bypass capacitors
❑ Describe the effect of internal transistor capacitances
  ◆ Identify the internal capacitance in BJTs and JFETs
❑ Explain Miller's theorem
  ◆ Calculate the Miller input and output capacitances

Effect of Coupling Capacitors

Recall from basic circuit theory that $X_C = 1/(2\pi fC)$. This formula shows that the capacitive reactance varies inversely with frequency. At lower frequencies the reactance is greater, and it decreases as the frequency increases. At lower frequencies—for example, audio frequencies below 10 Hz—capacitively coupled amplifiers such as those in Figure 10–1 have less voltage gain than they have at higher frequencies. The reason is that at lower frequencies more signal voltage is dropped across C_1 and C_3 because their reactances are higher. This higher signal voltage drop at lower frequencies reduces the voltage gain. Also, a phase shift is introduced by the coupling capacitors because C_1 forms a lead circuit with the R_{in} of the amplifier and C_3 forms a lead circuit with R_L in series with R_C or R_D. Recall that a *lead circuit* is an *RC* circuit in which the phase of the output voltage across *R* leads the input voltage.

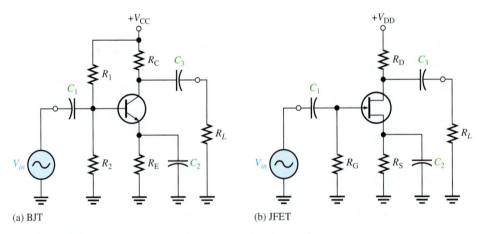

(a) BJT   (b) JFET

▲ FIGURE 10–1

Examples of capacitively coupled BJT and FET amplifiers.

Effect of Bypass Capacitors

At lower frequencies, the reactance of the bypass capacitor, C_2 in Figure 10–1, becomes significant and the emitter (or FET source terminal) is no longer at ac ground. The capacitive reactance X_{C2} in parallel with R_E (or R_S) creates an impedance that reduces the gain. This is illustrated in Figure 10–2.

For example, when the frequency is sufficiently high, $X_C \cong 0\ \Omega$ and the voltage gain of the CE amplifier is $A_v = R_C/r'_e$. At lower frequencies, $X_C \gg 0\ \Omega$ and the voltage gain is $A_v = R_C/(r'_e + Z_e)$.

Effect of Internal Transistor Capacitances

At high frequencies, the coupling and bypass capacitors become effective ac shorts and do not affect an amplifier's response. Internal transistor junction capacitances, however, do come into play, reducing an amplifier's gain and introducing phase shift as the signal frequency increases. These internal capacitances are referred to as parasitic capacitances. These are unwanted or stay capacitances that exist between circuit elements; the circuit elements can be any components that are in close proximity to each other.

In the case of transistors, **parasitic capacitance** is an unwanted capacitance that exists between terminals of the transistor (both BJTs and FETs).

Figure 10–3 shows the internal parasitic *pn* junction capacitances for both a bipolar junction transistor and a JFET. In the case of the BJT, C_{be} is the base-emitter junction capacitance and C_{bc} is the base-collector junction capacitance. In the case of the JFET, C_{gs} is the capacitance between gate and source and C_{gd} is the capacitance between gate and drain.

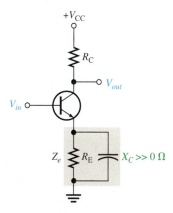

▲ **FIGURE 10–2**

Nonzero reactance of the bypass capacitor in parallel with R_E creates an emitter impedance, (Z_e), which reduces the voltage gain.

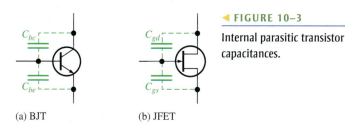

◀ **FIGURE 10–3**

Internal parasitic transistor capacitances.

(a) BJT　　　(b) JFET

Datasheets often refer to the BJT capacitance C_{bc} as the output capacitance, often designated C_{ob}. The capacitance C_{be} is often designated as the input capacitance C_{ib}. Datasheets for FETs normally specify input capacitance C_{iss} and reverse transfer capacitance C_{rss}. From these, C_{gs} and C_{gd} can be calculated, as you will see in Section 10–4.

At lower frequencies, the internal capacitances have a very high reactance because of their low capacitance value (usually only a few picofarads) and the low frequency value. Therefore, they look like opens and have no effect on the transistor's performance. As the frequency goes up, the internal capacitive reactances go down, and at some point they begin to have a significant effect on the transistor's gain. When the reactance of C_{be} (or C_{gs}) becomes small enough, a significant amount of the signal voltage is lost due to a voltage-divider effect of the signal source resistance and the reactance of C_{be}, as illustrated in Figure 10–4(a). When the reactance of C_{bc} (or C_{gd}) becomes small enough, a significant amount of output signal voltage is fed back out of phase with the input (negative feedback), thus effectively reducing the voltage gain. This is illustrated in Figure 10–4(b).

Miller's Theorem

Miller's theorem is used to simplify the analysis of inverting amplifiers at high frequencies where the internal transistor capacitances are important. The capacitance C_{bc} in BJTs (C_{gd} in FETs) between the input (base or gate) and the output (collector or drain) is shown in Figure 10–5(a) in a generalized form. A_v is the absolute voltage gain of the inverting amplifier at midrange frequencies, and C represents either C_{bc} or C_{gd}.

Miller's theorem states that C effectively appears as a capacitance from input to ground, as shown in Figure 10–5(b), that can be expressed as follows:

$$C_{in(Miller)} = C(A_v + 1)$$

Equation 10–1

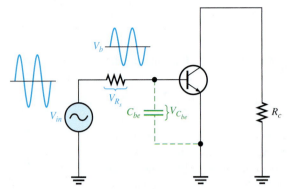

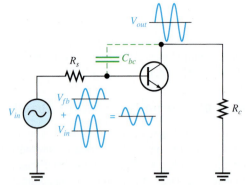

(a) Effect of C_{be}, where V_b is reduced by the voltage-divider action of R_s and $X_{C_{be}}$.

(b) Effect of C_{bc}, where part of V_{out} (V_{fb}) goes back through C_{bc} to the base and reduces the input signal because it is approximately 180° out of phase with V_{in}.

▲ **FIGURE 10–4**

AC equivalent circuit for a BJT amplifier showing effects of the internal capacitances C_{be} and C_{bc}.

▶ **FIGURE 10–5**

General case of Miller input and output capacitances. C represents C_{bc} or C_{gd}.

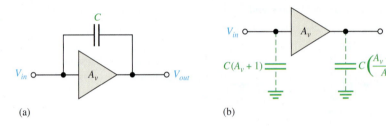

This formula shows that C_{bc} (or C_{gd}) has a much greater impact on input capacitance than its actual value. For example, if $C_{bc} = 6$ pF and the amplifier gain is 50, then $C_{in(Miller)} = 306$ pF. Figure 10–6 shows how this effective input capacitance appears in the actual ac equivalent circuit in parallel with C_{be} (or C_{gs}).

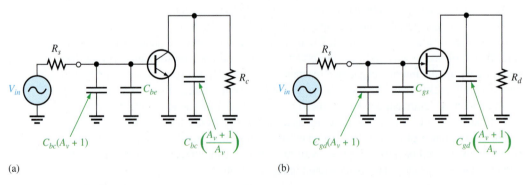

▲ **FIGURE 10–6**

Amplifier ac equivalent circuits showing internal capacitances and effective Miller capacitances.

Miller's theorem also states that C effectively appears as a capacitance from output to ground, as shown in Figure 10–5(b), that can be expressed as follows:

Equation 10–2

$$C_{out(Miller)} = C\left(\frac{A_v + 1}{A_v}\right)$$

This formula indicates that if the voltage gain is 10 or greater, $C_{out(Miller)}$ is approximately equal to C_{bc} or C_{gd} because $(A_v + 1)/A_v$ is approximately equal to 1. Figure 10–6 also shows how this effective output capacitance appears in the ac equivalent circuit for BJTs and FETs. Equations 10–1 and 10–2 are derived in "Derivations of Selected Equations" at www.pearsonhighered.com/floyd.

---

1. In an ac amplifier, which capacitors affect the low-frequency gain?
2. How is the high-frequency gain of an amplifier limited?
3. When can coupling and bypass capacitors be neglected?
4. Determine $C_{in(Miller)}$ if $A_v = -50$ and $C_{bc} = 5$ pF.
5. Determine $C_{out(Miller)}$ if $A_v = -25$ and $C_{bc} = 3$ pF.

10–2 THE DECIBEL

The decibel is a unit of logarithmic gain measurement and is commonly used to express amplifier response.

After completing this section, you should be able to

❏ **Use the decibel (dB) to express amplifier gain**
   ◆ Express power gain and voltage gain in dB
❏ Discuss the 0 dB reference
   ◆ Define *midrange gain*
❏ Define and discuss the critical frequency
❏ Discuss power measurement in dBm
   ◆ Identify the internal capacitance in BJTs and JFETs
❏ Explain Miller's theorem
   ◆ Calculate the Miller input and output capacitances

The use of decibels to express gain was introduced in Chapter 6. The decibel unit is important in amplifier measurements. The basis for the decibel unit stems from the logarithmic response of the human ear to the intensity of sound. Recall that the decibel was defined as a logarithmic measurement of the ratio of one power to another or one voltage to another. Power gain is expressed in decibels (dB) by the following formula:

$$A_{p(dB)} = 10 \log A_p$$ **Equation 10–3**

where A_p is the actual power gain, P_{out}/P_{in}. Voltage gain is expressed in decibels by the following formula:

$$A_{v(dB)} = 20 \log A_v$$ **Equation 10–4**

When working with decibel voltage gains, it is important that they are measured in the same impedance. In radio frequency and microwave systems, the impedance is generally 50 ohms; in audio systems, it is generally 600 ohms. If A_v is greater than 1, the dB gain is positive. If A_v is less than 1, the dB gain is negative and is usually called *attenuation*. You can use the LOG key on your calculator when working with these formulas.

EXAMPLE 10–1   Express each of the following ratios in dB:

(a) $\dfrac{P_{out}}{P_{in}} = 250$ (b) $\dfrac{P_{out}}{P_{in}} = 100$ (c) $A_v = 10$

(d) $A_p = 0.5$ (e) $\dfrac{V_{out}}{V_{in}} = 0.707$

Solution   (a) $A_{p(\text{dB})} = 10 \log (250) = \textbf{24 dB}$

(b) $A_{p(\text{dB})} = 10 \log (100) = \textbf{20 dB}$

(c) $A_{v(\text{dB})} = 20 \log (10) = \textbf{20 dB}$

(d) $A_{p(\text{dB})} = 10 \log (0.5) = \textbf{−3 dB}$

(e) $A_{v(\text{dB})} = 20 \log (0.707) = \textbf{−3 dB}$

Related Problem*   Express each of the following gains in dB: (a) $A_v = 1200$, (b) $A_p = 50$, (c) $A_v = 125{,}000$.

* Answers can be found at www.pearsonhighered.com/floyd.

FYI

The factor of 20 in Equation 10–4 is because power is proportional to voltage squared. Technically, the equation should be applied only when the voltages are measured in the same impedance. This is the case for many communication systems, such as in television or microwave systems.

0 dB Reference

It is often convenient in amplifiers to assign a certain value of gain as the 0 dB reference. This does not mean that the actual voltage gain is 1 (which is 0 dB); it means that the reference gain, no matter what its actual value, is used as a reference with which to compare other values of gain and is therefore assigned a 0 dB value. When an amplifier's response is shifted in the vertical axis to 0 dB, it is said to be **normalized**. To normalize an amplifier's response, all values are divided by the midband gain, forcing the reference level to be 0 dB. Note that the shape of the response is unaffected by this procedure. Normalization simplifies the comparison of different amplifier responses.

Many amplifiers exhibit a maximum gain over a certain range of frequencies and a reduced gain at frequencies below and above this range. The maximum gain occurs for the range of frequencies between the upper and lower critical frequencies and is called the **midrange gain**, which is assigned a 0 dB value. Any value of gain below midrange can be referenced to 0 dB and expressed as a negative dB value. For example, if the midrange voltage gain of a certain amplifier is 100 and the gain at a certain frequency below midrange is 50, then this reduced voltage gain can be expressed as $20 \log (50/100) = 20 \log (0.5) = -6$ dB. This indicates that it is 6 dB *below* the 0 dB reference. Halving the output voltage for a steady input voltage is always a 6 dB *reduction* in the gain. Correspondingly, a doubling of the output voltage is always a 6 dB *increase* in the gain. Figure 10–7 illustrates a normalized gain-versus-frequency curve showing several dB points.

▶ **FIGURE 10–7**

Normalized voltage gain versus frequency curve.

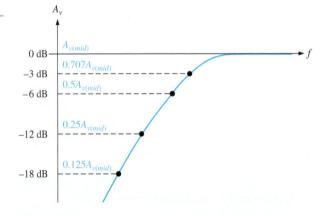

Table 10–1 shows how doubling or halving voltage gains that are measured in the same impedance translates into decibel values. Notice in the table that every time the voltage gain is doubled, the decibel value increases by 6 dB, and every time the gain is halved, the dB value decreases by 6 dB.

◀ TABLE 10–1

Decibel values corresponding to doubling and halving of the voltage gain that are measured in the same impedance.

VOLTAGE GAIN (A_v)	DECIBEL VALUE*
32	20 log (32) = 30 dB
16	20 log (16) = 24 dB
8	20 log (8) = 18 dB
4	20 log (4) = 12 dB
2	20 log (2) = 6 dB
1	20 log (1) = 0 dB
0.707	20 log (0.707) = −3 dB
0.5	20 log (0.5) = −6 dB
0.25	20 log (0.25) = −12 dB
0.125	20 log (0.125) = −18 dB
0.0625	20 log (0.0625) = −24 dB
0.03125	20 log (0.03125) = −30 dB

*Decibel values are with respect to zero reference.

Critical Frequency

A critical frequency (also known as **cutoff frequency** or *corner frequency*) is a frequency at which the output power drops to one-half of its midrange value. This corresponds to a 3 dB reduction in the power gain, as expressed in dB by the following formula:

$$A_{p(dB)} = 10 \log (0.5) = -3 \text{ dB}$$

Also, at the critical frequencies the voltage gain is 70.7% of its midrange value and is expressed in dB as

$$A_{v(dB)} = 20 \log (0.707) = -3 \text{ dB}$$

EXAMPLE 10–2

A certain amplifier has a midrange rms output voltage of 10 V. What is the rms output voltage for each of the following dB gain reductions with a constant rms input voltage?

(a) −3 dB   **(b)** −6 dB   **(c)** −12 dB   **(d)** −24 dB

Solution   Multiply the midrange output voltage by the voltage gain corresponding to the specified decibel value in Table 10–1.

(a) At −3 dB, $V_{out} = 0.707(10 \text{ V}) = \textbf{7.07 V}$

(b) At −6 dB, $V_{out} = 0.5(10 \text{ V}) = \textbf{5 V}$

(c) At −12 dB, $V_{out} = 0.25(10 \text{ V}) = \textbf{2.5 V}$

(d) At −24 dB, $V_{out} = 0.0625(10 \text{ V}) = \textbf{0.625 V}$

Related Problem   Determine the output voltage at the following decibel levels for a midrange value of 50 V:

(a) 0 dB   **(b)** −18 dB   **(c)** −30 dB

Power Measurement in dBm

The **dBm** is a unit for measuring power levels referenced to 1 mW. Positive dBm values represent power levels above 1 mW, and negative dBm values represent power levels below 1 mW.

Because the decibel (dB) can be used to represent only power *ratios*, not actual power, the dBm provides a convenient way to express actual power output of an amplifier or other device. Each 3 dBm increase corresponds to a doubling of the power, and a 3 dBm decrease corresponds to a halving of the power.

To state that an amplifier has a 3 dB power gain indicates only that the output power is twice the input power and nothing about the actual output power. To indicate actual output power, the dBm can be used. For example, 3 dBm is equivalent to 2 mW because 2 mW is twice the 1 mW reference. 6 dBm is equivalent to 4 mW, and so on. Likewise, −3 dBm is the same as 0.5 mW. Table 10–2 shows several values of power in terms of dBm.

▶ TABLE 10–2

Power in terms of dBm.

POWER	dBM
32 mW	15 dBm
16 mW	12 dBm
8 mW	9 dBm
4 mW	6 dBm
2 mW	3 dBm
1 mW	0 dBm
0.5 mW	−3 dBm
0.25 mW	−6 dBm
0.125 mW	−9 dBm
0.0625 mW	−12 dBm
0.03125 mW	−15 dBm

SECTION 10–2 CHECKUP

1. How much increase in actual voltage gain corresponds to +12 dB?
2. Convert a power gain of 25 to decibels.
3. What power corresponds to 0 dBm?
4. What is the standard impedance in audio systems that is assumed with decibel measurements?

10–3 LOW FREQUENCY AMPLIFIER RESPONSE

The voltage gain and phase shift of capacitively coupled amplifiers are affected when the signal frequency is below a critical value. At low frequencies, the reactance of the coupling capacitors becomes significant, resulting in a reduction in voltage gain and an increase in phase shift. Frequency responses of both BJT and FET capacitively coupled amplifiers are discussed.

After completing this section, you should be able to

❑ **Analyze the low-frequency response of an amplifier**
❑ Analyze a BJT amplifier
   ◆ Calculate the midrange voltage gain ◆ Identify the parts of the amplifier that affect low-frequency response
❑ Identify and analyze the BJT amplifier's input *RC* circuit
   ◆ Calculate the lower critical frequency and gain roll-off ◆ Sketch a Bode plot
   ◆ Define *decade* and *octave* ◆ Determine the phase shift
❑ Identify and analyze the BJT amplifier's output *RC* circuit
   ◆ Calculate the lower critical frequency ◆ Determine the phase shift

◻ Identify and analyze the BJT amplifier's bypass *RC* circuit
  ◆ Calculate the lower critical frequency   ◆ Explain the effect of a swamping resistor
◻ Analyze a FET amplifier
◻ Identify and analyze the D-MOSFET amplifier's input *RC* circuit
  ◆ Calculate the lower critical frequency   ◆ Determine the phase shift
◻ Identify and analyze the D-MOSFET amplifier's output *RC* circuit
  ◆ Calculate the lower critical frequency   ◆ Determine the phase shift
◻ Explain the total low-frequency response of an amplifier
  ◆ Illustrate the response with Bode plots
◻ Simulate the frequency response using Multisim
  ◆ Calculate the lower critical frequency   ◆ Determine the phase shift

BJT Amplifiers

A typical capacitively coupled common-emitter amplifier is shown in Figure 10–8. Assuming that the coupling and bypass capacitors are ideal shorts at the midrange signal frequency, you can determine the midrange voltage gain using Equation 10–5, where $R_c = R_C \| R_L$.

$$A_{v(mid)} = \frac{R_c}{r'_e}$$

Equation 10–5

If a swamping resistor (R_{E1}) is used, it appears in series with r'_e and the equation becomes

$$A_{v(mid)} = \frac{R_c}{r'_e + R_{E1}}$$

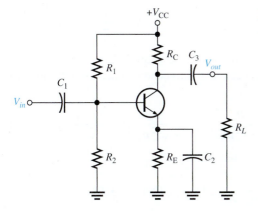

◀ FIGURE 10–8

A capacitively coupled BJT amplifier.

The BJT amplifier in Figure 10–8 has three high-pass *RC* circuits that affect its gain as the frequency is reduced below midrange. These are shown in the low-frequency ac equivalent circuit in Figure 10–9. Unlike the ac equivalent circuit used in previous chapters, which represented midrange response ($X_C \cong 0\ \Omega$), the low-frequency equivalent circuit retains the coupling and bypass capacitors because X_C is not small enough to neglect when the signal frequency is sufficiently low.

One *RC* circuit is formed by the input coupling capacitor C_1 and the input resistance of the amplifier. The second *RC* circuit is formed by the output coupling capacitor C_3, the resistance looking in at the collector (R_{out}), and the load resistance. The third *RC* circuit that affects the low-frequency response is formed by the emitter-bypass capacitor C_2 and the resistance looking in at the emitter.

▶ **FIGURE 10–9**

The low-frequency ac equivalent circuit of the amplifier in Figure 10–8 consists of three high-pass *RC* circuits.

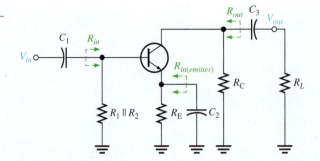

The Input *RC* Circuit

The input *RC* circuit for the BJT amplifier in Figure 10–8 is formed by C_1 and the amplifier's input resistance and is shown in Figure 10–10. (Input resistance was discussed in Chapter 6.) As the signal frequency decreases, X_{C1} increases. This causes less voltage across the input resistance of the amplifier at the base because more voltage is dropped across C_1 and because of this, the overall voltage gain of the amplifier is reduced. The base voltage for the input *RC* circuit in Figure 10–10 (neglecting the internal resistance of the input signal source) can be stated as

$$V_{base} = \left(\frac{R_{in}}{\sqrt{R_{in}^2 + X_{C1}^2}} \right) V_{in}$$

▶ **FIGURE 10–10**

Input *RC* circuit formed by the input coupling capacitor and the amplifier's input resistance.

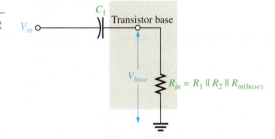

As previously mentioned, a critical point in the amplifier's response occurs when the output voltage is 70.7% of its midrange value. This condition occurs in the input *RC* circuit when $X_{C1} = R_{in}$.

$$V_{base} = \left(\frac{R_{in}}{\sqrt{R_{in}^2 + R_{in}^2}} \right) V_{in} = \left(\frac{R_{in}}{\sqrt{2R_{in}^2}} \right) V_{in} = \left(\frac{R_{in}}{\sqrt{2}R_{in}} \right) V_{in} = \left(\frac{1}{\sqrt{2}} \right) V_{in} = 0.707 V_{in}$$

In terms of measurement in decibels,

$$20 \log \left(\frac{V_{base}}{V_{in}} \right) = 20 \log (0.707) = -3 \text{ dB}$$

Lower Critical Frequency The condition where the gain is down 3 dB is logically called the −3 *dB point* of the amplifier response; the overall gain is 3 dB less than at midrange frequencies because of the attenuation (gain less than 1) of the input *RC* circuit. The frequency, f_{cl}, at which this condition occurs is called the *lower critical frequency* (also known as the *lower cutoff frequency, lower corner frequency*, or *lower break frequency*) and can be calculated as follows:

$$X_{C1} = \frac{1}{2\pi f_{cl(input)} C_1} = R_{in}$$

Equation 10–6

$$f_{cl(input)} = \frac{1}{2\pi R_{in} C_1}$$

If the resistance of the input source is taken into account, Equation 10–6 becomes

$$f_{cl(input)} = \frac{1}{2\pi(R_s + R_{in})C_1}$$

EXAMPLE 10–3

For the circuit in Figure 10–11, calculate the lower critical frequency due to the input RC circuit. Assumed $r'_e = 9.6\ \Omega$ and $\beta = 200$. Notice that a swamping resistor, R_{E1}, is used.

▶ **FIGURE 10–11**

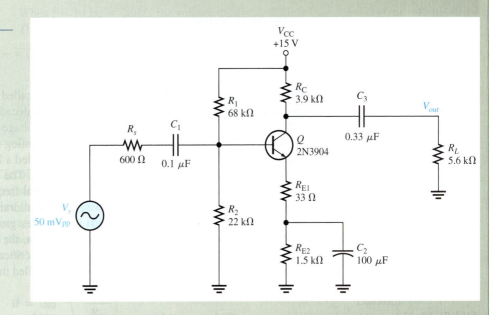

Solution   The input resistance is

$$R_{in} = R_1 \parallel R_2 \parallel (\beta(r'_e + R_{E1})) = 68\ \text{k}\Omega \parallel 22\ \text{k}\Omega \parallel (200(9.6\ \Omega + 33\ \Omega)) = 5.63\ \text{k}\Omega$$

The lower critical frequency is

$$f_{cl(input)} = \frac{1}{2\pi R_{in}C_1} = \frac{1}{2\pi(5.63\ \text{k}\Omega)(0.1\ \mu\text{F})} = \mathbf{282\ Hz}$$

Related Problem   What value of input capacitor will move the lower cutoff frequency to 130 Hz?

Open the Multisim file EXM10-03 or the LT Spice file EXS10-03 in the Examples folder on the website and read the critical frequency on the Bode plotter. The Bode plotter is not an actual instrument available, but allows the user to see the response of a circuit in the frequency domain (frequency is the independent variable). Notice that C_2 and C_3 are taken out of the calculation by making their value huge (1 F!). While this is unrealistic, it works nicely for the computer simulation to isolate the input response.

Voltage Gain Roll-Off at Low Frequencies   As you have seen, the input RC circuit reduces the overall voltage gain of an amplifier by 3 dB when the frequency is reduced to the critical value f_c. As the frequency continues to decrease below f_c, the overall voltage gain also continues to decrease. The rate of decrease in voltage gain with frequency is called **roll-off**. For each ten times reduction in frequency below f_c, *there is a 20 dB reduction in voltage gain.*

Related Problem Explain why C_2 is larger than C_1 or C_3.

Open the Multisim file EXM10-06 or the LT Spice file EXS10-06 in the Examples folder on the website and read the critical frequency on the Bode plotter. Notice that C_1 and C_3 are taken out of the calculation by making their value huge as before (1 F!).

FET Amplifiers

A zero-biased D-MOSFET amplifier with capacitive coupling on the input and output is shown in Figure 10–20. As you learned in Chapter 9, the midrange voltage gain of a zero-biased amplifier is

$$A_{v(mid)} = g_m R_d$$

This is the gain at frequencies high enough so that the capacitive reactances are approximately zero.

▶ **FIGURE 10–20**

Zero-biased D-MOSFET amplifier.

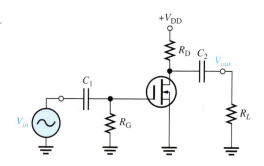

The amplifier in Figure 10–20 has only two high-pass *RC* circuits that influence its low-frequency response. One *RC* circuit is formed by the input coupling capacitor C_1 and the input resistance. The other circuit is formed by the output coupling capacitor C_2 and the output resistance looking in at the drain.

The Input *RC* Circuit

The input *RC* circuit for the FET amplifier in Figure 10–20 is shown in Figure 10–21. As in the case for the BJT amplifier, the reactance of the input coupling capacitor increases as the frequency decreases. When $X_{C1} = R_{in}$, the gain is down 3 dB below its midrange value.

▶ **FIGURE 10–21**

Input *RC* circuit.

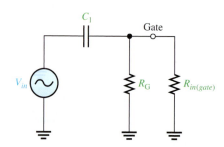

The lower critical frequency is

$$f_{cl(input)} = \frac{1}{2\pi R_{in} C_1}$$

The input resistance is

$$R_{in} = R_G \parallel R_{in(gate)}$$

where $R_{in(gate)}$ is determined from datasheet information as

$$R_{in(gate)} = \left| \frac{V_{GS}}{I_{GSS}} \right|$$

Therefore, the lower critical frequency is

$$f_{cl(input)} = \frac{1}{2\pi(R_G \| R_{in(gate)})C_1}$$

Equation 10–12

For practical work, the value of $R_{in(gate)}$ is so large it can be ignored, as will be illustrated in Example 10–7.

The gain rolls off below f_c at 20 dB/decade, as previously shown. The phase angle in the low-frequency input *RC* circuit is

$$\theta = \tan^{-1}\left(\frac{X_{C1}}{R_{in}}\right)$$

Equation 10–13

EXAMPLE 10–7

What is the lower critical frequency of the input *RC* circuit in the FET amplifier of Figure 10–22?

▶ **FIGURE 10–22**

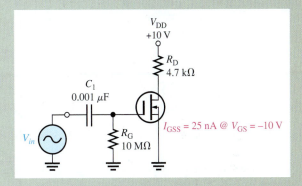

Solution   First determine R_{in} and then calculate f_c.

$$R_{in(gate)} = \left| \frac{V_{GS}}{I_{GSS}} \right| = \frac{10 \text{ V}}{25 \text{ nA}} = 400 \text{ M}\Omega$$

$$R_{in} = R_G \| R_{in(gate)} = 10 \text{ M}\Omega \| 400 \text{ M}\Omega = 9.8 \text{ M}\Omega$$

$$f_{cl(input)} = \frac{1}{2\pi R_{in}C_1} = \frac{1}{2\pi(9.8 \text{ M}\Omega)(0.001 \text{ }\mu\text{F})} = \textbf{16.2 Hz}$$

For all practical purposes,

$$R_{in} \cong R_G = 10 \text{ M}\Omega$$

and

$$f_{cl(input)} = \frac{1}{2\pi R_G C_1} = \frac{1}{2\pi(10 \text{ M}\Omega)(0.001 \text{ }\mu\text{F})} \cong 15.9 \text{ Hz}$$

There is very little difference in the two results.

The critical frequency of the input *RC* circuit of a FET amplifier is usually very low because of the very high input resistance and the high value of R_G.

Related Problem How much does the lower critical frequency of the input *RC* circuit change if the FET in Figure 10–22 is replaced by one with $I_{GSS} = 10$ nA @ $V_{GS} = -8$ V?

 Open the Multisim file EXM10-07 or the LT Spice file EXS10-07 in the Examples folder on the website and measure the low critical frequency for the input circuit. Compare to the calculated results.

The Output *RC* Circuit

The second *RC* circuit that affects the low-frequency response of the FET amplifier in Figure 10–20 is formed by a coupling capacitor C_2 and the output resistance looking in at the drain, as shown in Figure 10–23(a). The load resistor, R_L, is also included. As in the case of the BJT, the FET is treated as a current source, and the upper end of R_D is effectively ac ground, as shown in Figure 10–23(b). The Thevenin equivalent of the circuit to the left of C_2 is shown in Figure 10–23(c). The lower critical frequency for this *RC* circuit is

Equation 10–14

$$f_{cl(output)} = \frac{1}{2\pi(R_D + R_L)C_2}$$

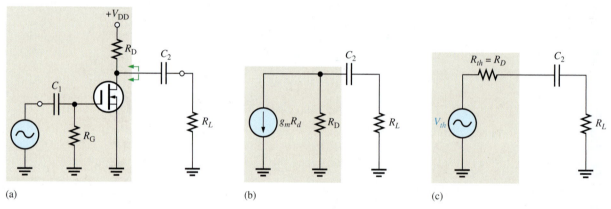

▲ FIGURE 10–23

Development of the equivalent low-frequency output *RC* circuit.

The effect of the output *RC* circuit on the amplifier's voltage gain below the midrange is similar to that of the input *RC* circuit. The circuit with the highest critical frequency dominates because it is the one that first causes the gain to roll off as the frequency drops below its midrange values. The phase angle in the low-frequency output *RC* circuit is

Equation 10–15

$$\theta = \tan^{-1}\left(\frac{X_{C2}}{R_D + R_L}\right)$$

Again, at the critical frequency, the phase angle is 45° and approaches 90° as the frequency approaches zero. However, starting at the critical frequency, the phase angle decreases from 45° and becomes very small as the frequency goes higher.

EXAMPLE 10–8

Determine the lower critical frequencies for the FET amplifier in Figure 10–24. Assume that the load is another identical amplifier with the same R_{in}. The datasheet shows $I_{GSS} = 100$ nA at $V_{GS} = -12$ V.

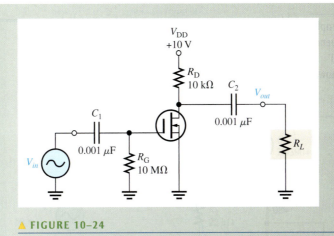

▲ FIGURE 10–24

Solution First, find the lower critical frequency for the input *RC* circuit.

$$R_{in(gate)} = \left| \frac{V_{GS}}{I_{GSS}} \right| = \frac{12\text{ V}}{100\text{ nA}} = 120\text{ M}\Omega$$

$$R_{in} = R_G \| R_{in(gate)} = 10\text{ M}\Omega \| 120\text{ M}\Omega = 9.2\text{ M}\Omega$$

$$f_{cl(input)} = \frac{1}{2\pi R_{in} C_1} = \frac{1}{2\pi (9.2\text{ M}\Omega)(0.001\ \mu\text{F})} = 17.3\text{ Hz}$$

The output *RC* circuit has a lower critical frequency of

$$f_{cl(output)} = \frac{1}{2\pi (R_D + R_L)C_2} = \frac{1}{2\pi (9.21\text{ M}\Omega)(0.001\ \mu\text{F})} \cong 17.3\text{ Hz}$$

Related Problem If the circuit in Figure 10–24 were operated with no load, how is the output low-frequency response affected?

 Open the Multisim file EXM10-08 or the LT Spice file EXS10-08 in the Examples folder on the website. Determine the total low-frequency response of the amplifier.

Total Low-Frequency Response of an Amplifier

Now that we have individually examined the high-pass *RC* circuits that affect a BJT or FET amplifier's voltage gain at low frequencies, let's look at the combined effect of the three *RC* circuits in a BJT amplifier. Each circuit has a critical frequency determined by the *R* and *C* values. The critical frequencies of the three *RC* circuits are not necessarily all equal. If one of the *RC* circuits has a critical (break) frequency higher than the other two, then it is the *dominant RC* circuit. The dominant circuit determines the frequency at which the overall voltage gain of the amplifier begins to drop at −20 dB/decade. The other circuits each cause an additional −20 dB/decade roll-off below their respective critical (break) frequencies.

To get a better picture of what happens at low frequencies, refer to the Bode plot in Figure 10–25, which shows the superimposed ideal responses for the three *RC* circuits (green lines) of a BJT amplifier. In this example, each *RC* circuit has a different critical frequency. The input *RC* circuit is dominant (highest f_c) in this case, and the bypass *RC* circuit has the lowest f_c. The ideal overall response is shown as the blue line.

Here is what happens. As the frequency is reduced from midrange, the first "break point" occurs at the critical frequency of the input *RC* circuit, $f_{cl(input)}$, and the gain begins to drop at −20 dB/decade. This constant roll-off rate continues until the critical frequency of the output *RC* circuit, $f_{cl(output)}$, is reached. At this break point, the output *RC* circuit adds

3. A certain RC circuit has an $f_{cl} = 235$ Hz, above which the attenuation is 0 dB. What is the dB attenuation at 23.5 Hz?

4. What is the amount of phase shift contributed by an input circuit when $X_C = 0.5R_{in}$ at a certain frequency below f_{cl1}?

5. What is the critical frequency when $R_D = 1.5$ kΩ, $R_L = 5$ kΩ, and $C_2 = 0.0022$ μF in a circuit like Figure 10–24?

10–4 HIGH-FREQUENCY AMPLIFIER RESPONSE

You have seen how the coupling and bypass capacitors affect the voltage gain of an amplifier at lower frequencies where the reactances of the coupling and bypass capacitors are significant. In the midrange of an amplifier, the effects of the capacitors are minimal and can be neglected. If the frequency is increased sufficiently, a point is reached where the transistor's internal capacitances begin to have a significant effect on the gain. The basic differences between BJTs and FETs are the specifications of the internal capacitances and the input resistance.

After completing this section, you should be able to

❑ **Analyze the high-frequency response of an amplifier**
❑ Analyze a BJT amplifier
  ◆ Apply Miller's theorem
❑ Identify and analyze the BJT amplifier's input RC circuit
  ◆ Calculate the upper critical frequency and gain roll-off ◆ Determine the phase shift
❑ Identify and analyze the BJT amplifier's output RC circuit
  ◆ Calculate the upper critical frequency ◆ Determine the phase shift
❑ Analyze a FET amplifier
❑ Identify and analyze a JFET amplifier
  ◆ Determine internal capacitances on a datasheet ◆ Apply Miller's theorem
❑ Identify and analyze the JFET amplifier's input RC circuit
  ◆ Calculate the upper critical frequency ◆ Determine the phase shift
❑ Identify and analyze the JFET amplifier's output RC circuit
  ◆ Calculate the upper critical frequency ◆ Determine the phase shift
❑ Discuss the total high-frequency response of an amplifier
  ◆ Use Bode plots to illustrate the high-frequency response

BJT Amplifiers

A high-frequency ac equivalent circuit for the BJT amplifier in Figure 10–31(a) is shown in Figure 10–31(b). Notice that the coupling and bypass capacitors are treated as effective shorts and do not appear in the equivalent circuit. The internal capacitances, C_{be} and C_{bc}, which are significant only at high frequencies, do appear in the diagram. As previously mentioned. C_{be} is sometimes called the input capacitance C_{ib}, and C_{bc} is sometimes called the output capacitance C_{ob}. C_{be} is specified on datasheets at a certain value of V_{BE}. Often, a datasheet will list C_{ib} as C_{ibo} and C_{ob} as C_{obo}. The o as the last letter in the subscript indicates the capacitance is measured with the base open. For example, a 2N2222A transistor has a C_{be} of 25 pF at $V_{BE} = 0.5$ V dc, $I_C = 0$, and $f = 1$ MHz. Also, C_{bc} is specified at a certain value of V_{BC}. The 2N2222A has a maximum C_{bc} of 8 pF at $V_{BC} = 10$ V dc.

Miller's Theorem in High-Frequency Analysis By applying Miller's theorem to the inverting amplifier in Figure 10–31(b) and using the midrange voltage gain, you have

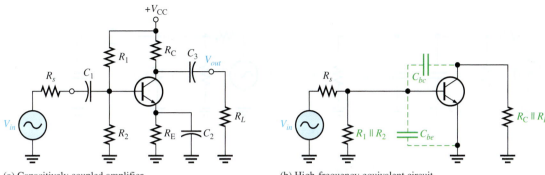

(a) Capacitively coupled amplifier   (b) High-frequency equivalent circuit

▲ FIGURE 10–31

Capacitively coupled amplifier and its high-frequency equivalent circuit.

a circuit that can be analyzed for high-frequency response. Looking in from the signal source, the capacitance C_{bc} appears in the Miller input capacitance from base to ground.

$$C_{in(Miller)} = C_{bc}(A_v + 1)$$

C_{be} simply appears as a capacitance to ac ground, as shown in Figure 10–32, in parallel with $C_{in(Miller)}$. Looking in at the collector, C_{bc} appears in the Miller output capacitance from collector to ground. As shown in Figure 10–32, the Miller output capacitance appears in parallel with R_c.

$$C_{out(Miller)} = C_{bc}\left(\frac{A_v + 1}{A_v}\right)$$

These two Miller capacitances create a high-frequency input RC circuit and a high-frequency output RC circuit. These two circuits differ from the low-frequency input and output circuits, which act as high-pass filters, because the capacitances go to ground and therefore act as low-pass filters. The equivalent circuit in Figure 10–32 is an ideal model because stray capacitances that are due to circuit interconnections are neglected.

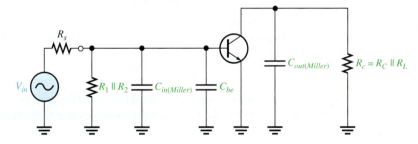

◄ FIGURE 10–32

High-frequency equivalent circuit after applying Miller's theorem.

The Input *RC* Circuit

At high frequencies, the input circuit is as shown in Figure 10–33(a), where $\beta_{ac}r'_e$ is the input resistance at the base of the transistor because the bypass capacitor effectively shorts the emitter to ground. By combining C_{be} and $C_{in(Miller)}$ in parallel and repositioning, you get the simplified circuit shown in Figure 10–33(b). Next, by thevenizing the circuit to the left of the capacitor, as indicated, the input RC circuit is reduced to the equivalent form shown in Figure 10–33(c).

As the frequency increases, the capacitive reactance becomes smaller. This causes the signal voltage at the base to decrease, so the amplifier's voltage gain decreases. The reason for this is that the capacitance and resistance act as a voltage divider and, as the frequency increases, more voltage is dropped across the resistance and less across the capacitance. At the critical frequency, the gain is 3 dB less than its midrange value. The upper critical high where $C_{tot} = C_{gs} + C_{in(Miller)}$. The input RC circuit produces a phase angle of

Equation 10–24

$$\theta = \tan^{-1}\left(\frac{R_s}{X_{C_{tot}}}\right)$$

The effect of the input RC circuit is to reduce the midrange gain of the amplifier by 3 dB at the critical frequency and to cause the gain to decrease at –20 dB/decade above f_c.

EXAMPLE 10–14

Find the upper critical frequency of the input RC circuit for the FET amplifier in Figure 10–42. $C_{iss} = 8$ pF, $C_{rss} = 3$ pF, and $g_m = 6500\ \mu$S.

▶ **FIGURE 10–42**

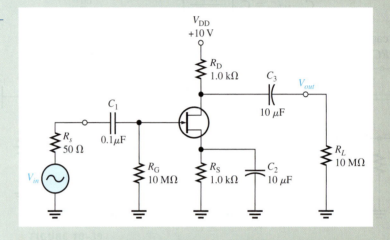

Solution   Determine C_{gd} and C_{gs}.

$$C_{gd} = C_{rss} = 3\ \text{pF}$$
$$C_{gs} = C_{iss} - C_{rss} = 8\ \text{pF} - 3\ \text{pF} = 5\ \text{pF}$$

Determine the upper critical frequency for the input RC circuit as follows:

$$A_v = g_m R_d = g_m(R_D \parallel R_L) \cong (6500\ \mu\text{S})(1\ \text{k}\Omega) = 6.5$$
$$C_{in(Miller)} = C_{gd}(A_v + 1) = (3\ \text{pF})(7.5) = 22.5\ \text{pF}$$

The total input capacitance is

$$C_{in(tot)} = C_{gs} + C_{in(Miller)} = 5\ \text{pF} + 22.5\ \text{pF} = 27.5\ \text{pF}$$

The upper critical frequency is

$$f_{cu(input)} = \frac{1}{2\pi R_s C_{in(tot)}} = \frac{1}{2\pi(50\ \Omega)(27.5\ \text{pF})} = \textbf{116 MHz}$$

Related Problem   If the gain of the amplifier in Figure 10–42 is increased to 10, what happens to f_c?

The Output RC Circuit

The high-frequency output RC circuit is formed by the Miller output capacitance and the output resistance looking in at the drain, as shown in Figure 10–43(a). As in the case of the BJT, the FET is treated as a current source. When you apply Thevenin's theorem, you get an equivalent output RC circuit consisting of R_D in parallel with R_L and an equivalent output capacitance.

$$C_{out(Miller)} = C_{gd}\left(\frac{A_v + 1}{A_v}\right)$$

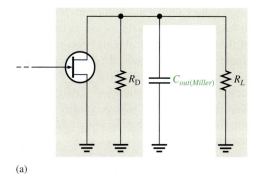

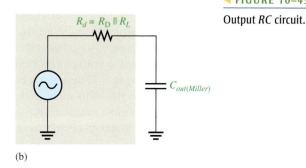

(a)   (b)

This equivalent output circuit is shown in Figure 10–43(b). The critical frequency of the output RC lag circuit is

$$f_{cu(output)} = \frac{1}{2\pi R_d C_{out(Miller)}}$$

Equation 10–25

The output circuit produces a phase shift of

$$\theta = \tan^{-1}\left(\frac{R_d}{X_{C_{out(Miller)}}}\right)$$

Equation 10–26

EXAMPLE 10–15

Determine the upper critical frequency of the output RC circuit for the FET amplifier in Figure 10–42. What is the phase shift introduced by this circuit at the critical frequency? Which RC circuit is dominant, that is, which one has the lower value of upper critical frequency?

Solution Since R_L is very large compared to R_D, it can be neglected, and the equivalent output resistance is

$$R_d \cong R_D = 1.0\ k\Omega$$

The equivalent output capacitance is

$$C_{out(Miller)} = C_{gd}\left(\frac{A_v + 1}{A_v}\right) = (3\ pF)\left(\frac{7.5}{6.5}\right) = 3.46\ pF$$

Therefore, the upper critical frequency is

$$f_{cu(output)} = \frac{1}{2\pi R_d C_{out(Miller)}} = \frac{1}{2\pi(1.0\ k\Omega)(3.46\ pF)} = 46\ MHz$$

Although it has been neglected, any stray wiring capacitance could significantly affect the frequency response because $C_{out(Miller)}$ is very small.

The phase angle is always **45°** at f_c for an RC circuit and the output lags.

In Example 10–14, the upper critical frequency of the input RC circuit was found to be 116 MHz. Therefore, the upper critical frequency for the output circuit is dominant because it is the lower of the two.

Related Problem If A_v of the amplifier in Figure 10–42 is increased to 10, what is the upper critical frequency of the output circuit?

Total High-Frequency Response of an Amplifier

As you have seen, the two RC circuits created by the internal transistor capacitances influence the high-frequency response of both BJT and FET amplifiers. As the frequency increases and reaches the high end of its midrange values, one of the *RC* circuits will cause the amplifier's gain to begin dropping off. The frequency at which this occurs is the dominant upper critical frequency; it is the lower of the two upper critical high frequencies. An ideal high-frequency Bode plot is shown in Figure 10–44(a). It shows the first break point at $f_{cu(input)}$ where the voltage gain begins to roll off at −20 dB/decade. At $f_{cu(output)}$, the gain begins dropping at −40 dB/decade because each *RC* circuit is providing a −20 dB/decade roll-off. Figure 10–44(b) shows a nonideal Bode plot where the voltage gain is actually −3 dB/decade below midrange at $f_{cu(input)}$. Other possibilities are that the output *RC* circuit is dominant or that both circuits have the same critical frequency.

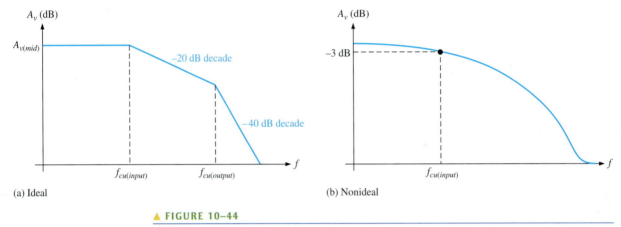

(a) Ideal   (b) Nonideal

▲ **FIGURE 10–44**

High-frequency Bode plots.

SECTION 10–4 CHECKUP	1. What determines the high-frequency response of an amplifier?

SECTION 10–4 CHECKUP

1. What determines the high-frequency response of an amplifier?
2. If an amplifier has a midrange voltage gain of 80, the transistor's C_{bc} is 4 pF, and C_{be} = 8 pF, what is the total input capacitance?
3. A certain amplifier has $f_{cu(input)}$ = 3.5 MHz and $f_{cu(output)}$ = 8.2 MHz. Which circuit dominates the high-frequency response?
4. What are the capacitances that are usually specified on a FET datasheet?
5. If C_{gs} = 4 pF and C_{gd} = 3 pF, what is the total input capacitance of a FET amplifier whose voltage gain is 25?

10–5 TOTAL AMPLIFIER FREQUENCY RESPONSE

In the previous sections, you learned how each *RC* circuit in an amplifier affects the frequency response. In this section, we will bring these concepts together and examine the total response of typical amplifiers and the specifications relating to their performance.

After completing this section, you should be able to

◻ **Analyze an amplifier for total frequency response**
◻ Discuss bandwidth
  ◆ Define the dominant critical frequencies
◻ Explain gain-bandwidth product
  ◆ Define *unity-gain frequency*

Figure 10–45(b) shows a generalized ideal response curve (Bode plot) for the BJT amplifier shown in Figure 10–45(a). As previously discussed, the three break points at the lower critical frequencies (f_{cl1}, f_{cl2}, and f_{cl3}) are produced by the three low-frequency RC circuits formed by the coupling and bypass capacitors. The break points at the upper critical frequencies, f_{cu1} and f_{cu2}, are produced by the two high-frequency RC circuits formed by the transistor's internal capacitances.

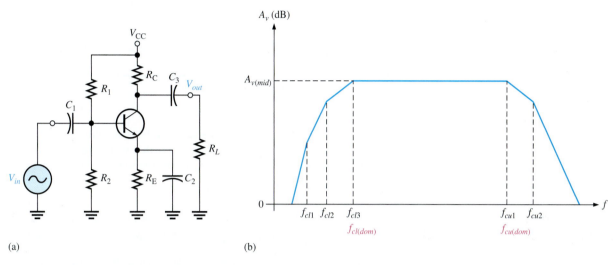

(a)   (b)

▲ **FIGURE 10–45**

A BJT amplifier and its generalized ideal response curve (Bode plot).

Of particular interest are the two dominant critical frequencies, f_{cl3} and f_{cu1}, in Figure 10–45(b). These two frequencies are where the voltage gain of the amplifier is 3 dB below its midrange value. These dominant frequencies are designated $f_{cl(dom)}$ and $f_{cu(dom)}$.

The upper and lower dominant critical frequencies are sometimes called the *half-power frequencies*. This term is derived from the fact that the output power of an amplifier at its critical frequencies is one-half of its midrange power, as previously mentioned. This can be shown as follows, starting with the fact that the output voltage is 0.707 of its midrange value at the dominant critical frequencies.

$$V_{out(f_c)} = 0.707 V_{out(mid)}$$

$$P_{out(f_c)} = \frac{V_{out(f_c)}^2}{R_{out}} = \frac{(0.707 V_{out(mid)})^2}{R_{out}} = \frac{0.5 V_{out(mid)}^2}{R_{out}} = 0.5 P_{out(mid)}$$

Bandwidth

An amplifier normally operates with signal frequencies between $f_{cl(dom)}$ and $f_{cu(dom)}$. As you know, when the input signal frequency is at $f_{cl(dom)}$ or $f_{cu(dom)}$, the output signal voltage level is 70.7% of its midrange value or −3 dB. If the signal frequency drops below $f_{cl(dom)}$, the gain and thus the output signal level drops at 20 dB/decade until the next critical frequency is reached. The same occurs when the signal frequency goes above $f_{cu(dom)}$.

The range (band) of frequencies lying between $f_{cl(dom)}$ and $f_{cu(dom)}$ is defined as the **bandwidth** of the amplifier, as illustrated in Figure 10–46. Only the dominant critical frequencies appear in the response curve because they determine the bandwidth. Also, sometimes the other critical frequencies are far enough away from the dominant frequencies that they play no significant role in the total amplifier response and can be neglected. The amplifier's bandwidth is expressed in units of hertz as

$$BW = f_{cu(dom)} - f_{cl(dom)}$$

Equation 10–27

▶ **FIGURE 10–46**

Response curve illustrating the bandwidth of an amplifier.

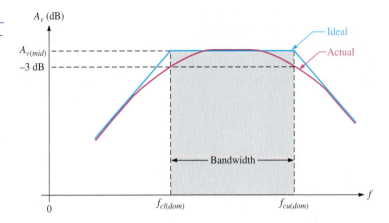

Ideally, all signal frequencies lying in an amplifier's bandwidth are amplified equally. For example, if a 10 mV rms signal is applied to an amplifier with a voltage gain of 20, it is ideally amplified to 200 mV rms for all frequencies in the bandwidth. In actually, the gain is down 3 dB at $f_{cl(dom)}$ and $f_{cu(dom)}$.

EXAMPLE 10–16

What is the bandwidth of an amplifier having an $f_{cl(dom)}$ of 200 Hz and an $f_{cu(dom)}$ of 2 kHz?

Solution

$$BW = f_{cu(dom)} - f_{cl(dom)} = 2000 \text{ Hz} - 200 \text{ Hz} = \textbf{1800 Hz}$$

Notice that bandwidth has the unit of hertz.

Related Problem

If $f_{cl(dom)}$ is increased, does the bandwidth increase or decrease? If $f_{cu(dom)}$ is increased, does the bandwidth increase or decrease?

Gain-Bandwidth Product

One characteristic of amplifiers is that the product of the voltage gain and the bandwidth is always constant but only when the roll-off is -20 dB/decade. This characteristic is called the **gain-bandwidth product**. Let's assume that the dominant lower critical frequency of a particular amplifier is much less than the dominant upper critical frequency.

$$f_{cl(dom)} \ll f_{cu(dom)}$$

The bandwidth can then be approximated as

$$BW = f_{cu(dom)} - f_{cl(dom)} \cong f_{cu}$$

Unity-Gain Frequency The simplified Bode plot for this condition is shown in Figure 10–47. Notice that $f_{cl(dom)}$ is neglected because it is so much smaller than $f_{cu(dom)}$, and the bandwidth approximately equals $f_{cu(dom)}$. Further, assume there is only one upper cutoff frequency (to meet the requirement of a -20 dB/decade roll-off rate). This assumption is valid for many amplifiers in which one upper cutoff frequency is significantly higher than the other, ensuring that the roll-off rate is -20 dB to the unity gain frequency. This is a useful assumption for many amplifiers including many operational amplifiers (op-amps) as you will see in Chapter 12. Beginning at $f_{cu(dom)}$, the gain rolls off until unity gain (0 dB) is reached. The frequency at which the amplifier's gain is 1 is called the *unity-gain frequency*, f_T. The significance of f_T is that it always equals the midrange voltage gain times the bandwidth and is constant for a given transistor.

Equation 10–28

$$f_T = A_{v(mid)}BW$$

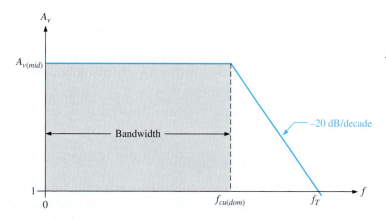

Simplified response curve where $f_{cl(dom)}$ is negligible (assumed to be zero) compared to $f_{cu(dom)}$ and one dominant upper frequency controls the roll-off.

For the case shown in Figure 10–47, $f_T = A_{v(mid)}f_{cu(dom)}$. For example, if a transistor datasheet specifies $f_T = 100$ MHz, this means that the transistor is capable of producing a voltage gain of 1 up to 100 MHz, or a gain of 100 up to 1 MHz, or any combination of gain and bandwidth that produces a product of 100 MHz.

EXAMPLE 10–17

A certain transistor has an f_T of 175 MHz. When this transistor is used in an amplifier with a midrange voltage gain of 50 and one dominant upper cutoff frequency, what bandwidth can be achieved ideally?

Solution

$$f_T = A_{v(mid)}BW$$

$$BW = \frac{f_T}{A_{v(mid)}} = \frac{175 \text{ MHz}}{50} = \textbf{3.5 MHz}$$

Related Problem

An amplifier has a midrange voltage gain of 20 and a bandwidth of 1 MHz. What is the f_T of the transistor?

SECTION 10–5 CHECKUP

1. What is the voltage gain of an amplifier at f_T?
2. What is the bandwidth of an amplifier when $f_{cu(dom)} = 25$ kHz and $f_{cl(dom)} = 100$ Hz?
3. The f_T of a certain transistor is 130 MHz. What is the maximum voltage gain that can be achieved with a bandwidth of 50 MHz?

10–6 FREQUENCY RESPONSE OF MULTISTAGE AMPLIFIERS

To this point, you have seen how the voltage gain of a single-stage amplifier changes over frequency. When two or more stages are cascaded to form a multistage amplifier, the overall frequency response is determined by the frequency response of each stage depending on the relationships of the critical frequencies.

After completing this section, you should be able to

❏ **Analyze multistage amplifiers for frequency response**
❏ Analyze the case where the stages have different critical frequencies
  ◆ Determine the overall bandwidth
❏ Analyze the case where the stages have equal critical frequencies
  ◆ Determine the overall bandwidth
❏ Simulate a two-stage amplifier using Multisim

When amplifier stages are cascaded to form a multistage amplifier, the dominant frequency response is determined by the responses of the individual stages. There are two cases to consider:

1. Each stage has a different dominant lower critical frequency and a different dominant upper critical frequency.

2. Each stage has the same dominant lower critical frequency and the same dominant upper critical frequency.

Different Critical Frequencies

Ideally, when the dominant lower critical frequency, $f_{cl(dom)}$, of each amplifier stage is different from the other stages, the overall dominant lower critical frequency, $f'_{cl(dom)}$, equals the dominant critical frequency of the stage with the highest $f_{cl(dom)}$.

Ideally, when the dominant upper critical frequency, $f_{cu(dom)}$, of each amplifier stage is different from the other stages, the overall dominant upper critical frequency, $f'_{cu(dom)}$, equals the dominant critical frequency of the stage with the lowest $f_{cu(dom)}$.

In practice, the critical frequencies interact, so these calculated values should be considered approximations that are useful for troubleshooting or estimating the response. When more accuracy is required, a computer simulation is the best solution.

Overall Bandwidth The bandwidth of a multistage amplifier is the difference between the overall dominant lower critical frequency and the overall dominant upper critical frequency.

$$BW = f'_{cu(dom)} - f'_{cl(dom)}$$

EXAMPLE 10–18

In a certain 2-stage amplifier, one stage has a dominant lower critical frequency of 850 Hz and a dominant upper critical frequency of 100 kHz. The other has a dominant lower critical frequency of 1 kHz and a dominant upper critical frequency of 230 kHz. Determine the overall bandwidth of the 2-stage amplifier.

Solution

$$f'_{cl(dom)} = 1 \text{ kHz}$$
$$f'_{cu(dom)} = 100 \text{ kHz}$$
$$BW = f'_{cu(dom)} - f'_{cl(dom)} = 100 \text{ kHz} - 1 \text{ kHz} = \textbf{99 kHz}$$

Related Problem

A certain 3-stage amplifier has the following dominant lower critical frequencies for each stage: $f_{cl(dom)(1)} = 50$ Hz, $f_{cl(dom)(2)} = 980$ Hz, and $f_{cl(dom)(3)} = 130$ Hz. What is the overall dominant lower critical frequency?

Equal Critical Frequencies

When each amplifier stage in a multistage arrangement has equal dominant critical frequencies, you may think that the overall dominant critical frequency is equal to the critical frequency of each stage. This is not the case, however.

When the dominant lower critical frequencies of each stage in a multistage amplifier are all the same, the overall dominant lower critical frequency is increased by a factor of $1/\sqrt{2^{1/n} - 1}$ as shown by the following formula (n is the number of stages in the multistage amplifier):

Equation 10–29

$$f'_{cl(dom)} = \frac{f_{cl(dom)}}{\sqrt{2^{1/n} - 1}}$$

When the dominant upper critical frequencies of each stage are all the same, the overall dominant upper critical frequency is reduced by a factor of $\sqrt{2^{1/n} - 1}$.

Equation 10–30

$$f'_{cu(dom)} = f_{cu(dom)}\sqrt{2^{1/n} - 1}$$

The proofs of these formulas are given in "Derivations of Selected Equations" at www.pearsonhighered.com/floyd.

EXAMPLE 10–19

Both stages in a 2-stage amplifier have a dominant lower critical frequency of 50 Hz and a dominant upper critical frequency of 80 kHz. Determine the overall bandwidth.

Solution

$$f'_{cl(dom)} = \frac{f_{cl(dom)}}{\sqrt{2^{1/n} - 1}} = \frac{50 \text{ Hz}}{\sqrt{2^{0.5} - 1}} = \frac{50 \text{ Hz}}{0.644} = 77.7 \text{ Hz}$$

$$f'_{cu(dom)} = f_{cu(dom)}\sqrt{2^{1/n} - 1} = (80 \text{ kHz})(0.644) = 51.5 \text{ kHz}$$

$$BW = f'_{cu(dom)} - f'_{cl(dom)} = 51.5 \text{ kHz} - 77.7 \text{ Hz} = \mathbf{50.7 \text{ kHz}}$$

Related Problem

If a third identical stage is connected in cascade to the 2-stage amplifier in this example, what is the resulting overall bandwidth?

Computer Simulation for Multistage Amplifiers

With multistage amplifiers, the detailed calculation of the frequency response is greatly simplified by computer simulation. There are several interactions within each stage and other interactions between the stages that affect the overall response. When you need more accuracy, a computer simulation is used. This is particularly useful in design work because you can change a component and see the effect immediately on the frequency response. The following example illustrates the application of computer analysis to a multistage amplifier.

EXAMPLE 10–20

A dc coupled two-stage amplifier is simulated with Multisim in Figure 10–48 to determine the overall frequency response.

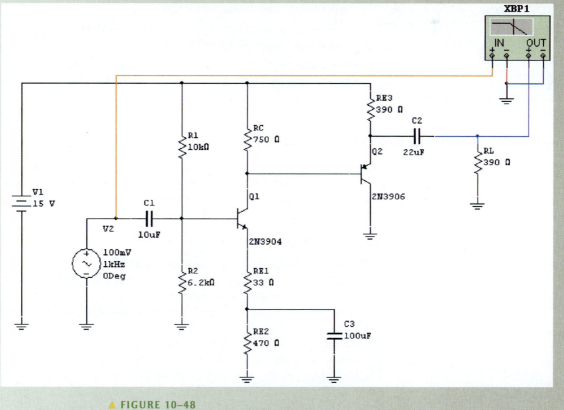

▲ FIGURE 10–48

Solution The circuit was constructed in Multisim by dragging the parts needed onto the simulated workbench and connecting them. Connect the Bode plotter and adjust it to show the complete response curve with upper and lower critical frequencies. Figure 10–49 shows the display. When the cursor is moved to the lower critical frequency (3 dB below midrange), a reading of approximately 56 Hz is observed. When the cursor is moved to the upper critical frequency, a reading of approximately 34 MHz is observed.

▶ FIGURE 10–49

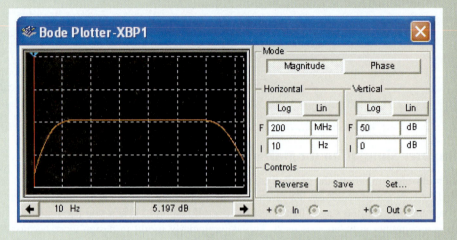

Related Problem Determine the gain of the amplifier in Figure 10–48.

SECTION 10–6 CHECKUP

1. One stage in an amplifier has $f_{cl} = 1$ kHz and the other stage has $f_{cl} = 325$ Hz. What is the dominant lower critical frequency?

2. In a certain 3-stage amplifier $f_{cu(1)} = 50$ kHz, $f_{cu(2)} = 55$ kHz, and $f_{cu(3)} = 49$ kHz. What is the dominant upper critical frequency?

3. When more identical stages are added to a multistage amplifier with each stage having the same critical frequency, does the bandwidth increase, decrease, or stay the same?

10–7 FREQUENCY RESPONSE MEASUREMENTS

Two basic methods are used to measure the frequency response of an amplifier. The methods apply to both BJT and FET amplifiers although a BJT amplifier is used as an example. You will concentrate on determining the two dominant critical frequencies. From these values, you can get the bandwidth.

After completing this section, you should be able to

❑ **Measure the frequency response of an amplifier**
❑ Analyze the case where the stages have different critical frequencies
   ◆ Determine the overall bandwidth
❑ Analyze the case where the stages have equal critical frequencies
   ◆ Determine the overall bandwidth
❑ Simulate a two-stage amplifier using Multisim
❑ Measure the frequency response of an amplifier
   ◆ Describe a general measurement procedure
❑ Apply frequency/amplitude measurement to determine critical frequencies
❑ Use step-response measurement
   ◆ Determine the upper critical frequency   ◆ Determine the lower critical frequency

Frequency/Amplitude Measurement

Figure 10–50(a) shows the test setup for an amplifier circuit board. The schematic for the circuit board is also shown. The amplifier is driven by a sinusoidal voltage source with a

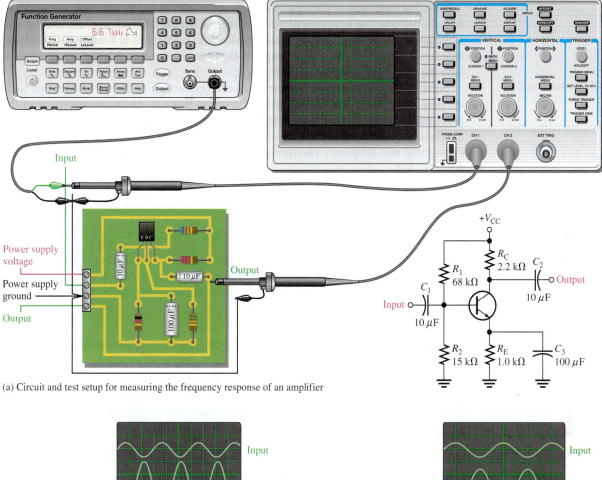

(a) Circuit and test setup for measuring the frequency response of an amplifier

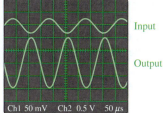

Ch1 50 mV   Ch2 0.5 V   50 μs

Input frequency control
on function generator   Amplifier input and output voltages

(b) Frequency is set to a midrange value (6.67 kHz in this case).
Input voltage adjusted for an output of 1 V peak.

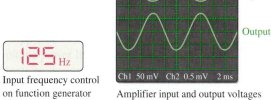

Ch1 50 mV   Ch2 0.5 mV   2 ms

Input frequency control
on function generator   Amplifier input and output voltages

(c) Frequency is reduced until the output is 0.707 V peak.
This is the lower critical frequency.

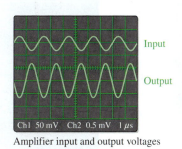

Ch1 50 mV   Ch2 0.5 mV   1 μs

Input frequency control
on function generator   Amplifier input and output voltages

(d) Frequency is increased until the output is again 0.707 V peak.
This is the upper critical frequency.

▲ FIGURE 10–50

A general procedure for measuring an amplifier's frequency response.

▶ **FIGURE 10–52**

The audio preamplifier with original capacitor values.

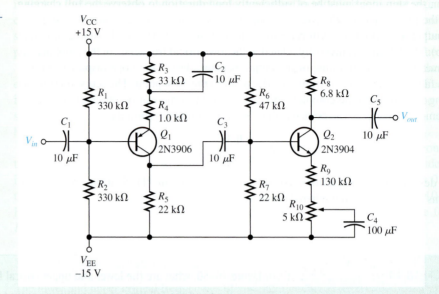

A frequency analysis of the original amplifier is as follows. For the Q_1 stage, the *input circuit* consists of C_1 and $R_1 \parallel R_2 \parallel \beta_{ac}R_4$. r_e' is neglected. The critical frequency is (assuming $\beta_{ac} = 100$)

$$f_{cl(input)} = \frac{1}{2\pi(R_1 \parallel R_2 \parallel \beta_{ac}R_4)C_1} = \frac{1}{2\pi(62.3\ k\Omega)10\ \mu F} = 0.255\ Hz$$

The *bypass circuit* consists of C_2 and

$$\left(R_4 + \left(\frac{R_1 \parallel R_2 \parallel R_{source}}{\beta_{ac}}\right)\right) \parallel R_3 \cong R_4$$

The expression reduces to approximately R_4 because R_{source} is assumed to be 300 Ω (microphone impedance) and R_3 is much greater than R_4.

$$f_{cl(bypass)} = \frac{1}{2\pi R_4 C_2} = \frac{1}{2\pi(1\ k\Omega)10\ \mu F} = 15.9\ Hz$$

The *output circuit* consists of C_3 and $R_5 + R_6 \parallel R_7 \parallel \beta_{ac}(R_9 + R_{10})$. r_e' is neglected. Assuming that R_{10} is set at 1 kΩ,

$$f_{cl(output)} = \frac{1}{2\pi(R_5 + R_6 \parallel R_7 \parallel \beta_{ac}(R_9 + R_{10}))C_3} = \frac{1}{2\pi(35.2\ k\Omega)10\ \mu F} = 0.452\ Hz$$

For the Q_2 stage, the *input circuit* is the same as the output circuit of the Q_1 stage.

$$f_{cl(input)} = \frac{1}{2\pi(R_5 + R_6 \parallel R_7 \parallel \beta_{ac}(R_9 + R_{10}))C_3} = \frac{1}{2\pi(35.2\ k\Omega)10\ \mu F} = 0.452\ Hz$$

The *bypass circuit* consists of C_4 and approximately $R_9 + R_{10} + (R_6 \parallel R_7)/\beta_{ac}$. The resistance is partially dependent on the setting of R_{10}. We will assume that the gain setting is such that R_{10} has negligible effect on the frequency.

$$f_{cl(bypass)} = \frac{1}{2\pi\left(R_9 + \dfrac{R_6 \parallel R_7}{\beta_{ac}}\right)C_4} = \frac{1}{2\pi(280\ \Omega)100\ \mu F} = 5.68\ Hz$$

The *output circuit* consists of C_5 and $R_8 + R_L$. The load is the 29 kΩ input resistance of the power amplifier.

$$f_{cl(output)} = \frac{1}{2\pi(R_8 + R_L)C_5} = \frac{1}{2\pi(35.8\ k\Omega)10\ \mu F} = 0.445\ Hz$$

The dominant critical frequency of the amplifier is established by the Q_1 stage bypass circuit and is $f_{cl(bypass)} = 15.9$ Hz, which is in very close agreement with the simulation.

Simulation of Original Circuit

The Multisim preamp with the original capacitor values is shown in Figure 10–53(a). A Bode plotter is connected to measure the frequency response. Figure 10–53(b) shows the logarithmic response curve with a midrange gain at 5 kHz of 33.3 dB. Moving the Bode

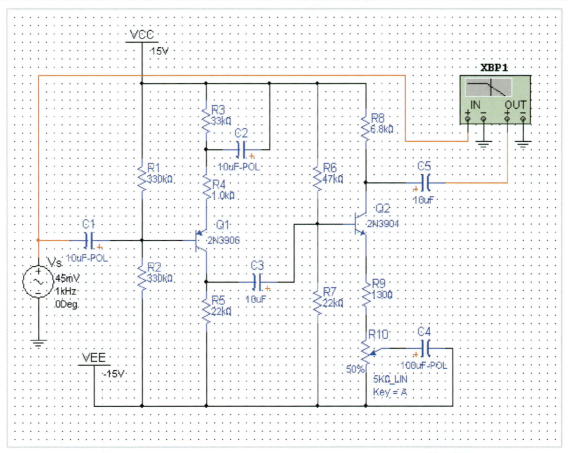

(a) Circuit screen with original capacitor values

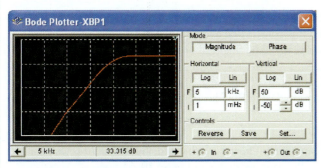

(b) At 5 kHz gain is 33.3 dB

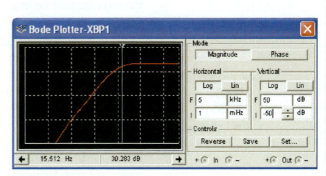

(c) Approximate f_c is 16 Hz at 30.3 dB (down 3 dB)

▲ FIGURE 10–53

Preamp frequency response with original capacitor values.

plotter cursor down until the gain is approximately 3 dB below midrange, or 30.3 dB, results in a lower critical frequency of 16 Hz at this gain setting (note that there is a small effect on the response for different gains due to a different path for C_4 to charge and discharge). This verifies that the response of the preamp includes the potentially troublesome 60 Hz interference.

Modification to Increase the Overall Lower Critical Frequency

Capacitor values must be reduced to achieve a critical frequency of 300 Hz $\pm$ 10%. The approach, in this case, will be to use C_1 and C_3 to set the new dominant critical frequency. C_2 and C_5 will be used to produce a faster roll-off below 60 Hz. C_4 will be left at 100 μF to avoid a change in frequency response when the gain is changed.

C_1 is part of the stage 1 input circuit, and C_3 is part of the stage 2 input circuit. These capacitor values will determine the proper dominant lower critical frequencies required to achieve an overall dominant critical frequency of 300 Hz.

Multistage Frequency Response   When the lower critical frequencies of each stage are equal, Equation 10–29 applies. The overall dominant lower critical frequency, $f'_{cl(dom)}$, is 300 Hz. Solving the equation for the dominant lower critical frequency of each stage, $f_{cl(dom)}$, you get

$$f_{cl(dom)} = f'_{cl(dom)} \sqrt{(2^{1/2} - 1)} = 300 \text{ Hz} \sqrt{(1.414 - 1)} = 300 \text{ Hz}(0.643) = 193 \text{ Hz}$$

Setting the dominant critical frequency of both stages of the amplifier to 193 Hz will produce an overall dominant lower critical frequency of 300 Hz. Using the frequency analysis that was done for the original circuit as a guide, do the following calculations.

1. Calculate the value of C_1 to produce a lower critical frequency of 193 Hz.
2. Calculate the value of C_3 to produce a lower critical frequency of 193 Hz.

The results of your calculation should agree with the values shown in Figure 10–54. The value for C_2 is the next lower available value in Multisim.

The Multisim circuit with reduced capacitor values is shown in Figure 10–54(a). As you can see in part (c), the new critical frequency is 276.604 Hz, which is within the specified 10% tolerance of 300 Hz. The gain is 9.744 dB for a frequency near 60 Hz with the volume setting at 85%, as shown in part (d).

3. From the Bode plots in Figure 10–54, determine how much the gain at 60 Hz is down from the midrange gain.

Simulate the preamp circuit using your Multisim or LT Spice software. Observe the operation with the Bode plotter.

Prototyping and Testing

Now that the revised circuit has been simulated and its operation verified, the modifications are made to the circuit and it is constructed and tested. After the circuit is successfully tested on a protoboard, it is ready to be finalized on a printed circuit board.

Lab Experiment

To build and test a similar circuit, go to Experiment 10 in your lab manual (*Laboratory Exercises for Electronic Devices* by David Buchla and Steven Wetterling).

Circuit Board

The capacitor values on the preamp circuit board are changed and the board is tested at 5 kHz and at 60 Hz using an oscilloscope, as shown in Figure 10–55.

4. What is the measured rms output voltage at 5 kHz in Figure 10–55?
5. What is the measured rms output voltage at 60 Hz in Figure 10–55?

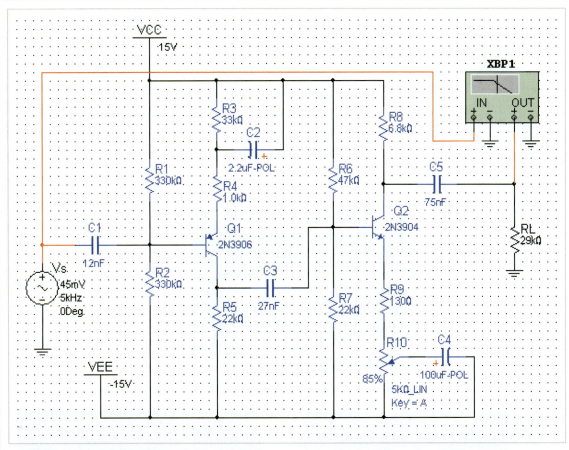

(a) Circuit screen with reduced capacitor values

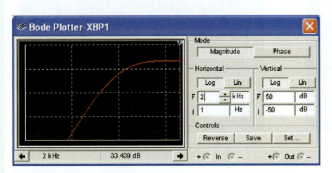

(b) Midrange gain is 33.439 Hz

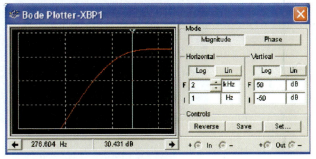

(c) f_c is 276.604 Hz at 30.431 dB (-3 dB)

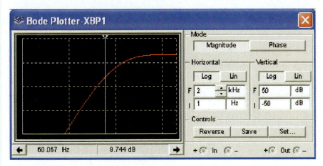

(d) At 60.067 Hz the gain is 9.744 dB (down 23.7 dB)

▲ FIGURE 10–54

Preamp frequency response with reduced capacitor values.

6. What would be the approximate rms amplitude of the output waveform at 300 Hz?
7. Based on the oscilloscope measurement in Figure 10–55, express the voltage gain at 5 kHz in dB.
8. Based on the oscilloscope measurement in Figure 10–55, express the voltage gain at 60 Hz in dB.

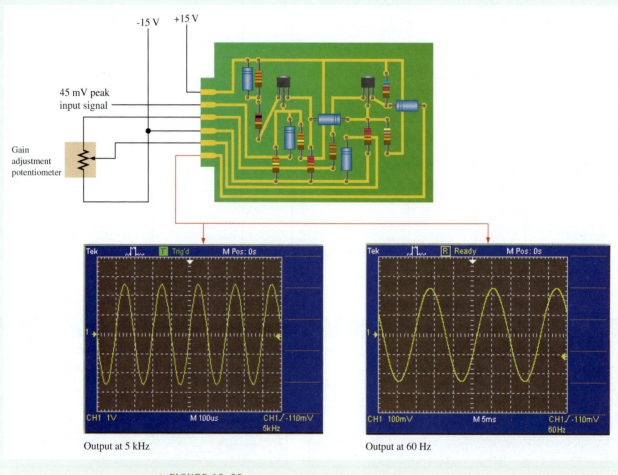

Output at 5 kHz   Output at 60 Hz

▲ FIGURE 10–55

Frequency test of new preamp board using an oscilloscope.

SUMMARY

Section 10–1 ◆ The coupling and bypass capacitors of an amplifier affect the low-frequency response.

◆ The internal transistor capacitances affect the high-frequency response.

Section 10–2 ◆ The decibel is a logarithmic unit of measurement for power gain and voltage gain.

◆ A decrease in voltage gain to 70.7% of midrange value is a reduction of 3 dB.

◆ A halving of the voltage gain corresponds to a reduction of 6 dB.

◆ The dBm is a unit for measuring power levels referenced to 1 mW.

◆ Critical frequencies are values of frequency at which the *RC* circuits reduce the voltage gain to 70.7% of its midrange value.

Section 10–3 ◆ Each *RC* circuit causes the gain to drop at a rate of –20 dB/decade.

◆ For the low-frequency *RC* circuits, the *highest* critical frequency is the dominant critical frequency.

◆ A decade of frequency change is a ten-times change (increase or decrease).

◆ An octave of frequency change is a two-times change (increase or decrease).

Section 10–4 ◆ For the high-frequency *RC* circuits, the *lowest* critical frequency is the dominant critical frequency.

Section 10–5 ◆ The bandwidth of an amplifier is the range of frequencies between the dominant lower critical frequency and the dominant upper critical frequency.

◆ The gain-bandwidth product is an amplifier parameter that is the product of gain and bandwidth. For an amplifier with a single dominant upper critical frequency the gain-bandwidth product is a constant.

Section 10–6 ◆ The dominant critical frequencies of a multistage amplifier establish the bandwidth.

Section 10–7 ◆ Two frequency response measurement methods are frequency/amplitude and step.

KEY TERMS

Key terms and other bold terms in the chapter are defined in the end-of-book glossary.

Bandwidth The characteristic of certain types of electronic circuits that specifies the usable range of frequencies that pass from input to output.

Bode plot An idealized graph of the gain in dB versus frequency used to graphically illustrate the response of an amplifier or filter.

Critical frequency The frequency at which the response of an amplifier or filter is 3 dB less than at midrange.

Decade A ten-times increase or decrease in the value of a quantity such as frequency.

Midrange gain The gain that occurs for the range of frequencies between the lower and upper critical frequencies.

Normalized adjusting the values of a quantity to produce a standardized response. For amplifiers, it refers to adjusting the midrange voltage gain to assign it a value of 1 or 0 dB by dividing all gain values by the midrange voltage gain.

Roll-off The rate of decrease in the gain of an amplifier above or below the critical frequencies; usually it is specified in dB/decade.

Parasitic capacitance an unavoidable and unwanted capacitance that exists between circuit elements; the circuit elements can be any components that are in close proximity to each other.

Gain-bandwidth product The product of an amplifier's gain and bandwidth; for amplifiers with one dominant upper cutoff frequency, the gain-bandwidth product is constant.

KEY FORMULAS

Miller's Theorem

10–1 $C_{in(Miller)} = C(A_v + 1)$ Miller input capacitance, where $C = C_{bc}$ or C_{gd}

10–2 $C_{out(Miller)} = C\left(\dfrac{A_v + 1}{A_v}\right)$ Miller output capacitance, where $C = C_{bc}$ or C_{gd}

The Decibel

10–3 $A_{p(dB)} = 10 \log A_p$ Power gain in decibels

10–4 $A_{v(dB)} = 20 \log A_v$ Voltage gain in decibels

BJT Amplifier Low-Frequency Response

10–5 $A_{v(mid)} = \dfrac{R_c}{r'_e}$ Midrange voltage gain for a CE amplifier with a fully bypassed emitter resistance

10–6 $f_{cl(input)} = \dfrac{1}{2\pi R_{in}C_1}$ Lower critical frequency, input *RC* circuit

10–7 $\quad \theta = \tan^{-1}\left(\dfrac{X_{C1}}{R_{in}}\right)$ $\qquad$ Phase angle, input RC circuit

10–8 $\quad f_{cl(output)} = \dfrac{1}{2\pi(R_C + R_L)C_3}$ $\qquad$ Lower critical frequency, output RC circuit

10–9 $\quad \theta = \tan^{-1}\left(\dfrac{X_{C3}}{R_C + R_L}\right)$ $\qquad$ Phase angle, output RC circuit

10–10 $\quad R_{in(emitter)} = r'_e + \dfrac{R_{th}}{\beta_{ac}}$ $\qquad$ Resistance looking in at emitter

10–11 $\quad f_{cl(bypass)} = \dfrac{1}{2\pi[(r'_e + R_{th}/\beta_{ac})\|R_E]C_2}$ $\qquad$ Lower critical frequency, bypass RC circuit

FET Amplifier Low-Frequency Response

10–12 $\quad f_{cl(input)} = \dfrac{1}{2\pi(R_G \| R_{in(gate)})C_1}$ $\qquad$ Lower critical frequency, input RC circuit

10–13 $\quad \theta = \tan^{-1}\left(\dfrac{X_{C1}}{R_{in}}\right)$ $\qquad$ Phase angle, input RC circuit

10–14 $\quad f_{cl(output)} = \dfrac{1}{2\pi(R_D + R_L)C_2}$ $\qquad$ Lower critical frequency, output RC circuit

10–15 $\quad \theta = \tan^{-1}\left(\dfrac{X_{C2}}{R_D + R_L}\right)$ $\qquad$ Phase angle, output RC circuit

BJT Amplifier High-Frequency Response

10–16 $\quad f_{cu(input)} = \dfrac{1}{2\pi(R_s \| R_1 \| R_2 \| \beta_{ac}r'_e)C_{tot}}$ $\qquad$ Upper critical frequency, input RC circuit

10–17 $\quad \theta = \tan^{-1}\left(\dfrac{R_s \| R_1 \| R_2 \| \beta_{ac}r'_e}{X_{C_{tot}}}\right)$ $\qquad$ Phase angle, input RC circuit

10–18 $\quad f_{cu(output)} = \dfrac{1}{2\pi R_c C_{out(Miller)}}$ $\qquad$ Upper critical frequency, output RC circuit

10–19 $\quad \theta = \tan^{-1}\left(\dfrac{R_c}{X_{C_{out(Miller)}}}\right)$ $\qquad$ Phase angle, output RC circuit

FET Amplifier High-Frequency Response

10–20 $\quad C_{gd} = C_{rss}$ $\qquad$ Gate-to-drain capacitance

10–21 $\quad C_{gs} = C_{iss} - C_{rss}$ $\qquad$ Gate-to-source capacitance

10–22 $\quad C_{ds} = C_{oss} - C_{rss}$ $\qquad$ Drain-to-source capacitance

10–23 $\quad f_{cu(input)} = \dfrac{1}{2\pi R_s C_{tot}}$ $\qquad$ Upper critical frequency, input RC circuit

10–24 $\quad \theta = \tan^{-1}\left(\dfrac{R_s}{X_{C_{tot}}}\right)$ $\qquad$ Phase angle, input RC circuit

10–25 $\quad f_{cu(output)} = \dfrac{1}{2\pi R_d C_{out(Miller)}}$ $\qquad$ Upper critical frequency, output RC circuit

10–26 $\quad \theta = \tan^{-1}\left(\dfrac{R_d}{X_{C_{out(Miller)}}}\right)$ $\qquad$ Phase angle, output RC circuit

Total Response

10–27 $\quad BW = f_{cu} - f_{cl}$ $\qquad$ Bandwidth

10–28 $\quad f_T = A_{v(mid)}BW$ $\qquad$ Unity-gain bandwidth

Multistage Response

10–29 $\quad f'_{cl(dom)} = \dfrac{f_{cl(dom)}}{\sqrt{2^{1/n} - 1}}$ $\qquad$ Overall dominant lower critical frequency for case of equal dominant critical frequencies

10–30 $f'_{cu(dom)} = f_{cu(dom)}\sqrt{2^{1/n} - 1}$ Overall dominant upper critical frequency for case of equal dominant critical frequencies

Measurement Techniques

10–31 $f_{cu} = \dfrac{0.35}{t_r}$ Upper critical frequency

10–32 $f_{cl} = \dfrac{0.35}{t_f}$ Lower critical frequency

TRUE/FALSE QUIZ   Answers can be found at www.pearsonhighered.com/floyd.

1. Coupling capacitors in an amplifier determine the low-frequency response.
2. Bypass capacitors in an amplifier determine the high-frequency response.
3. Internal transistor capacitance has no effect on an amplifier's frequency response.
4. Miller's theorem states that both gain and internal capacitances influence high-frequency response.
5. The midrange gain is between the upper and lower critical frequencies.
6. The critical frequency is one of two frequencies at which the gain is 6 dB less than the midrange gain.
7. dBm is a unit for measuring power levels.
8. A ten-times change in frequency is called a decade.
9. An octave corresponds to a doubling or halving of the frequency.
10. The input and output RC circuits have no effect on the frequency response.
11. A Bode plot shows the voltage gain versus frequency on a logarithmic scale.
12. Phase shift is part of an amplifier's frequency response.
13. An amplifier's low frequency cutoff can be measured by a rise-time measurement.
14. The gain-bandwith product is constant for amplifiers with a roll-off rate of -20 dB/decade.

CIRCUIT-ACTION QUIZ   Answers can be found at www.pearsonhighered.com/floyd.

1. If the value of R_1 in Figure 10–8 is increased, the signal voltage at the base will
   (a) increase   **(b)** decrease   **(c)** not change
2. If the value of C_1 in Figure 10–27 is decreased, the critical frequency associated with the input circuit will
   (a) increase   **(b)** decrease   **(c)** not change
3. If the value of R_L in Figure 10–27 is increased, the voltage gain will
   (a) increase   **(b)** decrease   **(c)** not change
4. If the value of R_C in Figure 10–27 is decreased, the voltage gain will
   (a) increase   **(b)** decrease   **(c)** not change
5. If V_{CC} in Figure 10–34 is increased, the dc emitter voltage will
   (a) increase   **(b)** decrease   **(c)** not change
6. If the transistor in Figure 10–34 is replaced with one having a higher β_{ac}, the critical frequency will
   (a) increase   **(b)** decrease   **(c)** not change
7. If the transistor in Figure 10–34 is replaced with one having a lower β_{ac}, the midrange voltage gain will
   (a) increase   **(b)** decrease   **(c)** not change
8. If the value of R_D in Figure 10–42 is increased, the voltage gain will
   (a) increase   **(b)** decrease   **(c)** not change
9. If the value of R_L in Figure 10–42 is increased, the critical frequency will
   (a) increase   **(b)** decrease   **(c)** not change
10. If the FET in Figure 10–42 is replaced with one having a higher g_m, the critical frequency will
    (a) increase   **(b)** decrease   **(c)** not change

| SELF-TEST | Answers can be found at www.pearsonhighered.com/floyd. |

Section 10–1

1. The low-frequency response of an amplifier is determined in part by

   (a) the voltage gain   (b) the type of transistor

   (c) the supply voltage   (d) the coupling capacitors

2. The high-frequency response of an amplifier is determined in part by

   (a) the gain-bandwidth product   (b) the bypass capacitor

   (c) the internal transistor capacitances   (d) the roll-off

3. The Miller input capacitance of an amplifier is dependent, in part, on

   (a) the input coupling capacitor   (b) the voltage gain

   (c) the bypass capacitor   (d) none of these

Section 10–2

4. The decibel is used to express

   (a) power gain   (b) voltage gain   (c) attenuation   (d) all of these

5. When the voltage gain is 70.7% of its midrange value, it is said to be

   (a) attenuated   (b) down 6 dB   (c) down 3 dB   (d) down 1 dB

6. In an amplifier, the gain that occurs between the lower and upper critical frequencies is called the

   (a) critical gain   (b) midrange gain   (c) bandwidth gain   (d) decibel gain

7. A certain amplifier has a voltage gain of 100 at midrange. If the gain decreases by 6 dB, it is equal to

   (a) 50   (b) 70.7   (c) 0   (d) 20

Section 10–3

8. The gain of a certain amplifier decreases by 6 dB when the frequency is reduced from 1 kHz to 10 Hz. The roll-off is

   (a) −3 dB/decade   (b) −6 dB/decade   (c) −3 dB/octave   (d) −6 dB/octave

9. The gain of a particular amplifier at a given frequency decreases by 6 dB when the frequency is doubled. The roll-off is

   (a) −12 dB/decade   (b) −20 dB/decade   (c) −6 dB/octave   (d) answers (b) and (c)

10. The lower critical frequency of a direct-coupled amplifier with no bypass capacitor is

    (a) variable   (b) 0 Hz   (c) dependent on the bias   (d) none of these

Section 10–4

11. At the upper critical frequency, the peak output voltage of a certain amplifier is 10 V. The peak voltage in the midrange of the amplifier is

    (a) 7.07 V   (b) 6.37 V   (c) 14.14 V   (d) 10 V

12. The high-frequency response of an amplifier is determined by the

    (a) coupling capacitors   (b) bias circuit

    (c) transistor capacitances   (d) all of these

13. The Miller input and output capacitances for a BJT inverting amplifier depend on

    (a) C_{bc} (b) β_{ac} (c) Av (d) answers (a) and (c)

Section 10–5

14. The bandwidth of an amplifier is determined by

    (a) the midrange gain   (b) the critical frequencies

    (c) the roll-off rate   (d) the input capacitance

15. A two-stage amplifier has the following lower critical frequencies: 35 Hz and 68 Hz, The upper critical frequencies are 140 kHz and 1.5 MHz. The midrange bandwidth is

    (a) 35 Hz   (b) 68 Hz   (c) 140 kHz   (d) 1.5 MHz

16. Ideally, the midrange gain of an amplifier

    (a) increases with frequency

    (b) decreases with frequency

    (c) remains constant with frequency

    (d) depends on the coupling capacitors

17. The frequency at which an amplifier's gain is 1 is called the

  (a) unity-gain frequency (b) midrange frequency

  (c) corner frequency (d) break frequency

18. When the voltage gain of an amplifier is increased, the bandwidth

  (a) is not affected (b) increases (c) decreases (d) becomes distorted

19. If the f_T of the transistor used in an amplifier with a constant -20 dB/decade roll off is 75 MHz and the bandwidth is 10 MHz. The midrange voltage gain is

  (a) 750 (b) 7.5 (c) 10 (d) 1

20. In the midrange of an amplifier's bandwidth, the peak output voltage is 6 V. At the lower critical frequency, the peak output voltage is

  (a) 3 V (b) 3.82 V (c) 8.48 V (d) 4.24 V

Section 10–6 21. The dominant lower critical frequency of a multistage amplifier is the

  (a) lowest f_{cl} (b) highest f_{cl} (c) average of all the f_{cl}'s (d) none of these

22. When the critical frequencies of all of the stages are the same, the dominant critical frequency is

  (a) higher than any individual f_{cl} (b) lower than any individual f_{cl}

  (c) equal to the individual f_{cl}'s (d) the sum of all individual f_{cl}'s

Section 10–7 23. In the step response of a noninverting amplifier, a longer rise time means

  (a) a narrower bandwidth (b) a lower f_{cl}

  (c) a higher f_{cu} (d) answers (a) and (b)

PROBLEMS

Answers to all odd-numbered problems are at the end of the book.

BASIC PROBLEMS

Section 10–1 **Basic Concepts**

1. (a) What part of the frequency response is affected by a transistor's parasitic capacitances?

  (b) What can a designer do to minimize the affect of parasitic capacitance in a CE amplifier?

2. Explain why the coupling capacitors do not have a significant effect on gain at sufficiently high-signal frequencies.

3. List the capacitances that affect high-frequency gain in both BJT and FET amplifiers.

4. In the amplifier of Figure 10–56, list the capacitances that affect the low-frequency response of the amplifier and those that affect the high-frequency response.

▶ **FIGURE 10–56**

Multisim and LT Spice file circuits are identified with a logo and are in the Problems folder on the companion website. Filenames correspond to figure numbers (e.g., FGM10-56 and FGS10-56).

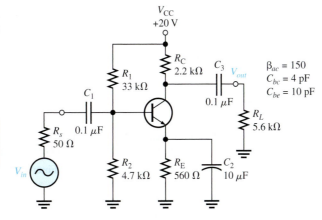

26. If the midrange gain of a given amplifier with one dominant upper frequency pole is 50 dB and therefore 47 dB at f_{cu}, what is the gain at $2f_{cu}$? At $4f_{cu}$? At $10f_{cu}$?

Section 10–6 Frequency Response of Multistage Amplifiers

27. In a certain two-stage amplifier, the first stage has critical frequencies of 230 Hz and 1.2 MHz. The second stage has critical frequencies of 195 Hz and 2 MHz. What are the dominant critical frequencies?

28. What is the bandwidth of the two-stage amplifier in Problem 27?

29. Determine the bandwidth of a two-stage amplifier in which each stage has a lower critical frequency of 400 Hz and an upper critical frequency of 800 kHz.

30. What is the dominant lower critical frequency of a three-stage amplifier in which $f_{cl} = 50$ Hz for each stage.

31. In a certain two-stage amplifier, the lower critical frequencies are $f_{cl(1)} = 125$ Hz and $f_{cl(2)} = 125$ Hz, and the upper critical frequencies are $f_{cu(1)} = 3$ MHz and $f_{cu(2)} = 2.5$ MHz. Determine the bandwidth.

Section 10–7 Frequency Response Measurements

32. In a step-response test of a certain amplifier, $t_r = 20$ ns and $t_f = 1$ ms. Determine f_{cl} and f_{cu}.

33. Suppose you are measuring the frequency response of an amplifier with a signal source and an oscilloscope. Assume that the signal level and frequency are set such that the oscilloscope indicates an output voltage level of 5 V rms in the midrange of the amplifier's response. If you wish to determine the upper critical frequency, indicate what you would do and what scope indication you would look for.

34. Determine the approximate bandwidth of an amplifier from the indicated results of the step-response test in Figure 10–61.

▶ FIGURE 10–61

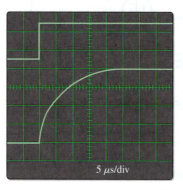
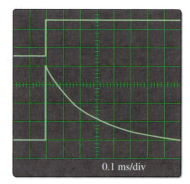

5 μs/div   0.1 ms/div

APPLICATION ACTIVITY PROBLEMS

35. Determine the dominant lower critical frequency for the amplifier in Figure 10–52 if the coupling capacitors are changed to 1 μF. Assume $R_L = 29$ kΩ and $\beta_{ac} = 100$.

36. Does the change in Problem 35 significantly affect the overall bandwidth?

37. How does a change from 29 kΩ to 100 kΩ in load resistance on the final output of the amplifier in Figure 10–52 affect the dominant lower critical frequency?

38. If the transistors in the modified preamp in the Device Application have a β_{ac} of 300, determine the effect on the dominant lower critical frequency.

DATASHEET PROBLEMS

39. Referring to the partial datasheet for a 2N3904 in Figure 10–62, determine the total input capacitance for an amplifier if the voltage gain is 25.

40. A certain amplifier uses a 2N3904 and has a midrange voltage gain of 50. Referring to the partial datasheet in Figure 10–62, determine its minimum bandwidth.

41. The datasheet for a 2N4351 MOSFET specifies the maximum values of internal capacitances as follows: $C_{iss} = 5$ pF, $C_{rss} = 1.3$ pF, and $C_{d(sub)} = 5$ pF. Determine C_{gd}, C_{gs}, and C_{ds}.

Electrical Characteristics $T_A = 25°C$ unless otherwise noted

Symbol	Parameter	Test Conditions	Min	Max	Units
OFF CHARACTERISTICS					
$V_{(BR)CEO}$	Collector-Emitter Breakdown Voltage	$I_C = 1.0$ mA, $I_B = 0$	40		V
$V_{(BR)CBO}$	Collector-Base Breakdown Voltage	$I_C = 10$ µA, $I_E = 0$	60		V
$V_{(BR)EBO}$	Emitter-Base Breakdown Voltage	$I_E = 10$ µA, $I_C = 0$	6.0		V
I_{BL}	Base Cutoff Current	$V_{CE} = 30$ V, $V_{EB} = 3$V		50	nA
I_{CEX}	Collector Cutoff Current	$V_{CE} = 30$ V, $V_{EB} = 3$V		50	nA
ON CHARACTERISTICS*					
h_{FE}	DC Current Gain	$I_C = 0.1$ mA, $V_{CE} = 1.0$ V	40		
		$I_C = 1.0$ mA, $V_{CE} = 1.0$ V	70		
		$I_C = 10$ mA, $V_{CE} = 1.0$ V	100	300	
		$I_C = 50$ mA, $V_{CE} = 1.0$ V	60		
		$I_C = 100$ mA, $V_{CE} = 1.0$ V	30		
$V_{CE(sat)}$	Collector-Emitter Saturation Voltage	$I_C = 10$ mA, $I_B = 1.0$ mA		0.2	V
		$I_C = 50$ mA, $I_B = 5.0$ mA		0.3	V
$V_{BE(sat)}$	Base-Emitter Saturation Voltage	$I_C = 10$ mA, $I_B = 1.0$ mA	0.65	0.85	V
		$I_C = 50$ mA, $I_B = 5.0$ mA		0.95	V
SMALL SIGNAL CHARACTERISTICS					
f_T	Current Gain - Bandwidth Product	$I_C = 10$ mA, $V_{CE} = 20$ V, f = 100 MHz	300		MHz
C_{obo}	Output Capacitance	$V_{CB} = 5.0$ V, $I_E = 0$, f = 1.0 MHz		4.0	pF
C_{ibo}	Input Capacitance	$V_{EB} = 0.5$ V, $I_C = 0$, f = 1.0 MHz		8.0	pF
NF	Noise Figure	$I_C = 100$ µA, $V_{CE} = 5.0$ V, $R_S = 1.0$kΩ,f=10 Hz to 15.7kHz		5.0	dB
SWITCHING CHARACTERISTICS					
t_d	Delay Time	$V_{CC} = 3.0$ V, $V_{BE} = 0.5$ V,		35	ns
t_r	Rise Time	$I_C = 10$ mA, $I_{B1} = 1.0$ mA		35	ns
t_s	Storage Time	$V_{CC} = 3.0$ V, $I_C = 10$mA		200	ns
t_f	Fall Time	$I_{B1} = I_{B2} = 1.0$ mA		50	ns

▲ **FIGURE 10–62**

Partial datasheet for the 2N3904. Copyright Fairchild Semiconductor Corporation. Used by permission.

ADVANCED PROBLEMS

42. Two single-stage capacitively coupled amplifiers like the one in Figure 10–56 are connected as a two-stage amplifier (R_L is removed from the first stage). Determine whether or not this configuration will operate as a linear amplifier with an input voltage of 10 mV rms. If not, modify the design to achieve maximum gain without distortion.

43. Two stages of the amplifier in Figure 10–60 are connected in cascade. Determine the overall bandwidth.

44. Redesign the amplifier in Figure 10–52 for an adjustable voltage gain of 50 to 500 and a lower critical frequency of 1 kHz.

MULTISIM TROUBLESHOOTING PROBLEMS

These file circuits are in the Troubleshooting Problems folder on the companion website.

45. Open file TPM10-45 and determine the fault.

46. Open file TPM10-46 and determine the fault.

47. Open file TPM10-47 and determine the fault.

48. Open file TPM10-48 and determine the fault.

▶ FIGURE 11–5

The silicon-controlled rectifier (SCR).

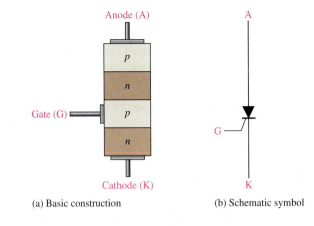

(a) Basic construction   (b) Schematic symbol

(c) Typical packages; the "hockey puck" style third from right is for switching very high currents.

▶ FIGURE 11–6

SCR equivalent circuit.

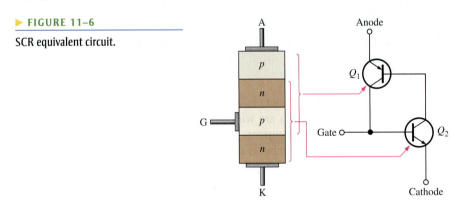

of Q_1 by providing a path for I_{B1}; in turn, Q_1 sustains the saturated conduction of Q_2 by providing I_{B2}. Thus, the device stays on (latches) once it is triggered on, as shown in Figure 11–7(c). Current in the SCR is entirely controlled by the impedance of the external circuit. In this state, the very low resistance between the anode and cathode can be approximated by a closed switch, as indicated. In practice, a small voltage (<1 V) will appear across the SCR when it is on due to the base-emitter drop of the equivalent Q_2 and the saturation drop in the equivalent Q_1.

Like the four-layer diode, an SCR can also be turned on without gate triggering by increasing the anode-to-cathode voltage to a value exceeding the forward-breakover voltage $V_{BR(F)}$, as shown on the characteristic curve in Figure 11–8(a). The forward-breakover voltage decreases as I_G is increased above 0 V, as shown by the set of curves in Figure 11–8(b). Eventually, a value of I_G is reached at which the SCR turns on at a very low anode-to-cathode voltage. So, as you can see, the gate current controls the value of forward breakover voltage, $V_{BR(F)}$, required for turn-on.

Although anode-to-cathode voltages in excess of $V_{BR(F)}$ will not damage the device if current is limited, this situation should be avoided because the normal control of the SCR is lost. It should normally be triggered on only with a pulse at the gate.

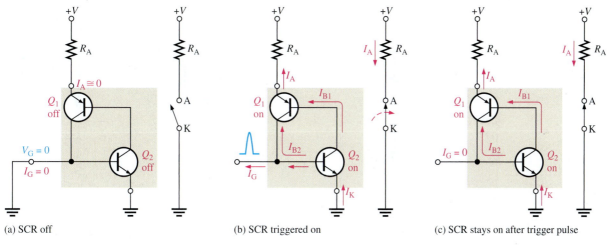

(a) SCR off   (b) SCR triggered on   (c) SCR stays on after trigger pulse

▲ **FIGURE 11–7**

The SCR turn-on process with the switch equivalents shown.

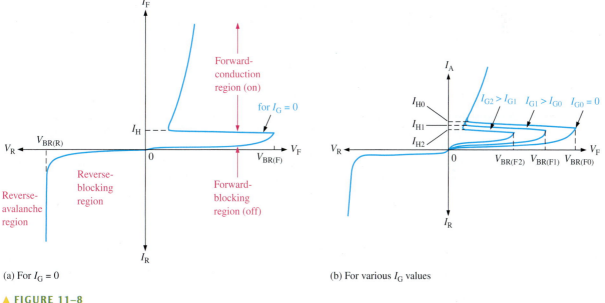

(a) For $I_G = 0$ (b) For various I_G values

▲ **FIGURE 11–8**

SCR characteristic curves.

Turning the SCR Off

When the gate returns to 0 V after the trigger pulse is removed, the SCR cannot turn off; it stays in the forward-conduction region. The anode current must drop below the value of the holding current, I_H, in order for turn-off to occur. The holding current is indicated in Figure 11–8.

There are two basic methods for turning off an SCR: *anode current interruption* and *forced commutation*. The anode current can be interrupted by either a momentary series or parallel switching arrangement, as shown in Figure 11–9. The series switch in part (a) simply reduces the anode current to zero and causes the SCR to turn off. The parallel switch in part (b) routes part of the total current away from the SCR, thereby reducing the anode current to a value less than I_H.

The **forced commutation** method basically requires momentarily forcing current through the SCR in the direction opposite to the forward conduction so that the net forward

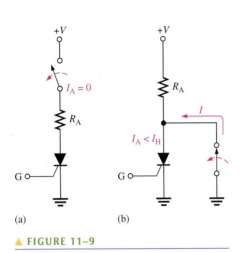

▲ **FIGURE 11–9**

SCR turn-off by anode current interruption.

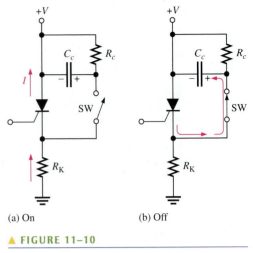

▲ **FIGURE 11–10**

SCR turn-off by forced commutation.

current is reduced below the holding value. The basic circuit, as shown in Figure 11–10, consists of a switch (normally a transistor switch) and a capacitor. While the SCR is conducting, the switch is open and C_c is charged to the supply voltage through R_c, as shown in part (a). To turn off the SCR, the switch is closed, placing the capacitor across the SCR and forcing current through it opposite to the forward current, as shown in part (b). Typically, turn-off times for SCRs range from a few microseconds up to about 30 μs.

SCR Characteristics and Ratings

Several of the most important SCR characteristics and ratings are defined as follows. Be aware that specifications are given for a specific temperature (generally 25°C); some specifications such as minimum gate trigger current will change at different temperatures. Use the curve in Figure 11–8(a) for reference where appropriate.

Forward-breakover voltage, $V_{BR(F)}$ This is the voltage at which the SCR enters the forward-conduction region. The value of $V_{BR(F)}$ is maximum when $I_G = 0$ and is designated $V_{BR(F0)}$. When the gate current in increased, $V_{BR(F)}$ decreases and is designated $V_{BR(F1)}$, $V_{BR(F2)}$, and so on, for increasing steps in gate current (I_{G1}, I_{G2}, and so on).

Holding current, I_H   This is the value of anode current below which the SCR switches from the forward-conduction region to the forward-blocking region. The value increases with decreasing values of I_G and is maximum for $I_G = 0$.

Gate trigger current, I_{GT}   This is the value of gate current necessary to switch the SCR from the forward-blocking region to the forward-conduction region under specified conditions.

Average forward current, $I_{F(avg)}$ This is the maximum continuous anode current (dc) that the device can withstand in the conduction state under specified conditions.

Forward-conduction region   This region corresponds to the *on* condition of the SCR where there is forward current from cathode to anode through the very low resistance (approximate short) of the SCR.

Forward-blocking and reverse-blocking regions   These regions correspond to the *off* condition of the SCR where the forward current from cathode to anode is blocked by the effective open circuit of the SCR.

Reverse-breakdown voltage, $V_{BR(R)}$ This parameter specifies the value of reverse voltage from cathode to anode at which the device breaks into the avalanche region and begins to conduct heavily (the same as in a *pn* junction diode).

The Light-Activated SCR (LASCR)

The light-activated silicon-controlled rectifier (**LASCR**) (also called photo SCR) is a four-layer semiconductor device (thyristor) that operates essentially as does the conventional SCR except that it can also be light-triggered using a small lens to focus light onto a light-sensitive gate. Once triggered, the LASCR will continue to conduct even if the light is no longer present. The LASCR conducts current in one direction when activated by a sufficient amount of light and continues to conduct until the current falls below a specified value. Figure 11–11 shows a LASCR schematic symbol. The LASCR is most sensitive to light when the gate terminal is open. It is also temperature sensitive; leakage current increases with temperature, so the LASCR can turn on at lower light levels when temperature increases. If necessary, a resistor from the gate to the cathode can be used to reduce the sensitivity.

Figure 11–12 shows a LASCR used to energize a latching relay. The input source turns on the lamp; the resulting incident light triggers the LASCR. The anode current energizes the relay and closes the contact. Notice that the input source is electrically isolated from the rest of the circuit. As is evident in the circuit diagram, a very small energy is all that is required to close the relay, which can switch power for an alarm, motor, or other device. LASCRs are extremely efficient and provide electrical isolation between the control circuit and the load. For these reasons, LASCRs are also used in high-voltage, high-current switching systems.

▲ FIGURE 11–11

LASCR symbol.

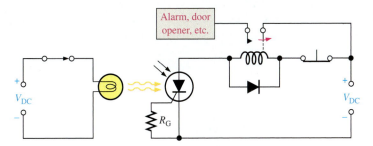

◄ FIGURE 11–12

A LASCR circuit.

The solid state relay (SSR) using E-MOSFETs was introduced in Section 9–6. Recall that a solid state relay is an interfacing circuit that can open or close one or more switches that are connected to a load. Solid state relays can be constructed with an internal opto-isolator and LASCR forming a photo SCR optocoupler. These devices are commonly used to interface logic circuits to a load and isolate the logic from the load. (An example is the H11CX series from Fairchild.) Figure 11–13 shows a basic circuit for interfacing TTL logic to a 25 W load (in this case a small lamp) An advantage for this is that the two circuits are completely isolated and the logic is not subjected to conductive noise generated by the load.

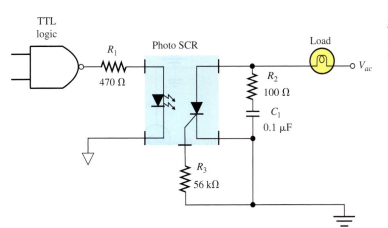

◄ FIGURE 11–13

Interfacing TTL logic with a load using a photo SCR

By adjusting R_2, the SCR can be made to trigger at any point on the positive half-cycle of the ac waveform between 0° and 90°, as shown in Figure 11–17.

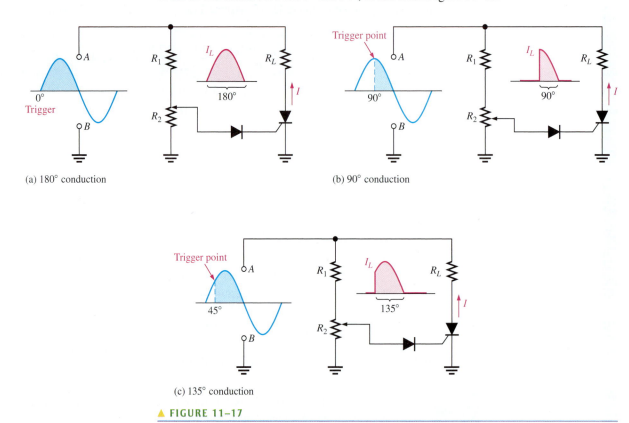

(a) 180° conduction

(b) 90° conduction

(c) 135° conduction

▲ FIGURE 11–17

Operation of the phase-control circuit.

When the SCR triggers near the beginning of the cycle (approximately 0°), as in Figure 11–17(a), it conducts for approximately 180° and maximum power is delivered to the load. When it triggers near the peak of the positive half-cycle (90°), as in Figure 11–17(b), the SCR conducts for approximately 90° and less power is delivered to the load. By adjusting R_2, triggering can be made to occur anywhere between these two extremes, and therefore, a variable amount of power can be delivered to the load. Figure 11–17(c) shows triggering at the 45° point as an example. When the ac input goes negative, the SCR turns off and does not conduct again until the trigger point on the next positive half-cycle. The diode prevents the negative ac voltage from being applied to the gate of the SCR.

EXAMPLE 11–2

Show the voltage waveform across the SCR in Figure 11–18 from anode to cathode (ground) in relation to the load current for 180°, 45°, and 90° conduction. Assume an ideal SCR.

▶ FIGURE 11–18

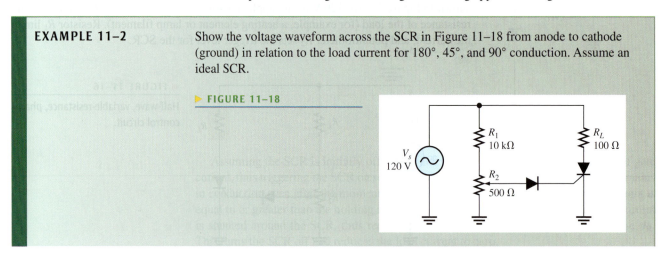

Solution   When there is load current, the SCR is conducting and the voltage across it is ideally zero. When there is no load current, the voltage across the SCR is the same as the applied voltage. The waveforms are shown in Figure 11–19.

Related Problem   What is the voltage across the SCR if it is never triggered?

 Open the Multisim file EXM11-02 or the LT Spice file EXS11-02 in the Examples folder on the website. View the voltage across the SCR with the oscilloscope. Vary the potentiometer setting and observe how V_{AK} changes.

▶ **FIGURE 11–19**

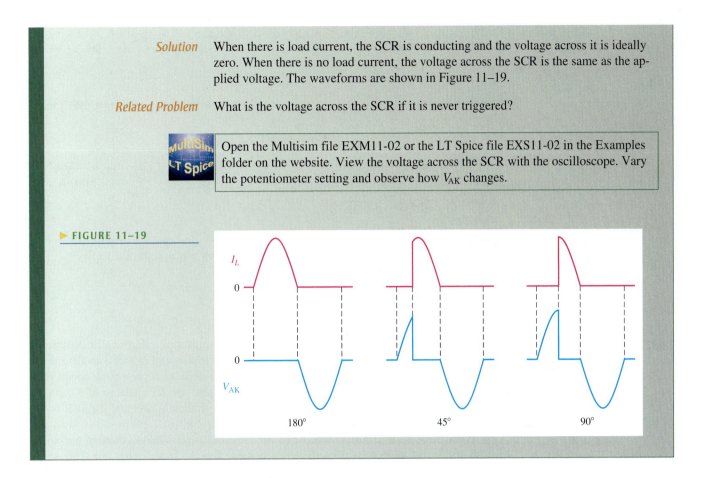

A useful circuit that uses a similar idea to the phase control circuit in Figure 11–16 is a motor speed controller for a dc motor, which can run on ac using an SCR to convert the ac to pulsed dc. A dc motor can even out pulses from the rectified ac, so there is no need to add a filter. The circuit is shown in Figure 11–20. To reference the motor to ground, it is connected to the cathode side of the SCR. Diode D_1 sets a minimum gate voltage of 0.7 V; R_2 adjusts the firing point of the SCR, and hence the average current in the motor. The circuit can be adapted to other loads that can operated with pulsed dc such as a lamp. Another variation of a speed control circuit uses a programmable unijunction transistor (PUT) and is introduced in the Device Application at the end of the chapter.

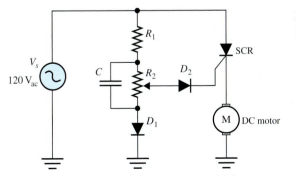

◀ **FIGURE 11–20**

Speed control for a dc motor.

Backup Lighting for Power Interruptions

As another example of SCR applications, let's examine a circuit that will maintain lighting by using a backup battery when there is an ac power failure. Figure 11–21 shows a center-tapped full-wave rectifier used for providing ac power to a low-voltage lamp. As long as the ac power is available, the battery charges through diode D_3 and R_1.

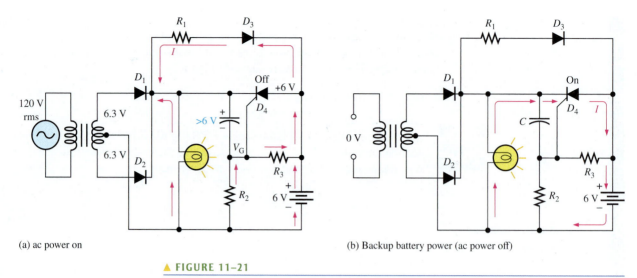

(a) ac power on

(b) Backup battery power (ac power off)

▲ **FIGURE 11–21**

Automatic backup lighting circuit.

The SCR's cathode voltage is established when the capacitor charges to the peak value of the full-wave rectified ac (6.3 V rms less the drops across R_2 and D_1). The anode is at the 6 V battery voltage, making it less positive than the cathode, thus preventing conduction. The SCR's gate is at a voltage established by the voltage divider made up of R_2 and R_3. Under these conditions the lamp is illuminated by the ac input power and the SCR is off, as shown in Figure 11–21(a).

When there is an interruption of ac power, the capacitor discharges through the closed path R_1, D_3, and R_3, making the cathode less positive than the anode or the gate. This action establishes a triggering condition, and the SCR begins to conduct. Current from the battery is through the SCR and the lamp, thus maintaining illumination, as shown in Figure 11–21(b). When ac power is restored, the capacitor recharges and the SCR turns off. The battery begins recharging.

An Over-Voltage Protection Circuit

Figure 11–22 shows a simple over-voltage protection circuit, sometimes called a "crowbar" circuit, in a dc power supply. (The name "crowbar" came from the suggestion that the circuit was like putting a crowbar across the output.) The dc output voltage from the

▶ **FIGURE 11–22**

A basic SCR over-voltage protection circuit (shown in blue).

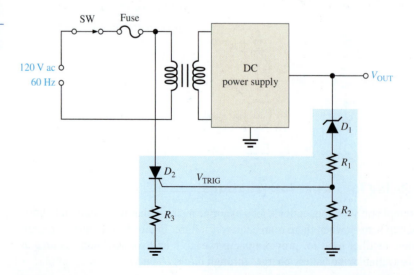

regulator is monitored by the zener diode (D_1) and the resistive voltage divider (R_1 and R_2). The upper limit of the output voltage is set by the zener voltage. If this voltage is exceeded, the zener conducts and the voltage divider produces an SCR trigger voltage. The trigger voltage turns on the SCR, which is connected across the line voltage. The SCR current causes the fuse to blow, thus disconnecting the line voltage from the power supply.

SECTION 11–3 CHECKUP	1. If the potentiometer in Figure 11–17 is set at its midpoint, during what part of the input cycle will the SCR conduct?
	2. In Figure 11–21, what is the purpose of diode D_3?

11–4 THE DIAC AND TRIAC

Both the diac and the triac are types of thyristors that can conduct current in both directions (bilateral). The difference between the two devices is that a diac has two terminals, while a triac has a third terminal, which is the gate for triggering. The diac functions basically like two parallel four-layer diodes turned in opposite directions. The triac functions basically like two parallel SCRs turned in opposite directions with a common gate terminal.

After completing this section, you should be able to

❏ **Describe the basic structure and operation of the diac and triac**
❏ Explain the operation of the diac
  ◆ Identify the schematic symbol ◆ Describe the characteristic curve
  ◆ Discuss the equivalent circuit
❏ Explain the operation of the triac
  ◆ Identify the schematic symbol ◆ Describe the characteristic curve
  ◆ Discuss the equivalent circuit
❏ Describe an application of the triac

The Diac

A **diac** (short for <u>di</u>ode <u>ac</u>) is a two-terminal thyristor that is equivalent to back-to-back inverse four-layer diodes that can conduct current in either direction when activated. The basic construction and schematic symbol for a diac are shown in Figure 11–23. Notice that the two terminals are labelled A_1 and A_2, which stand for Anode 1 and Anode 2. Neither terminal is referred to as a cathode. The top and bottom layers contain both n and p materials. The right side of the stack can be regarded as a *pnpn* structure with the same characteristics as a four-layer diode, while the left side is an inverted four-layer diode having an *npnp* structure.

Conduction occurs in a diac when the breakover voltage is reached with either polarity across the two terminals. The curve in Figure 11–24 illustrates this characteristic. Once breakover occurs, current is in a direction depending on the polarity of the voltage across the terminals. The device turns off when the current drops below the holding value.

The equivalent circuit of a diac consists of four transistors arranged as shown in Figure 11–25(a). When the diac is biased as in Figure 11–25(b), the *pnpn* structure from A_1 to A_2 provides the same operation as was described for the four-layer diode. In the equivalent circuit, Q_1 and Q_2 are forward-biased, and Q_3 and Q_4 are reverse-biased. The device operates on the upper right portion of the characteristic curve in Figure 11–24 under this bias condition. When the diac is biased as shown in Figure 11–25(c), the *pnpn* structure from A_2 and A_1 is used. In the equivalent circuit, Q_3 and Q_4 are forward-biased, and Q_1 and Q_2

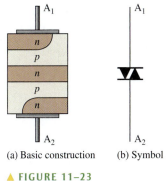

(a) Basic construction   (b) Symbol

▲ **FIGURE 11–23**

The diac.

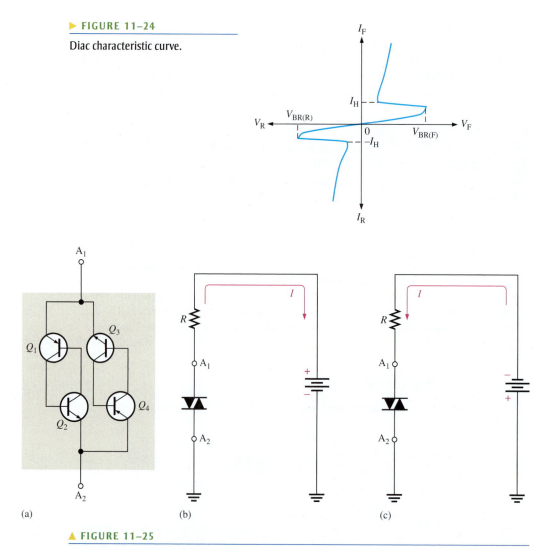

▶ FIGURE 11–24

Diac characteristic curve.

▲ FIGURE 11–25

Diac equivalent circuit and bias conditions.

are reverse-biased. Under this bias condition, the device operates on the lower left portion of the characteristic curve, as shown in Figure 11–24.

If the diac is driven by a sine wave as in Figure 11–26(a), it will fire on both the positive and negative half-cycles when the sine wave reaches V_{BR}. After firing, the diac will latch for remainder of the half cycle at which point the sine wave reverses direction and the diac drops out of conduction as the current drops below the hold current value, I_H. This action repeats on the opposite half cycle as shown in Fig 11–26(b). The voltage across the resistor (V_{R1}) has the same shape as the current in the circuit.

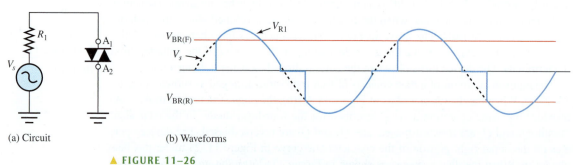

(a) Circuit (b) Waveforms

▲ FIGURE 11–26

Diac circuit with a sine wave input

The diac is designed for fast switching on each half of the waveform at exactly the same point because the internal junctions are doped to similar levels, making it useful as a triggering device. It is rarely used in circuits alone; the most common use is as a triggering device for triacs (covered next).

The Triac

A **triac** is like a diac with a gate terminal. A triac can be turned on by a pulse of gate current and does not require the breakover voltage to initiate conduction, as does the diac. Basically, a triac can be thought of simply as two SCRs connected in parallel and in opposite directions with a common gate terminal. Unlike the SCR, the triac can conduct current in either direction when it is triggered on, depending on the polarity of the voltage across its A_1 and A_2 terminals. Figure 11–27 shows the basic construction and schematic symbol for a triac.

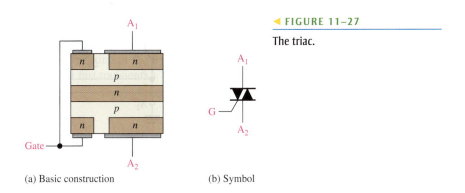

(a) Basic construction (b) Symbol

◀ **FIGURE 11–27**

The triac.

The characteristic curve is shown in Figure 11–28. Notice that the breakover potential decreases as the gate current increases, just as with the SCR. As with other thyristors, the triac ceases to conduct when the anode current drops below the specified value of the holding current, I_H. The only way to turn off the triac is to reduce the current to a sufficiently low level.

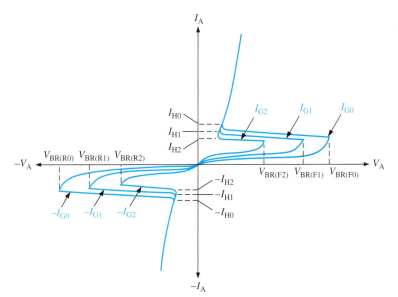

◀ **FIGURE 11–28**

Triac characteristic curves.

11–5 THE SILICON-CONTROLLED SWITCH (SCS)

The silicon-controlled switch (SCS) is similar in construction to the SCR. The SCS, however, has two gate terminals, the cathode gate and the anode gate. The SCS can be turned on and off using either gate terminal. Remember that the SCR can be only turned on using its gate terminal. Normally, the SCS is available in power ratings lower than those of the SCR.

After completing this section, you should be able to

❑ **Describe a silicon-controlled switch (SCS)**
❑ Explain the basic operation
  ◆ Identify the schematic symbol   ◆ Discuss the equivalent circuit
❑ Discuss SCS applications

Basic Operation

An **SCS** (silicon-controlled switch) is a four-terminal thyristor that has two gate terminals that are used to trigger the device on and off. The symbol and terminal identification for an SCS are shown in Figure 11–33. As in the case of an SCR, it is basically a rectifier (current in one direction).

As with the previous thyristors, the basic operation of the SCS can be understood by referring to the transistor equivalent, shown in Figure 11–34. To start, assume that both Q_1 and Q_2 are off, and therefore that the SCS is not conducting. A positive pulse on the cathode gate drives Q_2 into conduction and thus provides a path for Q_1 base current. When Q_1 turns on, its collector current provides base current for Q_2, thus sustaining the *on* state of the device. This regenerative action is the same as in the turn-on process of the SCR and the four-layer diode and is illustrated in Figure 11–34(a).

▲ **FIGURE 11–33**

The silicon-controlled switch (SCS).

▶ **FIGURE 11–34**

SCS operation.

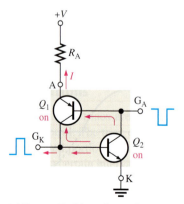

(a) *Turn-on*: Positive pulse on G_K *or* negative pulse on G_A

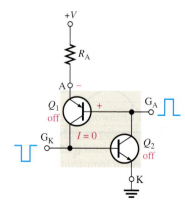

(b) *Turn-off*: Positive pulse on G_A *or* negative pulse on G_K

The SCS can also be turned on with a negative pulse on the anode gate, as indicated in Figure 11–34(a). This drives Q_1 into conduction which, in turn, provides base current for Q_2. Once Q_2 is on, it provides a path for Q_1 base current, thus sustaining the *on* state.

To turn the SCS off, the normal method is to apply a positive pulse to the anode gate. This reverse-biases the base-emitter junction of Q_1 and turns it off. Q_2, in turn, cuts off and the SCS ceases conduction, as shown in Figure 11–34(b). With a certain value of R_A, the device can also be turned off with a negative pulse on the cathode gate, as indicated in part (b). The SCS typically has a faster turn-off time than the SCR.

In addition to the positive pulse on the anode gate or the negative pulse on the cathode gate, there is another method for turning off an SCS. Figure 11–35(a) and (b) shows two switching methods to reduce the anode current below the holding value. In each case, the bipolar junction transistor (BJT) acts as a switch to interrupt the anode current.

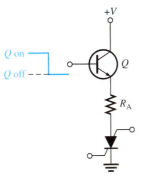

(a) Series switch turns off SCS

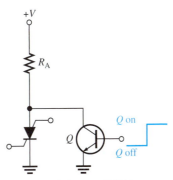

(b) Shunt switch turns off SCS

◀ **FIGURE 11–35**

The transistor switch in both series and shunt configurations reduces the anode current below the holding current and turns off the SCS.

Applications

The SCS and SCR are used in similar applications. The SCS has the advantage of faster turn-off with pulses on either gate terminal; however, it is more limited in terms of maximum current and voltage ratings. Also, the SCS is sometimes used in digital applications such as counters, registers, and timing circuits.

SECTION 11–5 CHECKUP	1. Explain the difference between an SCS and an SCR.
	2. How can an SCS be turned on?
	3. Describe three ways an SCS can be turned off.

11–6 THE UNIJUNCTION TRANSISTOR (UJT)

The unijunction transistor does not belong to the thyristor family because it does not have a four-layer type of construction. The term *unijunction* refers to the fact that the UJT has a single *pn* junction. The UJT is useful in certain oscillator applications and as a triggering device in thyristor circuits.

After completing this section, you should be able to

❑ **Describe the basic structure and operation of the unijunction transistor**
   ◆ Identify the schematic symbol
❑ Use the equivalent circuit to describe the basic operation
❑ Define and discuss the *standoff ratio*
❑ Discuss a UJT application

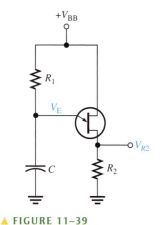

▲ FIGURE 11–39

Relaxation oscillator.

A UJT Application

The UJT can be used as a trigger device for SCRs and triacs. Other applications include nonsinusoidal oscillators, sawtooth generators, phase control, and timing circuits. Figure 11–39 shows a UJT relaxation oscillator as an example of one application.

The operation is as follows. When dc power is applied, the capacitor C charges exponentially through R_1 until it reaches the peak-point voltage V_P. At this point, the *pn* junction becomes forward-biased, and the emitter characteristic goes into the negative resistance region (V_E decreases and I_E increases). The capacitor then quickly discharges through the forward-biased junction, r'_B, and R_2. When the capacitor voltage decreases to the valley-point voltage V_V, the UJT turns off, the capacitor begins to charge again, and the cycle is repeated, as shown in the emitter voltage waveform in Figure 11–40 (top). During the discharge time of the capacitor, the UJT is conducting. Therefore, a voltage is developed across R_2, as shown in the waveform diagram in Figure 11–40 (bottom).

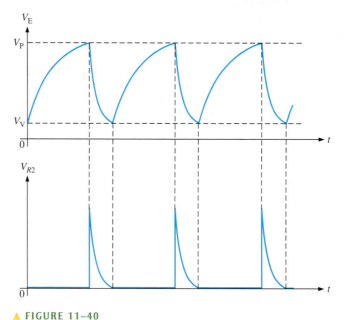

▲ FIGURE 11–40

Waveforms for UJT relaxation oscillator.

Conditions for Turn-On and Turn-Off In the relaxation oscillator of Figure 11–39, certain conditions must be met for the UJT to reliably turn on and turn off. First, to ensure turn-on, R_1 must not limit I_E at the peak point to less than I_P. To ensure this, the voltage drop across R_1 at the peak point should be greater than $I_P R_1$. Thus, the condition for turn-on is

$$V_{BB} - V_P > I_P R_1$$

or

$$R_1 < \frac{V_{BB} - V_P}{I_P}$$

To ensure turn-off of the UJT at the valley point, R_1 must be large enough that I_E (at the valley point) can decrease below the specified value of I_V. This means that the voltage across R_1 at the valley point must be less than $I_V R_1$. Thus, the condition for turn-off is

$$V_{BB} - V_V < I_V R_1$$

or

$$R_1 > \frac{V_{BB} - V_V}{I_V}$$

Therefore, for a proper turn-on and turn-off, R_1 must be in the range

$$\frac{V_{BB} - V_P}{I_P} > R_1 > \frac{V_{BB} - V_V}{I_V}$$

EXAMPLE 11–4

Determine a value of R_1 in Figure 11–41 that will ensure proper turn-on and turn-off of the UJT. The characteristic of the UJT exhibits the following values: $\eta = 0.5$, $V_V = 1$ V, $I_V = 10$ mA, $I_P = 20$ μA, and $V_P = 14$ V.

▶ **FIGURE 11–41**

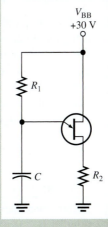

$$V_{BB}$$
$$+30 \text{ V}$$

Solution

$$\frac{V_{BB} - V_P}{I_P} > R_1 > \frac{V_{BB} - V_V}{I_V}$$

$$\frac{30 \text{ V} - 14 \text{ V}}{20 \ \mu A} > R_1 > \frac{30 \text{ V} - 1 \text{ V}}{10 \text{ mA}}$$

$$\mathbf{800 \ k\Omega > R_1 > 2.9 \ k\Omega}$$

As you can see, R_1 has quite a wide range of possible values that will work.

Related Problem

Determine a value of R_1 in Figure 11–41 that will ensure proper turn-on and turn-off for the following values: $\eta = 0.33$, $V_V = 0.8$ V, $I_V = 15$ mA, $I_P = 35$ μA, and $V_P = 18$ V.

SECTION 11–6 CHECKUP

1. Name the UJT terminals.

2. What is the intrinsic standoff ratio?

3. In a basic UJT relaxation oscillator such as in Figure 11–39, what three factors determine the period of oscillation?

11–7 THE PROGRAMMABLE UNIJUNCTION TRANSISTOR (PUT)

The programmable unijunction transistor (PUT) is actually a type of thyristor and not like the UJT at all in terms of structure. The only similarity to a UJT is that the PUT can be used in some oscillator applications to replace the UJT. The PUT is similar to an SCR except that its anode-to-gate voltage can be used to both turn on and turn off the device.

After completing this section, you should be able to

❑ **Describe the basic structure and operation of the programmable UJT**
  ◆ Identify the schematic symbol   ◆ Describe how a PUT differs from an SCR
  ◆ Compare a PUT and a UJT
❑ Explain how to set the trigger voltage
❑ Discuss a PUT application

A **PUT** (programmable unijunction transistor) is a type of three-terminal thyristor that is triggered into conduction when the voltage at the anode exceeds the voltage at the gate. The structure of the PUT is more similar to that of an SCR (four-layer) than to a UJT. The exception is that the gate is brought out as shown in Figure 11–42. Notice that the gate is connected to the *n* region adjacent to the anode. This *pn* junction controls the *on* and *off* states of the device. The gate is always biased positive with respect to the cathode. When the anode voltage exceeds the gate voltage by approximately 0.7 V, the *pn* junction is forward-biased and the PUT turns on. The PUT stays on until the anode voltage falls back below this level, then the PUT turns off.

▶ **FIGURE 11–42**

The programmable unijunction transistor (PUT).

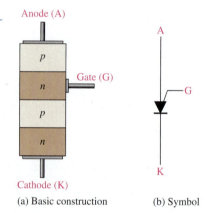

(a) Basic construction   (b) Symbol

Setting the Trigger Voltage

The gate can be biased to a desired voltage with an external voltage divider, as shown in Figure 11–43(a), so that when the anode voltage exceeds this "programmed" level, the PUT turns on.

▶ **FIGURE 11–43**

PUT biasing.

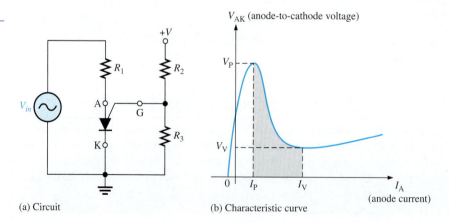

(a) Circuit   (b) Characteristic curve

An Application

A plot of the anode-to-cathode voltage, V_{AK}, versus anode current, I_A, in Figure 11–43(b) reveals a characteristic curve similar to that of the UJT. Therefore, the PUT replaces the UJT in many applications. One such application is the relaxation oscillator in Figure 11–44(a).

The basic operation of the PUT is as follows. The gate is biased at +9 V by the voltage divider consisting of resistors R_2 and R_3. When dc power is applied, the PUT is off and the capacitor charges toward +18 V through R_1. When the capacitor reaches $V_G + 0.7$ V, the PUT turns on and the capacitor rapidly discharges through the low *on* resistance of the PUT and R_4. A voltage spike is developed across R_4 during the discharge. As soon as the capacitor discharges, the PUT turns off and the charging cycle starts over, as shown by the waveforms in Figure 11–44(b).

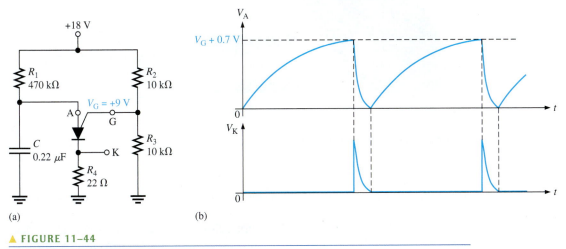

(a)   (b)

▲ **FIGURE 11–44**

PUT relaxation oscillator.

SECTION 11–7 CHECKUP

1. What does the term *programmable* mean as used in programmable unijunction transistor (PUT)?

2. Compare the structure and the operation of a PUT to those of other devices such as the UJT and SCR.

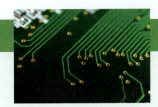

Device Application: *Motor Speed Control*

In this application, an SCR and a PUT are used to control the speed of a conveyor belt motor. The circuit controls the speed of the conveyor so that a predetermined average number of randomly spaced parts flow past a point on the production line in a specified period of time. This is to allow an adequate amount of time for the production line workers to perform certain tasks on each part. A basic diagram of the conveyor speed-control system is shown in Figure 11–45.

Each time a part on the moving conveyor belt passes the infrared (IR) detector and interrupts the IR beam, a digital counter in the processing circuits is advanced by one. The count of the passing parts is accumulated over a specified period of time and converted to

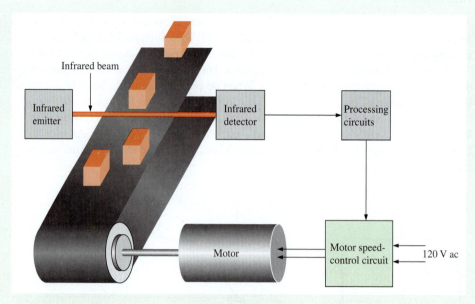

▲ FIGURE 11–45

Block diagram of conveyor speed-control system.

a proportional voltage by the processing circuits. The more parts that pass the IR detector during the specified time, the higher the voltage. The proportional voltage is applied to the motor speed-control circuit which, in turn, adjusts the speed of the electric motor that drives the conveyor belt in order to maintain the desired number of parts in a specified period of time.

The Motor Speed-Control Circuit

The proportional voltage from the processing circuits is applied to the gate of a PUT. This voltage determines the point in the ac cycle at which the SCR is triggered *on*. For a higher PUT gate voltage, the SCR turns on later in the half-cycle and therefore delivers less average power to the motor to decrease its speed. For a lower PUT gate voltage, the SCR turns on earlier in the half-cycle and delivers more average power to the motor to increase its speed. This process continually adjusts the motor speed to maintain the required number of parts per unit time moving on the conveyor. A potentiometer is used for calibration of the SCR trigger point. The motor speed-control circuit is shown in Figure 11–46.

▶ FIGURE 11–46

Motor speed-control circuit.

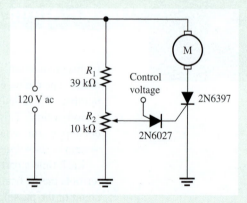

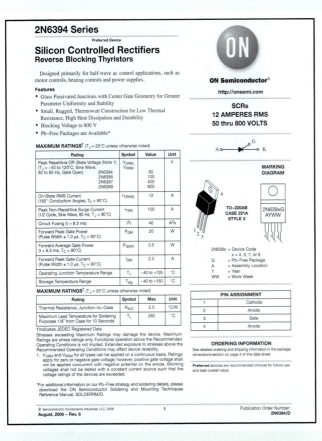

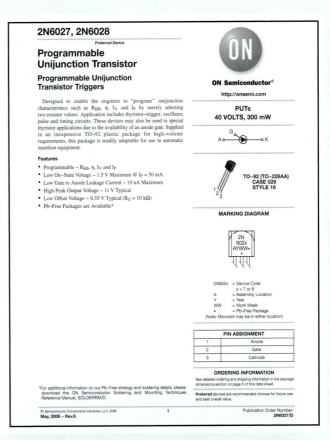

▲ FIGURE 11–47

Partial datasheets for the 2N6397 silicon-controlled rectifier and for the 2N6027 programmable unijunction transistor. Copyright of Semiconductor Component Industries, LLC. Used by permission.

The SCR used in the motor speed control is the 2N6397 *n*-channel. The partial datasheet is shown in Figure 11–47. The PUT is the 2N6027 and its partial datasheet is also shown in Figure 11–47.

Answer the following questions using the partial datasheets in Figure 11–47. If sufficient information doesn't appear on these datasheets, go to *onsemi.com* and download the complete datasheet(s).

1. How much peak voltage can the SCR withstand in the *off* state?
2. What is the maximum SCR current when it is turned on?
3. What is the maximum power dissipation of the PUT?

Simulation

The motor speed-control circuit is simulated in Multisim with a resistive/inductive load in place of the motor and a dc voltage source in place of the input from the processing circuit, as shown in Figure 11–48. The diode is placed across the motor for transient suppression.

4. On the scope display in Figure 11–48 identify when the SCR is conducting.
5. If the control voltage is reduced, will the SCR conduct more or less?
6. If the control voltage is reduced, will the motor speed increase or decrease?

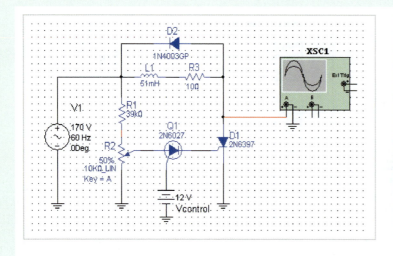

(a) Motor speed-control circuit

(b) Voltage across the SCR at V_{cont} = 12 V

▲ **FIGURE 11–48**

Simulation results for the motor speed-control circuit.

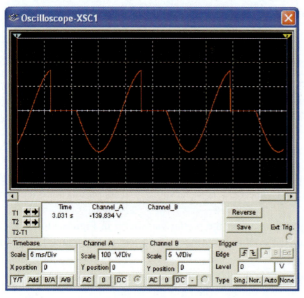

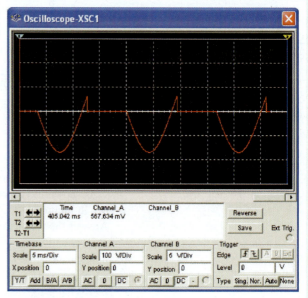

(a) Voltage across the SCR at V_{cont} = 14.2 V

(b) Voltage across the SCR at V_{cont} = 4 V

▲ **FIGURE 11–49**

SCR waveforms for two control voltages.

Figure 11–49 shows the results of varying $V_{control}$. You can see that as the control voltage is decreased, the SCR conducts for more of the cycle and therefore delivers more power to the motor to increase its speed.

Simulate the motor speed-control circuit using your Multisim or LT Spice software. Observe how the SCR voltage changes with changes in $V_{control}$.

Prototyping and Testing

Now that the circuit has been simulated, the prototype circuit is constructed and tested. After the circuit is successfully tested on a protoboard, it is ready to be finalized on a printed circuit board.

Lab Experiment

To build and test a similar circuit, go to Experiment 11 in your lab manual (*Laboratory Exercises for Electronic Devices* by David Buchla and Steven Wetterling).

Circuit Board

The motor speed-control circuit board is shown in Figure 11–50. The heat sink is for power dissipation in the SCR.

▶ **FIGURE 11–50**

Motor speed-control circuit board.

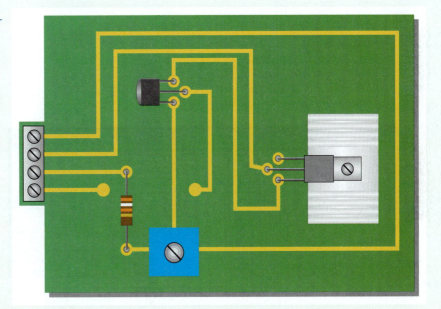

7. Check the printed circuit board for correctness by comparing with the schematic in Figure 11–46.
8. Label each input and output pin according to function.

Troubleshooting

Three circuit boards are tested, and the results are shown in Figure 11–51.

9. Determine the problem, if any, in each of the board tests in Figure 11–51.
10. List possible causes of any problem from item 9.

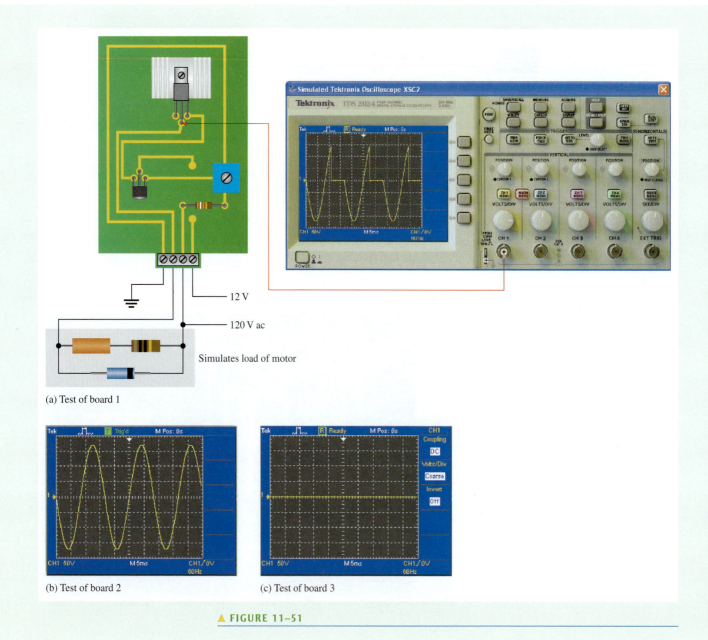

(a) Test of board 1

(b) Test of board 2

(c) Test of board 3

▲ FIGURE 11–51

SUMMARY OF THYRISTOR SYMBOLS

(a) 4-layer diode   (b) SCR   (c) LASCR   (d) Diac   (e) Triac   (f) SCS   (g) UJT   (h) PUT

SUMMARY

Section 11–1 ◆ Thyristors are devices constructed with four semiconductor layers (*pnpn*).

◆ Thyristors include four-layer diodes, SCRs, LASCRs, diacs, triacs, SCSs, and PUTs.

◆ The four-layer diode is a thyristor that conducts when the voltage across its terminals exceeds the breakover potential.

Section 11–2 ◆ The silicon-controlled rectifier (SCR) can be triggered on by a pulse at the gate and turned off by reducing the anode current below the specified holding value.

◆ Light acts as the trigger source in light-activated SCRs (LASCRs).

Section 11–3 ◆ The SCR has many applications including on/off current control, half-wave power control, back-up lighting, and over-voltage protection.

Section 11–4 ◆ The diac can conduct current in either direction and is turned on when a breakover voltage is exceeded. It turns off when the current drops below the holding value.

◆ The triac, like the diac, is a bidirectional device. It can be turned on by a pulse at the gate and conducts in a direction depending on the voltage polarity across the two anode terminals.

Section 11–5 ◆ The silicon-controlled switch (SCS) has two gate terminals and can be turned on by a pulse at the cathode gate and turned off by a pulse at the anode gate.

Section 11–6 ◆ The intrinsic standoff ratio of a unijunction transistor (UJT) determines the voltage at which the device will trigger on.

Section 11–7 ◆ The programmable unijunction transistor (PUT) can be externally programmed to turn on at a desired anode-to-gate voltage level.

KEY TERMS

Key terms and other bold terms in the chapter are defined in the end-of-book glossary.

Diac A two-terminal four-layer semiconductor device (thyristor) that can conduct current in either direction when properly activated.

Forward-breakover voltage ($V_{BR(F)}$) The voltage at which a device enters the forward-blocking region.

Four-layer diode The type of two-terminal thyristor that conducts current when the anode-to-cathode voltage reaches a specified "breakover" value.

Holding current (I_H) The value of the anode current below which a device switches from the forward-conduction region to the forward-blocking region.

LASCR Light-activated silicon-controlled rectifier; a four-layer semiconductor device (thyristor) that conducts current in one direction when activated by a sufficient amount of light and continues to conduct until the current falls below a specified value.

PUT Programmable unijunction transistor; a type of three-terminal thyristor (more like an SCR than a UJT) that is triggered into conduction when the voltage at the anode exceeds the voltage at the gate.

SCR Silicon-controlled rectifier; a type of three-terminal thyristor that conducts current when triggered on by a voltage at the single gate terminal and remains on until the anode current falls below a specified value.

SCS Silicon-controlled switch; a type of four-terminal thyristor that has two gate terminals that are used to trigger the device on and off.

Standoff ratio The characteristic of a UJT that determines its turn-on point.

Thyristor A class of four-layer (*pnpn*) semiconductor devices.

Triac A three-terminal thyristor that can conduct current in either direction when properly activated.

UJT Unijunction transistor; a three-terminal single *pn* junction device that exhibits a negative resistance characteristic.

KEY FORMULAS

11–1 $\eta = \dfrac{r'_{B1}}{r'_{BB}}$ UJT intrinsic standoff ratio

11–2 $V_P = \eta V_{BB} + V_{pn}$ UJT peak-point voltage

PROBLEMS

Answers to all odd-numbered problems are at the end of the book.

BASIC PROBLEMS

Section 11–1 **The Four-Layer Diode**

1. The four-layer diode in Figure 11–52 is biased such that it is in the forward-conduction region. Determine the anode current for $V_{BR(F)} = 20$ V, $V_{BE} = 0.7$, and $V_{CE(sat)} = 0.2$ V.

▶ FIGURE 11–52

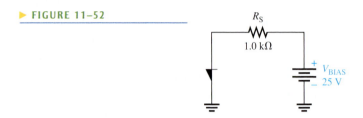

2. (a) Determine the resistance of a certain four-layer diode in the forward-blocking region if $V_{AK} = 15$ V and $I_A = 1$ μA.

   (b) If the forward-breakover voltage is 50 V, how much must V_{AK} be increased to switch the diode into the forward-conduction region?

Section 11–2 **The Silicon-Controlled Rectifier (SCR)**

3. Explain the operation of an SCR in terms of its transistor equivalent.

4. To what value must the variable resistor be adjusted in Figure 11–53 in order to turn the SCR off? Assume $I_H = 10$ mA and $V_{AK} = 0.7$ V.

▶ FIGURE 11–53

Multisim and LT Spice file circuits are identified with a logo and are in the Problems folder on the website. Filenames correspond to figure numbers (e.g., FGM11-53 or FGS11-53).

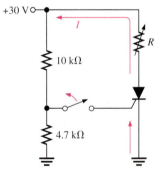

5. By examination of the circuit in Figure 11–54, explain its purpose and basic operation.

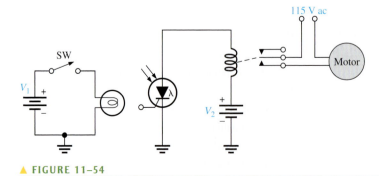

▲ FIGURE 11–54

6. Determine the voltage waveform across R_K in Figure 11–55.

▶ FIGURE 11–55

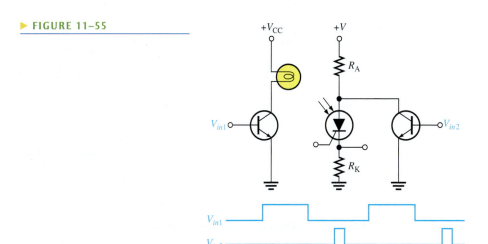

Section 11–3 SCR Applications

7. Describe how you would modify the circuit in Figure 11–16 so that the SCR triggers and conducts on the negative half-cycle of the input.

8. What is the purpose of diodes D_1 and D_2 in Figure 11–21?

9. Sketch the V_R waveform for the circuit in Figure 11–56, given the indicated relationship of the input waveforms.

▶ FIGURE 11–56

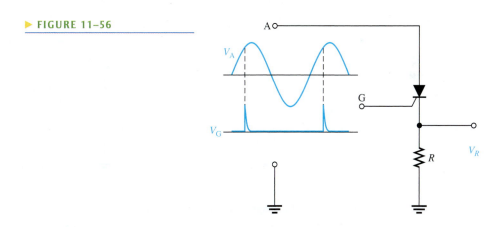

Section 11–4 The Diac and Triac

10. Sketch the current waveform for the circuit in Figure 11–57. The diac has a breakover potential of 20 V. $I_H = 20$ mA.

▶ FIGURE 11–57

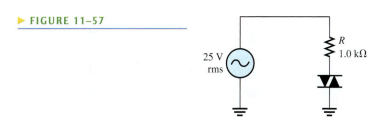

11. Repeat Problem 10 for the triac circuit in Figure 11–58. The breakover potential is 25 V and $I_H = 1$ mA.

▶ FIGURE 11–58

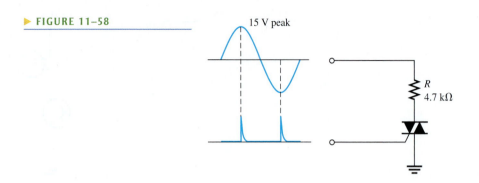

15 V peak

Section 11–5   The Silicon-Controlled Switch (SCS)

12. Explain the turn-on and turn-off operation of an SCS in terms of its transistor equivalent.

13. Name the terminals of an SCS.

Section 11–6   The Unijunction Transistor (UJT)

14. In a certain UJT, $r'_{B1} = 2.5$ kΩ and $r'_{B2} = 4$ kΩ. What is the intrinsic standoff ratio?

15. Determine the peak-point voltage for the UJT in Problem 14 if $V_{BB} = 15$ V.

16. Find the range of values of R_1 in Figure 11–59 that will ensure proper turn-on and turn-off of the UJT. $\eta = 0.68$, $V_V = 0.8$ V, $I_V = 15$ mA, $I_p = 10$ μA, and $V_P = 10$ V.

▶ FIGURE 11–59

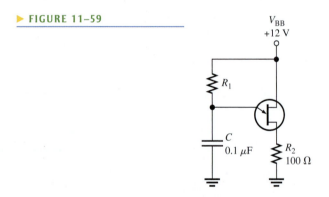

V_{BB}
+12 V

R_1

C
0.1 μF

R_2
100 Ω

Section 11–7   The Programmable Unijunction Transistor (PUT)

17. At what anode voltage (V_A) will each PUT in Figure 11–60 begin to conduct?

▶ FIGURE 11–60

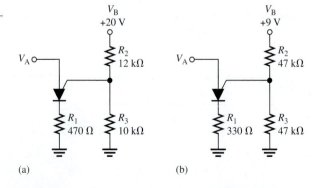

(a)   (b)

18. Draw the current waveform for each circuit in Figure 11–60 when there is a 10 V peak sinusoidal voltage at the anode. Neglect the forward voltage of the PUT.

19. Sketch the voltage waveform across R_1 in Figure 11–61 in relation to the input voltage waveform.

▶ FIGURE 11–61

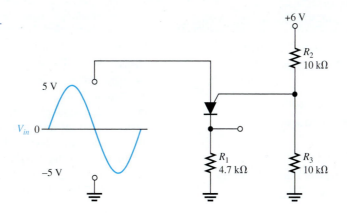

20. Repeat Problem 19 if R_3 is increased to 15 kΩ.

DEVICE APPLICATION PROBLEMS

21. In the motor speed-control circuit of Figure 11–46, at which PUT gate voltage does the electric motor run at the fastest speed: 0 V, 2 V, or 5 V?

22. Does the SCR in the motor speed-control circuit turn on earlier or later in the ac cycle if the resistance of the rheostat is reduced?

23. Describe the SCR action as the PUT gate voltage is increased in the motor speed-control circuit.

ADVANCED PROBLEMS

24. Refer to the SCR over-voltage protection circuit in Figure 11–22. For a +12 V output dc power supply, specify the component values that will provide protection for the circuit if the output voltage exceeds +15 V. Assume the fuse is rated at 1 A.

25. Design an SCR crowbar circuit to protect electronic circuits against a voltage from the power supply in excess of 6.2 V.

26. Design a relaxation oscillator to produce a frequency of 2.5 kHz using a UJT with $\eta = 0.75$ and a valley voltage of 1 V. The circuit must operate from a +12 V dc source. Design values of $I_V = 10$ mA and $I_P = 20$ μA are to be used.

MULTISIM TROUBLESHOOTING PROBLEMS

These file circuits are in the Troubleshooting Problems folder on the website.

27. Open file TPM11-27 and determine the fault.

28. Open file TPM11-28 and determine the fault.

29. Open file TPM11-29 and determine the fault.

12

THE OPERATIONAL AMPLIFIER

CHAPTER OBJECTIVES

◆ Describe the basic operational amplifier and its characteristics

◆ Discuss op-amp modes and several parameters

◆ Explain negative feedback in op-amps

◆ Analyze op-amps with negative feedback

◆ Describe how negative feedback affects op-amp impedances

◆ Discuss bias current and offset voltage

◆ Analyze the open-loop frequency response of an op-amp

◆ Analyze the closed-loop frequency response of an op-amp

◆ Troubleshoot op-amp circuits

KEY TERMS

◆ Operational amplifier (op-amp)
◆ Differential amplifier
◆ Differential mode
◆ Open-loop voltage gain
◆ Slew rate
◆ Negative feedback
◆ Closed-loop voltage gain
◆ Noninverting amplifier
◆ Voltage-follower
◆ Inverting amplifier
◆ Phase shift
◆ Gain-bandwidth product

DEVICE APPLICATION PREVIEW

For the Device Application in this chapter, the audio amplifier from the PA system in Chapter 7 is modified. The two-stage preamp portion of the amplifier is replaced by an op-amp circuit. The power amplifier portion is retained in its original configuration with the exception of the drive circuit so that the new design consists of an op-amp driving a push-pull power stage. In the original system there are two PC boards—one for the preamp and one for the power amplifier. The new design will allow both the preamp and the power amplifier to be on a single PC board.

VISIT THE WEBSITE

Study aids, Multisim files, and LT Spice files for this chapter are available at https://www.pearsonhighered.com /careersresources/

INTRODUCTION

In the previous chapters, you have studied a number of important electronic devices. These devices, such as the diode and the transistor, are separate devices that are individually packaged and interconnected in a circuit with other devices to form a complete, functional unit. Such devices are referred to as *discrete components*.

Now you will begin the study of linear integrated circuits (ICs), where many transistors, diodes, resistors, and capacitors are fabricated on a single tiny chip of semiconductive material and packaged in a single case to form a functional circuit. An integrated circuit, such as an operational amplifier (op-amp), is treated as a single device. This means that you will be concerned with what the circuit does more from an external viewpoint than from an internal, component-level viewpoint.

In this chapter, you will learn the basics of op-amps, which are the most versatile and widely used of all linear integrated circuits. You will also learn about open-loop and closed-loop frequency responses, bandwidth, phase shift, and other frequency-related parameters. The effects of negative feedback will be examined.

12–1 INTRODUCTION TO OPERATIONAL AMPLIFIERS

Early operational amplifiers (op-amps) were used primarily to perform mathematical operations such as addition, subtraction, integration, and differentiation—thus the term *operational*. These early devices were constructed with vacuum tubes and worked with high voltages. Today's op-amps are linear integrated circuits (ICs) that use relatively low dc supply voltages and are reliable and inexpensive.

After completing this section, you should be able to

❏ **Describe the basic operational amplifier and its characteristics**
  ◆ Identify the schematic symbol and IC package terminals
❏ Discuss the ideal op-amp
❏ Discuss the practical op-amp
  ◆ Draw the internal block diagram

The standard **operational amplifier (op-amp)** symbol is shown in Figure 12–1(a). It has two input terminals, the inverting (−) input and the noninverting (+) input, and one output terminal. Most op-amps operate with two dc supply voltages, one positive and the other negative, as shown in Figure 12–1(b), although some have a single dc supply. Usually these dc voltage terminals are left off the schematic symbol for simplicity but are understood to be there. Some typical op-amp IC packages are shown in Figure 12–1(c).

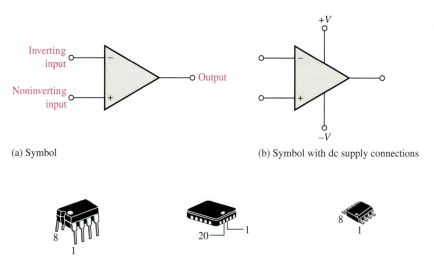

(a) Symbol

(b) Symbol with dc supply connections

(c) Typical packages. Pin 1 is indicated by a notch or dot on dual in-line (DIP) and surface-mount technology (SMT) packages, as shown.

◀ FIGURE 12–1

Op-amp symbols and packages.

The Ideal Op-Amp

To illustrate what an op-amp is, let's consider its ideal characteristics. A practical op-amp, of course, falls short of these ideal standards, but it is much easier to understand and analyze the device from an ideal point of view.

First, the ideal op-amp has *infinite voltage gain* and *infinite bandwidth*. Also, it has an *infinite input impedance* (open) so that it does not load the driving source. Finally, it has a *zero output impedance*. Op-amp characteristics are illustrated in Figure 12–2(a). The input voltage, V_{in}, appears between the two input terminals, and the output voltage is $A_v V_{in}$, as indicated by the internal voltage source symbol. The concept of infinite input impedance is a particularly valuable analysis tool for the various op-amp configurations, which will be discussed in Section 12–4.

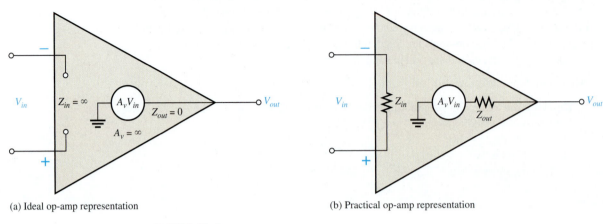

(a) Ideal op-amp representation

(b) Practical op-amp representation

▲ FIGURE 12–2

Basic op-amp representations.

The Practical Op-Amp

Although **integrated circuit (IC)** op-amps approach parameter values that can be treated as ideal in many cases, the ideal device can never be made. Any device has limitations, and the IC op-amp is no exception. Op-amps have both voltage and current limitations. Peak-to-peak output voltage, for example, is usually limited to slightly less than the two supply voltages. Output current is also limited by internal restrictions such as power dissipation and component ratings.

Characteristics of a practical op-amp are *very high voltage gain, very high input impedance, and very low output impedance*. These are labelled in Figure 12–2(b). Another practical consideration is that there is always noise generated within the op-amp. **Noise** is an undesired signal that affects the quality of a desired signal. Today, circuit designers are using smaller voltages that require high accuracy, so low-noise components are in greater demand. All circuits generate noise; op-amps are no exception, but the amount can be minimized.

Internal Block Diagram of an Op-Amp   A typical op-amp is made up of three types of amplifier circuits: a differential amplifier, a voltage amplifier, and a push-pull amplifier, as shown in Figure 12–3. The **differential amplifier** is the input stage for the op-amp. It provides amplification of the difference voltage between the two inputs. The second stage is usually a class A amplifier that provides additional gain. Some op-amps may have more than one voltage amplifier stage. A push-pull class B amplifier is typically used for the output stage.

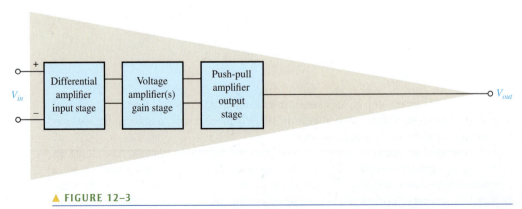

▲ FIGURE 12–3

Basic internal arrangement of an op-amp.

The differential amplifier was introduced in Chapter 6. The term *differential* comes from the amplifier's ability to amplify the difference of two input signals applied to its inputs. Only the difference in the two signals is amplified; if there is no difference, the output is zero. The differential amplifier exhibits two modes of operation based on the type of input signals. These modes are *differential* and *common,* which are described in the next section. Since the differential amplifier is the input stage of the op-amp, the op-amp exhibits the same modes.

SECTION 12–1 CHECKUP Answers can be found at www .pearsonhighered.com/floyd.	1. What are the connections to a basic op-amp? 2. Describe some of the characteristics of a practical op-amp. 3. List the amplifier stages in a typical op-amp. 4. What does a differential amplifier amplify?

12–2 OP-AMP INPUT MODES AND PARAMETERS

In this section, important op-amp input modes and several parameters are defined. Also several common IC op-amps are compared in terms of these parameters.

After completing this section, you should be able to

❑ **Discuss op-amp modes and several parameters**
  ◆ Identify the schematic symbol and IC package terminals
❑ Describe the input signal modes
  ◆ Explain the differential mode ◆ Explain the common mode
❑ Define and discuss op-amp parameters
  ◆ Define *common-mode rejection ratio (CMRR)* ◆ Calculate the CMRR
  ◆ Express the CMRR in decibels ◆ Define open-loop voltage gain
  ◆ Explain maximum output voltage swing ◆ Explain input offset voltage
  ◆ Explain input bias current ◆ Explain input impedance ◆ Explain input offset current ◆ Explain output impedance ◆ Explain slew rate
  ◆ Explain frequency response
❑ Compare op-amp parameters for several devices

Input Signal Modes

Recall that the input signal modes are determined by the differential amplifier input stage of the op-amp.

Differential Mode In the differential mode, either one signal is applied to an input with the other input grounded or two opposite-polarity signals are applied to the inputs. When an op-amp is operated in the single-ended differential mode, one input is grounded and a signal voltage is applied to the other input, as shown in Figure 12–4. In the case where the signal voltage is applied to the inverting input as in part (a), an inverted, amplified signal voltage appears at the output. In the case where the signal is applied to the noninverting input with the inverting input grounded, as in Figure 12–4(b), a noninverted, amplified signal voltage appears at the output.

▶ FIGURE 12–4

► **FIGURE 12–4**

Single-ended differential mode.

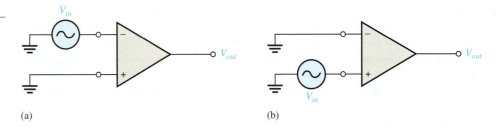

(a)   (b)

In the double-ended differential mode, two opposite-polarity (out-of-phase) signals are applied to the inputs, as shown in Figure 12–5(a). The amplified difference between the two inputs appears on the output. Equivalently, the double-ended differential mode can be represented by a single source connected between the two inputs, as shown in Figure 12–5(b).

► **FIGURE 12–5**

Double-ended differential mode.

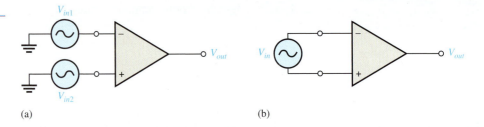

(a)   (b)

Common Mode   Recall that common mode and CMRR are terms that were introduced in Section 6–7 in connection with differential amplifiers. Because the front end of an op-amp is a differential amplifier, common mode and CMRR are important terms with op-amps and are reviewed here.

In the common mode, two signal voltages of the same phase, frequency, and amplitude are applied to the two inputs, as shown in Figure 12–6. When equal input signals are applied to both inputs, they tend to cancel, resulting in a zero output voltage.

► **FIGURE 12–6**

Common-mode operation.

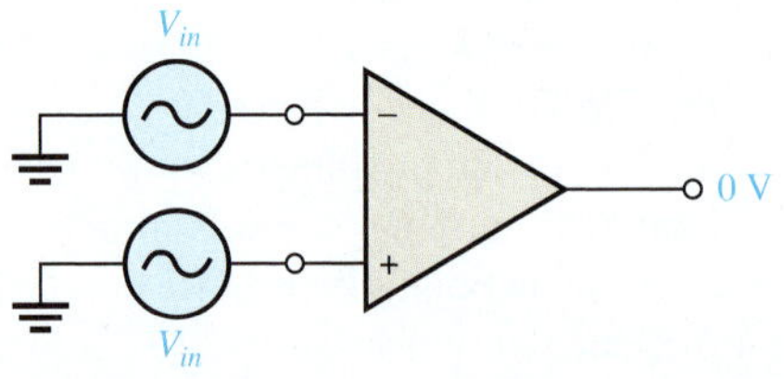

This action is called *common-mode rejection*. Its importance lies in the situation where an unwanted signal appears commonly on both op-amp inputs. Common-mode rejection means that this unwanted signal will not appear on the output and distort the desired signal. Common-mode signals (noise) generally are the result of the pick-up of radiated energy on the input lines, from adjacent lines, the 60 Hz power line, or other sources.

Op-Amp Parameters

Common-Mode Rejection Ratio   Desired signals can appear on only one input or with opposite polarities on both input lines. These desired signals are amplified and appear on the output as previously discussed. Unwanted signals (noise) appearing with the same polarity on both input lines are essentially cancelled by the op-amp and do not appear on the output. The measure of an amplifier's ability to reject common-mode signals is a parameter called the **CMRR (common-mode rejection ratio)**.

Ideally, an op-amp provides a very high gain for differential-mode signals and zero gain for common-mode signals. Practical op-amps, however, do exhibit a very small common-mode gain (usually much less than 1), while providing a high open-loop differential voltage gain (commonly from 100,000 to 1,000,000 or more for high-precision op-amps). The **open-loop voltage gain**, A_{ol}, of an op-amp is the internal voltage gain of the device and represents the ratio of output voltage to input voltage when there are no external components. The higher the open-loop gain with respect to the common-mode gain, the better the performance of the op-amp in terms of rejection of common-mode signals. This suggests that a good measure of the op-amp's performance in rejecting unwanted common-mode signals is the ratio of the open-loop differential voltage gain, A_{ol}, to the common-mode gain, A_{cm}. This ratio is the common-mode rejection ratio, CMRR.

$$\text{CMRR} = \frac{A_{ol}}{A_{cm}}$$

Equation 12–1

The higher the CMRR, the better. A very high value of CMRR means that the open-loop gain, A_{ol}, is high and the common-mode gain, A_{cm}, is low.

The CMRR is often expressed in decibels (dB) as

$$\text{CMRR} = 20 \log\left(\frac{A_{ol}}{A_{cm}}\right)$$

Equation 12–2

The open-loop voltage gain is set entirely by the internal design. Open-loop voltage gain can range up to 1,000,000,000 or more (120 dB) and is not a well-controlled parameter. Generally, a very high open-loop gain is better, but some very fast op-amps have values that are lower (a few thousand). Datasheets often refer to the open-loop voltage gain as the *large-signal voltage gain*. Even though open-loop gain is dimensionless, datasheets will often show it as V/mV or V/μV to express the very large values. Thus a gain of 200,000 can be expressed as 200 V/mV.

A CMRR of 100,000, for example, means that the desired input signal (differential) is amplified 100,000 times more than the unwanted noise (common-mode). If the amplitudes of the differential input signal and the common-mode noise are equal, the desired signal will appear on the output 100,000 times greater in amplitude than the noise. Thus, the noise or interference has been essentially eliminated.

CMRR is dependent on the frequency of the common-mode signal; as the frequency of the common-mode signal goes up, the CMRR is degraded. Manufacturers will publish a graph of the CMRR as a function of frequency. Figure 12–7 shows the response of CMRR as a function of the common-mode frequency for a high-quality op-amp. As you can see, the rejection is much better at very low frequencies.

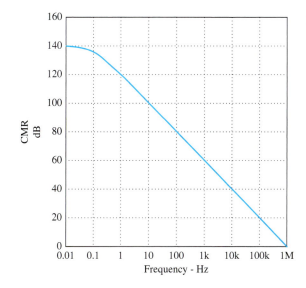

◄ **FIGURE 12–7**

CMR as a function of frequency.

After completing this section, you should be able to

❑ **Explain negative feedback in op-amps**
❑ Discuss why negative feedback is used
  ◆ Describe the effects of negative feedback on certain op-amp parameters

Negative feedback is illustrated in Figure 12–15. The inverting (−) input effectively makes the feedback signal 180° out of phase with the input signal.

▶ **FIGURE 12–15**

Illustration of negative feedback.

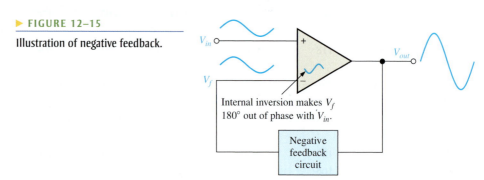

Internal inversion makes V_f 180° out of phase with V_{in}.

Negative feedback circuit

Why Use Negative Feedback?

As you can see in Table 12–1, the inherent open-loop voltage gain of a typical op-amp is very high (usually greater than 100,000). Therefore, an extremely small input voltage drives the op-amp into its saturated output states. In fact, even the input offset voltage of the op-amp can drive it into saturation. For example, assume $V_{IN} = 1$ mV and $A_{ol} = 100,000$. Then,

$$V_{IN} A_{ol} = (1 \text{ mV})(100,000) = 100 \text{ V}$$

Since the output level of an op-amp can never reach 100 V, it is driven deep into saturation and the output is limited to its maximum output levels, as illustrated in Figure 12–16 for both a positive and a negative input voltage of 1 mV.

▶ **FIGURE 12–16**

Without negative feedback, a small input voltage drives the op-amp to its output limits and it becomes nonlinear.

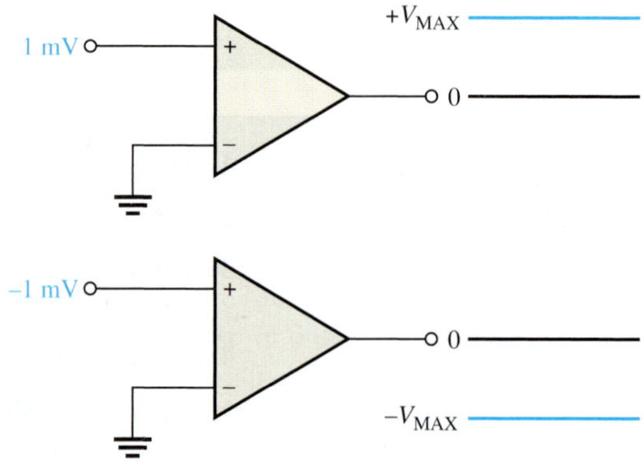

1 mV

$+V_{MAX}$

0

-1 mV

0

$-V_{MAX}$

The usefulness of an op-amp operated without negative feedback is generally limited to comparator applications (to be studied in Chapter 13). With negative feedback, the closed-loop voltage gain (A_{cl}) can be reduced and controlled so that the op-amp can function as a linear amplifier. In addition to providing a controlled, stable voltage gain, negative feedback also provides for control of the input and output impedances and amplifier bandwidth. Table 12–2 summarizes the general effects of negative feedback on op-amp performance.

▼ TABLE 12–2

	VOLTAGE GAIN	INPUT Z	OUTPUT Z	BANDWIDTH
Without negative feedback	A_{ol} is too high for linear amplifier applications	Relatively high (see Table 12–1)	Relatively low	Relatively narrow (because the gain is so high)
With negative feedback	A_{cl} is set to desired value by the feedback circuit	Can be increased or reduced to a desired value depending on type of circuit	Can be reduced to a desired value	Significantly wider

SECTION 12–3 CHECKUP

1. What are the benefits of negative feedback in an op-amp circuit?
2. Why is it generally necessary to reduce the gain of an op-amp from its open-loop value?

12–4 OP-AMPS WITH NEGATIVE FEEDBACK

An op-amp can be connected using negative feedback to stabilize the gain and increase frequency response. Negative feedback takes a portion of the output and applies it back out of phase with the input, creating an effective reduction in gain. This closed-loop gain is usually much less than the open-loop gain and independent of it.

After completing this section, you should be able to

❑ **Analyze op-amps with negative feedback**
❑ Discuss closed-loop voltage gain
❑ Identify and analyze the noninverting op-amp configuration
❑ Identify and analyze the voltage-follower configuration
❑ Identify and analyze the inverting amplifier configuration

Closed-Loop Voltage Gain, A_{cl}

The closed-loop voltage gain is the voltage gain of an op-amp with external feedback. The amplifier configuration consists of the op-amp and an external negative feedback circuit that connects the output to the inverting input. The closed-loop voltage gain is determined by the external component values and can be precisely controlled by them.

Noninverting Amplifier

An op-amp connected in a **closed-loop** configuration as a noninverting amplifier with a controlled amount of voltage gain is shown in Figure 12–17. The input signal is applied to the noninverting (+) input. The output is applied back to the inverting (−) input through the feedback circuit (closed loop) formed by the input resistor R_i and the feedback resistor R_f. This creates negative feedback as follows. Resistors R_i and R_f form a voltage-divider circuit, which reduces V_{out} and connects the reduced voltage V_f to the inverting input. The feedback voltage is expressed as

$$V_f = \left(\frac{R_i}{R_i + R_f}\right)V_{out}$$

Thus,

Equation 12–12

$$Z_{out(NI)} = \frac{Z_{out}}{1 + A_{ol}B}$$

This equation shows that the output impedance of the noninverting amplifier configuration with negative feedback is much less than the internal output impedance, Z_{out}, of the op-amp itself (without feedback) because Z_{out} is divided by the factor $1 + A_{ol}B$. In practical cases, the output impedance can be assumed to be zero as the next example illustrates.

EXAMPLE 12–5

(a) Determine the input and output impedances of the amplifier in Figure 12–26. The op-amp datasheet gives $Z_{in} = 2\ M\Omega$, $Z_{out} = 75\ \Omega$, and $A_{ol} = 200{,}000$.

(b) Find the closed-loop voltage gain.

▶ **FIGURE 12–26**

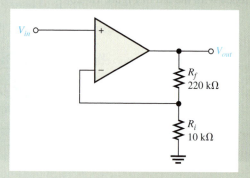

Solution

(a) The attenuation, B, of the feedback circuit is

$$B = \frac{R_i}{R_i + R_f} = \frac{10\ k\Omega}{230\ k\Omega} = 0.0435$$

$$Z_{in(NI)} = (1 + A_{ol}B)Z_{in} = [1 + (200{,}000)(0.0435)](2\ M\Omega)$$
$$= (1 + 8700)(2\ M\Omega) = \mathbf{17.4\ G\Omega}$$

This is such a large number that, for all practical purposes, it can be assumed to be infinite as in the ideal case.

$$Z_{out(NI)} = \frac{Z_{out}}{1 + A_{ol}B} = \frac{75\ \Omega}{1 + 8700} = \mathbf{8.6\ m\Omega}$$

This is such a small number that, for all practical purposes, it can be assumed to be zero as in the ideal case.

(b) $A_{cl(NI)} = 1 + \dfrac{R_f}{R_i} = 1 + \dfrac{220\ k\Omega}{10\ k\Omega} = \mathbf{23.0}$

Related Problem

(a) Determine the input and output impedances in Figure 12–26 for op-amp datasheet values of $Z_{in} = 3.5\ M\Omega$, $Z_{out} = 82\ \Omega$, and $A_{ol} = 135{,}000$.

(b) Find A_{cl}.

Open the Multisim file EXM12-05 or LT Spice file EXS12-05 in the Examples folder on the website. Measure the closed-loop voltage gain and compare with the calculated value.

Voltage-Follower Impedances

Since a voltage-follower is a special case of the noninverting amplifier configuration, the same impedance formulas are used but with $B = 1$.

$$Z_{in(VF)} = (1 + A_{ol})Z_{in}$$

Equation 12–13

$$Z_{out(VF)} = \frac{Z_{out}}{1 + A_{ol}}$$

Equation 12–14

As you can see, the voltage-follower input impedance is greater for a given A_{ol} and Z_{in} than for the noninverting amplifier configuration with the voltage-divider feedback circuit. Also, its output impedance is much smaller.

EXAMPLE 12–6

The op-amp in Example 12–5 is used in a voltage-follower configuration. Determine the input and output impedances.

Solution Since $B = 1$,

$$Z_{in(VF)} = (1 + A_{ol})Z_{in} = (1 + 200{,}000)(2\ \text{M}\Omega) \cong \mathbf{400\ G\Omega}$$

$$Z_{out(VF)} = \frac{Z_{out}}{1 + A_{ol}} = \frac{75\ \Omega}{1 + 200{,}000} = \mathbf{375\ \mu\Omega}$$

Notice that $Z_{in(VF)}$ is much greater than $Z_{in(NI)}$, and $Z_{out(VF)}$ is much less than $Z_{out(NI)}$ from Example 12–5. Again for all practical purposes, the ideal values can be assumed.

Related Problem If the op-amp in this example is replaced with one having a higher open-loop gain, how are the input and output impedances affected?

Impedances of the Inverting Amplifier

The input and output impedances of an inverting op-amp configuration are developed with the aid of Figure 12–27. Both the input signal and the negative feedback are applied, through resistors, to the inverting ($-$) terminal as shown.

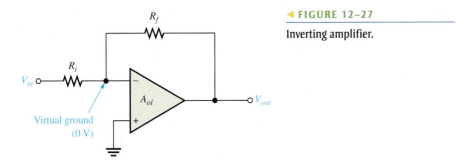

◀ **FIGURE 12–27**

Inverting amplifier.

Input Impedance The input impedance for an inverting amplifier is

$$Z_{in(I)} \cong R_i$$

Equation 12–15

This is because the inverting input of the op-amp is at virtual ground (0 V), and the input source simply sees R_i to ground, as shown in Figure 12–28.

Output Impedance As with a noninverting amplifier, the output impedance of an inverting amplifier is decreased by the negative feedback. In fact, the expression is the same as for the noninverting case.

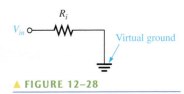

▲ **FIGURE 12–28**

Equation 12–16

$$Z_{out(I)} = \frac{Z_{out}}{1 + A_{ol}B}$$

Notice that Equation 12–16 for the inverting amplifier is the same as Equation 12–12 for the noninverting amplifier. The output impedance of both configurations is very low; in fact, it is almost zero in practical cases where A_{ol} is very large. Because of this near zero output impedance, any load impedance within limits can be connected to the op-amp output and not change the gain or the output voltage. The limits for the load impedance are determined by the maximum peak-to-peak swing of the output ($V_{O(p\text{-}p)}$) and the current limit of the op-amp.

EXAMPLE 12–7

Find the values of the input and output impedances in Figure 12–29. Also, determine the closed-loop voltage gain. The op-amp has the following parameters: $A_{ol} = 50,000$; $Z_{in} = 4$ MΩ; and $Z_{out} = 50$ Ω.

▶ **FIGURE 12–29**

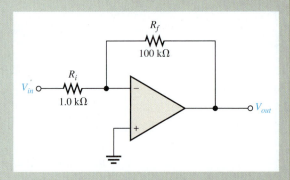

Solution

$$Z_{in(I)} \cong R_i = \mathbf{1.0\ k\Omega}$$

The feedback attenuation, B, is

$$B = \frac{R_i}{R_i + R_f} = \frac{1.0\ k\Omega}{101\ k\Omega} = 0.001$$

Then

$$Z_{out(I)} = \frac{Z_{out}}{1 + A_{ol}B} = \frac{50\ \Omega}{1 + (50,000)(0.001)}$$

$$= \mathbf{980\ m\Omega}\ (\text{zero for all practical purposes})$$

The closed-loop voltage gain is

$$A_{cl(I)} = -\frac{R_f}{R_i} = -\frac{100\ k\Omega}{1.0\ k\Omega} = \mathbf{-100}$$

Related Problem

Determine the input and output impedances and the closed-loop voltage gain in Figure 12–29. The op-amp parameters and circuit values are as follows: $A_{ol} = 100,000$; $Z_{in} = 5$ MΩ; $Z_{out} = 75$ Ω; $R_i = 560$ Ω; and $R_f = 82$ kΩ.

Open the Multisim file EXM12-07 or the LT Spice file EXS12-05 in the Examples folder on the website and measure the closed-loop voltage gain. Compare to the calculated result.

12–6 BIAS CURRENT AND OFFSET VOLTAGE

Certain deviations from the ideal op-amp must be recognized because of their effects on its operation. Transistors within the op-amp must be biased so that they have the correct values of base and collector currents and collector-to-emitter voltages. The ideal op-amp has no input current at its terminals; but in fact, the practical op-amp has small input bias currents typically in the nA range. Also, small internal imbalances in the transistors effectively produce a small offset voltage between the inputs. These nonideal parameters were described in Section 12–2.

After completing this section, you should be able to

◻ **Discuss bias current and offset voltage**
◻ Describe the effect of input bias current
◻ Discuss bias current compensation
  ◆ Explain bias current compensation in the voltage-follower   ◆ Explain bias current compensation in the noninverting and inverting amplifiers   ◆ Discuss the use of a BIFET
◻ Describe the effect of input offset voltage
◻ Discuss input offset voltage compensation

Effect of Input Bias Current

Figure 12–30(a) is an inverting amplifier with zero input voltage. Ideally, the current through R_i is zero because the input voltage is zero and the voltage at the inverting (−) terminal is zero. The small input bias current, I_1, is through R_f to the output terminal. I_1 creates a voltage across R_f, as indicated. The positive side of R_f is the output terminal, and therefore, the output error voltage is I_1R_f when it should be zero.

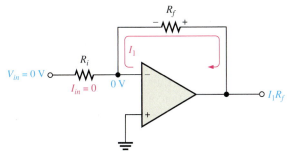

(a) Input bias current creates output error voltage (I_1R_f) in an inverting amplifier.

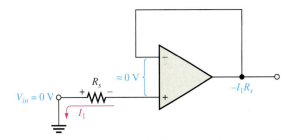

(b) Input bias current creates output error voltage in a voltage-follower.

▲ FIGURE 12–30

Effects of bias currents.

Figure 12–30(b) is a voltage-follower with zero input voltage and a source resistance, R_s. In this case, an input bias current, I_1, produces a voltage across R_s and creates an output voltage error as shown. The voltage at the inverting input terminal decreases to $-I_1R_s$ because the negative feedback tends to maintain a differential voltage of zero, as indicated. Since the inverting terminal is connected directly to the output terminal, the output error voltage is $-I_1R_s$.

Figure 12–31 is a noninverting amplifier with zero input voltage. Ideally, the voltage at the inverting terminal is also zero, as indicated. The input bias current, I_1, produces a voltage across R_f and thus creates an output error voltage of I_1R_f, just as with the inverting amplifier.

▶ **FIGURE 12–31**

Input bias current creates output error voltage in a noninverting amplifier.

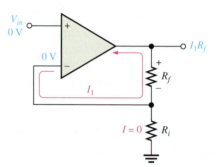

Bias Current Compensation

Voltage-Follower The output error voltage due to bias currents in a voltage-follower can be sufficiently reduced by adding a resistor, R_f, equal to the source resistance, R_s, in the feedback path, as shown in Figure 12–32. The voltage created by I_1 across the added resistor subtracts from the $-I_2R_s$ output error voltage. If $I_1 = I_2$, then the output voltage is zero. Usually I_1 does not quite equal I_2; but even in this case, the output error voltage is reduced as follows because I_{OS} is less than I_2.

$$V_{OUT(error)} = |I_1 - I_2|R_s = I_{OS}R_s$$

where I_{OS} is the input offset current.

▶ **FIGURE 12–32**

Bias current compensation in a voltage-follower.

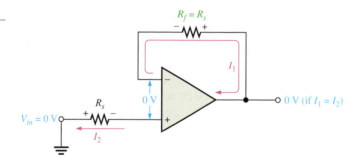

Noninverting and Inverting Amplifiers To compensate for the effect of bias current in the noninverting amplifier, a resistor R_c is added, as shown in Figure 12–33(a). The compensating resistor value equals the parallel combination of R_i and R_f. The input current creates a voltage drop across R_c that offsets the voltage across the combination of R_i and R_f, thus sufficiently reducing the output error voltage. The inverting amplifier is similarly compensated, as shown in Figure 12–33(b).

Use of a BIFET Op-Amp to Eliminate the Need for Bias Current Compensation The BIFET op-amp uses both BJTs and JFETs in its internal circuitry. The JFETs are used as the input devices to achieve a higher input impedance than is possible with standard BJT amplifiers. Because of their very high input impedance, BIFETs typically have input bias

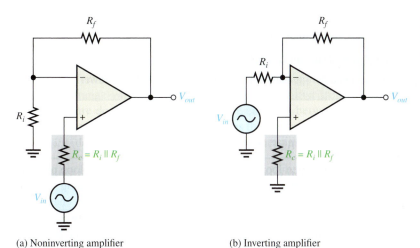

◀ FIGURE 12–33

Bias current compensation in the noninverting and inverting amplifier configurations.

(a) Noninverting amplifier

(b) Inverting amplifier

currents that are much smaller than in BJT op-amps, thus reducing or eliminating the need for bias current compensation.

Effect of Input Offset Voltage

The output voltage of an op-amp should be zero when the differential input is zero. However, there is always a small output error voltage present whose value typically ranges from microvolts to millivolts. This is due to unavoidable imbalances within the internal op-amp transistors aside from the bias currents previously discussed. In a negative feedback configuration, the input offset voltage V_{IO} can be visualized as an equivalent small dc voltage source, as illustrated in Figure 12–34 for a voltage-follower. Generally, the output error voltage due to the input offset voltage is

$$V_{OUT(error)} = A_{cl}V_{IO}$$

For the case of the voltage-follower, $A_{cl} = 1$, so

$$V_{OUT(error)} = V_{IO}$$

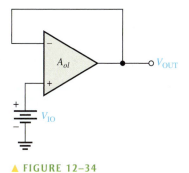

▲ FIGURE 12–34

Input offset voltage equivalent.

Input Offset Voltage Compensation

Most integrated circuit op-amps provide a means of compensating for offset voltage. This is usually done by connecting an external potentiometer to designated pins on the IC package, as illustrated in Figure 12–35(a) and (b) for a LM741 op-amp. The two terminals are labeled *offset null*. With no input, the potentiometer is simply adjusted until the output voltage reads 0, as shown in Figure 12–35(c).

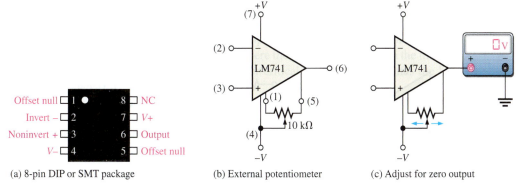

(a) 8-pin DIP or SMT package

(b) External potentiometer

(c) Adjust for zero output

▲ FIGURE 12–35

Input offset voltage compensation for an LM741 op-amp.

Unity-Gain Bandwidth Notice in Figure 12–37 that the gain steadily decreases to a point where it is equal to unity (1 or 0 dB). The value of the frequency at which this unity gain occurs is the *unity-gain frequency* designated f_T. f_T is also called the *unity-gain bandwidth*.

Gain-Versus-Frequency Analysis

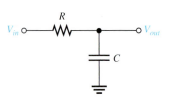

▲ FIGURE 12–38

RC lag circuit.

The *RC* lag (low-pass) circuits within an op-amp are responsible for the roll-off in gain as the frequency increases, just as was discussed for the discrete amplifiers in Chapter 10. From basic ac circuit theory, the attenuation of an *RC* lag circuit, such as in Figure 12–38, is expressed as

$$\frac{V_{out}}{V_{in}} = \frac{X_C}{\sqrt{R^2 + X_C^2}}$$

Dividing both the numerator and denominator to the right of the equals sign by X_C,

$$\frac{V_{out}}{V_{in}} = \frac{1}{\sqrt{1 + R^2/X_C^2}}$$

The critical frequency of an *RC* circuit is

$$f_c = \frac{1}{2\pi RC}$$

Dividing both sides by f gives

$$\frac{f_c}{f} = \frac{1}{2\pi RCf} = \frac{1}{(2\pi fC)R}$$

Since $X_C = 1/(2\pi fC)$, the previous expression can be written as

$$\frac{f_c}{f} = \frac{X_C}{R}$$

Substituting this result in the previous equation for V_{out}/V_{in} produces the following expression for the attenuation of an *RC* lag circuit in terms of frequency:

Equation 12–18

$$\frac{V_{out}}{V_{in}} = \frac{1}{\sqrt{1 + f^2/f_c^2}}$$

If an op-amp is represented by a voltage gain element with a gain of $A_{ol(mid)}$ plus a single *RC* lag circuit, as shown in Figure 12–39, it is known as a compensated op-amp. The total open-loop gain of the op-amp is the product of the midrange open-loop gain, $A_{ol(mid)}$, and the attenuation of the *RC* circuit.

▶ FIGURE 12–39

Op-amp represented by a gain element and an internal *RC* circuit.

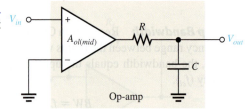

Equation 12–19

$$A_{ol} = \frac{A_{ol(mid)}}{\sqrt{1 + f^2/f_c^2}}$$

As you can see from Equation 12–19, the open-loop gain equals the midrange gain when the signal frequency f is much less than the critical frequency f_c and drops off as the frequency increases. Since f_c is part of the open-loop response of an op-amp, we will refer to it as $f_{c(ol)}$.

The following example demonstrates how the open-loop gain decreases as the frequency increases above $f_{c(ol)}$.

EXAMPLE 12–8

Determine A_{ol} for the following values of f. Assume $f_{c(ol)} = 100$ Hz and $A_{ol(mid)} = 100{,}000$.

(a) $f = 0$ Hz   **(b)** $f = 10$ Hz   **(c)** $f = 100$ Hz   **(d)** $f = 1000$ Hz

Solution

(a) $A_{ol} = \dfrac{A_{ol(mid)}}{\sqrt{1 + f^2/f_{c(ol)}^2}} = \dfrac{100{,}000}{\sqrt{1 + 0}} = \mathbf{100{,}000}$

(b) $A_{ol} = \dfrac{100{,}000}{\sqrt{1 + (0.1)^2}} = \mathbf{99{,}503}$

(c) $A_{ol} = \dfrac{100{,}000}{\sqrt{1 + (1)^2}} = \dfrac{100{,}000}{\sqrt{2}} = \mathbf{70{,}710}$

(d) $A_{ol} = \dfrac{100{,}000}{\sqrt{1 + (10)^2}} = \mathbf{9950}$

Related Problem

Find A_{ol} for the following frequencies. Assume $f_{c(ol)} = 200$ Hz and $A_{ol(mid)} = 80{,}000$.

(a) $f = 2$ Hz   **(b)** $f = 10$ Hz   **(c)** $f = 2500$ Hz

Phase Shift

As you know from Chapter 10, an *RC* circuit causes a propagation delay from input to output, thus creating a **phase shift** between the input signal and the output signal. An *RC* lag circuit such as found in an op-amp stage causes the output signal voltage to lag the input, as shown in Figure 12–40. From basic ac circuit theory, the phase shift, θ, is

$$\theta = -\tan^{-1}\left(\frac{R}{X_C}\right)$$

Since $R/X_C = f/f_c$,

$$\theta = -\tan^{-1}\left(\frac{f}{f_c}\right)$$

Equation 12–20

The negative sign indicates that the output lags the input. This equation shows that the phase shift increases with frequency and approaches $-90°$ as f becomes much greater than f_c.

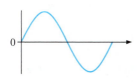 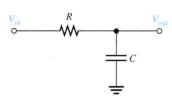 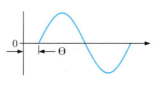

◀ **FIGURE 12–40**

Output voltage lags input voltage.

EXAMPLE 12–9

Calculate the phase shift for an *RC* lag circuit for each of the following frequencies, and then plot the curve of phase shift versus frequency. Assume $f_c = 100$ Hz.

(a) $f = 1$ Hz   **(b)** $f = 10$ Hz   **(c)** $f = 100$ Hz

(d) $f = 1000$ Hz   **(e)** $f = 10,000$ Hz

Solution

(a) $\theta = -\tan^{-1}\left(\dfrac{f}{f_c}\right) = -\tan^{-1}\left(\dfrac{1 \text{ Hz}}{100 \text{ Hz}}\right) = \mathbf{-0.573°}$

(b) $\theta = -\tan^{-1}\left(\dfrac{10 \text{ Hz}}{100 \text{ Hz}}\right) = \mathbf{-5.71°}$

(c) $\theta = -\tan^{-1}\left(\dfrac{100 \text{ Hz}}{100 \text{ Hz}}\right) = \mathbf{-45°}$

(d) $\theta = -\tan^{-1}\left(\dfrac{1000 \text{ Hz}}{100 \text{ Hz}}\right) = \mathbf{-84.3°}$

(e) $\theta = -\tan^{-1}\left(\dfrac{10,000 \text{ Hz}}{100 \text{ Hz}}\right) = \mathbf{-89.4°}$

The phase shift-versus-frequency curve is plotted in Figure 12–41. Note that the frequency axis is logarithmic.

▶ **FIGURE 12–41**

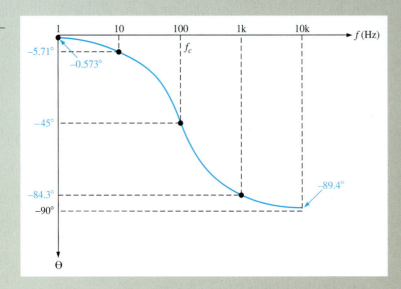

Related Problem   At what frequency, in this example, is the phase shift 60°?

Overall Frequency Response

Previously, an op-amp was defined to have a constant roll-off of -20 dB/decade above its critical frequency. For most op-amps this is the case; for some, however, the situation is more complex. This occurs more frequently when the op-amp is designed for higher speed, including fast slew rate or (in some cases) very low noise. The more complex IC operational amplifier may consist of several cascaded amplifier stages. The gain of each stage is frequency dependent and rolls off at -20 dB/decade above its critical frequency. Therefore, the total response of an op-amp is a composite of the individual responses of the internal stages. As an example, a three-stage op-amp is represented in Figure 12–42(a), and the frequency response of each stage is shown in Figure 12–42(b). As you know, dB gains are added so that the total op-amp frequency response is as shown in Figure 12–42(c). Since the roll-off rates are additive, the total roll-off rate increases by −20 dB/decade (−6 dB/decade) as each critical frequency is reached. When an op-amp with a response like this is used in a circuit, care must be taken to avoid oscillations.

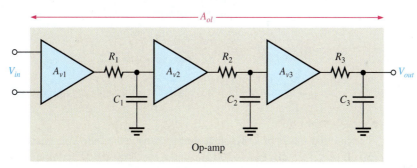

(a) Representation of an op-amp with three internal stages

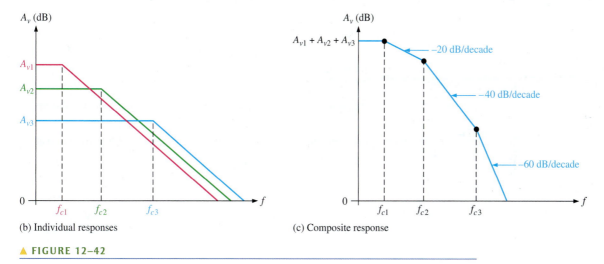

(b) Individual responses

(c) Composite response

▲ **FIGURE 12–42**

Op-amp open-loop frequency response.

Overall Phase Response

In a multistage amplifier, each stage contributes to the total phase lag. As you have seen, each RC lag circuit can produce up to a −90° phase shift. Since each stage in an op-amp includes an RC lag circuit, a three-stage op-amp, for example, can have a maximum phase lag of −270°. Also, the phase lag of each stage is less than −45° when the frequency is below the critical frequency, equal to −45° at the critical frequency, and greater than −45° when the frequency is above the critical frequency. The phase lags of the stages of an op-amp are added to produce a total phase lag, according to the following formula for three stages:

$$\theta_{tot} = -\tan^{-1}\left(\frac{f}{f_{c1}}\right) - \tan^{-1}\left(\frac{f}{f_{c2}}\right) - \tan^{-1}\left(\frac{f}{f_{c3}}\right)$$

EXAMPLE 12–10

A certain op-amp has three internal amplifier stages with the following gains and critical frequencies:

Stage 1: $A_{v1} = 40$ dB, $f_{c1} = 2$ kHz

Stage 2: $A_{v2} = 32$ dB, $f_{c2} = 40$ kHz

Stage 3: $A_{v3} = 20$ dB, $f_{c3} = 150$ kHz

Determine the open-loop midrange gain in decibels and the total phase lag when $f = f_{c1}$.

Solution $A_{ol(mid)} = A_{v1} + A_{v2} + A_{v3} = 40$ dB $+ 32$ dB $+ 20$ dB $= 92$ dB

$$\theta_{tot} = -\tan^{-1}\left(\frac{f}{f_{c1}}\right) - \tan^{-1}\left(\frac{f}{f_{c2}}\right) - \tan^{-1}\left(\frac{f}{f_{c3}}\right)$$

$$= -\tan^{-1}(1) - \tan^{-1}\left(\frac{2}{40}\right) - \tan^{-1}\left(\frac{2}{150}\right) = -45° - 2.86° - 0.76° = \mathbf{-48.6°}$$

Related Problem   The internal stages of a two-stage amplifier have the following characteristics: $A_{v1} = 50$ dB, $A_{v2} = 25$ dB, $f_{c1} = 1500$ Hz, and $f_{c2} = 3000$ Hz. Determine the open-loop midrange gain in decibels and the total phase lag when $f = f_{c1}$.

SECTION 12–7 CHECKUP

1. How do the open-loop voltage gain and the closed-loop voltage gain of an op-amp differ?

2. The upper critical frequency of a particular op-amp is 100 Hz. What is its open-loop 3 dB bandwidth?

3. Does the open-loop gain increase or decrease with frequency above the critical frequency?

4. If the individual stage gains of an op-amp are 20 dB and 30 dB, what is the total gain in decibels?

5. If the individual phase lags are $-49°$ and $-5.2°$, what is the total phase lag?

12–8   CLOSED-LOOP FREQUENCY RESPONSE

Op-amps are normally used in a closed-loop configuration with negative feedback in order to achieve precise control of the gain and bandwidth. In this section, you will see how feedback affects the gain and frequency response of an op-amp.

After completing this section, you should be able to

❏ **Analyze the closed-loop frequency response of an op-amp**
   ◆ Review the closed-loop voltage gain for each op-amp configuration
❏ Analyze the effect of negative feedback on bandwidth
❏ Define and discuss the gain-bandwidth product

Recall that midrange gain of an op-amp is reduced by negative feedback, as indicated by the following closed-loop gain expressions for the three amplifier configurations previously covered, where B is the feedback attenuation. For a non-inverting amplifier,

$$A_{cl(\text{NI})} = \frac{A_{ol}}{1 + A_{ol}B} \cong \frac{1}{B} = 1 + \frac{R_f}{R_i}$$

For an inverting amplifier,

$$A_{cl(\text{I})} \cong -\frac{R_f}{R_i}$$

For a voltage-follower,

$$A_{cl(\text{VF})} = 1$$

Effect of Negative Feedback on Bandwidth

You know how negative feedback affects the gain; now you will learn how it affects the amplifier's bandwidth. The closed-loop critical frequency of an op-amp is

$$f_{c(cl)} = f_{c(ol)}(1 + BA_{ol(mid)})$$

Equation 12–21

This expression shows that the closed-loop critical frequency, $f_{c(cl)}$, is higher than the open-loop critical frequency $f_{c(ol)}$ by the factor $1 + BA_{ol(mid)}$. You will find a derivation of Equation 12–21 in "Derivations of Selected Equations" at www.pearsonhighered.com/floyd.

Since $f_{c(cl)}$ equals the bandwidth for the closed-loop amplifier, the closed-loop bandwidth (BW_{cl}) is also increased by the same factor.

$$BW_{cl} = BW_{ol}(1 + BA_{ol(mid)})$$

Equation 12–22

EXAMPLE 12–11

A certain amplifier has an open-loop midrange gain of 150,000 and an open-loop 3 dB bandwidth of 200 Hz. The attenuation (B) of the feedback loop is 0.002. What is the closed-loop bandwidth?

Solution $BW_{cl} = BW_{ol}(1 + BA_{ol(mid)}) = 200\ \text{Hz}[1 + (0.002)(150,000)] = \textbf{60.2 kHz}$

Related Problem   If $A_{ol(mid)} = 200,000$ and $B = 0.05$, what is the closed-loop bandwidth?

Figure 12–43 graphically illustrates the concept of closed-loop response for a compensated op-amp. When the open-loop gain of an op-amp is reduced by negative feedback,

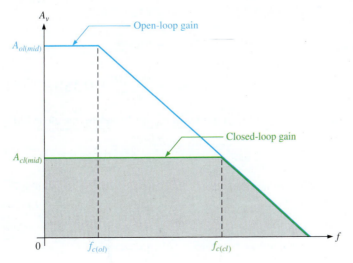

◀ **FIGURE 12–43**

Closed-loop gain compared to open-loop gain.

the bandwidth is increased. The closed-loop gain is independent of the open-loop gain up to the point of intersection of the two gain curves. This point of intersection is the critical frequency, $f_{c(cl)}$, for the closed-loop response. Notice that the closed-loop gain has the same roll-off rate as the open-loop gain, beyond the closed-loop critical frequency.

Gain-Bandwidth Product

As you saw in Figure 12–37, an increase in closed-loop gain causes a decrease in the bandwidth and vice versa, such that the product of gain and bandwidth is constant. This is true as long as the roll-off rate is constant as in the case of a compensated op-amp. If you let A_{cl} represent the gain of any of the closed-loop configurations and $f_{c(cl)}$ represent the closed-loop critical frequency (same as the bandwidth), then

$$A_{cl}f_{c(cl)} = A_{ol}f_{c(ol)}$$

The **gain-bandwidth product** for a compensated op-amp is always equal to the frequency at which the op-amp's open-loop gain is unity or 0 dB (unity-gain bandwidth, f_T).

Equation 12–23

$$f_T = A_{cl}f_{c(cl)}$$

EXAMPLE 12–12

Determine the bandwidth of each of the amplifiers in Figure 12–44. Both op-amps have an open-loop gain of 100 dB and a unity-gain bandwidth (f_T) of 3 MHz.

▶ **FIGURE 12–44**

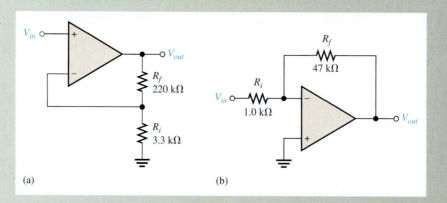

(a)   (b)

Solution   **(a)** For the noninverting amplifier in Figure 12–44(a), the closed-loop gain is

$$A_{cl} = 1 + \frac{R_f}{R_i} = 1 + \frac{220 \text{ k}\Omega}{3.3 \text{ k}\Omega} = 67.7$$

Use Equation 12–23 and solve for $f_{c(cl)}$ (where $f_{c(cl)} = BW_{cl}$).

$$f_{c(cl)} = BW_{cl} = \frac{f_T}{A_{cl}}$$

$$BW_{cl} = \frac{3 \text{ MHz}}{67.7} = \textbf{44.3 kHz}$$

(b) For the inverting amplifier in Figure 12–44(b), the closed-loop gain is

$$A_{cl} = -\frac{R_f}{R_i} = -\frac{47 \text{ k}\Omega}{1.0 \text{ k}\Omega} = -47$$

Using the absolute value of A_{cl}, the closed-loop bandwidth is

$$BW_{cl} = \frac{3 \text{ MHz}}{47} = \textbf{63.8 kHz}$$

Related Problem Determine the bandwidth of each of the amplifiers in Figure 12–44. Both op-amps have an A_{ol} of 90 dB and a unity-gain bandwidth of 2 MHz.

Open the Multisim file EXM12-12 or the LT Spice file EXS12-12 in the Examples folder on the website. Measure the bandwidth of each amplifier using the Bode plotter and compare with the calculated values.

SECTION 12–8 CHECKUP

1. Is the closed-loop gain always less than the open-loop gain?
2. A certain compensated op-amp is used in a feedback configuration having a gain of 30 and a bandwidth of 100 kHz. If the external resistor values are changed to increase the gain to 60, what is the new bandwidth?
3. What is the unity-gain bandwidth of the op-amp in Question 2?

12–9 TROUBLESHOOTING

As a technician, you may encounter situations in which an op-amp or its associated circuitry has malfunctioned. The op-amp is a complex integrated circuit with many types of internal failures possible. However, since you cannot troubleshoot the op-amp internally, treat it as a single device with only a few connections to it. If it fails, replace it just as you would a resistor, capacitor, or transistor.

After completing this section, you should be able to

❑ **Troubleshoot op-amp circuits**
❑ Determine faults in the noninverting amplifier
❑ Determine faults in the voltage-follower
❑ Determine faults in the inverting amplifier

In the basic op-amp configurations, there are only a few external components that can fail. These are the feedback resistor, the input resistor, and the resistor or potentiometer used for offset voltage compensation. Also, of course, the op-amp itself can fail or there can be faulty contacts in the circuit. Let's examine the three basic configurations for possible faults and the associated symptoms.

The first thing to do when you suspect a faulty circuit is to check for the proper supply voltage and ground at the pins of the op-amp. Having done that, several other possible faults are as follows. A visual inspection should also be done.

Faults in the Noninverting Amplifier

Open Feedback Resistor If the feedback resistor, R_f, in Figure 12–45 opens, the op-amp is operating with its very high open-loop gain, which causes the input signal to drive the device into nonlinear operation and results in a severely clipped output signal as shown in part (a).

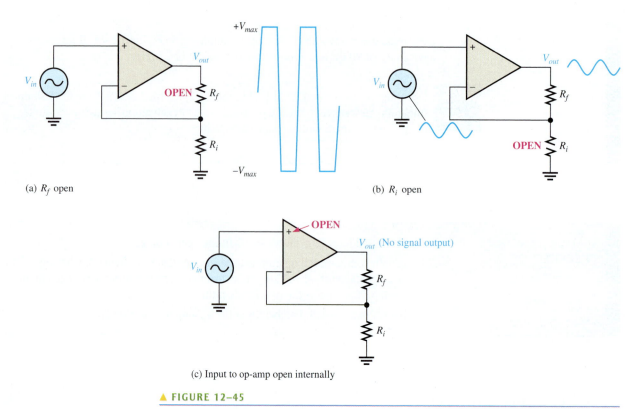

(a) R_f open

(b) R_i open

(c) Input to op-amp open internally

Faults in the noninverting amplifier.

Open Input Resistor   In this case, you still have a closed-loop configuration. Since R_i is open and effectively equal to infinity (∞), the closed-loop gain from Equation 12–8 is

$$A_{cl(NI)} = 1 + \frac{R_f}{R_i} = 1 + \frac{R_f}{\infty} = 1 + 0 = 1$$

This shows that the amplifier acts like a voltage-follower. You would observe an output signal that is the same as the input, as indicated in Figure 12–45(b).

Internally Open Noninverting Op-Amp Input   In this situation, because the input voltage is not applied to the op-amp, the output is zero. This is indicated in Figure 12–45(c).

Other Op-Amp Faults   In general, an internal failure will result in a loss or distortion of the output signal. The best approach is to first make sure that there are no external failures or faulty conditions. If everything else is good, then the op-amp must be bad.

Faults in the Voltage-Follower

The voltage-follower is a special case of the noninverting amplifier. Except for a faulty op-amp, an open or shorted external connection, or a problem with the offset null potentiometer, about the only thing that can happen in a voltage-follower circuit is an open feedback loop. This would have the same effect as an open feedback resistor as previously discussed.

Faults in the Inverting Amplifier

Open Feedback Resistor   If R_f opens, as indicated in Figure 12–46(a), the input signal still feeds through the input resistor and is amplified by the high open-loop gain of the op-amp. This forces the device to be driven into nonlinear operation, and you will see an output something like that shown. This is a similar result as in the noninverting amplifier configuration.

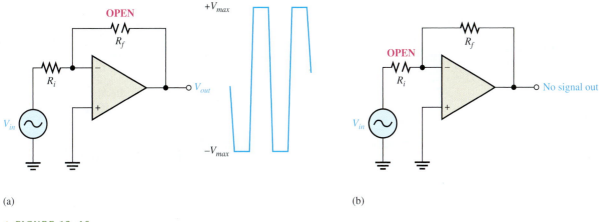

(a)   (b)

▲ **FIGURE 12–46**

Faults in the inverting amplifier.

Open Input Resistor   This prevents the input signal from getting to the op-amp input, so there will be no output signal, as indicated in Figure 12–46(b).

Failures in the op-amp itself or the offset null potentiometer have the same effects as previously discussed for the noninverting amplifier configuration.

Multisim Troubleshooting Exercises

These file circuits are in the Troubleshooting Exercises folder on the website. Open each file and determine if the circuit is working properly. If it is not working properly, determine the fault.

1. Multisim file TSM12-01

2. Multisim file TSM12-02

3. Multisim file TSM12-03

4. Multisim file TSM12-04

5. Multisim file TSM12-05

**SECTION 12–9
CHECKUP**

1. If you notice that the op-amp output signal is beginning to clip on one peak as you increase the input signal, what should you check?

2. For a noninverting amplifier, if there is no op-amp output signal when there is a verified input signal at the input pin, what would you suspect as being faulty?

Device Application: *Op-Amp Audio Amplifier*

The company that manufactures the PA system developed in Chapters 6 and 7 wants to replace the audio amplifier with a new design using an op-amp instead of the discrete transistor preamp circuit to reduce parts and cost. The power amplifier will still retain its basic design with only a few changes. The block diagram of the PA system is shown in Figure 12–47.

▶ FIGURE 12–47

Block diagram of public address system.

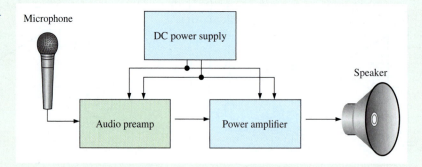

The Circuit

The schematic of the new op-amp design is shown in Figure 12–48. A KA741 operational amplifier is used for the preamp stage. The power amplifier retains the original push-pull complementary Darlington configuration with the exception of the driver stage. This has been eliminated because the op-amp, with its very low output resistance, is capable of driving the push-pull power amplifier stage without a buffer interface. The rheostat, R_{gain}, is for adjusting the voltage gain, and the potentiometer, R_{null}, is for output nulling (making the dc output 0 V).

▶ FIGURE 12–48

Audio amplifier.

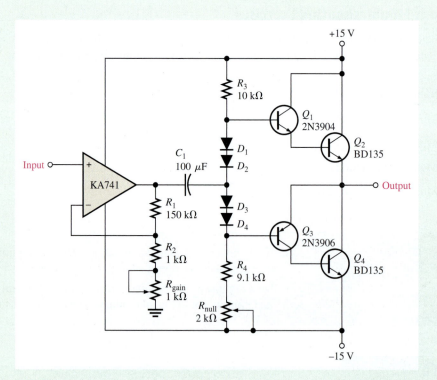

1. Identify the op-amp configuration.
2. Calculate the maximum and minimum voltage gains of the op-amp.
3. What is the maximum rms output of the op-amp stage if the input is 50 mV rms?
4. Determine the ideal maximum power delivered by the audio amplifier to an 8 Ω speaker.

A partial datasheet for a KA741 op-amp is shown in Figure 12–49.

FAIRCHILD
SEMICONDUCTOR®

www.fairchildsemi.com

KA741

Single Operational Amplifier

Features

- Short circuit protection
- Excellent temperature stability
- Internal frequency compensation
- High Input voltage range
- Null of offset

Description

The KA741 series are general purpose operational amplifiers. It is intended for a wide range of analog applications. The high gain and wide range of operating voltage provide superior performance in intergrator, summing amplifier, and general feedback applications.

8-DIP

8-SOP

Internal Block Diagram

Rev. 1.0.1

©2001 Fairchild Semiconductor Corporation

Absolute Maximum Ratings (T$_A$ = 25°C)

Parameter	Symbol	Value	Unit
Supply Voltage	V$_{CC}$	±18	V
Differential Input Voltage	V$_{I(DIFF)}$	30	V
Input Voltage	V$_I$	±15	V
Output Short Circuit Duration	-	Indefinite	
Power Dissipation	P$_D$	500	mW
Operating Temperature Range KA741 KA741I	T$_{OPR}$	0 ~ + 70 -40 ~ +85	°C
Storage Temperature Range	T$_{STG}$	-65 ~ + 150	°C

Electrical Characteristics

(V$_{CC}$ = 15V, V$_{EE}$ = - 15V. T$_A$ = 25° C, unless otherwise specified)

Parameter	Symbol	Conditions		KA741/KA741I Min.	Typ.	Max.	Unit
Input Offset Voltage	V$_{IO}$	R$_S$ ≤ 10 KΩ		-	2.0	6.0	mV
		R$_S$ ≤ 50 Ω		-	-	-	
Input Offset Voltage Adjustment Range	V$_{IO(R)}$	V$_{CC}$ = ±20 V		-	±15	-	mV
Input Offset Current	I$_{IO}$			-	20	200	nA
Input Bias Current	I$_{BIAS}$			-	80	500	nA
Input Resistance (Note1)	R$_I$	V$_{CC}$ = ±20 V		0.3	2.0	-	MΩ
Input Voltage Range	V$_{I(R)}$			±12	±13	-	V
Large Signal Voltage Gain	G$_V$	R$_L$ ≥ 2 KΩ	V$_{CC}$ = ± 20 V, V$_{O(P-P)}$ = ± 15 V	-	-	-	V/mV
			V$_{CC}$ = ± 15 V, V$_{O(P-P)}$ = ±10 V	20	200		
Output Short Circuit Current	I$_{SC}$			-	25	-	mA
Output Voltage Swing	V$_{O(P-P)}$	V$_{CC}$ = ±20V	R$_L$ ≥ 10 KΩ	-	-	-	V
			R$_L$ ≥ 2 KΩ	-	-	-	
		V$_{CC}$ = ±15V	R$_L$ ≥ 10 KΩ	±12	±14	-	
			R$_L$ ≥ 2 KΩ	±10	±13	-	
Common Mode Rejection Ratio	CMRR	R$_S$ ≤ 10 KΩ, V$_{CM}$ = ±12V		70	90	-	dB
		R$_S$ ≤ 50 Ω, V$_{CM}$ = ±12V		-	-	-	
Power Supply Rejection Ratio	PSRR	V$_{CC}$ = ±15V to V$_{CC}$ = ±15V R$_S$ ≤ 50 Ω		-	-	-	dB
		V$_{CC}$ = ±15V to V$_{CC}$ = ±15V R$_S$ ≤ 50 Ω		77	96	-	
Transient Response	Rise Time	T$_R$	Unity Gain	-	0.3	-	μs
	Overshoot	OS		-	10	-	%
Bandwidth	BW			-	-	-	MHz
Slew Rate	SR	Unity Gain		-	0.5	-	V/μs
Supply Current	I$_{CC}$	R$_L$ = ∞Ω		-	1.5	2.8	mA
Power Consumption	P$_C$	V$_{CC}$ = ±20V		-	-	-	mW
		V$_{CC}$ = ±15V		-	50	85	

Electrical Characteristics

(V$_{CC}$ = ±15V, unless otherwise specified)
The following specification apply over the range of 0°C ≤ T$_A$ ≤ +70 °C for the KA741; and the -40°C ≤ T$_A$ ≤+85 °C for the KA741I

Parameter	Symbol	Conditions		KA741/KA741I Min.	Typ.	Max.	Unit
Input Offset Voltage	V$_{IO}$	R$_S$ ≤ 50 Ω		-	-	-	mV
		R$_S$ ≤ 10 KΩ		-	-	7.5	
Input Offset Voltage Drift	ΔV$_{IO}$/ΔT			-	-	-	μV/°C
Input Offset Current	I$_{IO}$			-	-	300	nA
Input Offset Current Drift	ΔI$_{IO}$/ΔT			-	-	-	nA/°C
Input Bias Current	I$_{BIAS}$			-	-	0.8	μA
Input Resistance (Note1)	R$_I$	V$_{CC}$ = ±20 V		-	-	-	MΩ
Input Voltage Range	V$_{I(R)}$			±12	±13	-	V
Output Voltage Swing	V$_{O(P-P)}$	V$_{CC}$ = ± 20 V	R$_S$ ≥ 10 KΩ	-	-	-	V
			R$_S$ ≥ 2 KΩ	-	-	-	
		V$_{CC}$ = ± 15 V	R$_S$ ≥ 10 KΩ	±12	±14	-	
			R$_S$ ≥ 2 KΩ	±10	±13	-	
Output Short Circuit Current	I$_{SC}$			10	-	40	mA
Common Mode Rejection Ratio	CMRR	R$_S$ ≤ 10 KΩ, V$_{CM}$ = ±12 V		70	90	-	dB
		R$_S$ ≤ 50 Ω, V$_{CM}$ = ±12 V		-	-	-	
Power Supply Rejection Ratio	PSRR	V$_{CC}$ = ±20V to ±5 V	R$_S$ ≤ 50 Ω	-	-	-	dB
			R$_S$ ≤ 10 KΩ	77	96	-	
Large Signal Voltage Gain	G$_V$	R$_S$ ≥ 2 KΩ	V$_{CC}$ = ±20V, V$_{O(P-P)}$ = ±15V	-	-	-	V/mV
			V$_{CC}$ = ±15V, V$_{O(P.P)}$ = ±10V	15	-	-	
			V$_{CC}$ = ±15V, V$_{O(P-P)}$ = ±2V	-	-	-	

▲ **FIGURE 12–49**

Partial datasheet for the KA741 op-amp. Copyright Fairchild Semiconductor Corporation. Used by permission.

5. Using the datasheet, assign pin numbers to the op-amp in Figure 12–48.
6. Determine the maximum power consumption of the op-amp with the ± 15 V supply voltages.
7. To what typical voltage can the output swing with ± 15 V supply voltages?

Simulation

The audio amplifier is simulated with an input signal of 50 mV using Multisim. The results are shown in Figure 12–50 where an 8.2 Ω resistor is used to simulate the speaker.

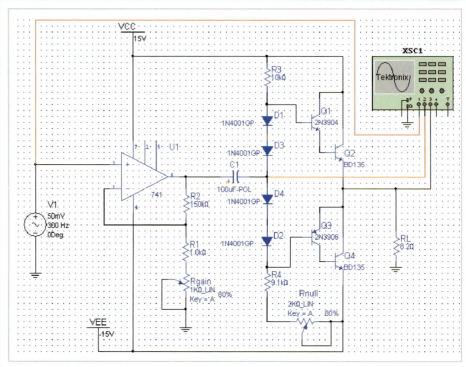

(a) Circuit screen

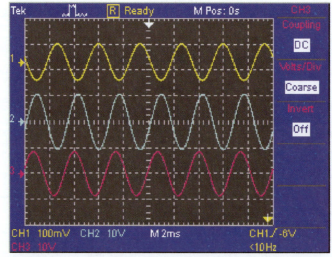

(b) Input signal, op-amp output signal, and final output signal

▲ FIGURE 12–50

Simulation of the audio amplifier.

8. From the scope display in Figure 12–50, determine the rms value of each voltage.
9. Determine the voltage gain of the op-amp stage from the measured signals.
10. Determine the overall voltage gain from the measured signals.

Simulate the op-amp audio amplifier using your Multisim or LT Spice software. Observe the signal voltages with the oscilloscope.

Prototyping and Testing

Now that the circuit has been simulated, the prototype circuit is constructed and tested. After the circuit is successfully tested on a protoboard, it is ready to be finalized on a printed circuit board.

Lab Experiment

To build and test a similar circuit, go to Experiment 12–A in your lab manual (*Laboratory Exercises for Electronic Devices* by David Buchla and Steven Wetterling).

The original system had two boards, the preamp board and the power amp board. By using the new op-amp design, the audio amplifier is simplified to one board, as shown in Figure 12–51.

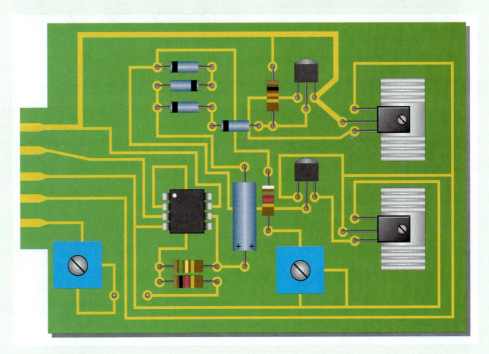

▲ FIGURE 12–51

Audio amplifier board.

11. Check the printed circuit board for correctness by comparing it with the schematic in Figure 12–48.
12. Label each input and output pin according to function.

Troubleshooting

Four circuit boards are tested, and the results are shown in Figure 12–52. [Parts (b), (c), and (d) are shown on the next page.]

13. Determine the problem, if any, in each of the board tests in Figure 12–52.

14. List possible causes of any problem from item 13.

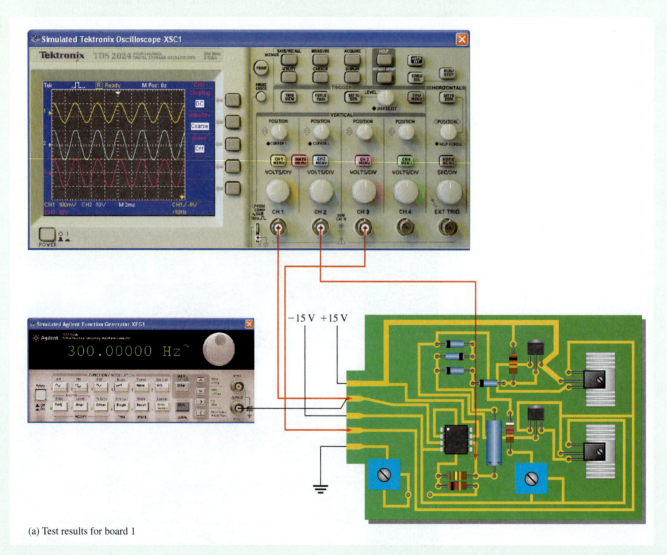

(a) Test results for board 1

▲ **FIGURE 12–52**

Results of audio amplifier board tests.

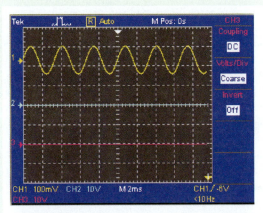

(b) Test results for board 2

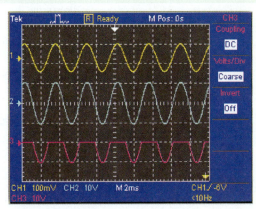

(c) Test results for board 3

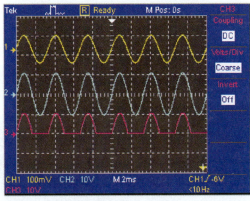

(d) Test results for board 4

▲ **FIGURE 12–52**

Continued

Programmable Analog Technology

FPAAs (field-programmable analog arrays) and dpASPs (dynamically programmable analog signal processors) are based on switched-capacitor technology that was covered in Chapter 9. These are integrated circuit devices that can be programmed with software for various types of analog functions and designs. Both the FPAA and the dpASP can be reprogrammed, but the FPAA is statically reconfigurable and must first be reset, whereas the dpASP can be **dynamically reconfigured** "on-the-fly" while operating in a system. The software allows you to experiment and design with various analog devices that are covered in this textbook by specifying parameters, observing operation, interconnecting devices for more complex circuits, and preparing the software for **downloading** to an actual device. Refer to the tutorial available with the AnadigmDesigner®2 software or the tutorial in the *Laboratory Exercises for Electronic Devices* lab manual.

The Designer2 software differs in its primary purpose from the Multisim software you have been using. Electronic Workbench Multisim is basically a simulation software that allows you to test circuits on the computer using simulated discrete and integrated components. The Multisim software is useful for verifying that a circuit actually works as intended, but it has limited hardware interface capability. AnadigmDesigner2 is both a simulation and a hardware interface tool that allows you to custom program an analog design and implement it in an integrated circuit chip. It is based on an extensive library of analog functions called configurable analog modules (**CAMs**) that can be connected using a simple "drag-and drop" format and tested with virtual instruments on the computer. The design can then be converted to hardware by downloading it to an actual FPAA or dpASP IC chip.

A free trial version of the AnadigmDesigner2 software is available for downloading at www .anadigm.com. The basic steps in implementing a design are as follows using a specific CAM for illustration. These steps apply to any CAM or multiple CAMs.

CAM Selection

The first step in implementing a programmable analog design is to open the AnadigmDesigner2 software. The representation of a blank FPAA or dpASP chip will appear as shown in Figure 12–53. Select the *Configure* icon to open a list of available CAMs, as shown in Figure 12–54.

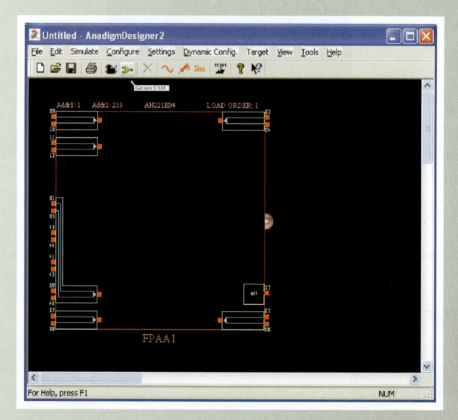

▲ **FIGURE 12–53**

Click on *Configure* icon.

Configure and Place CAM

Select the desired CAM and set up the parameters in the *Set CAM Parameters* screen shown in Figure 12–55. Next place the CAM in the chip outline and connect to an input and output, as shown in Figure 12–56. Several CAMs can be placed in one chip and interconnected.

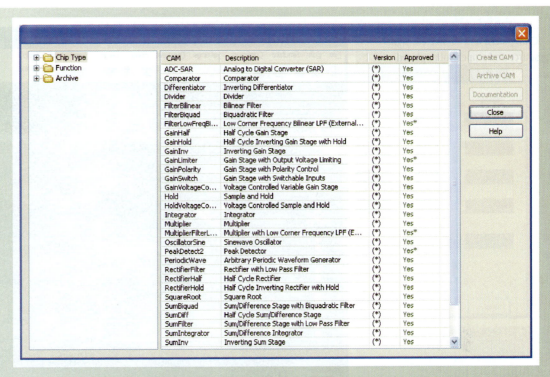

▲ FIGURE 12–54

List of available CAMs.

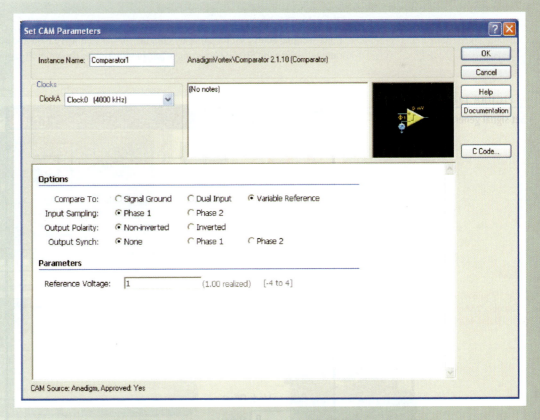

▲ FIGURE 12–55

Set CAM parameters.

Op-Amp Impedances

12–11	$Z_{in(NI)} = (1 + A_{ol}B)Z_{in}$	Input impedance (noninverting)
12–12	$Z_{out(NI)} = \dfrac{Z_{out}}{1 + A_{ol}B}$	Output impedance (noninverting)
12–13	$Z_{in(VF)} = (1 + A_{ol})Z_{in}$	Input impedance (voltage-follower)
12–14	$Z_{out(VF)} = \dfrac{Z_{out}}{1 + A_{ol}}$	Output impedance (voltage-follower)
12–15	$Z_{in(I)} \cong R_i$	Input impedance (inverting)
12–16	$Z_{out(I)} = \dfrac{Z_{out}}{1 + A_{ol}B}$	Output impedance (inverting)

Op-Amp Frequency Responses

12–17	$BW = f_{cu}$	Op-amp bandwidth
12–18	$\dfrac{V_{out}}{V_{in}} = \dfrac{1}{\sqrt{1 + f^2/f_c^2}}$	RC attenuation
12–19	$A_{ol} = \dfrac{A_{ol(mid)}}{\sqrt{1 + f^2/f_c^2}}$	Open-loop voltage gain
12–20	$\theta = -\tan^{-1}\left(\dfrac{f}{f_c}\right)$	RC phase shift
12–21	$f_{c(cl)} = f_{c(ol)}(1 + BA_{ol(mid)})$	Closed-loop critical frequency
12–22	$BW_{cl} = BW_{ol}(1 + BA_{ol(mid)})$	Closed-loop bandwidth
12–23	$f_T = A_{cl}f_{c(cl)}$	Unity-gain bandwidth

TRUE/FALSE QUIZ

Answers can be found at www.pearsonhighered.com/floyd.

1. An ideal op-amp has an infinite input impedance.
2. An ideal op-amp has a very high output impedance.
3. The op-amp can operate in either the differential mode or the common mode.
4. Common-mode rejection means that a signal appearing on both inputs is effectively cancelled.
5. CMRR stands for common-mode rejection reference.
6. Slew rate determines how fast the output can change in response to a step input.
7. Negative feedback reduces the gain of an op-amp from its open-loop value.
8. Negative feedback reduces the bandwidth of an op-amp from its open-loop value.
9. A noninverting amplifier uses negative feedback.
10. The gain of a voltage-follower is very high.
11. Negative feedback affects the input and output impedances of an op-amp.
12. A compensated op-amp has a gain roll-off of –20dB decade above the critical frequency.
13. The gain-bandwidth product equals the unity-gain frequency for a compensated op-amp.
14. If the feedback resistor in an inverting amplifier opens, the gain becomes zero.

CIRCUIT-ACTION QUIZ

Answers can be found at www.pearsonhighered.com/floyd.

1. If R_f is decreased in the circuit of Figure 12–19, the voltage gain will
   (a) increase   (b) decrease   (c) not change
2. If $V_{in} = 1$ mV and R_f opens in the circuit of Figure 12–19, the output voltage will
   (a) increase   (b) decrease   (c) not change

3. If R_i is increased in the circuit of Figure 12–19, the voltage gain will

(a) increase   (b) decrease   (c) not change

4. If 10 mV are applied to the input to the op-amp circuit of Figure 12–23 and R_f is increased, the output voltage will

(a) increase   (b) decrease   (c) not change

5. In Figure 12–29, if R_f is changed from 100 kΩ to 68 kΩ, the feedback attenuation will

(a) increase   (b) decrease   (c) not change

6. If the closed-loop gain in Figure 12–44(a) is increased by increasing the value of R_f, the closed-loop bandwidth will

(a) increase   (b) decrease   (c) not change

7. If R_f is changed to 470 kΩ and R_i is changed to 10 kΩ in Figure 12–44(b), the closed-loop bandwidth will

(a) increase   (b) decrease   (c) not change

8. If R_i in Figure 12–44(b) opens, the output voltage will

(a) increase   (b) decrease   (c) not change

SELF-TEST

Answers can be found at www.pearsonhighered.com/floyd.

Section 12–1

1. An integrated circuit (IC) op-amp has

(a) two inputs and two outputs   (b) one input and one output

(c) two inputs and one output

2. Which of the following characteristics does not necessarily apply to an op-amp?

(a) High gain   (b) Low power

(c) High input impedance   (d) Low output impedance

3. A differential amplifier

(a) is part of an op-amp   (b) has one input and one output

(c) has two outputs   (d) answers (a) and (c)

Section 12–2

4. When an op-amp is operated in the single-ended differential mode,

(a) the output is grounded

(b) one input is grounded and a signal is applied to the other

(c) both inputs are connected together

(d) the output is not inverted

5. In the double-ended differential mode,

(a) a signal is applied between the two inputs   (b) the gain is 1

(c) the outputs are different amplitudes   (d) only one supply voltage is used

6. In the common mode,

(a) both inputs are grounded   (b) the outputs are connected together

(c) an identical signal appears on both inputs   (d) the output signals are in-phase

7. Common-mode gain is

(a) very high   (b) very low

(c) always unity   (d) unpredictable

8. If $A_{ol} = 3500$ and $A_{cm} = 0.35$, the CMRR is

(a) 1225   (b) 10,000

(c) 80 dB   (d) answers (b) and (c)

9. With zero volts on both inputs, an op-amp ideally should have an output equal to

(a) the positive supply voltage   (b) the negative supply voltage

(c) zero   (d) the CMRR

8. Distinguish between input bias current and input offset current, and then calculate the input offset current in Problem 7.

9. Figure 12–62 shows the output voltage of an op-amp in response to a step input. What is the slew rate?

10. How long does it take the output voltage of an op-amp to go from -10 V to $+10$ V if the slew rate is 0.5 V/μS?

▶ **FIGURE 12–62**

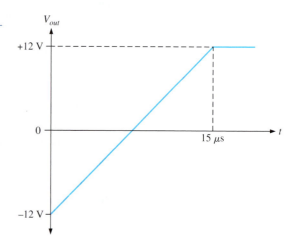

Section 12–4   Op-Amps with Negative Feedback

11. Identify each of the op-amp configurations in Figure 12–63.

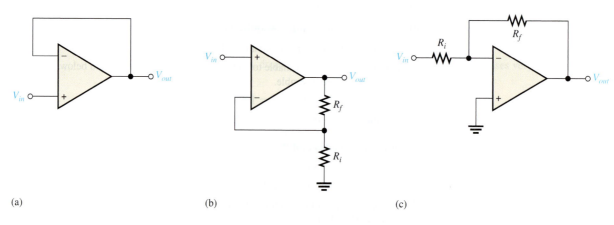

(a)   (b)   (c)

▲ **FIGURE 12–63**

12. A noninverting amplifier has an R_i of 1.0 kΩ and an R_f of 100 kΩ. Determine V_f and B if $V_{out} = 5$ V.

13. For the amplifier in Figure 12–64, determine the following:

    (a) $A_{cl(NI)}$ **(b)** V_{out} **(c)** V_f

▶ **FIGURE 12–64**

Multisim or LT Spice file circuits are identified with a logo and are in the Problems folder on the website. Filenames correspond to figure numbers (e.g., FGM12-64 or FGS12-64).

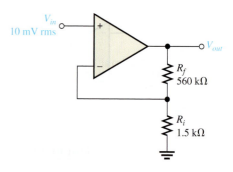

14. Determine the closed-loop gain of each amplifier in Figure 12–65.

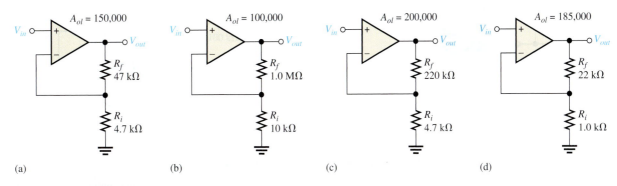

(a)   (b)   (c)   (d)

▲ FIGURE 12–65

15. Find the value of R_f that will produce the indicated closed-loop gain in each amplifier in Figure 12–66.

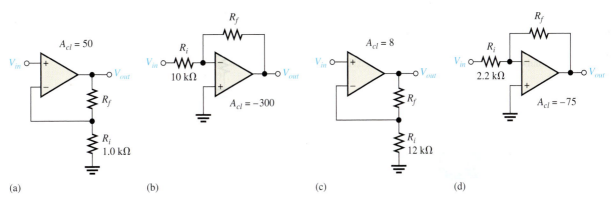

(a)   (b)   (c)   (d)

▲ FIGURE 12–66

16. Find the gain of each amplifier in Figure 12–67.

17. If a signal voltage of 10 mV rms is applied to each amplifier in Figure 12–67, what are the output voltages and what is their phase relationship with inputs?

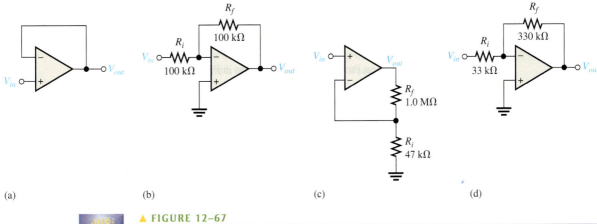

(a)   (b)   (c)   (d)

▲ FIGURE 12–67

Section 12–8   Closed-Loop Frequency Response

35. Determine the midrange gain in dB of each amplifier in Figure 12–73. Are these open-loop or closed-loop gains?

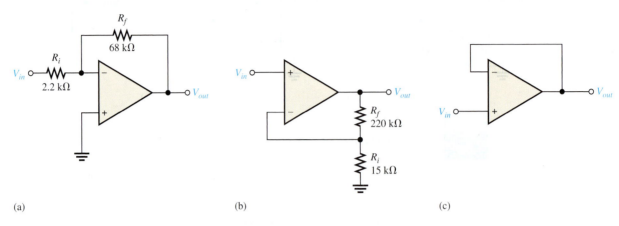

(a)   (b)   (c)

▲ **FIGURE 12–73**

36. A certain amplifier has an open-loop gain in midrange of 180,000 and an open-loop critical frequency of 1500 Hz. If the attenuation of the feedback path is 0.015, what is the closed-loop bandwidth?

37. Given that $f_{c(ol)}$ = 750 Hz, A_{ol} = 89 dB, and $f_{c(cl)}$ = 5.5 kHz, determine the closed-loop gain in decibels.

38. What is the unity-gain bandwidth in Problem 37?

39. For each amplifier in Figure 12–74, determine the closed-loop gain and bandwidth. The op-amps in each circuit exhibit an open-loop gain of 125 dB and a unity-gain bandwidth of 2.8 MHz.

▶ **FIGURE 12–74**

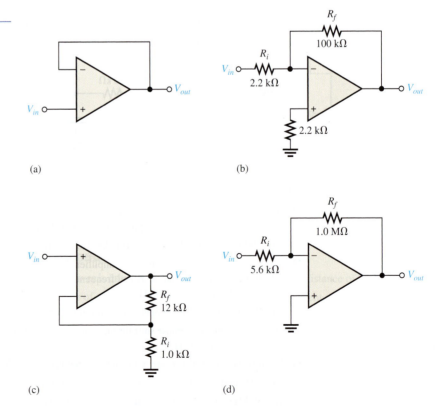

(a)   (b)

(c)   (d)

40. Which of the amplifiers in Figure 12–75 has the smaller bandwidth?

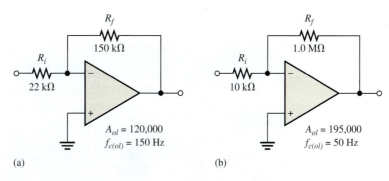

(a)   (b)

▲ **FIGURE 12–75**

Section 12–9   Troubleshooting

41. Determine the most likely fault(s) for each of the following symptoms in Figure 12–76 with a 100 mV signal applied.

   (a) No output signal.

   (b) Output severely clipped on both positive and negative swings.

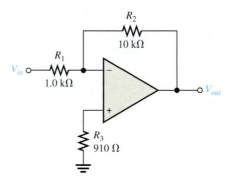

▲ **FIGURE 12–76**

42. Determine the effect on the output if the circuit in Figure 12–76 has the following fault (one fault at a time).

   (a) Output pin is shorted to the inverting input.

   (b) R_3 is open.

   (c) R_3 is 10 kΩ instead of 910 Ω.

   (d) R_1 and R_2 are swapped.

43. On the circuit board in Figure 12–77, what happens if the middle lead (wiper) of the 100 kΩ potentiometer is broken?

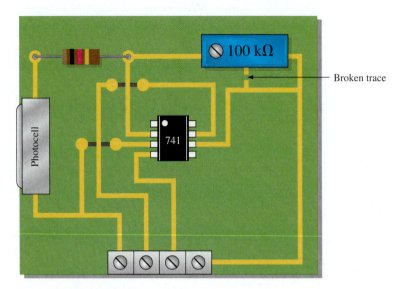

▲ FIGURE 12–77

DEVICE APPLICATION PROBLEMS

44. In the amplifier circuit of Figure 12–48, list the possible faults that will cause the push-pull stage to operate nonlinearly.

45. What indication would you observe if a 100 kΩ resistor is incorrectly installed for R_2 in Figure 12–48?

46. What voltage will you measure on the output of the amplifier in Figure 12–48 if diode D_1 opens?

DATASHEET PROBLEMS

47. Refer to the partial 741 datasheet (LM741) in Figure 12–78. Determine the input resistance (impedance) of a noninverting amplifier which uses a 741 op-amp with $R_f = 47$ kΩ and $R_i = 470$ Ω. Use typical values.

48. Refer to the partial datasheet in Figure 12–78. Determine the input impedances of an LM741 op-amp connected as an inverting amplifier with a closed-loop voltage gain of 100 and $R_f = 100$ kΩ.

49. Refer to Figure 12–78 and determine the minimum open-loop voltage gain for an LM741 expressed as a ratio of output volts to input volts.

50. Refer to Figure 12–78. How long does it typically take the output voltage of an LM741 to make a transition from −8 V to +8 V in response to a step input?

ADVANCED PROBLEMS

51. Design a noninverting amplifier with an appropriate closed-loop voltage gain of 150 and a minimum input impedance of 100 MΩ using a LM741 op-amp. Include bias current compensation.

52. Design an inverting amplifier using a LM741 op-amp. The voltage gain must be 68 ± 5% and the input impedance must be approximately 10 kΩ. Include bias current compensation.

Electrical Characteristics

Parameter	Conditions	LM741A			LM741			LM741C			Units
		Min	Typ	Max	Min	Typ	Max	Min	Typ	Max	
Input offset voltage	$T_A = 25°C$	—	—	—	—	—	—	—	—	—	
	$R_S \leq 10\ k\Omega$	—	—	—	—	—	5.0	—	2.0	6.0	mV
	$R_S \leq 50\ \Omega$	—	0.8	3.0	—	—	—	—	—	—	mV
	$T_{AMIN} \leq T_A \leq T_{AMAX}$	—	—	—	—	—	—	—	—	—	
	$R_S \leq 50\ \Omega$	—	—	4.0	—	—	—	—	—	—	mV
	$R_S \leq 10\ k\Omega$	—	—	—	—	—	6.0	—	—	7.5	mV
Average input offset voltage drift		—	—	15	—	—	—	—	—	—	μV/°C
		—	—	—	—	—	—	—	—	—	
Input offset voltage adjustment range	$T_A = 25°C, V_S = \pm 20\ V$	±10	—	—	—	±15	—	—	±15	—	mV
		—	—	—	—	—	—	—	—	—	
Input offset current	$T_A = 25°C$	—	3.0	30	—	20	200	—	20	200	nA
	$T_{AMIN} \leq T_A \leq T_{AMAX}$	—	—	70	—	85	500	—	—	300	nA
Average input offset current drift		—	—	0.5	—	—	—	—	—	—	nA/°C
		—	—	—	—	—	—	—	—	—	
Input bias current	$T_A = 25°C$	—	30	80	—	80	500	—	80	500	nA
	$T_{AMIN} \leq T_A \leq T_{AMAX}$	—	—	0.210	—	—	1.5	—	—	0.8	μA
Input resistance	$T_A = 25°C, V_S = \pm 20\ V$	1.0	6.0	—	0.3	2.0	—	0.3	2.0	—	MΩ
	$T_{AMIN} \leq T_A \leq T_{AMAX}$	0.5	—	—	—	—	—	—	—	—	MΩ
	$V_S = \pm 20\ V$	—	—	—	—	—	—	—	—	—	
Input voltage range	$T_A = 25°C$	—	—	—	—	—	—	—	±12	±13	V
	$T_{AMIN} \leq T_A \leq T_{AMAX}$	—	—	—	±12	±13	—	—	—	—	V
Large-signal voltage gain	$T_A = 25°C, R_L \geq 2\ k\Omega$	—	—	—	—	—	—	—	—	—	
	$V_S = \pm 20\ V, V_O = \pm 15\ V$	50	—	—	—	—	—	—	—	—	V/mV
	$V_S = \pm 15\ V, V_O = \pm 10\ V$	—	—	—	50	200	—	20	200	—	V/mV
	$T_{AMIN} \leq T_A \leq T_{AMAX}$	—	—	—	—	—	—	—	—	—	
	$R_L \geq 2\ k\Omega$	—	—	—	—	—	—	—	—	—	
	$V_S = \pm 20\ V, V_O = \pm 15\ V$	32	—	—	—	—	—	—	—	—	V/mV
	$V_S = \pm 15\ V, V_O = \pm 10\ V$	—	—	—	25	—	—	15	—	—	V/mV
	$V_S = \pm 5\ V, V_O = \pm 2\ V$	10	—	—	—	—	—	—	—	—	V/mV
Output voltage swing	$V_S = \pm 20\ V$	—	—	—	—	—	—	—	—	—	
	$R_L \geq 10\ k\Omega$	±16	—	—	—	—	—	—	—	—	V
	$R_L \geq 2\ k\Omega$	±15	—	—	—	—	—	—	—	—	V
	$V_S = \pm 15\ V$	—	—	—	—	—	—	—	—	—	
	$R_L \geq 10\ k\Omega$	—	—	—	±12	±14	—	±12	±14	—	V
	$R_L \geq 2\ k\Omega$	—	—	—	±10	±13	—	±10	±13	—	V
Output short circuit Current	$T_A = 25°C$	10	25	35	—	25	—	—	25	—	mA
	$T_{AMIN} \leq T_A \leq T_{AMAX}$	10	—	40	—	—	—	—	—	—	mA
Common-mode rejection ratio	$T_{AMIN} \leq T_A \leq T_{AMAX}$	—	—	—	—	—	—	—	—	—	
	$R_S \leq 10\ k\Omega, V_{CM} = \pm 12\ V$	—	—	—	70	90	—	70	90	—	dB
	$R_S \leq 50\ \Omega, V_{CM} = \pm 12\ V$	80	95	—	—	—	—	—	—	—	dB
Supply voltage rejection ratio	$T_{AMIN} \leq T_A \leq T_{AMAX},$	—	—	—	—	—	—	—	—	—	
	$V_S = \pm 20\ V$ to $V_S = \pm 5\ V$	—	—	—	—	—	—	—	—	—	
	$R_S \leq 50\ \Omega$	86	96	—	—	—	—	—	—	—	dB
	$R_S \leq 10\ k\Omega$	—	—	—	77	96	—	77	96	—	dB
Transient response	$T_A = 25°C$, Unity gain	—	—	—	—	—	—	—	—	—	
Rise time		—	0.25	0.8	—	0.3	—	—	0.3	—	μs
Overshoot		—	6.0	20	—	5	—	—	5	—	%
Bandwidth	$T_A = 25°C$	0.437	1.5	—	—	—	—	—	—	—	MHz
Slew rate	$T_A = 25°C$, Unity gain	0.3	0.7	—	—	0.5	—	—	0.5	—	V/μs
Supply current	$T_A = 25°C$			—	—	1.7	2.8	—	1.7	2.8	mA
Power consumption	$T_A = 25°C$	—	—	—	—	—	—	—	—	—	
	$V_S = \pm 20\ V$	—	80	150	—	—	—	—	—	—	mW
	$V_S = \pm 15\ V$	—	—	—	—	50	85	—	50	85	mW
LM741A	$V_S = \pm 20\ V$	—	—	—	—	—	—	—	—	—	
	$T_A = T_{AMIN}$	—	—	165	—	—	—	—	—	—	mW
	$T_A = T_{AMAX}$	—	—	135	—	—	—	—	—	—	mW
LM741	$V_S = \pm 15\ V$	—	—	—	—	—	—	—	—	—	
	$T_A = T_{AMIN}$	—	—	—	—	60	100	—	—	—	mW
	$T_A = T_{AMAX}$	—	—	—	—	45	75	—	—	—	mW

▲ FIGURE 12–78

53. Design a noninverting amplifier with an upper critical frequency, f_{cu}, of 10 kHz using an LM741 op-amp. The dc supply voltages are ± 15 V. Refer to Figure 12–79. Include bias current compensation.

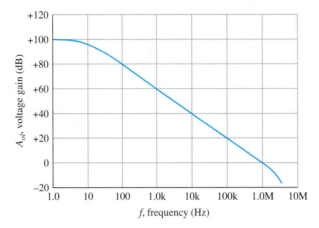

▲ **FIGURE 12–79**

54. For the circuit you designed in Problem 53, determine the minimum load resistance if the minimum output voltage swing is to be ± 10 V. Refer to the datasheet graphs in Figure 12–80.

55. Design an inverting amplifier using an LM741 op-amp if a midrange voltage gain of 50 and a bandwidth of 20 kHz is required. Include bias current compensation.

56. What is the maximum closed-loop voltage gain that can be achieved with an LM741 op-amp if the bandwidth must be no less than 5 kHz?

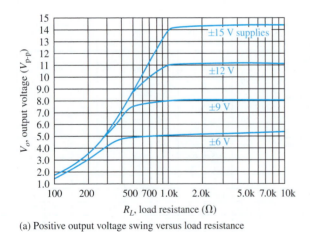

(a) Positive output voltage swing versus load resistance

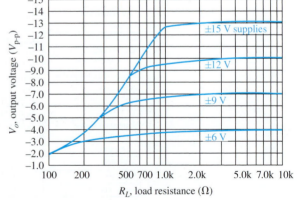

(b) Negative output voltage swing versus load resistance

▲ **FIGURE 12–80**

MULTISIM TROUBLESHOOTING PROBLEMS

These file circuits are in the Troubleshooting Problems folder on the website.

57. Open file TPM12-57 and determine the fault.

58. Open file TPM12-58 and determine the fault.

59. Open file TPM12-59 and determine the fault.

60. Open file TPM12-60 and determine the fault.

61. Open file TPM12-61 and determine the fault.

62. Open file TPM12-62 and determine the fault.

63. Open file TPM12-63 and determine the fault.

64. Open file TPM12-64 and determine the fault.

65. Open file TPM12-65 and determine the fault.

66. Open file TPM12-66 and determine the fault.

67. Open file TPM12-67 and determine the fault.

68. Open file TPM12-68 and determine the fault.

69. Open file TPM12-69 and determine the fault.

70. Open file TPM12-70 and determine the fault.

71. Open file TPM12-71 and determine the fault.

72. Open file TPM12-72 and determine the fault.

13

BASIC OP-AMP CIRCUITS

CHAPTER OBJECTIVES

◆ Describe and analyze the operation of several types of comparator circuits

◆ Describe and analyze the operation of several types of summing amplifiers

◆ Describe and analyze the operation of integrators and differentiators

◆ Troubleshoot op-amp circuits

KEY TERMS

◆ Comparator
◆ Hysteresis
◆ Schmitt trigger
◆ Bounding
◆ Summing amplifier
◆ Integrator
◆ Differentiator

DEVICE APPLICATION PREVIEW

In the Device Application in this chapter, an audio signal generator is modified to include a pulse generator to provide a signal source for digital circuits. A voltage comparator generates the pulse waveform from the sine wave output of the audio generator. The duty cycle of the pulse waveform can be varied and is compatible with +5 V logic circuits.

VISIT THE WEBSITE

Study aids and Multisim files for this chapter are available at https://www.pearsonhighered.com/careersresources/

INTRODUCTION

In the last chapter, you learned about the principles, operation, and characteristics of the operational amplifier. Op-amps are used in such a wide variety of circuits and applications that it is impossible to cover all of them in one chapter, or even in one book. Therefore, in this chapter, four fundamentally important circuits are covered to give you a foundation in op-amp circuits.

13–1 COMPARATORS

Operational amplifiers are often used as comparators to compare the amplitude of one voltage with another. In this application, the op-amp is used in the open-loop configuration, with the input voltage on one input and a reference voltage on the other.

After completing this section, you should be able to

❑ **Describe and analyze the operation of several types of comparator circuits**
❑ Discuss the operation of a zero-level detector
❑ Describe the operation of a nonzero-level detector
   ◆ Calculate the reference voltage   ◆ Analyze a nonzero-level detector
❑ Discuss how input noise affects comparator operation
   ◆ Define *hysteresis*   ◆ Explain how to reduce noise effects with hysteresis
   ◆ Calculate the upper and lower trigger points   ◆ Explain what a Schmitt trigger is
❑ Describe the operation of comparators with output bounding
   ◆ Define *bounding*   ◆ Analyze a comparator with both hysteresis and output bounding
❑ Discuss examples of comparator applications
   ◆ Explain the operation of an over-temperature sensing circuit   ◆ Describe analog-to-digital (A/D) conversion

A **comparator** is a specialized op-amp circuit that compares two input voltages and produces an output that is always at either one of two states, indicating the greater or less than relationship between the inputs. Comparators provide very fast switching times, and many have additional capabilities (such as fast propagation delay or internal reference voltages) to optimize the comparison function. For example, some ultra-high-speed comparators can have propagation delays of as little as 500 ps. Because the output is always in one of two states, comparators are often used to interface between an analog and digital circuit.

For less critical applications, an op-amp running without negative feedback (open-loop) is often used as a comparator. Although op-amps are much slower and lack other special features, they have very high open-loop gain, which enables them to detect very tiny differences in the inputs. In general, comparators cannot be used as op-amps, but op-amps can be used as comparators in noncritical applications. Because an op-amp without negative feedback is essentially a comparator, we will look at the comparison function using a typical op-amp.

Zero-Level Detection

One application of an op-amp used as a comparator is to determine when an input voltage exceeds a certain level. Figure 13–1(a) shows a zero-level detector. Notice that the inverting (−) input is grounded to produce a zero level and that the input signal voltage is applied to the noninverting (+) input. Because of the high open-loop voltage gain, a very small difference voltage between the two inputs drives the amplifier into saturation, causing the output voltage to go to its limit. For example, consider an op-amp having $A_{ol} = 100,000$. A voltage difference of only 0.25 mV between the inputs could produce an output voltage of $(0.25 \text{ mV})(100,000) = 25$ V *if* the op-amp were capable. However, since most op-amps have maximum output voltage limitations near the value of their dc supply voltages, the device would be driven into saturation.

Figure 13–1(b) shows the result of a sinusoidal input voltage applied to the noninverting (+) input of the zero-level detector. When the sine wave is positive, the output is at its maximum positive level. When the sine wave crosses 0, the amplifier is driven to its opposite state and the output goes to its maximum negative level, as shown. As you can see, the zero-level detector can be used as a squaring circuit to produce a square wave from a sine wave. The rise and fall time of the square wave is determined by the slew rate of the op-amp; a true comparator will have a shorter rise and fall time and better response.

The op-amp as a zero-level detector.

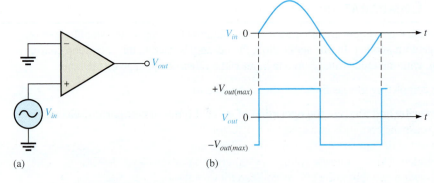

(a)

(b)

Nonzero-Level Detection

The zero-level detector in Figure 13–1 can be modified to detect positive and negative voltages by connecting a fixed reference voltage source to the inverting (−) input, as shown in Figure 13–2(a). A more practical arrangement is shown in Figure 13–2(b) using a voltage divider to set the reference voltage, V_{REF}, as follows:

$$V_{REF} = \frac{R_2}{R_1 + R_2}(+V)$$

where $+V$ is the positive op-amp dc supply voltage. The circuit in Figure 13–2(c) uses a zener diode to set the reference voltage ($V_{REF} = V_Z$). As long as V_{in} is less than V_{REF}, the

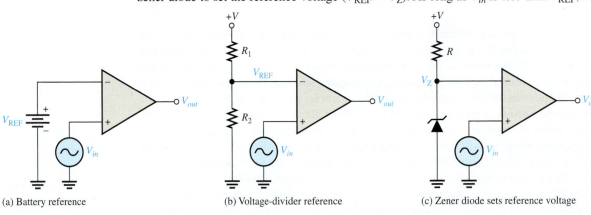

(a) Battery reference

(b) Voltage-divider reference

(c) Zener diode sets reference voltage

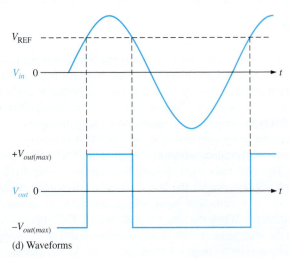

(d) Waveforms

▲ FIGURE 13–2

Nonzero-level detectors.

output remains at the maximum negative level. When the input voltage exceeds the reference voltage, the output goes to its maximum positive voltage, as shown in Figure 13–2(d) with a sinusoidal input voltage.

EXAMPLE 13–1

The input signal in Figure 13–3(a) is applied to the comparator in Figure 13–3(b). Draw the output showing its proper relationship to the input signal. Assume the maximum output levels of the comparator are ± 14 V.

▶ FIGURE 13–3

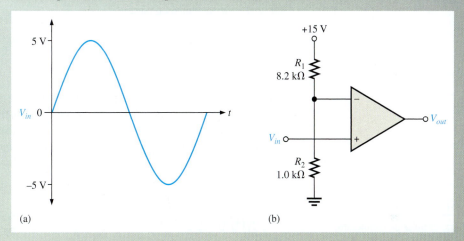

(a)   (b)

Solution   The reference voltage is set by R_1 and R_2 as follows:

$$V_{REF} = \frac{R_2}{R_1 + R_2}(+V) = \frac{1.0 \text{ k}\Omega}{8.2 \text{ k}\Omega + 1.0 \text{ k}\Omega}(+15 \text{ V}) = 1.63 \text{ V}$$

As shown in Figure 13–4, each time the input exceeds +1.63 V, the output voltage switches to its +14 V level, and each time the input goes below +1.63 V, the output switches back to its −14 V level.

▶ FIGURE 13–4

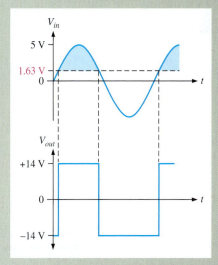

Related Problem*   Determine the reference voltage in Figure 13–3 if $R_1 = 22$ kΩ and $R_2 = 3.3$ kΩ.

***Answers can be found at www.pearsonhighered.com/floyd.**

 Open the Multisim file EXM13-01 or file LT Spice EXS13-01 in the Examples folder on the website. Compare the output waveform to the specified input at any arbitrary frequency and verify that the reference voltage agrees with the calculated value.

Effects of Input Noise on Comparator Operation

In many practical situations, noise (unwanted voltage fluctuations) appears on the input line. This noise voltage becomes superimposed on the input voltage, as shown in Figure 13–5 for the case of a sine wave, and can cause a comparator to erratically switch output states.

▶ **FIGURE 13–5**

Sine wave with superimposed noise.

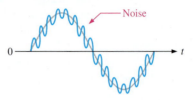

In order to understand the potential effects of noise voltage, consider a low-frequency sinusoidal voltage applied to the noninverting (+) input of an op-amp comparator used as a zero-level detector, as shown in Figure 13–6(a). Part (b) of the figure shows the input sine wave plus noise and the resulting output. When the sine wave approaches 0, the fluctuations due to noise may cause the total input to vary above and below 0 several times, thus producing an erratic output voltage.

▶ **FIGURE 13–6**

Effects of noise on comparator circuit.

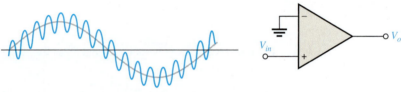

(a)

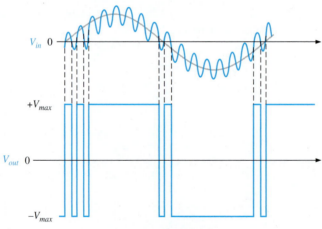

(b)

Reducing Noise Effects with Hysteresis An erratic output voltage caused by noise on the input occurs because the op-amp comparator switches from its negative output state to its positive output state at the same input voltage level that causes it to switch in the opposite direction, from positive to negative. This unstable condition occurs when the input voltage hovers around the reference voltage, and any small noise fluctuations cause the comparator to switch first one way and then the other.

In order to make the comparator less sensitive to noise, a technique incorporating positive feedback, called **hysteresis**, can be used. Basically, hysteresis means that there is a higher reference level when the input voltage goes from a lower to higher value than when it goes from a higher to a lower value. A good example of hysteresis is a common household thermostat that turns the furnace on at one temperature and off at another.

The two reference levels are referred to as the upper trigger point (UTP) and the lower trigger point (LTP). This two-level hysteresis is established with a positive feedback arrangement, as shown in Figure 13–7. Notice that the noninverting (+) input is connected to a resistive voltage divider such that a portion of the output voltage is fed back to the input. The input signal is applied to the inverting (−) input in this case.

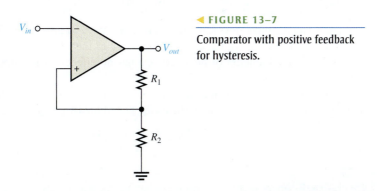

◀ **FIGURE 13–7**

Comparator with positive feedback for hysteresis.

The basic operation of the comparator with hysteresis is illustrated in Figure 13–8. Assume that the output voltage is at its positive maximum, $+V_{out(max)}$. The voltage fed back to the noninverting input is V_{UTP} and is expressed as

$$V_{UTP} = \frac{R_2}{R_1 + R_2}(+V_{out(max)})$$

Equation 13–1

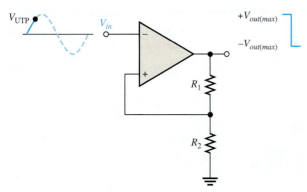

(a) When the output is at the maximum positive voltage and the input exceeds UTP, the output switches to the maximum negative voltage.

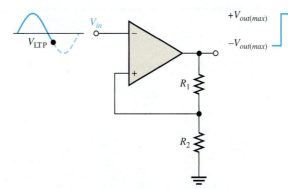

(b) When the output is at the maximum negative voltage and the input goes below LTP, the output switches back to the maximum positive voltage.

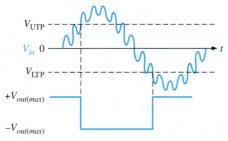

(c) Device triggers only once when UTP or LTP is reached; thus, there is immunity to noise that is riding on the input signal.

▲ **FIGURE 13–8**

Operation of a comparator with hysteresis.

When V_{in} exceeds V_{UTP}, the output voltage drops to its negative maximum, $-V_{out(max)}$, as shown in part (a). Now the voltage fed back to the noninverting input is V_{LTP} and is expressed as

Equation 13–2

$$V_{LTP} = \frac{R_2}{R_1 + R_2}(-V_{out(max)})$$

The input voltage must now fall below V_{LTP}, as shown in part (b), before the device will switch from the maximum negative voltage back to the maximum positive voltage. This means that a small amount of noise voltage has no effect on the output, as illustrated by Figure 13–8(c).

A comparator with built-in hysteresis is sometimes known as a **Schmitt trigger**. The amount of hysteresis is defined by the difference of the two trigger levels.

Equation 13–3

$$V_{HYS} = V_{UTP} - V_{LTP}$$

EXAMPLE 13–2

Determine the upper and lower trigger points for the comparator circuit in Figure 13–9. Assume that $+V_{out(max)} = +5$ V and $-V_{out(max)} = -5$ V.

▶ FIGURE 13–9

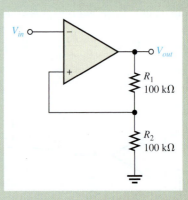

Solution

$$V_{UTP} = \frac{R_2}{R_1 + R_2}(+V_{out(max)}) = 0.5(5\ V) = +2.5\ V$$

$$V_{LTP} = \frac{R_2}{R_1 + R_2}(-V_{out(max)}) = 0.5(-5\ V) = -2.5\ V$$

Related Problem

Determine the upper and lower trigger points in Figure 13–9 for $R_1 = 68$ kΩ and $R_2 = 82$ kΩ. Also assume the maximum output voltage levels are now ± 7 V.

Open the Multisim file EXM13-02 or the LT Spice file EXS13-02 in the Examples folder on the website. Determine the upper and lower trigger points and compare with the calculated values using a 5 V rms, 60 Hz sine wave for the input.

Output Bounding

In some applications, it is necessary to limit the output voltage levels of a comparator to a value less than that provided by the saturated op-amp. A single zener diode can be used, as shown in Figure 13–10, to limit the output voltage to the zener voltage in one direction and to the forward diode voltage drop in the other. This process of limiting the output range is called **bounding**.

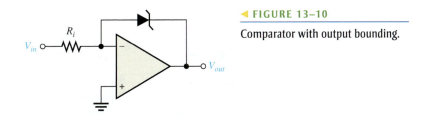

◀ FIGURE 13–10

Comparator with output bounding.

The operation is as follows. Since the anode of the zener is connected to the inverting (−) input, it is at virtual ground (≅ 0 V). Therefore, when the output voltage reaches a positive value equal to the zener voltage, it limits at that value, as illustrated in Figure 13–11(a). When the output switches negative, the zener acts as a regular diode and becomes forward-biased at 0.7 V, limiting the negative output voltage to this value, as shown in part (b). Turning the zener around limits the output voltage in the opposite direction.

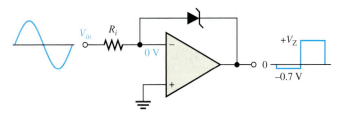

(a) Bounded at a positive value

◀ FIGURE 13–11

Operation of a bounded comparator.

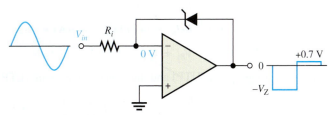

(b) Bounded at a negative value

Two zener diodes arranged as in Figure 13–12 limit the output voltage to the zener voltage plus the forward voltage drop (0.7 V) of the forward-biased zener, both positively and negatively, as shown.

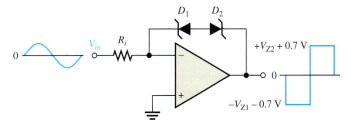

◀ FIGURE 13–12

Double-bounded comparator.

EXAMPLE 13–3

Determine the output voltage waveform for Figure 13–13.

Solution   This comparator has both hysteresis and zener bounding. The voltage across D_1 and D_2 in either direction is 4.7 V + 0.7 V = 5.4 V. This is because one zener is always forward-biased with a drop of 0.7 V when the other one is in breakdown.

is one of the drawbacks of the simultaneous ADC, but IC technology has reduced the problem somewhat by combining multiple comparators and associated circuits on a single IC chip. For example, 6- or 8-bit flash converters are readily available. These ADCs are useful in applications that require the fastest possible conversion times, such as video processing.

▶ FIGURE 13–16

A simplified simultaneous (flash) analog-to-digital converter (ADC) using op-amps as comparators.

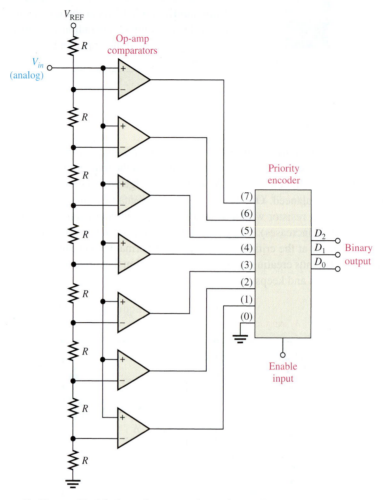

In Figure 13–16, the reference voltage for each comparator is set by the resistive voltage-divider circuit and V_{REF}. The output of each comparator is connected to an input of the priority encoder. The *priority encoder* is a digital device that produces a binary number on its output representing the highest value input.

The encoder samples its input when a pulse occurs on the enable line (sampling pulse), and a three-digit binary number proportional to the value of the analog input signal appears on the encoder's outputs.

The sampling rate determines the accuracy with which the sequence of binary numbers represents the changing input signal. The more samples taken in a given unit of time, the more accurately the analog signal is represented in digital form. Sample rates of over 1 GHz are available with flash converters. The output of the priority encoder is latched (held) during the interval between samples.

The following example illustrates the basic operation of the simultaneous ADC in Figure 13–16.

EXAMPLE 13–4

Determine the binary number sequence of the three-digit simultaneous ADC in Figure 13–16 for the input signal in Figure 13–17 and the sampling pulses (encoder enable) shown. Assume the output is latched (held) after each sample pulse. Draw the resulting digital output waveforms.

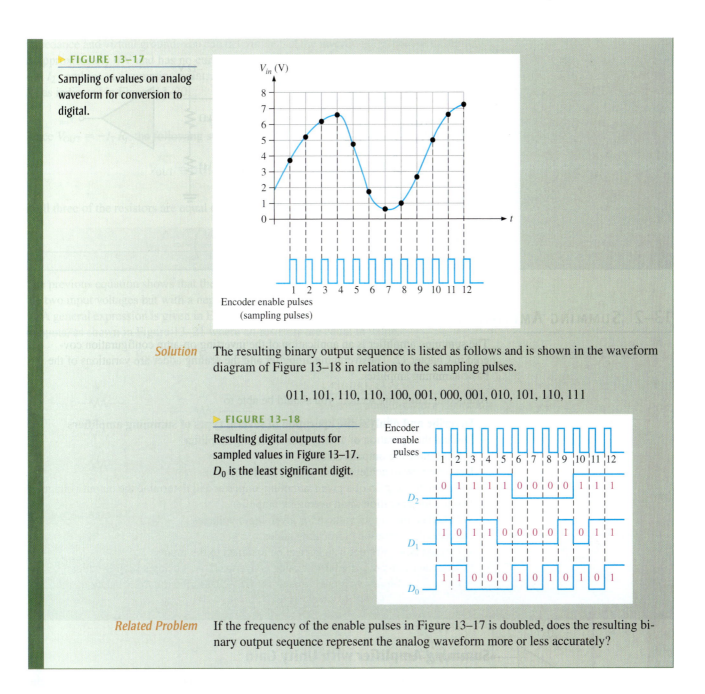

▶ FIGURE 13–17

Sampling of values on analog waveform for conversion to digital.

Encoder enable pulses
(sampling pulses)

Solution   The resulting binary output sequence is listed as follows and is shown in the waveform diagram of Figure 13–18 in relation to the sampling pulses.

011, 101, 110, 110, 100, 001, 000, 001, 010, 101, 110, 111

▶ FIGURE 13–18

Resulting digital outputs for sampled values in Figure 13–17. D_0 is the least significant digit.

Related Problem   If the frequency of the enable pulses in Figure 13–17 is doubled, does the resulting binary output sequence represent the analog waveform more or less accurately?

Specific Comparators

The LM111-N and LM311-N are examples of specific comparators that exhibit high switching speeds (200 ns) and other features not normally found on the general type of op-amp. These comparators can operate with supply voltages from ±15 V to a single +5 V. The open collector output provides the capability of driving loads that require voltages up to 50 V referenced to ground or to the supply voltages. An offset balancing input and a strobe input allow the output to be turned on or off regardless of the differential input.

SECTION 13–1 CHECKUP Answers can be found at www .pearsonhighered.com/floyd.	1. What is the reference voltage for each comparator in Figure 13–19? 2. What is the purpose of hysteresis in a comparator? 3. Define the term *bounding* in relation to a comparator's output.

From Figure 13–27(b), the first binary input code is 0000, which produces an output voltage of 0 V. The next input code is 0001 (it stands for decimal 1). For this, the output voltage is −0.25 V. The next code is 0010, which produces an output voltage of −0.5 V. The next code is 0011, which produces an output voltage of −0.25 V + (−0.5 V) = −0.75 V. Each successive binary code increases the output voltage by −0.25 V. So, for this particular straight binary sequence on the inputs, the output is a stairstep waveform going from 0 V to −3.75 V in −0.25 V steps, as shown in Figure 13–28. If the steps are very small, the output approximates a straight line (linear).

▶ FIGURE 13–28

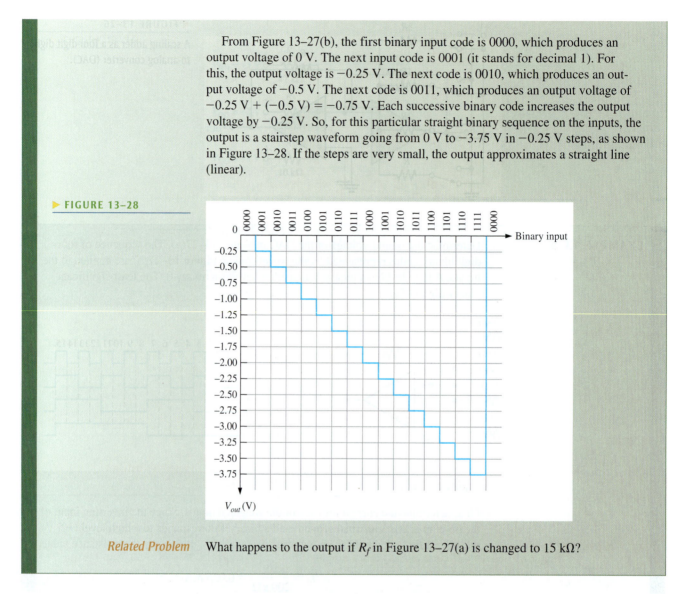

Related Problem What happens to the output if R_f in Figure 13–27(a) is changed to 15 kΩ?

As mentioned before, the *R/2R* ladder is more commonly used for D/A conversion than the scaling adder and is shown in Figure 13–29 for four bits. It overcomes one of the disadvantages of the binary-weighted-input DAC because it requires only two resistor values.

▶ FIGURE 13–29

An *R/2R* ladder DAC.

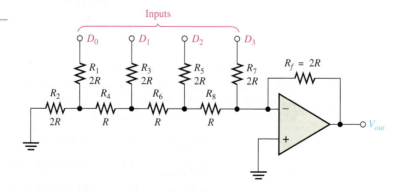

Assume that the D_3 input is HIGH (+5 V) and the others are LOW (ground, 0 V). This condition represents the binary number 1000. A circuit analysis will show that this reduces to the equivalent form shown in Figure 13–30(a). Essentially no current goes through the

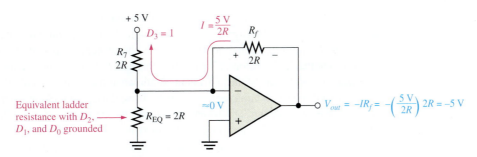

(a) Equivalent circuit for $D_3 = 1$, $D_2 = 0$, $D_1 = 0$, $D_0 = 0$

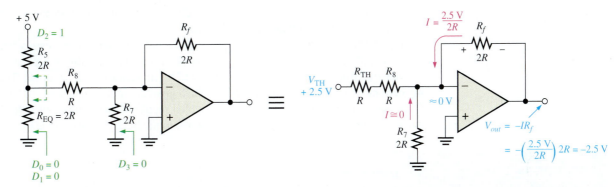

(b) Equivalent circuit for $D_3 = 0$, $D_2 = 1$, $D_1 = 0$, $D_0 = 0$

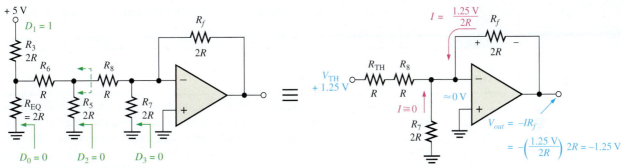

(c) Equivalent circuit for $D_3 = 0$, $D_2 = 0$, $D_1 = 1$, $D_0 = 0$

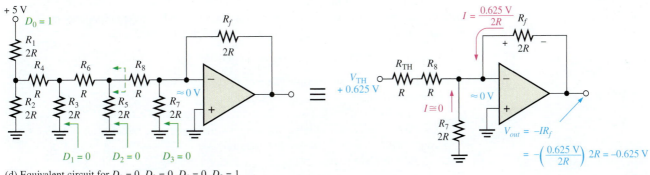

(d) Equivalent circuit for $D_3 = 0$, $D_2 = 0$, $D_1 = 0$, $D_0 = 1$

▲ FIGURE 13–30

Analysis of the $R/2R$ ladder DAC.

$2R$ equivalent resistance because the inverting input is at virtual ground. Thus, all of the current ($I = 5$ V/$2R$) through R_7 is also through R_f, and the output voltage is -5 V. The operational amplifier keeps the inverting ($-$) input near zero volts (≈ 0 V) because of negative feedback. Therefore, all current is through R_f rather than into the inverting input.

Figure 13–30(b) shows the equivalent circuit when the D_2 input is at $+5$ V and the others are at ground. This condition represents 0100. If we thevenize looking from R_8, we get 2.5 V in series with R, as shown. This results in a current through R_f of $I = 2.5$ V/$2R$, which gives an output voltage of -2.5 V. Keep in mind that there is no current from the op-amp inverting input and that there is no current through R_7 because it has 0 V across it, due to the virtual ground.

Figure 13–30(c) shows the equivalent circuit when the D_1 input is at $+5$ V and the others are at ground. This condition represents 0010. Again thevenizing looking from R_8, you get 1.25 V in series with R as shown. This results in a current through R_f of $I = 1.25$ V/$2R$, which gives an output voltage of -1.25 V.

In part (d) of Figure 13–30, the equivalent circuit representing the case where D_0 is at $+5$ V and the other inputs are at ground is shown. This condition represents 0001. Thevenizing from R_8 gives an equivalent of 0.625 V in series with R as shown. The resulting current through R_f is $I = 0.625$ V/$2R$, which gives an output voltage of -0.625 V.

Notice that each successively lower-weighted input produces an output voltage that is halved, so that the output voltage is proportional to the binary weight of the input bits.

SECTION 13–2 CHECKUP

1. Define *summing point*.
2. What is the value of R_f/R for a five-input averaging amplifier?
3. A certain scaling adder has two inputs, one having twice the weight of the other. If the resistor value for the lower-weighted input is 10 kΩ, what is the value of the other input resistor?

13–3 INTEGRATORS AND DIFFERENTIATORS

An op-amp **integrator** simulates mathematical integration, which is basically a summing process that determines the total area under the curve of a function. An op-amp **differentiator** simulates mathematical differentiation, which is a process of determining the instantaneous rate of change of a function. It is not necessary for you to understand mathematical integration or differentiation, at this point, in order to learn how an integrator and differentiator work. Ideal integrators and differentiators are used to show basic principles. Practical integrators often have an additional resistor in parallel with the feedback capacitor to prevent saturation. Practical differentiators may include a resistor in series with the comparator to reduce high frequency noise.

After completing this section, you should be able to

❑ **Describe and analyze the operation of integrators and differentiators**
❑ Describe and identify the op-amp integrator
  ♦ Discuss the ideal integrator   ♦ Explain how a capacitor charges
  ♦ Discuss the capacitor voltage, the output voltage, and the rate of change of the output voltage   ♦ Describe the practical integrator
❑ Describe and identify the op-amp differentiator
  ♦ Discuss the ideal differentiator   ♦ Discuss the practical differentiator

The Op-Amp Integrator

The Ideal Integrator   An ideal integrator is shown in Figure 13–31. Notice that the feedback element is a capacitor that forms an RC circuit with the input resistor.

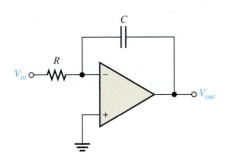

◀ FIGURE 13–31

An ideal op-amp integrator.

How a Capacitor Charges   To understand how an integrator works, it is important to review how a capacitor charges. Recall that the charge Q on a capacitor is proportional to the charging current (I_C) and the time (t).

$$Q = I_C t$$

Also, in terms of the voltage, the charge on a capacitor is

$$Q = CV_C$$

From these two relationships, the capacitor voltage can be expressed as

$$V_C = \left(\frac{I_C}{C}\right)t$$

This expression has the form of an equation for a straight line that begins at zero with a constant slope of I_C/C. Remember from algebra that the general formula for a straight line is $y = mx + b$. In this case, $y = V_C$, $m = I_C/C$, $x = t$, and $b = 0$.

Recall that the capacitor voltage in a simple RC circuit with a constant input voltage is not linear but is exponential. This is because the charging current continuously decreases as the capacitor charges and causes the rate of change of the voltage to continuously decrease. The key thing about using an op-amp with an RC circuit to form an integrator is that if the capacitor's charging current is made constant, the output will be a straight-line (linear) voltage rather than an exponential voltage. Now let's see why this is true.

In Figure 13–32, the inverting input of the op-amp is at virtual ground (0 V), so the voltage across R_i equals V_{in}. Therefore, the input current is

$$I_{in} = \frac{V_{in}}{R_i}$$

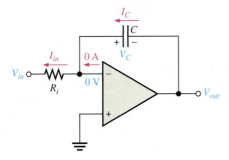

◀ FIGURE 13–32

Currents in an integrator.

If V_{in} is a constant voltage, then I_{in} is also a constant because the inverting input always remains at 0 V, keeping a constant voltage across R_i. Because of the very high input impedance of the op-amp, there is negligible current at the inverting input. This makes the constant input current charge the capacitor, as indicated in Figure 13–32, so

$$I_C = I_{in}$$

The Capacitor Voltage   Since I_{in} is constant, so is I_C. The constant I_C charges the capacitor linearly and produces a linear voltage across C. The positive side of the capacitor is held at 0 V by the virtual ground of the op-amp. The voltage on the negative side of the capacitor, which is the op-amp output voltage, decreases linearly from zero as the capacitor charges, as shown in Figure 13–33. This voltage, V_C, is called a *negative ramp* and is the consequence of a constant positive input.

▶ FIGURE 13–33

A linear ramp voltage is produced across the capacitor by the constant charging current.

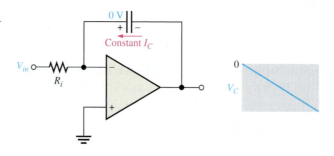

The Output Voltage V_{out} is the same as the voltage on the negative side of the capacitor. When a constant positive input voltage in the form of a step or pulse (a pulse has a constant amplitude when high) is applied, the output ramp decreases negatively until the op-amp saturates at its maximum negative level. This is indicated in Figure 13–34.

▶ FIGURE 13–34

A constant input voltage produces a ramp on the output of the integrator.

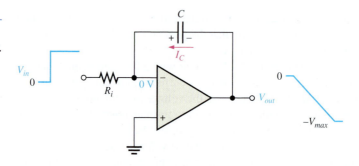

Rate of Change of the Output Voltage   The rate at which the capacitor charges, and therefore the slope of the output ramp, is set by the ratio I_C/C, as you have seen. Since $I_C = V_{in}/R_i$, the rate of change or slope of the integrator's output voltage is $\Delta V_{out}/\Delta t$.

Equation 13–7

$$\frac{\Delta V_{out}}{\Delta t} = -\frac{V_{in}}{R_iC}$$

Integrators are especially useful in triangular-wave oscillators as you will see in Chapter 16.

The Practical Integrator   The ideal integrator uses only a capacitor in the feedback path, which is open to dc. This implies that the gain at dc is the open-loop gain of the op-amp. In a practical integrator, any dc error voltage due to offset error will cause the output to produce a ramp that moves toward either positive or negative saturation (depending on the offset), even when no signal is present. Also, if the signal source is not perfectly centered with no offset, the output will move toward saturation.

Practical integrators must have some means of overcoming the effects of offset and bias current and other small differences in the circuit. Various solutions are available, such as chopper stabilized amplifiers; however, the simplest effective solution is to use a resistor in parallel with the capacitor in the feedback path, as shown in Figure 13–35. The feedback resistor, R_f, should be large compared to the input resistor R_{in}, in order to have a negligible effect on the output waveform. In addition, a compensating resistor, R_c, may be added to the noninverting input to balance the effects of bias current.

One useful application for integrators is in waveshaping of periodic inputs. If the input is a square wave, the integrator can convert it to a triangle waveform. It takes a few cycles for the output to reach steady state, but after a few cycles a square-wave input will be a triangle on the output with a dc level equal to the average of the dc input times the gain.

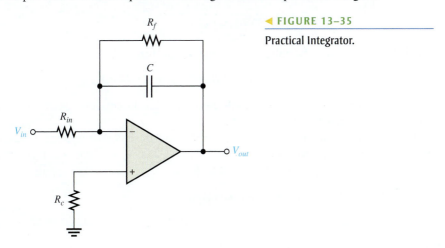

◄ FIGURE 13–35

Practical Integrator.

EXAMPLE 13–10

(a) Determine the rate of change of the output voltage in response to the input square wave, as shown for the practical integrator in Figure 13–36(a). Assume steady state conditions have been reached. The input is a 1.0 kHz, 5 V_{pp} square wave centered at 0 V.

(b) Draw the output waveform.

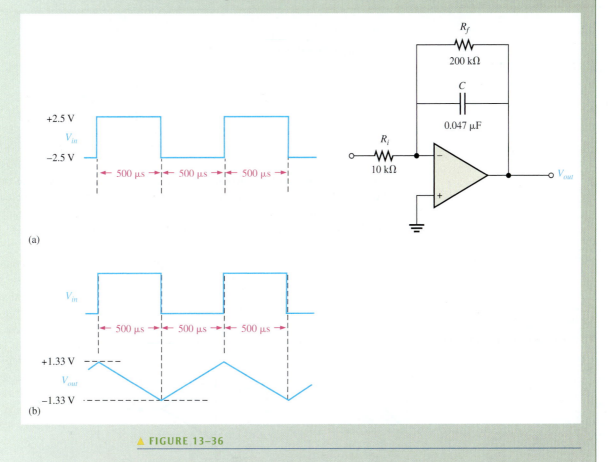

▲ FIGURE 13–36

Solution　**(a)** The rate of change of the output voltage during the time that the input is at $+2.5$ V (capacitor charging) is

$$\frac{\Delta V_{out}}{\Delta t} = -\frac{V_{in}}{R_i C} = -\frac{2.5 \text{ V}}{(10 \text{ k}\Omega)(0.047 \text{ }\mu\text{F})} = -5.32 \text{ kV/s} = \mathbf{-5.32 \text{ mV/}\mu\text{s}}$$

(b) In 500 μs (the time the pulse is high), the output changes by

$$\Delta V_{out} = (-5.32 \text{ mV/}\mu\text{s})(500 \text{ }\mu\text{s}) = -2.66 \text{ V}$$

Because the output has had time to reach steady state conditions, it is centered on 0 V and thus goes from $+1.33$ V to -1.33 V.

The time that the input is -2.5 V is also 500 μs and the charging rate is the same as before but of opposite sign. Therefore, the $\Delta V_{out} = +2.66$ V. The output will go from -1.33 V to $+1.33$ V as shown in Figure 13–36(b).

Related Problem　What happens to the rate of change of the output voltage if R_f is doubled?

The Op-Amp Differentiator

The Ideal Differentiator　An ideal differentiator is shown in Figure 13–37. Notice how the placement of the capacitor and resistor differ from the integrator. The capacitor is now the input element, and the resistor is the feedback element. A differentiator produces an output that is proportional to the rate of change of the input voltage.

▶ **FIGURE 13–37**

An ideal op-amp differentiator.

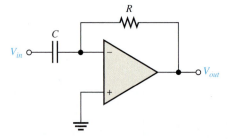

To see how the differentiator works, apply a positive-going ramp voltage to the input as indicated in Figure 13–38. In this case, $I_C = I_{in}$ and the voltage across the capacitor is equal to V_{in} at all times ($V_C = V_{in}$) because of virtual ground on the inverting input.

From the basic formula, $V_C = (I_C/C)t$, the capacitor current is

$$I_C = \left(\frac{V_C}{t}\right)C$$

▶ **FIGURE 13–38**

A differentiator with a ramp input.

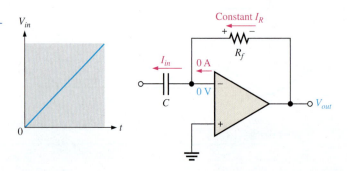

Since the current at the inverting input is negligible, $I_R = I_C$. Both currents are constant because the slope of the capacitor voltage (V_C/t) is constant. The output voltage is also constant and equal to the voltage across R_f because one side of the feedback resistor is always 0 V (virtual ground).

$$V_{out} = I_R R_f = I_C R_f$$

$$\boldsymbol{V_{out} = -\left(\frac{V_C}{t}\right)R_f C}$$

Equation 13–8

The output is negative when the input is a positive-going ramp and positive when the input is a negative-going ramp, as illustrated in Figure 13–39. During the positive slope of the input, the capacitor is charging from the input source and the constant current through the feedback resistor is in the direction shown. During the negative slope of the input, the current is in the opposite direction because the capacitor is discharging.

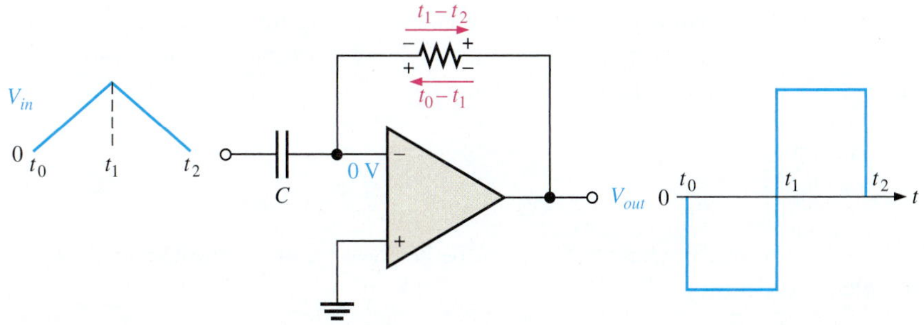

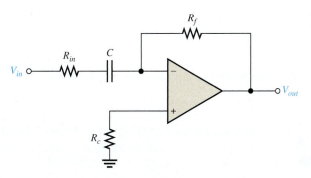

▲ FIGURE 13–39

Output of a differentiator with a series of positive and negative ramps (triangle wave) on the input.

Notice in Equation 13–8 that the term V_C/t is the slope of the input. If the slope increases, V_{out} increases. If the slope decreases, V_{out} decreases. The output voltage is proportional to the slope (rate of change) of the input. The constant of proportionality is the time constant, $R_f C$.

The Practical Differentiator   The ideal differentiator uses a capacitor in series with the inverting input. Because a capacitor has very low impedance at high frequencies, the combination of R_f and C form a very high gain amplifier at high frequencies. This means that a differentiator circuit tends to be noisy because electrical noise mainly consists of high frequencies. The solution to this problem is simply to add a resistor, R_{in}, in series with the capacitor to act as a low-pass filter and reduce the gain at high frequencies. The resistor should be small compared to the feedback resistor in order to have a negligible effect on the desired signal. Figure 13–40 shows a practical differentiator. A bias compensating resistor may also be used on the noninverting input.

◀ FIGURE 13–40

Practical Differentiator.

EXAMPLE 13–11

Determine the output voltage of the practical op-amp differentiator in Figure 13–42 for the triangular-wave input shown. The input resistor can be ignored as it is small compared to R_f.

▶ **FIGURE 13–41**

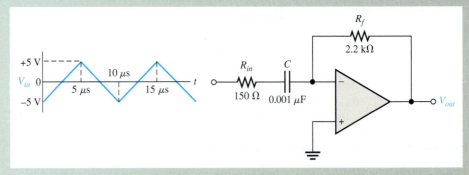

Solution

Starting at $t = 0$, the input voltage is a positive-going ramp ranging from -5 V to $+5$ V (a $+10$ V change) in 5 μs. Then it changes to a negative-going ramp ranging from $+5$ V to -5 V (a -10 V change) in 5 μs.

The time constant is

$$R_f C = (2.2 \text{ k}\Omega)(0.001 \text{ }\mu\text{F}) = 2.2 \text{ }\mu\text{s}$$

Determine the slope or rate of change (V_C/t) of the positive-going ramp and calculate the output voltage as follows:

$$\frac{V_C}{t} = \frac{10 \text{ V}}{5 \text{ }\mu\text{s}} = 2 \text{ V/}\mu\text{s}$$

$$V_{out} = -\left(\frac{V_C}{t}\right)R_f C = -(2 \text{ V/}\mu\text{s})2.2 \text{ }\mu\text{s} = -4.4 \text{ V}$$

Likewise, the slope of the negative-going ramp is -2 V/μs, and the output voltage is

$$V_{out} = -(-2 \text{ V/}\mu\text{s})2.2 \text{ }\mu\text{s} = +4.4 \text{ V}$$

Figure 13–42 shows a graph of the output voltage waveform relative to the input.

▶ **FIGURE 13–42**

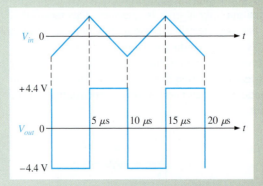

Related Problem

What would the output voltage be if the feedback resistor in Figure 13–40 is changed to 3.3 kΩ?

  Open the Multisim file EXM13-11 in the Examples folder on the website. Compare the output waveform to with the calculated value.

SECTION 13–3
CHECKUP

1. What is the feedback element in an ideal op-amp integrator?
2. For a constant input voltage to an integrator, why is the voltage across the capacitor linear?
3. What is the feedback element in an op-amp differentiator?
4. How is the output of a differentiator related to the input?

13–4 TROUBLESHOOTING

Although integrated circuit op-amps are extremely reliable and trouble-free, failures do occur from time to time. Before trying to isolate the failure, it is a good idea to analyze the symptoms to see if they point to the problem. For example, a circuit with no output could point to a failure of a power supply. One type of internal failure mode is a condition where the op-amp output is in a saturated state, resulting in a constant high or constant low level, regardless of the input. Also, external component failures will produce various types of failure modes in op-amp circuits. Some examples are presented in this section.

After completing this section, you should be able to

◻ **Troubleshoot op-amp circuits**
◻ Describe and explain symptoms of several component failures in a bounded comparator
◻ Describe symptoms of component failures in a summing amplifier

Figure 13–43 illustrates an internal failure of a comparator circuit that results in a "stuck" output.

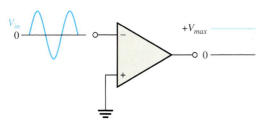

(a) Output failed in the HIGH state   (b) Output failed in the LOW state

▲ **FIGURE 13–43**

Internal comparator failures typically result in the output being "stuck" in the HIGH or LOW state.

Symptoms of External Component Failures in Comparator Circuits

A comparator with zener-bounding and hysteresis is shown in Figure 13–44. In addition to a failure of the op-amp itself, a zener diode or one of the resistors could be faulty. For

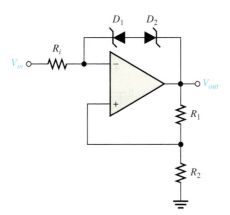

◀ **FIGURE 13–44**

A bounded comparator with hysteresis.

A new version of the signal generator is being developed that adds a pulse wave-form generator to the audio signal generator in a single unit. The pulse generator will produce an output with a variable duty cycle that can be used to drive 5 V digital logic circuits. The sine wave generator will remain the same, but the frequency control and the output terminals will be common to both the sine wave generator and the pulse genera-tor. The output function will be switch-selectable, and the pulse waveform will require an additional front panel control for adjusting the duty cycle. The minimum specifica-tions are given in Table 13–1. The front panel for the sine/pulse generator is shown in Figure 13–49.

▶ TABLE 13–1

	OUTPUT VOLTAGE RANGE	FREQUENCY RANGE	DUTY CYCLE RANGE
Sine	0.1 V–20 V p-p	20 Hz–20 kHz	
Pulse	5 V amplitude	20 Hz–20 kHz	15%–85%

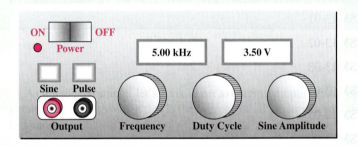

▲ FIGURE 13–49

Front panel of the sine/pulse generator.

The Circuit

The schematic of the new design is shown in Figure 13–50. The pulse waveform is de-rived from the 10 V peak sine wave that is available internally in the existing signal gen-erator. An LM111H comparator is used for producing the pulse waveform using the sine wave as the driving source. The variable reference voltage at the inverting input of the comparator provides the duty cycle control. The duty cycle adjustment range is from 10% to 90%.

The LM111H comparator has an open collector output that is pulled up to +5 V with a 1 kΩ resistor, and the emitter of the output transistor is connected to ground, as shown. As a result, the output pulses vary between 0 V and +5 V.

1. Which components determine the comparator's variable reference voltage?
2. Calculate the minimum reference voltage.
3. Calculate the maximum reference voltage.
4. What sets the amplitude of the output pulses?
5. Explain how the duty cycle control works.

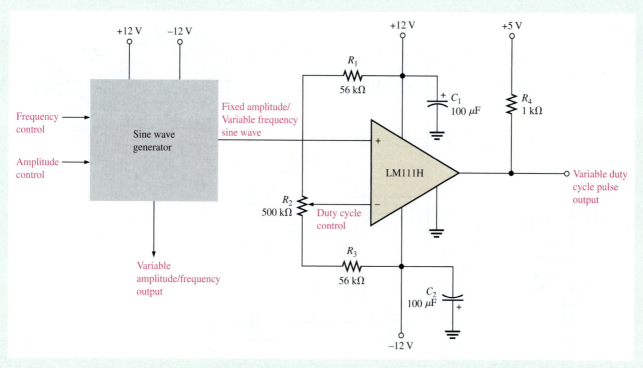

▲ FIGURE 13–50

Sine/pulse generator.

The pin diagram from the LM111H datasheet is shown in Figure 13–51. Pins 5 and 6 are unused in this application.

▶ FIGURE 13–51

Pin diagram for the LM111H comparator.

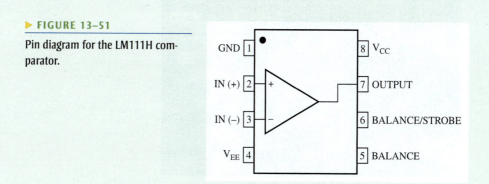

6. Referencing the pin diagram, assign pin numbers to the comparator in Figure 13–50.

Simulation

The sine/pulse generator is simulated using Multisim with an input signal of 7.07 V rms to represent the existing sine wave generator output. The results are shown in Figure 13–52 where the duty cycle of the pulse waveform is set to 50%.

7. From the scope display in Figure 13–52, verify the rms value of the sine wave.
8. Measure the amplitude of the pulse waveform on the display.
9. Verify the frequency of the waveforms on the display.

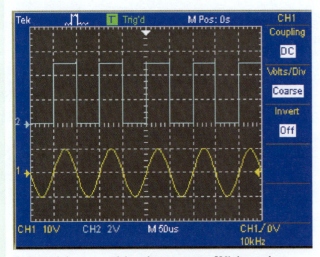

(a) Circuit screen

(b) Internal sine wave and the pulse output set at 50% duty cycle

▲ FIGURE 13–52

Simulation of the sine/pulse generator at a frequency of 10 kHz.

Figure 13–53 shows the simulation results for the pulse duty cycle measurement at test frequencies of 1 kHz and 10 kHz.

10. In Figure 13–53, determine if the minimum and maximum duty cycles meet or exceed specifications for the frequencies shown.

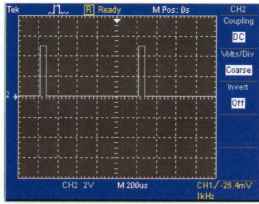

(a) Minimum duty cycle at 1 kHz

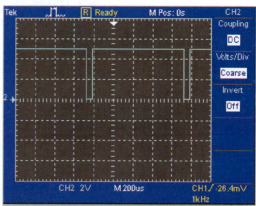

(b) Maximum duty cycle at 1 kHz

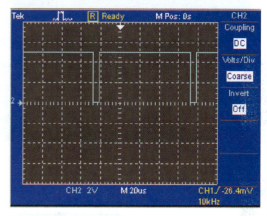

(c) Minimum duty cycle at 10 kHz

(d) Maximum duty cycle at 10 kHz

▲ FIGURE 13–53

Simulation results.

Simulate the sine/pulse generator using your Multisim or LT Spice software. Observe the output voltages with the oscilloscope as the duty cycle control is varied.

Hint: Before running a simulation that includes a comparator, it may be necessary to reset the default value of the relative error tolerance to avoid interpolation error, which results in slow transitions particularly at higher frequencies. To do this, select *Simulate* and click on *Interactive Simulation Settings*. Navigate to the *Analysis Options* tab and choose *Customize.* Change the *Relative Tolerance Error* to be le-005.

Prototyping and Testing

Now that the circuit has been simulated, the prototype circuit is constructed and tested. After the circuit is successfully tested on a protoboard, it is ready to be finalized on a printed circuit board.

Lab Experiment

To build and test a similar circuit, go to Experiment 13–A in your lab manual (*Laboratory Exercises for Electronic Devices* by David Buchla and Steven Wetterling).

Circuit Board

The pulse generator board is shown in Figure 13–54. This board will be added to the existing audio generator and connected to the front panel controls to complete the sine/pulse generator.

KEY FORMULAS

Comparator

$$13\text{--}1 \qquad V_{\text{UTP}} = \frac{R_2}{R_1 + R_2}(+V_{out(max)}) \qquad \text{Upper trigger point}$$

$$13\text{--}2 \qquad V_{\text{LTP}} = \frac{R_2}{R_1 + R_2}(-V_{out(max)}) \qquad \text{Lower trigger point}$$

$$13\text{--}3 \qquad V_{\text{HYS}} = V_{\text{UTP}} - V_{\text{LTP}} \qquad \text{Hysteresis voltage}$$

Summing Amplifier

$$13\text{--}4 \qquad V_{\text{OUT}} = -(V_{\text{IN1}} + V_{\text{IN2}} + \cdots + V_{\text{IN}n}) \qquad n\text{-input adder}$$

$$13\text{--}5 \qquad V_{\text{OUT}} = -\frac{R_f}{R}(V_{\text{IN1}} + V_{\text{IN2}} + \cdots + V_{\text{IN}n}) \qquad \text{Adder with gain}$$

$$13\text{--}6 \qquad V_{\text{OUT}} = -\left(\frac{R_f}{R_1}V_{\text{IN1}} + \frac{R_f}{R_2}V_{\text{IN2}} + \cdots + \frac{R_f}{R_n}V_{\text{IN}n}\right) \qquad \text{Scaling adder with gain}$$

Integrator and Differentiator

$$13\text{--}7 \qquad \frac{\Delta V_{out}}{\Delta t} = -\frac{V_{in}}{R_i C} \qquad \text{Integrator output rate of change}$$

$$13\text{--}8 \qquad V_{out} = -\left(\frac{V_C}{t}\right)R_f C \qquad \text{Differentiator output voltage with ramp input}$$

TRUE/FALSE QUIZ

Answers can be found at www.pearsonhighered.com/floyd.

1. The output of a comparator has two states.
2. The reference voltage on a comparator input establishes the gain.
3. Hysteresis incorporates positive feedback.
4. A comparator with hyteresis has two trigger points.
5. A summing amplifier can have more than two inputs.
6. The gain of a summing amplifier must always be unity (1).
7. DAC stands for digital-to-analog comparator.
8. An $R/2R$ ladder circuit is one form of DAC.
9. An integrator produces a ramp when a step input is applied.
10. In a practical integrator, a resistor is connected across the capacitor.
11. When a triangular waveform is applied to a differential, a sine wave appears on the output.
12. In a practical differentiator, a resistor is connected in series with the capacitor.

CIRCUIT-ACTION QUIZ

Answers can be found at www.pearsonhighered.com/floyd.

1. If R_2 opens in the comparator of Figure 13–3, the output voltage amplitude will
   (a) increase   (b) decrease   (c) not change
2. In the trigger circuit of Figure 13–9, if R_1 is decreased to 50 kΩ, the upper trigger-point voltage will
   (a) increase   (b) decrease   (b) not change

3. If the zener diodes in Figure 13–13 are changed to ones with a rating of 5.6 V, the output voltage amplitude will

    (a) increase   (b) decrease   (c) not change

4. If the top resistor in Figure 13–22 opens, the output voltage will

    (a) increase   (b) decrease   (c) not change

5. If V_{IN2} is changed to −1 V in Figure 13–22, the output voltage will

    (a) increase   (b) decrease   (c) not change

6. If V_{IN1} is increased to 0.4 V and V_{IN2} is reduced to 0.3 V in Figure 13–23, the output voltage will

    (a) increase   (b) decrease   (c) not change

7. If V_{IN3} is changed to −7 V in Figure 13–24, the output voltage will

    (a) increase   (b) decrease   (c) not change

8. If R_f in Figure 13–25 opens, the output voltage will

    (a) increase   (b) decrease   (c) not change

9. If the value of C in Figure 13–36 is reduced, the frequency of the output waveform will

    (a) increase   (b) decrease   (c) not change

10. If the frequency of the input waveform in Figure 13–40 is increased, the amplitude of the output voltage will

    (a) increase   (b) decrease   (c) not change

SELF-TEST

Answers can be found at www.pearsonhighered.com/floyd.

Section 13–1

1. In a zero-level detector, the output changes state when the input

    (a) is positive   (b) is negative   (c) crosses zero   (d) has a zero rate of change

2. The zero-level detector is one application of a

    (a) comparator   (b) differentiator   (c) summing amplifier   (d) diode

3. Noise on the input of a comparator can cause the output to

    (a) hang up in one state

    (b) go to zero

    (c) change back and forth erratically between two states

    (d) produce the amplified noise signal

4. The effects of noise can be reduced by

    (a) lowering the supply voltage   (b) using positive feedback

    (c) using negative feedback   (d) using hysteresis

    (e) answers (a) and (d)

5. A comparator with hysteresis

    (a) has one trigger point   (b) has two trigger points

    (c) has a variable trigger point   (d) is like a magnetic circuit

6. In a comparator with hysteresis,

    (a) a bias voltage is applied between the two inputs

    (b) only one supply voltage is used

    (c) a portion of the output is fed back to the inverting input

    (d) a portion of the output is fed back to the noninverting input

7. Using output bounding in a comparator

    (a) makes it faster   (b) keeps the output positive

    (c) limits the output levels   (d) stabilizes the output

Section 13–2

8. A summing amplifier can have

    (a) only one input   (b) only two inputs   (c) any number of inputs

temperature- or pressure-sensitive transducer, and the resulting small electrical signal is sent over a long line subject to electrical noise that produces common-mode voltages in the line. The instrumentation amplifier at the end of the line must amplify the small signal from the remote sensor and reject the large common-mode voltage. Figure 14–3 illustrates this.

▶ FIGURE 14–3

Illustration of the rejection of large common-mode voltages and the amplification of smaller signal voltages by an instrumentation amplifier.

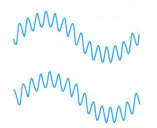 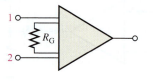

Small differential high-frequency signal riding on a larger low-frequency common-mode signal

Instrumentation amplifier

Amplified differential signal. No common-mode signal.

A Specific Instrumentation Amplifier

Now that you have the basic idea of how an instrumentation amplifier works, let's look at a specific device. A representative device, the AD622, is shown in Figure 14–4 where IC pin numbers are given for reference. This instrumentation amplifier is based on the design using three op-amps that was shown in Figure 14–1.

▶ FIGURE 14–4

The AD622 instrumentation amplifier.

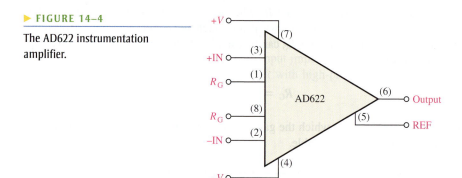

Some of the features of the AD622 are as follows. The voltage gain can be adjusted from 2 to 1000 with an external resistor R_G. There is unity gain with no external resistor. The input impedance is 10 GΩ. The common-mode rejection ratio (CMRR) has a minimum value of 66 dB. Recall that a higher CMRR means better rejection of common-mode voltages. The AD622 has a bandwidth of 800 kHz at a gain of 10 and a slew rate of 1.2 V/μs.

Setting the Voltage Gain For the AD622, an external resistor must be used to achieve a voltage gain greater than unity, as indicated in Figure 14–5. Resistor R_G is connected between the R_G terminals (pins 1 and 8). No resistor is required for unity. R_G is selected for the desired gain based on the following formula:

$$R_G = \frac{50.5 \text{ k}\Omega}{A_v - 1}$$

Notice that this formula is the same as Equation 14–2 for the three-op-amp configuration with an external R_G where the internal resistors R_1 and R_2 are each 25.25 kΩ.

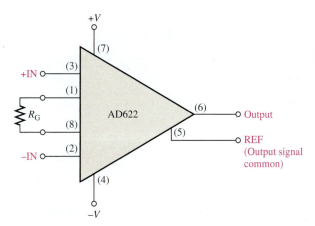

The AD622 with a gain-setting resistor.

Gain versus Frequency Figure 14–6 shows how the gain varies with frequency for gains of 1, 10, 100, and 1000. As the curves show, the bandwidth decreases as the gain increases.

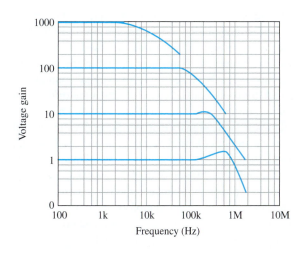

◀ FIGURE 14–6

Gain versus frequency for the AD622 instrumentation amplifier.

EXAMPLE 14–2

Calculate the voltage gain and determine the bandwidth using the graph in Figure 14–6 for the instrumentation amplifier in Figure 14–7.

▶ FIGURE 14–7

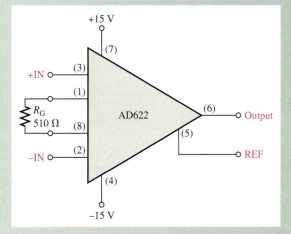

Solution   Determine the voltage gain as follows:

$$R_G = \frac{50.5 \text{ k}\Omega}{A_v - 1}$$

$$A_v - 1 = \frac{50.5 \text{ k}\Omega}{R_G}$$

$$A_v = \frac{50.5 \text{ k}\Omega}{510 \text{ }\Omega} + 1 = \mathbf{100}$$

Determine the approximate bandwidth from the graph at the point where the curve begins to drop in Figure 14–6.

$$BW \approx \mathbf{80 \text{ kHz}}$$

Related Problem   Modify the circuit in Figure 14–7 for a gain of approximately 45.

Noise Effects in Instrumentation Amplifier Applications

Various types of transducers are used to sense temperature, strain, pressure, and other parameters in many types of applications. Instrumentation amplifiers are generally used to process the small voltages produced by a transducer and often are used in noisy industrial environments where long cables connect the transducer output to the amplifier inputs. Noise in the form of common-mode signals picked up from external sources can be minimized, but not totally eliminated, by using coaxial cable in which the differential signal wires are surrounded by a metal mesh sheathing called a *shield.* As you know, in an electrically noisy environment any common-mode signals that are induced on the signal lines are rejected because both inputs to the amplifier have the same common-mode signal. However, when a shielded cable is used, there are stray capacitances distributed along its length between each signal line and the shield. The differences in these stray capacitances, particularly at higher frequencies, result in a phase shift between the two common-mode signals, as illustrated in Figure 14–8. The result is a degradation in the common-mode rejection of the amplifier because the two signals are no longer in phase and do not completely cancel so that a differential voltage is created at the amplifier inputs.

▶ **FIGURE 14–8**

Degradation of common-mode rejection in a shielded cable connection due to unwanted phase shifts.

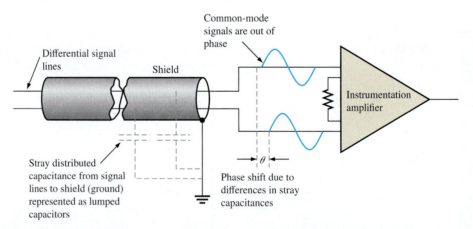

Shield Guard   **Guarding** is a technique to reduce the effects of noise on the common-mode operation of an instrumentation amplifier operating in critical environments by connecting the common-mode voltage to the shield of a coaxial cable. The common-mode signal is fed back to the shield by a voltage-follower stage, as shown in Figure 14–9. The purpose is to eliminate voltage differences between the signal lines and the shield, virtually eliminating leakage currents and cancelling the effects of the distributed capacitances so that the common-mode voltages are the same in both lines.

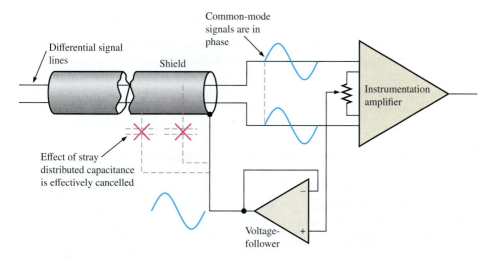

◀ FIGURE 14–9

Instrumentation amplifier with shield guard to prevent degradation of the common-mode rejection.

The voltage-follower is a low-impedance source that drives the common-mode signal onto the shield to eliminate the voltage difference between the signal lines and the shield. When the voltage between each signal line and the shield is zero, the leakage currents are also zero and the capacitive reactances become infinitely large. An infinitely large X_C implies a zero capacitance.

A Specific Instrumentation Amplifier with a Guard Output

Most instrumentation amplifiers can be configured externally to provide a shield guard driver. Certain IC amplifiers, however, provide an internally generated guard output that is intended for very critical environments. An example is the AD522, shown in Figure 14–10, which is a precision IC instrumentation amplifier designed for applications requiring high accuracy under worst-case operating conditions and with very small signals. The pin labeled DATA GUARD is the shield-guard output.

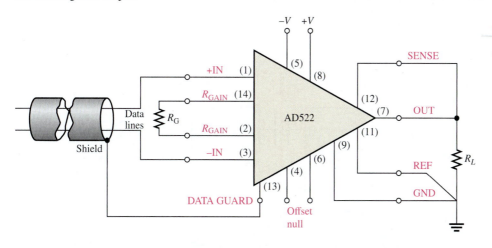

◀ FIGURE 14–10

The AD522 instrumentation amplifier in a typical configuration.

SECTION 14–1 CHECKUP

Answers can be found at www .pearsonhighered.com/floyd.

1. What is the main purpose of an instrumentation amplifier and what are three of its key characteristics?

2. What components do you need to construct a basic instrumentation amplifier?

3. How is the gain determined in an instrumentation amplifier?

4. In a certain AD622 configuration, $R_G = 10\ \text{k}\Omega$. What is the voltage gain?

5. Describe the purpose of a shield guard.

14–2 ISOLATION AMPLIFIERS

An isolation amplifier provides dc isolation between input and output. It is used for the protection of human life or sensitive equipment in those applications where hazardous power-line leakage or high-voltage transients are possible. The principal areas of application are in medical instrumentation, power plant instrumentation, industrial processing, and automated testing.

After completing this section, you should be able to

❑ **Explain and analyze the operation of an isolation amplifier**
❑ Describe a basic capacitor-coupled isolation amplifier
  ◆ Show the block diagram ◆ Define *modulation* ◆ Discuss the modulation process ◆ Describe the ISO124 as an example of an isolation amplifier
❑ Describe a transformer-coupled isolation amplifier
  ◆ Discuss the 3656KG ◆ Establish the voltage gain ◆ Describe a medical application

A Basic Capacitor-Coupled Isolation Amplifier

An **isolation amplifier** is a device that consists of two electrically isolated stages. The input stage and the output stage are separated from each other by an isolation barrier so that a signal must be processed in order to be coupled across the isolation barrier. Some isolation amplifiers use optical coupling or transformer coupling to provide isolation between the stages. However, many modern isolation amplifiers use capacitive coupling for isolation. Each stage has separate supply voltages and grounds so that there are no common electrical paths between them. A simplified block diagram for a typical isolation amplifier is shown in Figure 14–11. Notice two different ground symbols are used to reinforce the concept of stage separation.

▶ **FIGURE 14–11**

Simplified block diagram of a typical isolation amplifier.

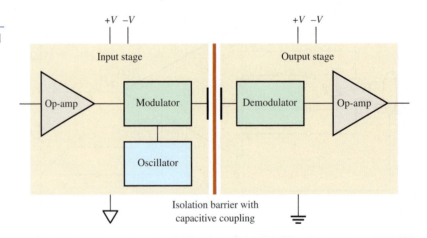

The input stage consists of an amplifier, an oscillator, and a modulator. **Modulation** is the process of allowing a signal containing information to modify a characteristic of another signal, such as amplitude, frequency, or pulse width, so that the information in the first signal is also contained in the second. In this case, the modulator uses a high-frequency square-wave oscillator to modify the original signal. A small-value capacitor (2 pF) in the isolation barrier is used to couple the lower-frequency modulated signal or dc voltage from the input to the output. Without modulation, prohibitively high-value capacitors would be necessary with a resulting degradation in the isolation between the stages.

The output stage consists of a demodulator that extracts the original input signal from the modulated signal so that the original signal from the input stage is back to its original form.

The high-frequency oscillator output in Figure 14–11 can be either amplitude or pulse-width modulated by the signal from the input amplifier (oscillators are covered in Chapter 16). In amplitude modulation, the amplitude of the oscillator output is varied corresponding to the variations of the input signal, as indicated in Figure 14–12(a), which uses one cycle of a sine wave for illustration. In pulse-width modulation, the duty cycle of the oscillator output is varied by changing the pulse width corresponding to the variations of the input signal. An isolation amplifier using pulse-width modulation is represented in Figure 14–12(b).

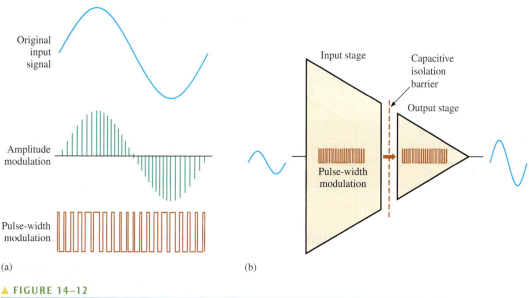

(a)   (b)

▲ FIGURE 14–12

Modulation.

Although it uses a relatively complex process internally, the isolation amplifier is still just an amplifier and is simple to use. When separate dc supply voltages and an input signal are applied, an amplified output signal is the result. The isolation function itself is an unseen process.

EXAMPLE 14–3

The ISO124 is an integrated circuit isolation amplifier. It has a voltage gain of 1 and operates on positive and negative dc supply voltages for both stages. This device uses pulse-width modulation (sometimes called duty cycle modulation) with a frequency of 500 kHz. It is recommended that the supply voltages be decoupled with external capacitors to reduce noise. Show the appropriate connections.

Solution The manufacturer recommends a 1 μF tantalum capacitor (for low leakage) from each dc power supply pin to ground. This is shown in Figure 14–13 where the supply voltages are ± 15 V.

Applications

As previously mentioned, the isolation amplifier is used in applications that require no common grounds between a transducer and the processing circuits where interfacing to sensitive equipment is required. In chemical, nuclear, and metal-processing industries, for example, millivolt signals typically exist in the presence of large common-mode voltages that can be in the kilovolt range. In this type of environment, the isolation amplifier can amplify small signals from very noisy equipment and provide a safe output to sensitive equipment such as computers.

Another important application is in various types of medical equipment. In medical applications where body functions such as heart rate and blood pressure are monitored, the very small monitored signals are combined with large common-mode signals, such as 60 Hz power-line pickup from the skin. In these situations, without isolation, dc leakage or equipment failure could be fatal. Figure 14–16 shows a simplified diagram of an isolation amplifier in a cardiac-monitoring application. In this situation, heart signals, which are very small, are combined with much larger common-mode signals caused by muscle noise, electrochemical noise, residual electrode voltage, and 60 Hz power-line pickup from the skin.

▶ **FIGURE 14–16**

Fetal heartbeat monitoring using an isolation amplifier.

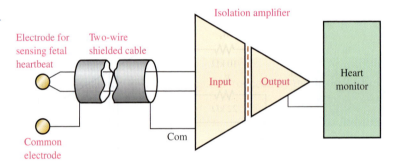

The monitoring of the fetal heartbeat, as illustrated, is the most demanding type of cardiac monitoring because in addition to the fetal heartbeat that typically generates 50 μV, there is also the mother's heartbeat that typically generates 1 mV. The common-mode voltages can run from about 1 mV to about 100 mV. The CMR (common-mode rejection) of the isolation amplifier separates the signal of the fetal heartbeat from that of the mother's heartbeat and from those common-mode signals. Therefore, the signal from the fetal heartbeat is essentially all that the amplifier sends to the monitoring equipment.

SECTION 14–2 CHECKUP

1. In what types of applications are isolation amplifiers used?
2. What are the two stages in a typical isolation amplifier and what is the purpose of having two stages?
3. How are the stages in an isolation amplifier connected?
4. What is the purpose of the oscillator in an isolation amplifier?

14–3 OPERATIONAL TRANSCONDUCTANCE AMPLIFIERS (OTAS)

Conventional op-amps are, as you know, primarily voltage amplifiers in which the output voltage equals the gain times the differential input voltage. The **operational transconductance amplifier (OTA)** is primarily a voltage-to-current amplifier in which the output current equals the gain times the differential input voltage. Because it can either source or sink output current, it can serve as an excellent current source for various applications and is often referred to as a voltage-controlled current source.

After completing this section, you should be able to

❑ **Explain and analyze the operation of an operation transconductance amplifier (OTA)**
   ◆ Identify the OTA schematic symbol
❑ Discuss the gain of an OTA
   ◆ Define *transconductance*   ◆ Explain how the transconductance is a function of bias current
❑ Describe some OTA circuits
   ◆ Discuss the OTA as an inverting amplifier   ◆ Discuss the OTA with resistance-controlled gain   ◆ Discuss the OTA with voltage-controlled gain
❑ Describe the LM13700 as an example of a specific OTA
   ◆ Describe how the input and output resistances change with bias current
❑ Discuss two OTA applications
   ◆ Describe an amplitude modulator   ◆ Describe a Schmitt trigger

Figure 14–17 shows the symbol for an OTA. The double circle symbol at the output represents an output current source that is dependent on a bias current. Like the conventional op-amp, the OTA has two differential input terminals, a high input impedance, and a high CMRR. Unlike the conventional op-amp, the OTA has a bias-current input terminal, a high output impedance, and no fixed open-loop voltage gain.

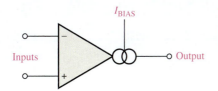

I_{BIAS}

Inputs

Output

◀ **FIGURE 14–17**

Symbol for an operational transconductance amplifier (OTA).

Transconductance

The **transconductance** of an electronic device is the ratio of the output current to the input voltage. For an OTA, the differential voltage is the input variable and current is the output variable; therefore, the ratio of output current to input voltage is also its gain. Because it is an operational amplifier, the input voltage that is amplified is actually the difference voltage between the inputs; hence, V_{in} represents a differential voltage in the equation. Consequently, the voltage-to-current gain of an OTA is the transconductance, g_m.

$$g_m = \frac{I_{out}}{V_{in}}$$

Equation 14–5

In an OTA, the transconductance is dependent on a constant (K) times the bias current (I_{BIAS}), as indicated in Equation 14–6. The value of the constant is dependent on the internal circuit design.

$$g_m = KI_{BIAS}$$

Equation 14–6

The output current is controlled by the input voltage and the bias current as shown by the following formula:

$$I_{out} = g_m V_{in} = KI_{BIAS}V_{in}$$

The relationship of the transconductance and the bias current in an OTA is an important characteristic. Figure 14–18 illustrates a typical relationship. Notice that the transconductance increases linearly with the bias current. The constant of proportionality, K, is the slope of the line. In this case, K is approximately 16 μS/μA. Also, K is somewhat temperature dependent and is lower at increasing temperature; this can affect how the circuit behaves.

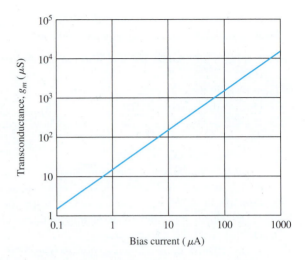

Example of a transconductance versus bias current graph for a typical OTA.

Unlike most operational amplifier circuits, the OTA is used without feedback. As shown in Figure 14–18, the transconductance and hence the output current can be adjusted within certain limits by the bias current.

EXAMPLE 14–5

If an OTA has a $g_m = 1000 \ \mu S$, what is the output current when the input differential voltage is 25 mV?

Solution

$$I_{out} = g_m V_{in} = (1000 \ \mu S)(25 \ mV) = \mathbf{25 \ \mu A}$$

Related Problem

Based on $K \cong 16 \ \mu S/\mu A$, calculate the approximate bias current required to produce $g_m = 1000 \ \mu S$.

Basic OTA Circuits

Figure 14–19 shows the OTA used as an inverting amplifier with a fixed voltage gain. The voltage gain is set by the transconductance and the load resistance as follows:

$$V_{out} = I_{out}R_L$$

Dividing both sides by V_{in},

$$\frac{V_{out}}{V_{in}} = \left(\frac{I_{out}}{V_{in}}\right)R_L$$

Since V_{out}/V_{in} is the voltage gain and $I_{out}/V_{in} = g_m$,

$$A_v = g_m R_L$$

An OTA as an inverting amplifier with a fixed voltage gain.

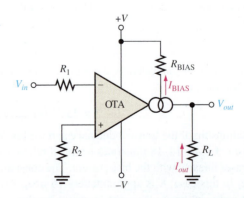

The transconductance of the amplifier in Figure 14–19 is determined by the amount of bias current, which is set by the dc supply voltages and the bias resistor R_{BIAS}.

One of the most useful features of an OTA is that the voltage gain can be controlled by the amount of bias current. This can be done manually, as shown in Figure 14–20(a), by using a variable resistor in series with R_{BIAS} in the circuit of Figure 14–19. By changing the resistance, you can produce a change in I_{BIAS}, which changes the transconductance. A change in the transconductance changes the voltage gain. The voltage gain can also be controlled with an externally applied variable voltage, as shown in Figure 14–20(b). A variation in the applied bias voltage causes a change in the bias current.

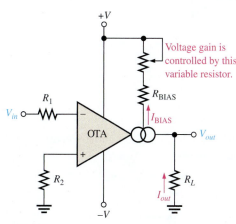

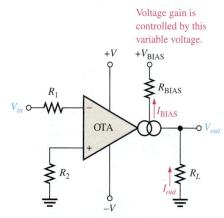

(a) Amplifier with resistance-controlled gain

(b) Amplifier with voltage-controlled gain

◀ FIGURE 14–20

An OTA as an inverting amplifier with a variable-voltage gain.

A Specific OTA

The LM13700 is a typical OTA and serves as a representative device. The LM13700 is a dual-device package containing two OTAs and output buffer circuits. Figure 14–21 shows the pin configuration using a single OTA in the package. The maximum dc supply voltages are ± 18 V, and its transconductance characteristic happens to be the same as indicated by the graph in Figure 14–18. For an LM13700, the bias current is determined by the following formula:

$$I_{BIAS} = \frac{+V_{BIAS} - (-V) - 1.4 \text{ V}}{R_{BIAS}}$$

The 1.4 V is due to the internal circuit where a base-emitter junction and a diode connect the external R_{BIAS} with the negative supply voltage $(-V)$. The positive bias voltage, $+V_{BIAS}$, may be obtained from the positive supply voltage, $+V$.

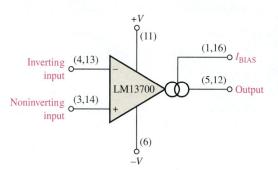

◀ FIGURE 14–21

An LM13700 OTA. There are two in an IC package. The buffer transistors are not shown. Pin numbers for both OTAs are given in parentheses.

Not only does the transconductance of an OTA vary with bias current, but so do the input and output resistances. Both the input and output resistances decrease as the bias current increases, as shown in Figure 14–22.

Example of input and output resistances versus bias current.

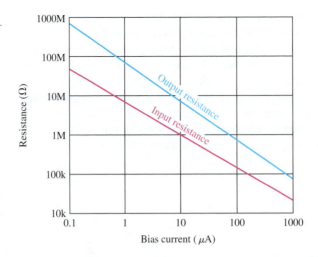

EXAMPLE 14–6

The OTA in Figure 14–23 is connected as an inverting fixed-gain amplifier where $+V_{BIAS} = +V$. Determine the approximate voltage gain.

▶ **FIGURE 14–23**

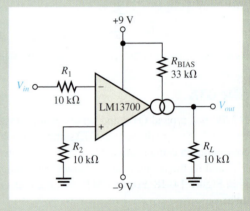

Solution   Calculate the bias current as follows:

$$I_{BIAS} = \frac{+V_{BIAS} - (-V) - 1.4 \text{ V}}{R_{BIAS}} = \frac{9 \text{ V} - (-9 \text{ V}) - 1.4 \text{ V}}{33 \text{ k}\Omega} = 503 \ \mu A$$

Using $K \cong 16 \ \mu S/\mu A$ from the graph in Figure 14–18, the value of transconductance corresponding to $I_{BIAS} = 503 \ \mu A$ is approximately

$$g_m = KI_{BIAS} \cong (16 \ \mu S/\mu A)(503 \ \mu A) = 8.05 \times 10^3 \ \mu S$$

Using this value of g_m, calculate the voltage gain.

$$A_v = g_m R_L \cong (8.05 \times 10^3 \ \mu S)(10 \text{ k}\Omega) = \textbf{80.5}$$

Related Problem   If the OTA in Figure 14–23 is operated with dc supply voltages of ± 12 V, will this change the voltage gain and, if so, to what value?

Two OTA Applications

Amplitude Modulator   Figure 14–24 illustrates an OTA connected as an amplitude modulator. The voltage gain is varied by applying a modulation voltage to the bias input. When a constant-amplitude input signal is applied, the amplitude of the output signal will

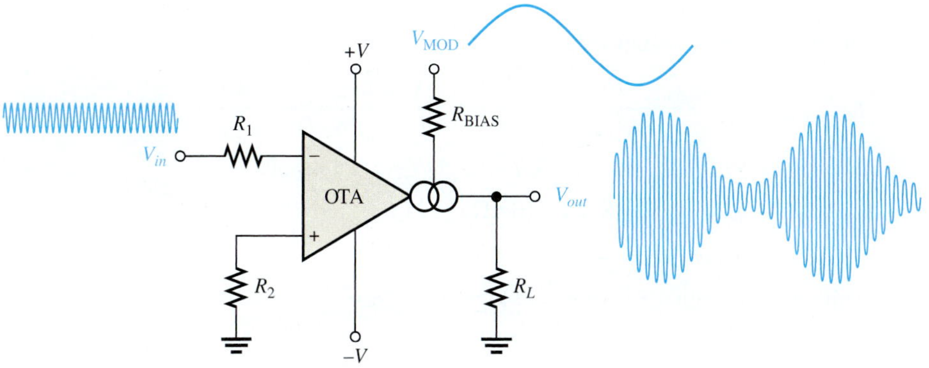

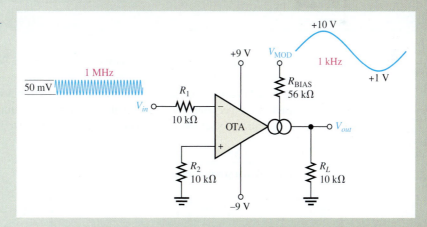

◀ **FIGURE 14–24**

The OTA as an amplitude modulator.

vary according to the modulation voltage on the bias input. The gain is dependent on bias current, and bias current is related to the modulation voltage by the following relationship:

$$I_{BIAS} = \frac{V_{MOD} - (-V) - 1.4\ V}{R_{BIAS}}$$

This modulating action is shown in Figure 14–24 for a higher-frequency sinusoidal input voltage and a lower-frequency sinusoidal modulating voltage.

EXAMPLE 14–7

The input to the OTA amplitude modulator in Figure 14–25 is a 50 mV peak-to-peak, 1 MHz sine wave. Determine the output signal, given the modulation voltage shown is applied to the bias input.

▶ **FIGURE 14–25**

Solution   The maximum voltage gain is when I_{BIAS}, and thus g_m, is maximum. This occurs at the maximum peak of the modulating voltage, V_{MOD}.

$$I_{BIAS(max)} = \frac{V_{MOD(max)} - (-V) - 1.4\ V}{R_{BIAS}} = \frac{10\ V - (-9\ V) - 1.4\ V}{56\ k\Omega} = 314\ \mu A$$

From the graph in Figure 14–18, the constant K is approximately 16 μS/μA.

$$g_m = KI_{BIAS(max)} \cong (16\ \mu S/\mu A)(314\ \mu A) = 5.02\ mS$$
$$A_{v(max)} = g_m R_L \cong (5.02\ mS)(10\ k\Omega) = 50.2$$
$$V_{out(max)} = A_{v(max)}V_{in} \cong (50.2)(50\ mV) = 2.51\ V$$

Calculate the minimum output voltage as follows:

$$I_{BIAS(min)} = \frac{V_{MOD(min)} - (-V) - 1.4 \text{ V}}{R_{BIAS}} = \frac{1 \text{ V} - (-9 \text{ V}) - 1.4 \text{ V}}{56 \text{ k}\Omega} = 154 \text{ } \mu A$$

$$g_m = K I_{BIAS(min)} \cong (16 \text{ } \mu S/\mu A) (154 \text{ } \mu A) = 2.46 \text{ mS}$$

$$A_{v(min)} = g_m R_L \cong (2.46 \text{ mS}) (10 \text{ k}\Omega) = 24.6$$

$$V_{out(min)} = A_{v(min)} V_{in} \cong (24.6) (50 \text{ mV}) = 1.23 \text{ V}$$

The resulting output voltage is shown in Figure 14–26.

▶ FIGURE 14–26

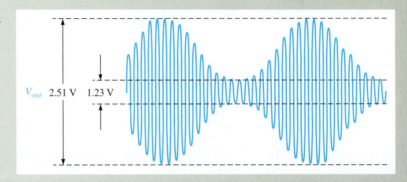

V_{out} 2.51 V   1.23 V

Related Problem   Repeat this example with the sinusoidal modulating signal replaced by a square wave with the same maximum and minimum levels and a bias resistor of 39 kΩ.

 Open the Multisim file EXM14-07 or the LT Spice file EXS14-07 in the Examples folder on the website and run the simulation and measure the output voltage.

Schmitt Trigger   Figure 14–27 shows an OTA used in a Schmitt-trigger configuration. Basically, a Schmitt trigger is a comparator with hysteresis where the input voltage is large enough to drive the device into its saturated states. When the input voltage exceeds a certain threshold value or trigger point, the device switches to one of its saturated output states. When the input falls below another threshold value, the device switches to its other saturated output state.

▶ FIGURE 14–27

The OTA as a Schmitt trigger.

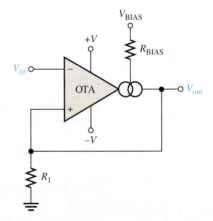

In the case of the OTA Schmitt trigger, the threshold levels are set by the current through resistor R_1. The maximum output current in an OTA equals the bias current. Therefore, in the saturated output states, $I_{out} = I_{BIAS}$. The maximum positive output voltage is $I_{out}R_1$, and this voltage is the positive threshold value or upper trigger point. When the input voltage exceeds this value, the output switches to its maximum negative voltage, which is $-I_{out}R_1$. Since $I_{out} = I_{BIAS}$, the trigger points can be controlled by the bias current. Figure 14–28 illustrates this operation.

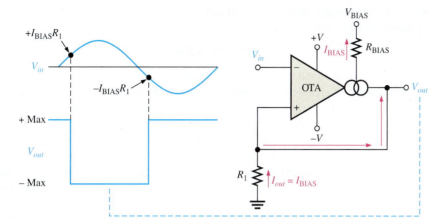

◄ FIGURE 14–28

Basic operation of the OTA Schmitt trigger.

SECTION 14–3 CHECKUP

1. What does OTA stand for?
2. If the bias current in an OTA is increased, does the transconductance increase or decrease?
3. What happens to the voltage gain if the OTA is connected as a fixed-voltage amplifier and the supply voltages are increased?
4. What happens to the voltage gain if the OTA is connected as a variable-gain voltage amplifier and the voltage at the bias terminal is decreased?

14–4 LOG AND ANTILOG AMPLIFIERS

Log and antilog amplifiers are used in applications that require compression of analog input data, linearization of transducers that have exponential outputs, and analog multiplication and division. They are often used in high-frequency communication systems, including fiber optic systems for processing wide dynamic range signals.

After completing this section, you should be able to

❏ **Explain and analyze the operation of log and antilog amplifiers**
  ◆ Define *logarithm*
❏ Describe the basic log amplifier
  ◆ Define *natural logarithm* ◆ Explain how a diode provides a logarithmic characteristic ◆ Describe the operation of a log amplifier with a diode in the feedback loop ◆ Describe the operation of a log amplifier with a BJT in the feedback loop
❏ Describe the basic antilog amplifier
  ◆ Define *antilogarithm* ◆ Explain how a diode or transistor is connected to form an antilog amplifier
❏ Discuss signal compression with log amplifiers
  ◆ Describe the difference between linear and logarithmic signal compression

The **logarithm** of a number is the power to which the base must be raised to get that number. A logarithmic (log) amplifier produces an output that is proportional to the logarithm of the input, and an antilogarithmic (antilog) amplifier takes the antilog or inverse log of the input.

The Basic Logarithmic Amplifier

The key element in a log amplifier is a device that exhibits a logarithmic characteristic that, when placed in the feedback loop of an op-amp, produces a logarithmic response. This means that the output voltage is a function of the logarithm of the input voltage, as expressed by the following general equation:

Equation 14–7

$$V_{out} = -K \ln(V_{in})$$

where K is a constant and ln is the natural logarithm to the base e. A **natural logarithm** is the exponent to which the base e must be raised in order to equal a given quantity. Although we will use natural logarithms in the formulas in this section, each expression can be converted to a logarithm to the base 10 ($\log_{10}$) using the relationship $\ln x = 2.3 \log_{10} x$.

The semiconductor *pn* junction in the form of either a diode or the base-emitter junction of a BJT provides a logarithmic characteristic. You may recall that a diode has a nonlinear characteristic up to a forward voltage of approximately 0.7 V. Figure 14–29 shows the characteristic curve, where V_F is the forward diode voltage and I_F is the forward diode current.

▶ **FIGURE 14–29**

A portion of a diode (*pn* junction) characteristic curve (V_F versus I_F).

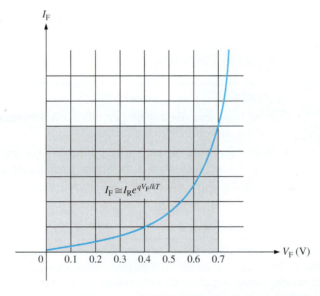

$$I_F \cong I_R e^{qV_F/kT}$$

As you can see on the graph, the diode curve is nonlinear. Not only is the characteristic curve nonlinear, it is logarithmic and is specifically defined by the following formula:

$$I_F \cong I_R e^{qV_F/kT}$$

where I_R is the reverse leakage current, q is the charge on an electron, k is Boltzmann's constant, and T is the absolute temperature in Kelvin. From the previous equation, the diode forward voltage, V_F, can be determined as follows. Take the natural logarithm (*ln* is the logarithm to the base e) of both sides.

$$\ln I_F = \ln I_R e^{qV_F/kT}$$

The ln of a product of two terms equals the sum of the ln of each term.

$$\ln I_F = \ln I_R + \ln e^{qV_F/kT} = \ln I_R + \frac{qV_F}{kT}$$

$$\ln I_F - \ln I_R = \frac{qV_F}{kT}$$

The difference of two ln terms equals the ln of the quotient of the terms.

$$\ln\left(\frac{I_F}{I_R}\right) = \frac{qV_F}{kT}$$

Solving for V_F,

$$V_F = \left(\frac{kT}{q}\right)\ln\left(\frac{I_F}{I_R}\right)$$

Log Amplifier with a Diode   When you place a diode in the feedback loop of an op-amp circuit, as shown in Figure 14–30, you have a basic log amplifier. Since the inverting input is at virtual ground (0 V), the output is at $-V_F$ when the input is positive. Since V_F is logarithmic, so is V_{out}. The output is limited to a maximum value of approximately -0.7 V because the diode's logarithmic characteristic is restricted to voltages below 0.7 V. Also, the input must be positive when the diode is connected in the direction shown in the figure. To handle negative inputs, you must turn the diode around.

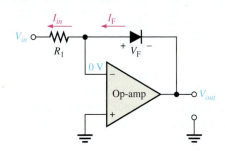

◀ **FIGURE 14–30**

A basic log amplifier using a diode as the feedback element.

An analysis of the circuit in Figure 14–30 is as follows, beginning with the facts that $V_{out} = -V_F$ and $I_F = I_{in}$ because there is no current at the inverting input.

$$V_{out} = -V_F$$

$$I_F = I_{in} = \frac{V_{in}}{R_1}$$

Substituting into the formula for V_F,

$$V_{out} = -\left(\frac{kT}{q}\right)\ln\left(\frac{V_{in}}{I_R R_1}\right)$$

The term kT/q is a constant equal to approximately 25 mV at 25°C. Therefore, the output voltage can be expressed as

$$V_{out} \cong -(0.025 \text{ V})\ln\left(\frac{V_{in}}{I_R R_1}\right)$$

Equation 14–8

From Equation 14–8, you can see that the output voltage is the negative of a logarithmic function of the input voltage. The value of the output is controlled by the value of the input voltage and the value of the resistor R_1. The other factor, I_R, is a constant for a given diode.

EXAMPLE 14–8

Determine the output voltage for the log amplifier in Figure 14–31. Assume $I_R = 50$ nA.

▶ **FIGURE 14–31**

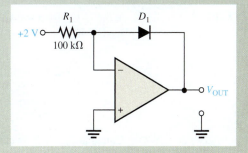

Solution The input voltage and the resistor value are given in Figure 14–31.

$$V_{OUT} = -(0.025 \text{ V})\ln\left(\frac{V_{in}}{I_R R_1}\right) = -(0.025 \text{ V})\ln\left(\frac{2 \text{ V}}{(50 \text{ nA})(100 \text{ k}\Omega)}\right)$$

$$= -(0.025 \text{ V})\ln(400) = -(0.025 \text{ V})(5.99) = \textbf{−0.150 V}$$

Related Problem Calculate the output voltage of the log amplifier with a +4 V input.

 Open the Multisim file EXM14-08 or the LT Spice file EXS14-08 in the Examples folder on the website. Apply the specified input voltage and measure the output voltage.

Log Amplifier with a BJT The base-emitter junction of a bipolar junction transistor exhibits the same type of logarithmic characteristic as a diode because it is also a *pn* junction. A log amplifier with a BJT connected in a common-base form in the feedback loop is shown in Figure 14–32. Notice that V_{out} with respect to ground is equal to $-V_{BE}$.

▶ **FIGURE 14–32**

A basic log amplifier using a transistor as the feedback element.

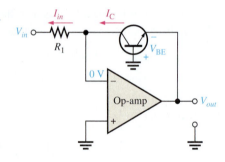

The analysis for this circuit is the same as for the diode log amplifier except that V_{BE} replaces V_F, I_C replaces I_F, and I_{EBO} replaces I_R. The expression for the V_{BE} versus I_C characteristic curve is

$$I_C = I_{EBO}e^{qV_{BE}/kT}$$

where I_{EBO} is the emitter-to-base leakage current. The expression for the output voltage is

Equation 14–9

$$V_{out} = -(0.025 \text{ V})\ln\left(\frac{V_{in}}{I_{EBO}R_1}\right)$$

EXAMPLE 14–9 What is V_{out} for a transistor log amplifier with $V_{in} = 3$ V and $R_1 = 68$ kΩ? Assume $I_{EBO} = 40$ nA.

Solution
$$V_{out} = -(0.025 \text{ V})\ln\left(\frac{V_{in}}{I_{EBO}R_1}\right) = -(0.025 \text{ V})\ln\left(\frac{3 \text{ V}}{(40 \text{ nA})(68 \text{ k}\Omega)}\right)$$

$$= -(0.025 \text{ V})\ln(1103) = \textbf{−175.1 mV}$$

Related Problem Calculate V_{out} if R_1 is changed to 33 kΩ.

The Basic Antilog Amplifier

The **antilogarithm** of a number is the result obtained when the base is raised to a power equal to the logarithm of that number. To get the antilogarithm, you must take the exponential of the logarithm (antilogarithm of $x = e^{\ln x}$).

An antilog amplifier is formed by connecting a transistor (or diode) as the input element as shown in Figure 14–33. The exponential formula still applies to the base-emitter *pn* junction. The output voltage is determined by the current (equal to the collector current) through the feedback resistor.

$$V_{out} = -R_f I_C$$

The characteristic equation of the *pn* junction is

$$I_C = I_{EBO}e^{qV_{BE}/kT}$$

Substituting into the equation for V_{out},

$$V_{out} = -R_f I_{EBO}e^{qV_{BE}/kT}$$

As you can see in Figure 14–33, $V_{in} = V_{BE}$.

$$V_{out} = -R_f I_{EBO}e^{qV_{in}/kT}$$

The exponential term can be expressed as an antilogarithm as follows:

$$V_{out} = -R_f I_{EBO}\text{antilog}\left(\frac{V_{in}q}{kT}\right)$$

Since kT/q is approximately 25 mV,

$$V_{out} = -R_f I_{EBO}\text{antilog}\left(\frac{V_{in}}{25\text{ mV}}\right)$$

Equation 14–10

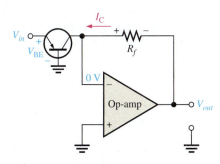

◄ **FIGURE 14–33**

A basic antilog amplifier.

EXAMPLE 14–10

For the antilog amplifier in Figure 14–34, find the output voltage. Assume $I_{EBO} = 40$ nA.

▶ **FIGURE 14–34**

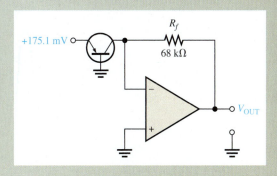

Solution First of all, notice that the input voltage in Figure 14–34 is the inverted output voltage of the log amplifier in Example 14–9, where the output voltage is proportional to the logarithm of the input voltage. In this case, the antilog amplifier reverses the process and produces an output that is proportional to the antilog of the input. Stated another way, the input of an antilog amplifier is proportional to the logarithm of the output. So, the output voltage of the antilog amplifier in Figure 14–34 should have the same magnitude as the input voltage of the log amplifier in Example 14–9 because all the constants are the same. Let's see if it does.

$$V_{out} = -R_f I_{EBO} \text{antilog}\left(\frac{V_{in}}{25 \text{ mV}}\right) = -(68 \text{ k}\Omega)(40 \text{ nA})\text{antilog}\left(\frac{175.1 \text{ mV}}{25 \text{ mV}}\right)$$

$$= -(68 \text{ k}\Omega)(40 \text{ nA})(1101) = -3 \text{ V}$$

Related Problem Determine V_{out} for the amplifier in Figure 14–34 if the feedback resistor is changed to 100 kΩ.

Signal Compression with Logarithmic Amplifiers

In certain applications, a signal may be too large in magnitude for a particular system to handle. The term *dynamic range* is often used to describe the range of voltages contained in a signal. In these cases, the signal voltage must be scaled down by a process called **signal compression** so that it can be properly handled by the system. If a linear circuit is used to scale a signal down in amplitude, the lower voltages are reduced by the same percentage as the higher voltages. Linear signal compression often results in the lower voltages becoming obscured by noise and difficult to accurately distinguish, as illustrated in Figure 14–35(a). To overcome this problem, a signal with a large dynamic range can be compressed using a logarithmic response, as shown in Figure 14–35(b). In logarithmic signal compression, the higher voltages are reduced by a greater percentage than the lower voltages, thus keeping the lower-voltage signals from being lost in noise.

▶ **FIGURE 14–35**

The basic concept of signal compression with a logarithmic amplifier.

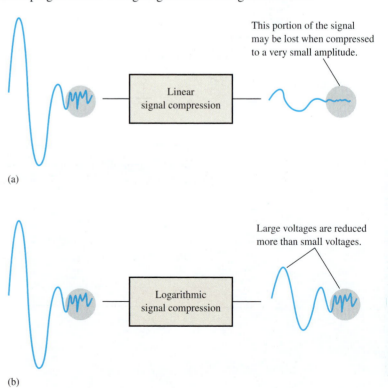

SECTION 14–4 CHECKUP	1. What purpose does the diode or transistor perform in the feedback loop of a log amplifier?
	2. Why is the output of a log amplifier limited to about 0.7 V?
	3. What are the factors that determine the output voltage of a basic log amplifier?
	4. In terms of implementation, how does a basic antilog amplifier differ from a basic log amplifier?

14–5 CONVERTERS AND OTHER INTEGRATED CIRCUITS

This section introduces a few more devices that represent basic applications of the op-amp and linear integrated circuits. You will learn about the constant-current source, the current-to-voltage converter, the voltage-to-current converter, the peak detector, and the LM386 audio amplifier. This is intended only to introduce you to some common basic applications.

After completing this section, you should be able to

❑ **Explain and analyze other types of integrated circuits**
❑ Describe a constant-current source
❑ Explain a current-to-voltage converter
❑ Discuss a voltage-to-current converter
❑ Explain how a peak detector works
❑ Discuss a particular audio amplifier

Constant-Current Source

Recall that a constant-current source, such as the OTA discussed in Section 14–3, delivers a load current that remains constant when the load resistance changes. Figure 14–36 shows a basic op-amp circuit in which a stable voltage source (V_{IN}) provides a constant current (I_i) through the input resistor (R_i). Since the inverting ($-$) input of the op-amp is at virtual ground (0 V), the value of I_i is determined by V_{IN} and R_i as

$$I_i = \frac{V_{IN}}{R_i}$$

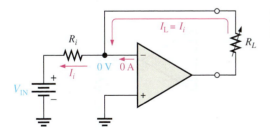

◄ **FIGURE 14–36**

A basic constant-current source.

Now, since the internal input impedance of the op-amp is extremely high (ideally infinite), practically all of I_i is through R_L, which is connected in the feedback path. Since $I_i = I_L$,

$$I_L = \frac{V_{IN}}{R_i}$$

Equation 14–11

If R_L changes, I_L remains constant as long as V_{IN} and R_i are held constant.

Current-to-Voltage Converter

A current-to-voltage converter converts a variable input current to a proportional output voltage. A basic circuit that accomplishes this is shown in Figure 14–37(a). Since practically all of I_i is through the feedback path, the voltage dropped across R_f is I_iR_f. Because the left side of R_f is at virtual ground (0 V), the output voltage equals the voltage across R_f, which is proportional to I_i.

Equation 14–12

$$V_{out} = I_iR_f$$

▶ FIGURE 14–37

Current-to-voltage converter.

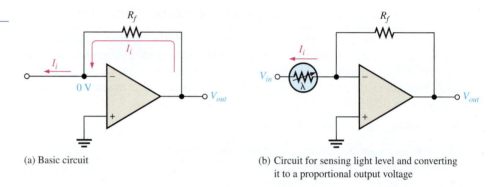

(a) Basic circuit

(b) Circuit for sensing light level and converting it to a proportional output voltage

A specific application of this circuit is illustrated in Figure 14–37(b), where a photoconductive cell is used to sense changes in light level. As the amount of light changes, the current through the photoconductive cell varies because of the cell's change in resistance. This change in resistance produces a proportional change in the output voltage ($\Delta V_{out} = \Delta I_iR_f$).

Voltage-to-Current Converter

A basic voltage-to-current converter is shown in Figure 14–38. Like the OTA, this circuit can be used in applications where it is necessary to have an output (load) current that is controlled by an input voltage. A drawback to this circuit is that the load is not grounded.

Neglecting the input offset voltage, both inverting and noninverting input terminals of the op-amp are at the same voltage, V_{in}. Therefore, the voltage across R_1 equals V_{in}. Since there is negligible current at the inverting input, the current through R_1 is the same as the current through R_L; thus

Equation 14–13

$$I_L = \frac{V_{in}}{R_1}$$

▶ FIGURE 14–38

Voltage-to-current converter.

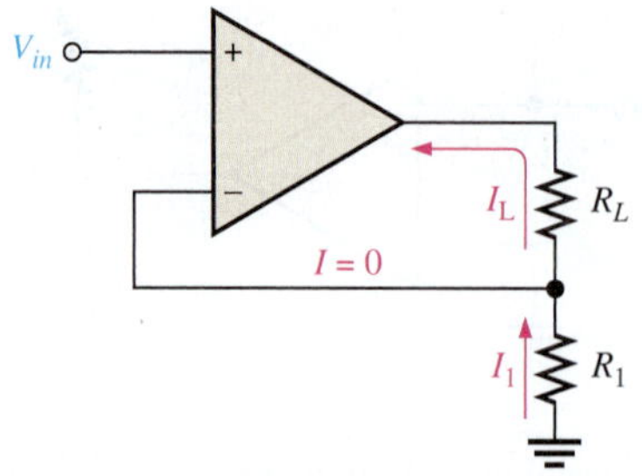

Peak Detector

An interesting application of the op-amp is in a peak detector circuit such as the one shown in Figure 14–39. In this case the op-amp is used as a comparator. This circuit is used to detect the peak of the input voltage and store that peak voltage on a capacitor. For example,

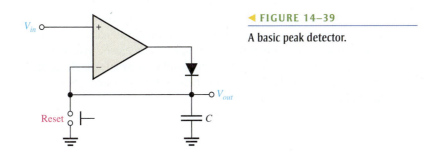

A basic peak detector.

this circuit can be used to detect and store the maximum value of a voltage surge; this value can then be measured at the output with a voltmeter or recording device. The basic operation is as follows. When a positive voltage is applied to the noninverting input of the op-amp, the high-level output voltage of the op-amp forward-biases the diode and charges the capacitor. The capacitor continues to charge until its voltage reaches a value equal to the input voltage, and thus both op-amp inputs are at the same voltage. At this point, the op-amp comparator switches, and its output goes to the low level. The diode is now reverse-biased, and the capacitor stops charging. It has reached a voltage equal to the peak of V_{in} and will hold this voltage until the charge eventually leaks off or until it is reset with a switch as indicated. If a greater input peak occurs, the capacitor charges to the new peak.

Audio Amplifiers

Audio amplifiers are used in numerous applications and are available as a complete system in integrated circuits. One common application is in receiver systems for radio or TV. The signal from a radio or TV is sent as an encoded signal embedded in the radio frequency signal. The receiver recovers the audio signal from the radio frequency signal. It is amplified with a small power amplifier and used to drive the speaker(s). Audio amplifiers typically have bandwidths of 3 kHz to 15 kHz depending on the requirements of the system. IC audio amplifiers are available with a range of capabilities.

The LM386 Audio Power Amplifier   This device is an example of a low-power audio amplifier that is capable of providing several hundred milliwatts to a speaker. It operates from any dc supply voltage in the 4 V to 12 V range, making it a good choice for portable or battery operation. The pin configuration of the LM386 is shown in Figure 14–40(a). The voltage gain of the LM386 is 20 without external connections to the gain terminals, as shown in Figure 14–40(b). A voltage gain of 200 is achieved by connecting a capacitor from pin 1 to pin 8, as shown in Figure 14–40(c). Voltage gains between 20 and 200 can be realized by a resistor and capacitor connected in series from pin 1 to pin 8 as shown in Figure 14–40(d). These external components are effectively placed in parallel with an internal gain-setting resistor.

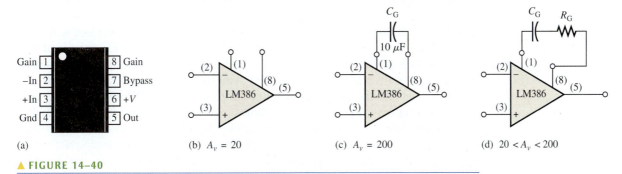

(a)

(b) $A_v = 20$

(c) $A_v = 200$

(d) $20 < A_v < 200$

▲ FIGURE 14–40

Pin configuration and gain connections for the LM386 audio amplifier.

A typical application of the LM386 as a power amplifier is shown in Figure 14–41, which is the last stage of a radio receiver. Here the audio signal is fed to the inverting input through the volume control potentiometer. In radio receivers, you may see an extra filter such as formed by R_2 and C_3 to remove any residual unwanted high frequency carrier signal. R_3 and C_6 provide additional filtering before the audio signal is applied to the speaker through the coupling capacitor C_7.

▶ **FIGURE 14–41**

The LM386 used as an audio power amplifier.

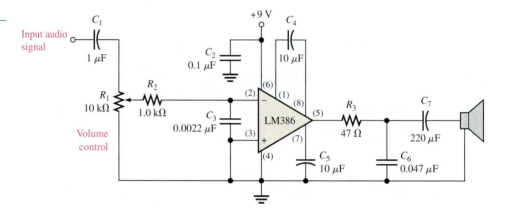

**SECTION 14–5
CHECKUP**

1. For the constant-current source in Figure 14–36, the input reference voltage is 6.8 V and R_i is 10 kΩ. What value of constant current does the circuit supply to a 1.0 kΩ load? To a 5 kΩ load?

2. What element determines the constant of proportionality that relates input current to output voltage in the current-to-voltage converter?

3. What is the typical bandwidth of an audio amplifier?

Device Application: *Liquid Level Control*

The system in this application is designed to maintain a constant liquid level in a tank. The level is kept constant by an electric pump and a pressure sensor (transducer) that detects a change in the level of the liquid by sensing the pressure in a tube.

Level-Sensing Method

A tube with both ends open is placed vertically in a liquid so that one end is above the surface of the liquid. The level of liquid in the tube will be the same as the level in the tank. Now, if the upper end is closed, the pressure of the air trapped in the tube will vary proportional to a change in level of the liquid. For example, if the liquid is water and it rises in the tank by 20 mm, then the pressure in the tube will increase by 20 mm of water. A pressure sensor is placed on the upper end of the tube when the liquid is at its reference level, and the other side is exposed to atmospheric pressure. When the water level decreases, a negative change in pressure is measured by the pressure sensor and a small proportional voltage is produced. The voltage from the pressure sensor is connected to an instrumentation amplifier, which amplifies the small voltage to drive a comparator with hysteresis (Schmitt trigger). The comparator reference voltage is adjusted to the desired

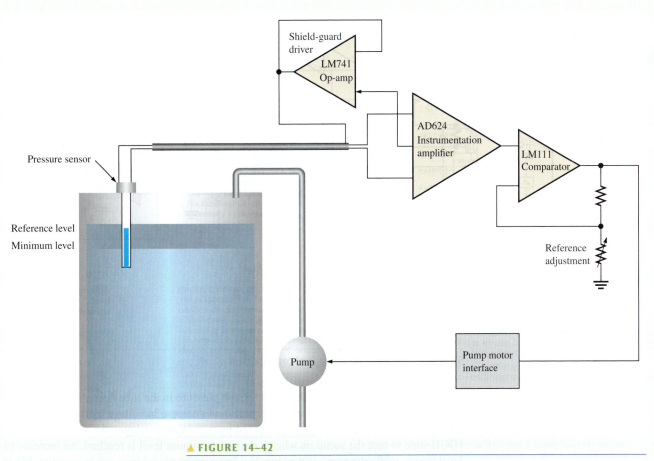

▲ **FIGURE 14–42**

Block diagram of the liquid-level control system.

value; when the level falls below the reference, the comparator switches states and turns the pump *on* to refill the tank to the reference level. The pressure sensor detects when the reference level of the liquid is reached, and the comparator switches back to its other state, turning the pump *off*. A basic diagram of the system is shown in Figure 14–42.

The Circuit

This system will operate in an industrial environment with exposure to mostly 60 Hz electrical noise. Also, the circuit will be located some distance from the tank and connected to the pressure sensor with a long coaxial cable. The output voltage of the pressure sensor is very small ($100 \, \mu\text{V} - 200 \, \mu\text{V}$). For these reasons, a shield-guard driver is incorporated to minimize the effects of noise on the small signal. The AD624 instrumentation amplifier is used to drive an LM111 comparator with hysteresis controlled by a rheostat in the feedback circuit. An LM741 op-amp connected as a voltage-follower is used for the guard driver. The circuit diagram is shown in Figure 14–43. Power supply connections are omitted to simplify the drawing. Resistors R_1 and R_2 provide a return path for bias currents to prevent output drift. R_3 is a pull-up resistor for the comparator output, and R_4 and R_5 provide for the adjustable reference levels by varying the hysteresis. R_6 provides a resistance in series with the shield-guard driver to limit current.

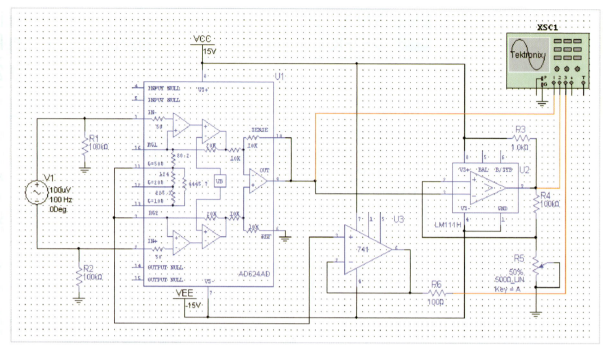

(a) Circuit screen

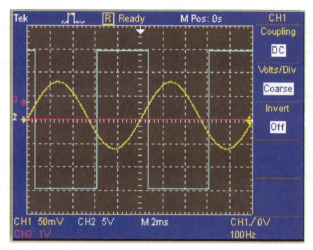

(b) Output of IA (yellow), output of comparator (blue), and output of guard driver (pink)

▲ **FIGURE 14–46**

Simulation with a differential input.

The input to the sensor circuit is simulated using a multisim with an input signal of 100 mV at 100 Hz to represent the sensor. Although the sensor output will contain very small signals, the sensor also picks up a high-frequency signal in order to observe the circuit operation. The simulated circuit is shown in Figure 14–46(a) for a differential input. The resulting outputs are shown in part (b). The comparator is triggered when the IA output goes to zero to test signal on the output of the comparator, which is equal to the input signal on the inputs.

Next, the input is changed to a 100 mV common-mode signal at a frequency of 60 Hz, and the simulation is run as shown in Figure 14–47. This simulates a low-frequency noise environment. Notice on the scope display that there is no output signal from the instrumentation amplifier, which indicates that it is rejecting the common-mode signal. The scope display also shows that the shield-guard driver correctly produces the common-mode signal.

5. Verify that the shield-guard driver output is equal to the common-mode signal.

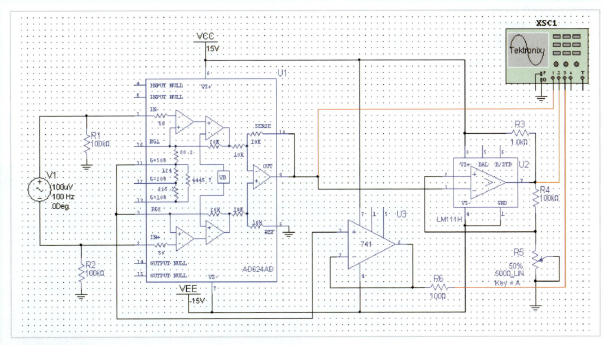

(a) Circuit screen

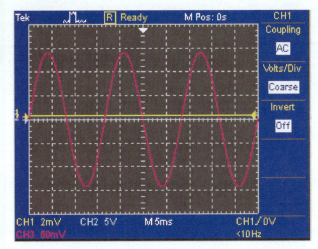

(b) Output of IA (yellow), output of comparator (blue), and output of guard driver (pink)

▲ FIGURE 14–47

Simulation with a common-mode input.

 Simulate the liquid-level control circuit using your Multisim or LT Spice software. Observe the operation with the virtual oscilloscope.

Prototyping and Testing

Now that the circuit has been simulated, the prototype circuit is constructed and tested. After the circuit is successfully tested on a protoboard, it is ready to be finalized on a printed circuit board.

Lab Experiment

To build and test a similar circuit, go to Experiment 14–A in your lab manual (*Laboratory Exercises for Electronic Devices* by David Buchla and Steven Wetterling).

Circuit Board

The liquid-level control circuit is implemented on a printed circuit board as shown in Figure 14–48. The dark gray lines represent backside connections.

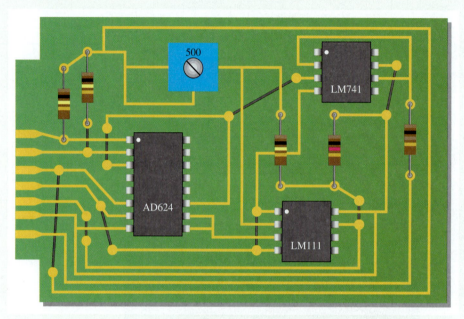

▲ **FIGURE 14–48**

Liquid-level control board.

6. Check the printed circuit board and verify that it agrees with the schematic.
7. Label each input and output pin according to function.

Programmable Analog Technology

Assignment

Create a circuit to provide a function similar to that of the level-control circuit in the Device Application.

Procedure: Open your Designer2 software and configure the CAMs as shown in Figure 14–49.

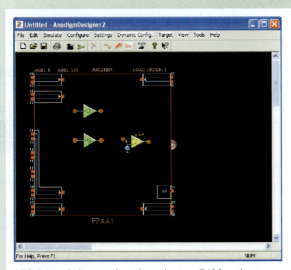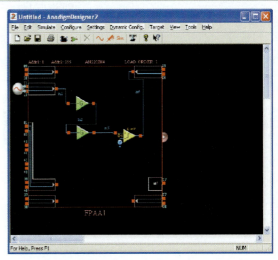

(a) Select and place two inverting gain stage CAMs and one comparator CAM.

(b) Connect the CAMs and add a differential signal source.

▲ **FIGURE 14–49**

Configure the signal generator as shown in Figure 14–50. Set the signal generator to represent a pressure sensor with a differential output and an amplitude of 100 μV. Note that the frequency is selected only to facilitate viewing.

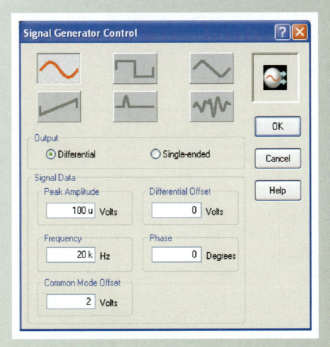

▲ **FIGURE 14–50**

Configure the gain stages for a total gain of 500, as shown in Figure 14–51.

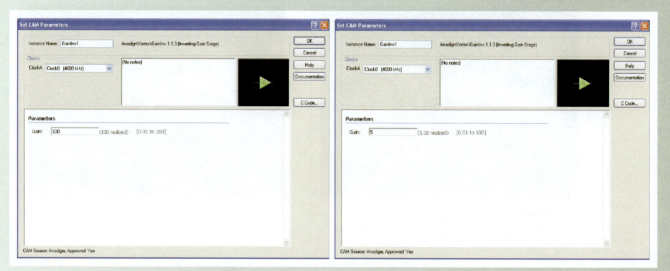

▲ FIGURE 14–51

First stage has gain of 100 and second stage has gain of 5.

Configure the comparator for a hysteresis of 40 mV, as shown in Figure 14–52.

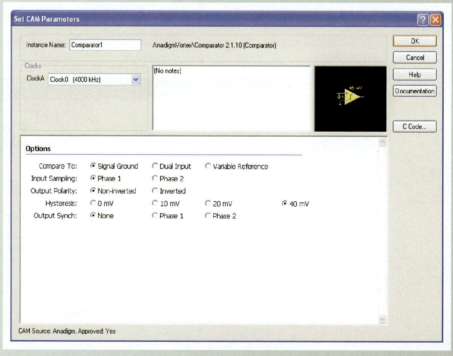

▲ FIGURE 14–52

Analysis: Place probes as shown in Figure 14–53 (top) and run a simulation. The results are shown in Figure 14–53 (bottom).

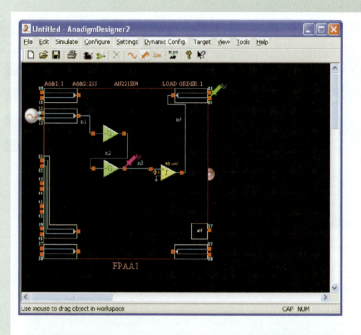

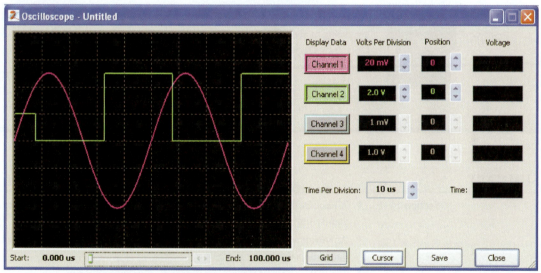

▲ **FIGURE 14–53**

Waveform measurement with a 40 mV comparator hysteresis.

Change the comparator hysteresis to 10 mV, and you get the waveform shown in Figure 14–54. Notice how the trigger points change.

The comparator hysteresis sets the trigger points on the signal so that a minimum and a maximum level can be set to control the level in a tank. Once an FPAA/dpASP is programmed with this design, the levels can be changed by programming a different hysteresis.

Programming Exercises

1. Open your Designer2 software.
2. Implement the level control circuit described.
3. Observe the output for a comparator hysteresis of 0, 10 mV, 20 mV, and 40 mV.

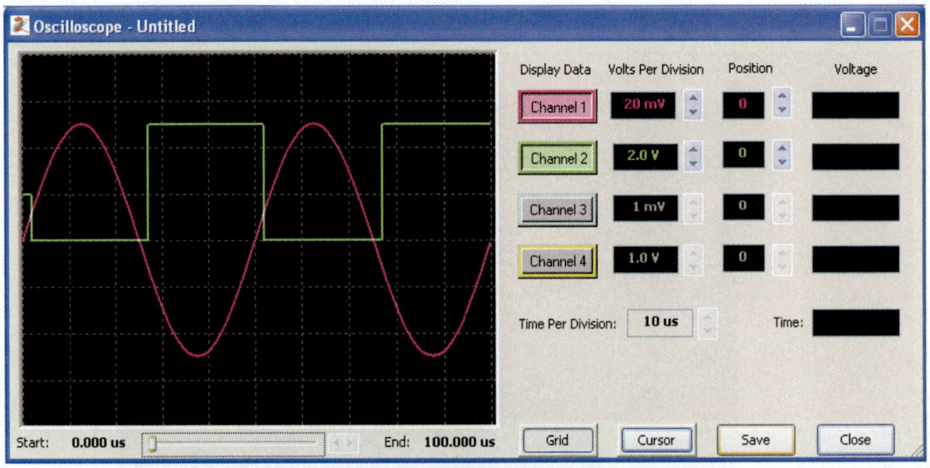

△ FIGURE 14–54

Waveform measurement with a 10 mV comparator hysteresis.

PAM Experiment

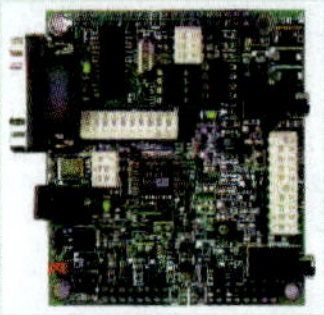

To program, download, and test a circuit using AnadigmDesigner2 software and the programmable analog module (PAM) board, go to Experiment 14–B in *Laboratory Exercises for Electronic Devices* by David Buchla and Steven Wetterling.

SUMMARY

Section 14–1
- A basic instrumentation amplifier is formed by three op-amps and seven resistors, including the gain-setting resistor R_G.
- An instrumentation amplifier has high input impedance, high CMRR, low output offset, and low output impedance.
- The voltage gain of a basic instrumentation amplifier is set by a single external resistor.
- An instrumentation amplifier is useful in applications where small signals are embedded in large common-mode noise.

Section 14–2
- A basic isolation amplifier has electrically isolated input and output stages.
- Isolation amplifiers use capacitive, optical, or transformer coupling for isolation.
- Isolation amplifiers are used to interface sensitive equipment with high-voltage environments and to provide protection from electrical shock in certain medical applications.

Section 14–3
- The operational transconductance amplifier (OTA) is a voltage-to-current amplifier.
- The output current of an OTA is the differential input voltage times the transconductance.
- In an OTA, transconductance varies with bias current; therefore, the gain of an OTA can be varied with a bias voltage or a variable resistor.

Section 14–4
- The operation of log and antilog amplifiers is based on the nonlinear (logarithmic) characteristics of a *pn* junction.
- A log amplifier has a *pn* junction in the feedback loop, and an antilog amplifier has a *pn* junction in series with the input.

Section 14–5 ◆ A constant-current source delivers the same load current regardless of load resistance (within limits).

◆ In a peak detector, an op-amp is used as a comparator to charge a capacitor through a diode to the peak value of the input voltage. It is useful in measuring peak voltage surges.

KEY TERMS

Key terms and other bold terms in the chapter are defined in the end-of-book glossary.

Instrumentation amplifier An amplifier used for amplifying small signals riding on large common-mode voltages.

Isolation amplifier An amplifier with electrically isolated internal stages.

Natural logarithm The exponent to which the base e ($e = 2.71828$) must be raised in order to equal a given quantity.

Operational transconductance amplifier (OTA) A voltage-to-current amplifier.

Transconductance In an electronic device, the ratio of the output current to the input voltage.

KEY FORMULAS

Instrumentation Amplifier

14–1 $$A_{cl} = 1 + \frac{2R}{R_G}$$

14–2 $$R_G = \frac{2R}{A_{cl} - 1}$$

Isolation Amplifier

14–3 $$A_{v1} = \frac{R_{f1}}{R_{i1}} + 1$$

14–4 $$A_{v2} = \frac{R_{f2}}{R_{i2}} + 1$$

Operational Transconductance Amplifier (OTA)

14–5 $$g_m = \frac{I_{out}}{V_{in}}$$

14–6 $$g_m = KI_{BIAS}$$

Log and Antilog Amplifiers

14–7 $$V_{out} = -K\ln(V_{in})$$

14–8 $$V_{out} \cong -(0.025\,\text{V})\ln\left(\frac{V_{in}}{I_R R_1}\right)$$

14–9 $$V_{out} = -(0.025\,\text{V})\ln\left(\frac{V_{in}}{I_{EBO} R_1}\right)$$

14–10 $$V_{out} = -R_f I_{EBO}\,\text{antilog}\left(\frac{V_{in}}{25\,\text{mV}}\right)$$

Converters and Other Op-Amp Circuits

14–11 $$I_L = \frac{V_{IN}}{R_i}$$ Constant-current source

14–12 $$V_{out} = I_i R_f$$ Current-to-voltage converter

14–13 $$I_L = \frac{V_{in}}{R_1}$$ Voltage-to-current converter

TRUE/FALSE QUIZ

Answers can be found at www.pearsonhighered.com/floyd.

1. Instrumentation amplifiers are particularly useful for amplifying small signals in a noisy environment.
2. The gain of an instrumentation amplifier cannot be changed.
3. A basic instrumentation amplifier consists of three op-amps.
4. An isolation amplifier prefers to operate alone.
5. An isolation amplifier consists of two electrically isolated stages.
6. All isolation amplifiers use transformer coupling.
7. OTA stands for operational transistor amplifier.
8. The transconductance of an OTA is dependent on a bias current.
9. A log amplifier can be used for compression of large dynamic range signals.
10. A peak detector is a circuit that uses a diode and a capacitor to produce a dc voltage equal to the peak of the input signal voltage.

CIRCUIT-ACTION QUIZ

Answers can be found at www.pearsonhighered.com/floyd.

1. If the value of R_G in Figure 14–7 is increased, the voltage gain will
   (a) increase   (b) decrease   (c) not change
2. If the voltage gain of the instrumentation amplifier in Figure 14–7 is set to 10 at 1 kHz and the frequency is increased to 100 kHz, the gain will
   (a) increase   (b) decrease   (c) not change
3. If the voltage gain of the instrumentation amplifier in Figure 14–7 is increased from 10 to 100, the bandwidth will
   (a) increase   (b) decrease   (c) not change
4. If R_{f1} in the isolation amplifier of Figure 14–15 is increased to 33 kΩ, the total voltage gain will
   (a) increase   (b) decrease   (c) not change
5. If the values of all the capacitors in Figure 14–15 are changed to 0.68 μF, the gain of the output stage will
   (a) increase   (b) decrease   (c) not change
6. If the value of R_L in the OTA of Figure 14–23 is reduced, the voltage gain will
   (a) increase   (b) decrease   (c) not change
7. If the bias current in the OTA of Figure 14–23 is increased, the voltage gain will
   (a) increase   (b) decrease   (c) not change
8. In the log amplifier of Figure 14–31, when the value of R_1 is decreased, the output voltage will
   (a) increase   (b) decrease   (c) not change

SELF-TEST

Answers can be found at www.pearsonhighered.com/floyd.

Section 14–1
1. To make a basic instrumentation amplifier, it takes
   (a) one op-amp with a certain feedback arrangement
   (b) two op-amps and seven resistors
   (c) three op-amps and seven capacitors
   (d) three op-amps and seven resistors
2. Typically, an instrumentation amplifier has an external resistor used for
   (a) establishing the input impedance   (b) setting the voltage gain
   (c) setting the current gain   (d) interfacing with an instrument

3. Instrumentation amplifiers are used primarily in
   (a) high-noise environments   (b) medical equipment
   (c) test instruments   (d) filter circuits

Section 14–2
4. Isolation amplifiers are used primarily in
   (a) remote, isolated locations
   (b) systems that isolate a single signal from many different signals
   (c) applications where there are high voltages and sensitive equipment
   (d) applications where human safety is a concern
   (e) answers (c) and (d)

5. The two parts of a basic isolation amplifier are
   (a) amplifier and filter   (b) input stage and coupling stage
   (c) input stage and output stage   (d) gain stage and offset stage

6. The stages of many isolation amplifiers are connected by
   (a) copper strips   (b) a capacitor   (c) microwave links   (d) current loops

7. The characteristic that allows an isolation amplifier to amplify small signal voltages in the presence of much greater noise voltages is its
   (a) CMRR   (b) high gain
   (c) high input impedance   (d) magnetic coupling between input and output

Section 14–3
8. The term *OTA* means
   (a) operational transistor amplifier   (b) operational transformer amplifier
   (c) operational transconductance amplifier   (d) output transducer amplifier

9. In an OTA, the transconductance is controlled by
   (a) the dc supply voltage   (b) the input signal voltage
   (c) the manufacturing process   (d) a bias current

10. The voltage gain of an OTA circuit is set by
    (a) a feedback resistor   (b) the transconductance only
    (c) the transconductance and the load resistor   (d) the bias current and supply voltage

11. An OTA is basically a
    (a) voltage-to-current amplifier   (b) current-to-voltage amplifier
    (c) current-to-current amplifier   (d) voltage-to-voltage amplifier

Section 14–4
12. The operation of a logarithmic amplifier is based on
    (a) the nonlinear operation of an op-amp
    (b) the logarithmic characteristic of a *pn* junction
    (c) the reverse breakdown characteristic of a *pn* junction
    (d) the logarithmic charge and discharge of an *RC* circuit

13. If the input to a log amplifier is x, the output is proportional to
    (a) e^x (b) $\ln x$ (c) $\log_{10}x$
    (d) $2.3 \log_{10}x$ (e) answers (a) and (c)   (f) answers (b) and (d)

14. If the input to an antilog amplifier is x, the output is proportional to
    (a) $e^{\ln x}$ (b) e^x (c) $\ln x$ (d) e^{-x}

Section 14–5
15. A constant-current source provides a nonchanging current to a load
    (a) for all values of current
    (b) for all values of load resistance
    (c) for all values of load resistance within defined limits

16. A peak detector consists of
    (a) a comparator, a transistor, and a capacitor   (b) a comparator, a diode, and a capacitor
    (c) a comparator, a diode, and an inductor   (d) an integrator, a diode, and a capacitor

PROBLEMS

Answers to all odd-numbered problems are at the end of the book.

BASIC PROBLEMS

Section 14–1   **Instrumentation Amplifiers**

1. Determine the voltage gains of op-amps A1 and A2 for the instrumentation amplifier configuration in Figure 14–55.

▶ **FIGURE 14–55**

Multisim and LT Spice file circuits are identified with a logo and are in the Problems folder on the website. File names correspond to Figure numbers (e.g., FGM14-55 or FGS14-55).

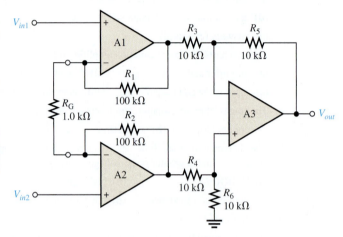

2. Find the overall voltage gain of the instrumentation amplifier in Figure 14–55.

3. The following voltages are applied to the instrumentation amplifier in Figure 14–55:
   $V_{in1} = 5$ mV, $V_{in2} = 10$ mV, and $V_{cm} = 225$ mV. Determine the final output voltage.

4. What value of R_G must be used to change the gain of the instrumentation amplifier in Figure 14–55 to 1000?

5. What is the voltage gain of the AD622 instrumentation amplifier in Figure 14–56?

▶ **FIGURE 14–56**

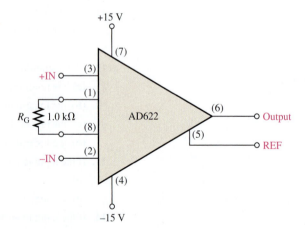

6. Determine the approximate bandwidth of the amplifier in Figure 14–56 if the voltage gain is set to 10. Use the graph in Figure 14–6.

7. Specify what you must do to change the gain of the amplifier in Figure 14–56 to approximately 24.

8. Determine the value of R_G in Figure 14–56 for a voltage gain of 20.

Section 14–2   **Isolation Amplifiers**

9. The op-amp in the input stage of a certain isolation amplifier has a voltage gain of 30. The output stage is set for a gain of 10. What is the total voltage gain of this device?

10. Determine the total voltage gain of each 3656KG in Figure 14–57.

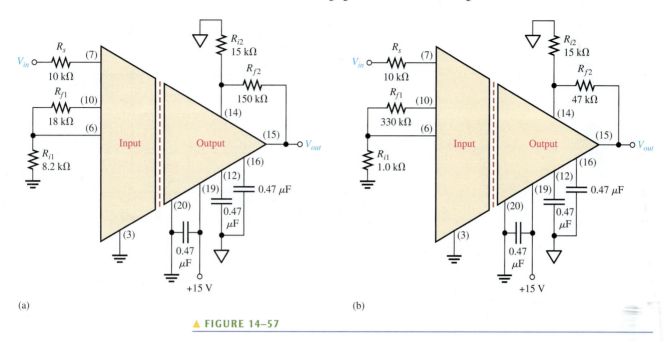

(a)　　　　　　　　　　　　　　　　　　　(b)

▲ FIGURE 14–57

11. Specify how you would change the total gain of the amplifier in Figure 14–57(a) to approximately 100 by changing only the gain of the input stage.

12. Specify how you would change the total gain in Figure 14–57(b) to approximately 440 by changing only the gain of the output stage.

13. Specify how you would connect each amplifier in Figure 14–57 for unity gain.

Section 14–3 **Operational Transconductance Amplifiers (OTAs)**

14. A certain OTA has an input voltage of 10 mV and an output current of 10 μA. What is the transconductance?

15. A certain OTA with a transconductance of 5000 μS has a load resistance of 10 kΩ. If the input voltage is 100 mV, what is the output current? What is the output voltage?

16. The output voltage of a certain OTA with a load resistance is determined to be 3.5 V. If its transconductance is 4000 μS and the input voltage is 100 mV, what is the value of the load resistance?

17. Determine the voltage gain of the OTA in Figure 14–58. Assume $K = 16$ μS/μA for the graph in Figure 14–59.

▶ FIGURE 14–58

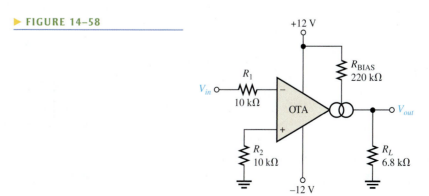

▶ **FIGURE 14–59**

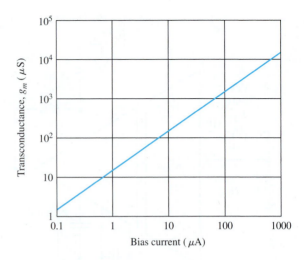

18. If a 10 kΩ rheostat is added in series with the bias resistor in Figure 14–58, what are the minimum and maximum voltage gains?

19. The OTA in Figure 14–60 functions as an amplitude modulation circuit. Determine the output voltage waveform for the given input waveforms assuming $K = 16\ \mu S/\mu A$.

▶ **FIGURE 14–60**

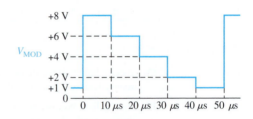

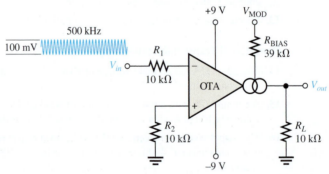

20. Determine the trigger points for the Schmitt trigger in Figure 14–61.

21. Determine the output voltage waveform for the Schmitt trigger in Figure 14–61 in relation to a 1 kHz sine wave with peak values of ±10 V.

▶ **FIGURE 14–61**

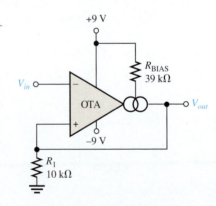

Section 14–4   Log and Antilog Amplifiers

22. Using your calculator, find the natural logarithm (ln) of each of the following numbers:

(a) 0.5   (b) 2   (c) 50   (d) 130

23. Repeat Problem 22 for $\log_{10}$.

24. What is the antilog of 1.6?

25. Explain why the output of a log amplifier is limited to approximately 0.7 V.

26. What is the output voltage of a certain log amplifier with a diode in the feedback path when the input voltage is 3 V? The input resistor is 82 kΩ and the reverse leakage current is 100 nA.

27. Determine the output voltage for the log amplifier in Figure 14–62. Assume I_{EBO} = 60 nA.

 ► **FIGURE 14–62**

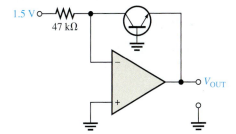

28. Determine the output voltage for the antilog amplifier in Figure 14–63. Assume I_{EBO} = 60 nA.

► **FIGURE 14–63**

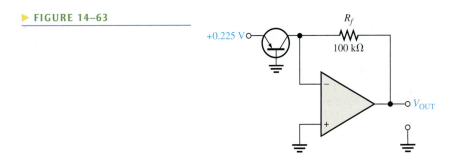

29. Signal compression is one application of logarithmic amplifiers. Suppose an audio signal with a maximum voltage of 1 V and a minimum voltage of 100 mV is applied to the log amplifier in Figure 14–62. What will be the maximum and minimum output voltages? What conclusion can you draw from this result?

Section 14–5   Converters and Other Integrated Circuits

30. Determine the load current in each circuit of Figure 14–64.

► **FIGURE 14–64**

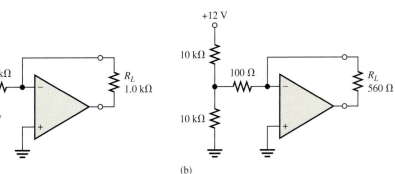

(a)   (b)

31. Devise a circuit for remotely sensing temperature and producing a proportional voltage that can then be converted to digital form for display. A thermistor can be used as the temperature-sensing element.

MULTISIM TROUBLESHOOTING PROBLEMS

These file circuits are in the Troubleshooting Problems folder on the website.

32. Open file TPM14-33 and determine the fault.
33. Open file TPM14-34 and determine the fault.
34. Open file TPM14-35 and determine the fault.
35. Open file TPM14-36 and determine the fault.
36. Open file TPM14-37 and determine the fault.

ACTIVE FILTERS

15

CHAPTER OBJECTIVES

◆ Describe and analyze the gain-versus-frequency responses of basic types of filters

◆ Describe three types of filter response characteristics and other parameters

◆ Identify and analyze active low-pass filters

◆ Identify and analyze active high-pass filters

◆ Analyze basic types of active band-pass filters

◆ Describe basic types of active band-stop filters

◆ Discuss two methods for measuring frequency response

KEY TERMS

◆ Filter
◆ Low-pass filter
◆ Pole
◆ Roll-off
◆ High-pass filter
◆ Band-pass filter
◆ Band-stop filter
◆ Damping factor

DEVICE APPLICATION PREVIEW

RFID stands for Radio Frequency Identification and is a technology that enables the tracking and/or identification of objects. Typically, an RFID system consists of an RF tag containing an IC chip that transmits data about the object, a reader that receives transmitted data from the tag, and a data-processing system that processes and stores the data passed to it by the reader. In this application, you will focus on the RFID reader. RFID systems are used in metering applications such as electronic toll collection, inventory control and tracking, merchandise control, asset tracking and recovery, tracking parts moving through a manufacturing process, and tracking goods in a supply chain.

VISIT THE WEBSITE

Study aids and Multisim files for this chapter are available at https://www.pearsonhighered.com/careersresources/

INTRODUCTION

Power-supply filters were introduced in Chapter 2. In this chapter, active filters that are used for signal processing are introduced. Filters are circuits that are capable of passing signals with certain selected frequencies while rejecting signals with other frequencies. This property is called *selectivity.*

Active filters use transistors or op-amps combined with passive *RC*, *RL*, or *RLC* circuits. The active devices provide voltage gain, and the passive circuits provide frequency selectivity. Although inductors are used in passive filters, they are avoided in active filters because inductors tend to be bulky, more expensive than capacitors, and not easily integrated. In terms of general response, the four basic categories of active filters are low-pass, high-pass, band-pass, and band-stop. In this chapter, you will study active filters using op-amps and *RC* circuits.

15–1   BASIC FILTER RESPONSES

Filters are usually categorized by the manner in which the output voltage varies with the frequency of the input voltage. The categories of active filters are low-pass, high-pass, band-pass, and band-stop. Each of these general responses are examined.

After completing this section, you should be able to

❑ **Describe and analyze the gain-versus-frequency responses of basic types of filters**
❑ Describe low-pass filter response
   ◆ Define *passband* and *critical frequency*   ◆ Determine the bandwidth
   ◆ Define *pole*   ◆ Explain roll-off rate and define its unit   ◆ Calculate the critical frequency
❑ Describe high-pass filter response
   ◆ Explain how the passband is limited   ◆ Calculate the critical frequency
❑ Describe band-pass filter response
   ◆ Determine the bandwidth   ◆ Determine the center frequency   ◆ Calculate the quality factor (Q)
❑ Describe band-stop filter response
   ◆ Determine the bandwidth

Low-Pass Filter Response

A **filter** is a circuit that passes certain frequencies and attenuates or rejects all other frequencies. The **passband** of a filter is the range of frequencies that are allowed to pass through the filter with minimum attenuation (usually defined as less than −3 dB of attenuation). The **critical frequency**, f_c, (also called the *cutoff frequency*) defines the end of the passband and is normally specified at the point where the response drops −3 dB (70.7%) from the passband response. Following the passband is a region called the *transition region* that leads into a region called the *stopband*. There is no precise point between the transition region and the stopband.

A **low-pass filter** is one that passes frequencies from dc to f_c and significantly attenuates all other frequencies. The passband of the ideal low-pass filter is shown in the blue-shaded area of Figure 15–1(a); the response drops to zero at frequencies beyond the passband. This ideal response is sometimes referred to as a "brick-wall" because nothing gets through beyond the wall. The bandwidth of an ideal low-pass filter is equal to f_c.

Equation 15–1

$$BW = f_c$$

The ideal response shown in Figure 15–1(a) is not attainable by any practical filter. Actual filter responses depend on the number of **poles**, a term used with filters to describe the number of *RC* circuits contained in the filter. (The term *pole* has a number of meanings in electrical work, but this definition is specifically for filters. In mathematics, it is a value for which a function goes to infinity, such as when you divide by zero.) The most basic low-pass filter is a simple *RC* circuit consisting of just one resistor and one capacitor; the output is taken across the capacitor as shown in Figure 15–1(b). This basic *RC* filter has a single pole, and it rolls off at −20 dB/decade beyond the critical frequency. The actual response is indicated by the blue line in Figure 15–1(a). The response is plotted on a standard log plot that is used for filters to show details of the curve as the gain drops. Notice that the gain drops off slowly until the frequency is at the critical frequency; after this, the gain drops rapidly.

The −20 dB/decade **roll-off** rate for the gain of a basic *RC* filter means that at a frequency of $10f_c$, the output will be −20 dB (10%) of the input. This roll-off rate is not a particularly good filter characteristic because too much of the unwanted frequencies (beyond the passband) are allowed through the filter.

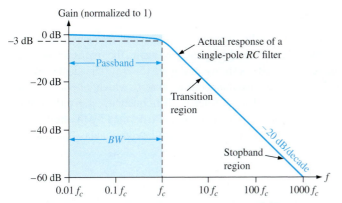

(a) Comparison of an ideal low-pass filter response (blue area) with actual response.

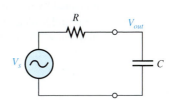

(b) Basic low-pass circuit

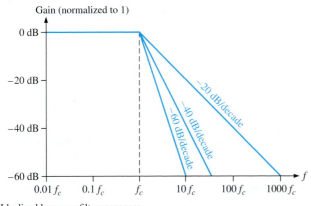

(c) Idealized low-pass filter responses

▲ **FIGURE 15–1**

Low-pass filter responses.

The critical frequency of a low-pass RC filter occurs when $X_C = R$, where

$$f_c = \frac{1}{2\pi RC}$$

Recall from your basic dc/ac studies that the output at the critical frequency is 70.7% of the input. This response is equivalent to an attenuation of −3 dB.

Figure 15–1(c) illustrates three idealized low-pass response curves including the basic one-pole response (−20 dB/decade). The approximations show a flat response to the cutoff frequency and a roll-off at a constant rate after the cutoff frequency. Actual filters do not have a perfectly flat response up to the cutoff frequency but drop to −3 dB at this point as described previously.

The simple one-pole RC filter in Figure 15–1 is a passive filter because it is composed only of passive components. More poles can be added to increase the steepness of the transition region, but the downside is that the accuracy of the filter is less due to loading effects. A better way to produce a filter that has a steeper transition region is to add active circuitry (an amplifier) to the basic filter. Responses that are steeper than −20 dB/decade in the transition region cannot be obtained by simply cascading identical RC stages (due to loading effects). However, by combining an op-amp with frequency-selective feedback circuits, filters can be designed with roll-off rates of −40, −60, or more dB/decade. Filters that include one or more op-amps in the design are called **active filters.** These filters can optimize the roll-off rate or other attribute (such as phase response) with a particular filter design. In general, the more poles the filter uses, the steeper its transition region will be. The exact response depends on the type of filter and the number of poles.

High-Pass Filter Response

A **high-pass filter** is one that significantly attenuates or rejects all frequencies below f_c and passes all frequencies above f_c. The critical frequency is, again, the frequency at which the output is 70.7% of the input (or -3 dB) as shown in Figure 15–2(a). The ideal response, indicated by the blue-shaded area, has an instantaneous drop at f_c, which, of course, is not achievable. Ideally, the passband of a high-pass filter is all frequencies above the critical frequency. The high-frequency response of practical circuits is limited by the finite band-width of active components and unwanted stray capacitance in components that make up the filter.

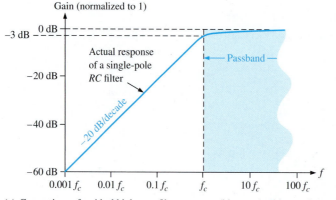

(a) Comparison of an ideal high-pass filter response (blue area) with actual response

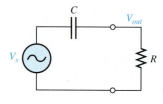

(b) Basic high-pass circuit

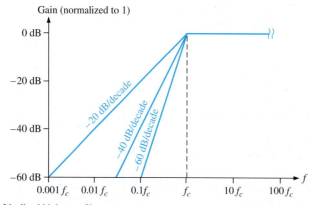

(c) Idealized high-pass filter responses

▲ **FIGURE 15–2**

High-pass filter responses.

A simple *RC* circuit consisting of a single resistor and capacitor can be configured as a high-pass filter by taking the output across the resistor as shown in Figure 15–2(b). As in the case of the low-pass filter, the basic *RC* circuit has a roll-off rate of -20 dB/decade, as indicated by the blue line in Figure 15–2(a). Also, the critical frequency for the basic high-pass filter occurs when $X_C = R$, where

$$f_c = \frac{1}{2\pi RC}$$

Figure 15–2(c) illustrates three idealized high-pass response curves, including the basic one-pole response (-20 dB/decade) for a high-pass *RC* circuit. As in the case of the low-pass filter, the approximations show a flat response to the cutoff frequency and a roll-off at a constant rate after the cutoff frequency. Actual high-pass filters do not have the perfectly flat response indicated or the precise roll-off rate shown. Responses that are steeper than −20 dB/decade in the transition region are also possible with both passive and active high-pass filters; the particular response depends on the type of filter and the number of poles. In general, active filters suffer from fewer loading effects.

Band-Pass Filter Response

A band-pass filter passes all signals lying within a band between a lower-frequency limit and an upper-frequency limit and essentially rejects all other frequencies that are outside this specified band. A generalized band-pass response curve is shown in Figure 15–3. The bandwidth (BW) is defined as the difference between the upper critical frequency (f_{c2}) and the lower critical frequency (f_{c1}).

$$BW = f_{c2} - f_{c1}$$

Equation 15–2

The critical frequencies are, of course, the points at which the response curve is 70.7% of its maximum. Recall from Chapter 12 that these critical frequencies are also called *3 dB frequencies*. The frequency about which the passband is centered is called the *center frequency*, f_0, defined as the geometric mean of the critical frequencies.

$$f_0 = \sqrt{f_{c1}f_{c2}}$$

Equation 15–3

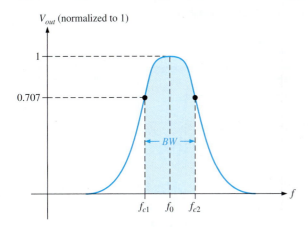

▲ FIGURE 15–3

General band-pass response curve.

Quality Factor The **quality factor** (Q) of a band-pass filter is the ratio of the center frequency to the bandwidth.

$$Q = \frac{f_0}{BW}$$

Equation 15–4

The value of Q is an indication of the selectivity of a band-pass filter. The higher the value of Q, the narrower the bandwidth and the better the selectivity for a given value of f_0. Band-pass filters are sometimes classified as narrow-band ($Q > 10$) or wide-band ($Q < 10$). The quality factor (Q) can also be expressed in terms of the damping factor (*DF*) of the filter as

$$Q = \frac{1}{DF}$$

You will study the damping factor in Section 15–2.

EXAMPLE 15–1

A certain band-pass filter has a center frequency of 15 kHz and a bandwidth of 1 kHz. Determine Q and classify the filter as narrow-band or wide-band.

Solution

$$Q = \frac{f_0}{BW} = \frac{15 \text{ kHz}}{1 \text{ kHz}} = \textbf{15}$$

Because $Q > 10$, this is a narrow-band filter.

*Related Problem**

If the quality factor of the filter is doubled, what will the bandwidth be?

*Answers can be found at www.pearsonhighered.com/floyd.

Band-Stop Filter Response

Another category of active filter is the **band-stop filter**, also known as *notch, band-reject,* or *band-elimination* filter. You can think of the operation as opposite to that of the band-pass filter because frequencies within a certain bandwidth are rejected, and frequencies outside the bandwidth are passed. A general response curve for a band-stop filter is shown in Figure 15–4. Notice that the bandwidth is the band of frequencies between the 3 dB points, just as in the case of the band-pass filter response.

▶ **FIGURE 15–4**

General band-stop filter response.

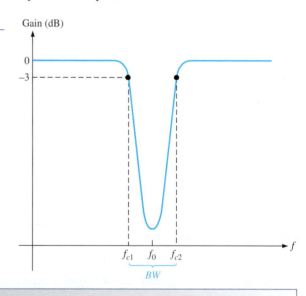

SECTION 15–1 CHECKUP

Answers can be found at www .pearsonhighered.com/floyd.

1. What determines the bandwidth of a low-pass filter?
2. What limits the passband of an active high-pass filter?
3. How are the Q and the bandwidth of a band-pass filter related? Explain how the selectivity is affected by the Q of a filter.

15–2 FILTER RESPONSE CHARACTERISTICS

Each type of filter response (low-pass, high-pass, band-pass, or band-stop) can be tailored by circuit component values to have either a Butterworth, Chebyshev, or Bessel characteristic. Each of these characteristics is identified by the shape of the response curve, and each has an advantage in certain applications.

After completing this section, you should be able to

❑ **Describe three types of filter response characteristics and other parameters**
❑ Discuss the Butterworth characteristic
❑ Describe the Chebyshev characteristic
❑ Discuss the Bessel characteristic
❑ Define *damping factor*
    ◆ Calculate the damping factor   ◆ Show the block diagram of an active filter
❑ Analyze a filter for critical frequency and roll-off rate
    ◆ Explain how to obtain multiorder filters   ◆ Describe the effects of cascading on roll-off rate

Butterworth, Chebyshev, or Bessel response characteristics can be realized with most active filter circuit configurations by proper selection of certain component values. A general comparison of the three response characteristics for a low-pass filter response curve is shown in Figure 15–5. High-pass and band-pass filters can also be designed to have any one of the characteristics.

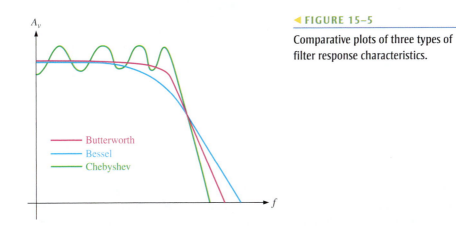

◀ **FIGURE 15–5**

Comparative plots of three types of filter response characteristics.

The Butterworth Characteristic   The **Butterworth** characteristic provides a very flat amplitude response in the passband and a roll-off rate of -20 dB/decade/pole. The phase response is not linear, however, and the phase shift (thus, time delay) of signals passing through the filter varies nonlinearly with frequency. Therefore, a pulse applied to a filter with a Butterworth response will cause overshoots on the output because each frequency component of the pulse's rising and falling edges experiences a different time delay. Filters with the Butterworth response are normally used when all frequencies in the passband must have the same gain. The Butterworth response is often referred to as a maximally flat response.

The Chebyshev Characteristic   Filters with the **Chebyshev** response characteristic are useful when a rapid roll-off is required because it provides a roll-off rate greater than -20 dB/decade/pole. This is a greater rate than that of the Butterworth, so filters can be implemented with the Chebyshev response with fewer poles and less complex circuitry for a given roll-off rate. This type of filter response is characterized by overshoot or ripples in the passband (depending on the number of poles) and an even less linear phase response than the Butterworth.

The Bessel Characteristic   The **Bessel** response exhibits a linear phase characteristic, meaning that the phase shift increases linearly with frequency. The result is almost no overshoot on the output with a pulse input. For this reason, filters with the Bessel response are used for filtering pulse waveforms without distorting the shape of the waveform.

The Damping Factor

As mentioned, an active filter can be designed to have either a Butterworth, Chebyshev, or Bessel response characteristic regardless of whether it is a low-pass, high-pass, band-pass, or band-stop type. The **damping factor (*DF*)** of an active filter circuit determines which response characteristic the filter exhibits. To explain the basic concept, a generalized active filter is shown in Figure 15–6. It includes an amplifier, a negative feedback circuit, and a filter section. The amplifier and feedback are connected in a noninverting configuration. The damping factor is determined by the negative feedback circuit and is defined by the following equation:

Equation 15–5

$$DF = 2 - \frac{R_1}{R_2}$$

▶ FIGURE 15–6

General diagram of an active filter.

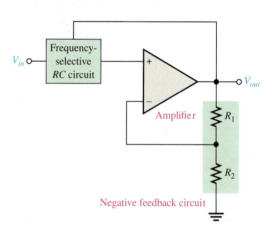

Basically, the damping factor affects the filter response by negative feedback action. Any attempted increase or decrease in the output voltage is offset by the opposing effect of the negative feedback. This tends to make the response curve flat in the passband of the filter if the value for the damping factor is precisely set. By advanced mathematics, which we will not cover, values for the damping factor have been derived for various orders of filters to achieve the maximally flat response of the Butterworth characteristic.

The value of the damping factor required to produce a desired response characteristic depends on the **order** (number of poles) of the filter. In mathematics, a pole is a point at which a function approaches infinity. For a filter, poles are determined by the resistors and capacitors present; for example a filter circuit with one resistor and one capacitor used to affect the frequency response is a one pole filter. The more poles a filter has, the faster its roll-off rate is. To achieve a second-order Butterworth response, for example, the damping factor must be 1.414. To implement this damping factor, the feedback resistor ratio must be

$$\frac{R_1}{R_2} = 2 - DF = 2 - 1.414 = 0.586$$

This ratio gives the closed-loop gain of the noninverting amplifier portion of the filter, $A_{cl(NI)}$, a value of 1.586, derived as follows:

$$A_{cl(NI)} = \frac{1}{B} = \frac{1}{R_2/(R_1 + R_2)} = \frac{R_1 + R_2}{R_2} = \frac{R_1}{R_2} + 1 = 0.586 + 1 = 1.586$$

EXAMPLE 15-2

If resistor R_2 in the feedback circuit of an active single-pole filter of the type in Figure 15–6 is 10 kΩ, what value must R_1 be to obtain a maximally flat Butterworth response?

Solution

$$\frac{R_1}{R_2} = 0.586$$

$$R_1 = 0.586R_2 = 0.586(10\ \text{k}\Omega) = \mathbf{5.86\ k\Omega}$$

Using the nearest standard 5% value of 5.6 kΩ will get very close to the ideal Butterworth response.

Related Problem   What is the damping factor for $R_2 = 10\ \text{k}\Omega$ and $R_1 = 5.6\ \text{k}\Omega$?

Critical Frequency and Roll-Off Rate

The critical frequency is determined by the values of the resistors and capacitors in the frequency-selective RC circuit shown in Figure 15–6. For a single-pole (first-order) filter, as shown in Figure 15–7, the critical frequency is

$$f_c = \frac{1}{2\pi RC}$$

Although we show a low-pass configuration, the same formula is used for the f_c of a single-pole high-pass filter. The number of poles determines the roll-off rate of the filter. A Butterworth response produces -20 dB/decade/pole. So, a first-order (one-pole) filter has a roll-off of -20 dB/decade; a second-order (two-pole) filter has a roll-off rate of -40 dB/decade; a third-order (three-pole) filter has a roll-off rate of -60 dB/decade; and so on.

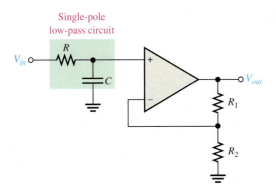

Single-pole
low-pass circuit

◀ **FIGURE 15–7**

First-order (one-pole) low-pass filter.

Generally, to obtain a filter with three poles or more, one-pole or two-pole filters are cascaded, as shown in Figure 15–8. To obtain a third-order filter, for example, cascade a second-order and a first-order filter; to obtain a fourth-order filter, cascade two second-order filters; and so on. Each filter in a cascaded arrangement is called a *stage* or *section*.

Because of its maximally flat response, the Butterworth characteristic is the most widely used. Therefore, we will limit our coverage to the Butterworth response to illustrate basic filter concepts. Table 15–1 lists the roll-off rates, damping factors, and feedback resistor ratios for up to sixth-order Butterworth filters. Resistor designations correspond to the gain-setting resistors in Figure 15–8 and may be different on other circuit diagrams.

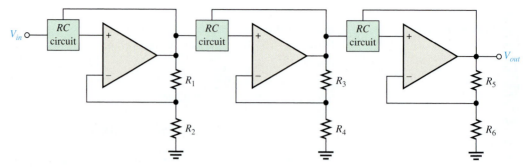

▲ FIGURE 15–8

The number of filter poles can be increased by cascading.

▼ TABLE 15–1

Values for the Butterworth response.

ORDER	ROLL-OFF DB/DECADE	1ST STAGE			2ND STAGE			3RD STAGE		
		POLES	DF	R_1/R_2	POLES	DF	R_3/R_4	POLES	DF	R_5/R_6
1	−20	1	Optional							
2	−40	2	1.414	0.586						
3	−60	2	1.00	1	1	1.00	1			
4	−80	2	1.848	0.152	2	0.765	1.235			
5	−100	2	1.00	1	2	1.618	0.382	1	0.618	1.382
6	−120	2	1.932	0.068	2	1.414	0.586	2	0.518	1.482

SECTION 15–2 CHECKUP

1. Explain how Butterworth, Chebyshev, and Bessel responses differ.
2. What determines the response characteristic of a filter?
3. Name the basic parts of an active filter.

15–3 ACTIVE LOW-PASS FILTERS

Filters that use op-amps as the active element provide several advantages over passive filters (R, L, and C elements only). The op-amp provides gain, so the signal is not attenuated as it passes through the filter. The high input impedance of the op-amp prevents excessive loading of the driving source, and the low output impedance of the op-amp prevents the filter from being affected by the load that it is driving. Active filters are also easy to adjust over a wide frequency range without altering the desired response.

After completing this section, you should be able to

❑ **Identify and analyze active low-pass filters**
❑ Identify a single-pole low-pass filter circuit
  ◆ Determine the closed-loop voltage gain ◆ Determine the critical frequency
❑ Identify a Sallen-Key low-pass filter circuit
  ◆ Describe the filter operation ◆ Calculate the critical frequency
❑ Analyze cascaded low-pass filters
  ◆ Explain how the roll-off rate is affected

A Single-Pole Filter

Figure 15–9(a) shows an active filter with a single low-pass RC frequency-selective circuit that provides a roll-off of -20 dB/decade above the critical frequency, as indicated by the response curve in Figure 15–9(b). The critical frequency of the single-pole filter is $f_c = 1/(2\pi RC)$. The op-amp in this filter is connected as a noninverting amplifier with the closed-loop voltage gain in the passband set by the values of R_1 and R_2.

$$A_{cl(NI)} = \frac{R_1}{R_2} + 1$$

Equation 15–6

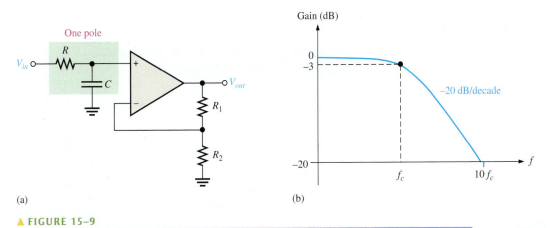

(a)   (b)

▲ **FIGURE 15–9**

Single-pole active low-pass filter and response curve.

The Sallen-Key Low-Pass Filter

The Sallen-Key is one of the most common configurations for a second-order (two-pole) filter. It is also known as a VCVS (voltage-controlled voltage source) filter. A low-pass version of the Sallen-Key filter is shown in Figure 15–10. Notice that there are two low-pass RC circuits that provide a roll-off of -40 dB/decade above the critical frequency (assuming a Butterworth characteristic). One RC circuit consists of R_A and C_A, and the second circuit consists of R_B and C_B. A unique feature of the Sallen-Key low-pass filter is the capacitor C_A that provides feedback for shaping the response near the edge of the passband. The critical frequency for the Sallen-Key filter is

$$f_c = \frac{1}{2\pi\sqrt{R_A R_B C_A C_B}}$$

Equation 15–7

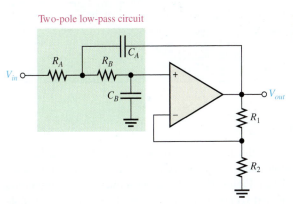

◀ **FIGURE 15–10**

Basic Sallen-Key low-pass filter.

The component values can be made equal so that $R_A = R_B = R$ and $C_A = C_B = C$. In this case, the expression for the critical frequency simplifies to

$$f_c = \frac{1}{2\pi RC}$$

As in the single-pole filter, the op-amp in the second-order Sallen-Key filter acts as a non-inverting amplifier with the negative feedback provided by resistors R_1 and R_2. As you have learned, the damping factor is set by the values of R_1 and R_2, thus making the filter response either Butterworth, Chebyshev, or Bessel. For example, from Table 15–1, the R_1/R_2 ratio must be 0.586 to produce the damping factor of 1.414 required for a second-order Butterworth response.

EXAMPLE 15–3

Determine the critical frequency of the Sallen-Key low-pass filter in Figure 15–11, and set the value of R_1 for an approximate Butterworth response.

▶ **FIGURE 15–11**

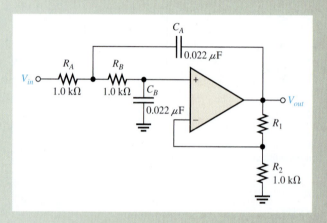

Solution   Since $R_A = R_B = R = 1.0\,\text{k}\Omega$ and $C_A = C_B = C = 0.022\,\mu\text{F}$,

$$f_c = \frac{1}{2\pi RC} = \frac{1}{2\pi(1.0\,\text{k}\Omega)(0.022\,\mu\text{F})} = \textbf{7.23 kHz}$$

For a Butterworth response, $R_1/R_2 = 0.586$.

$$R_1 = 0.586R_2 = 0.586(1.0\,\text{k}\Omega) = \textbf{586}\,\boldsymbol{\Omega}$$

Select a standard value as near as possible to this calculated value.

Related Problem   Determine f_c for Figure 15–11 if $R_A = R_B = R_2 = 2.2\,\text{k}\Omega$ and $C_A = C_B = 0.01\,\mu\text{F}$. Also determine the value of R_1 for a Butterworth response.

 Open the Multisim file EXM15-03 or the LT Spice file EXS15-03 in the Examples folder on the website. Determine the critical frequency and compare with the calculated value.

Cascaded Low-Pass Filters

A three-pole filter is required to get a third-order low-pass response (−60 dB/decade). This is done by cascading a two-pole Sallen-Key low-pass filter and a single-pole low-pass filter, as shown in Figure 15–12(a). Figure 15–12(b) shows a four-pole configuration obtained by cascading two Sallen-Key (two-pole) low-pass filters. In general, a four-pole filter is preferred because it uses the same number of op-amps to achieve a faster roll-off.

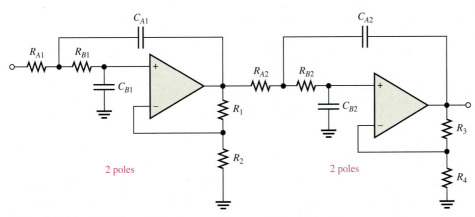

Cascaded low-pass filters.

(a) Third-order configuration

(b) Fourth-order configuration

EXAMPLE 15–4

For the four-pole filter in Figure 15–12(b), determine the capacitance values required to produce a critical frequency of 2680 Hz if all the resistors in the RC low-pass circuits are 1.8 kΩ. Also select values for the feedback resistors to get a Butterworth response.

Solution   Both stages must have the same f_c. Assuming equal-value capacitors,

$$f_c = \frac{1}{2\pi RC}$$

$$C = \frac{1}{2\pi Rf_c} = \frac{1}{2\pi(1.8\ \text{k}\Omega)(2680\ \text{Hz})} = 0.033\ \mu\text{F}$$

$$C_{A1} = C_{B1} = C_{A2} = C_{B2} = \textbf{0.033}\ \boldsymbol{\mu}\textbf{F}$$

Also select $R_2 = R_4 = 1.8$ kΩ for simplicity. Refer to Table 15–1. For a Butterworth response in the first stage, $DF = 1.848$ and $R_1/R_2 = 0.152$. Therefore,

$$R_1 = 0.152R_2 = 0.152(1800\ \Omega) = \textbf{274}\ \boldsymbol{\Omega}$$

Choose $R_1 = 270$ Ω.
In the second stage, DF = 0.765 and $R_3/R_4 = 1.235$. Therefore,

$$R_3 = 1.235R_4 = 1.235(1800\ \Omega) = \textbf{2.22 k}\boldsymbol{\Omega}$$

Choose $R_3 = 2.2$ kΩ.

Related Problem   For the filter in Figure 15–12(b), determine the capacitance values for $f_c = 1$ kHz if all the filter resistors are 680 Ω. Also specify the values for the feedback resistors to produce a Butterworth response.

1. How many poles does a second-order low-pass filter have? How many resistors and how many capacitors are used in the frequency-selective circuit?
2. Why is the damping factor of a filter important?
3. What is the primary purpose of cascading low-pass filters?

15–4 ACTIVE HIGH-PASS FILTERS

In high-pass filters, the roles of the capacitor and resistor are reversed in the *RC* circuits. Otherwise, the basic parameters are the same as for the low-pass filters.

After completing this section, you should be able to

❑ **Identify and analyze active high-pass filters**
❑ Identify a single-pole high-pass filter circuit
  ◆ Explain limitations at higher pass-band frequencies
❑ Identify a Sallen-Key high-pass filter circuit
  ◆ Describe the filter operation ◆ Calculate component values
❑ Discuss cascaded high-pass filters
  ◆ Describe a six-pole filter

A Single-Pole Filter

A high-pass active filter with a −20 dB/decade roll-off is shown in Figure 15–13(a). Notice that the input circuit is a single high-pass *RC* circuit. The negative feedback circuit is the same as for the low-pass filters previously discussed. The high-pass response curve is shown in Figure 15–13(b).

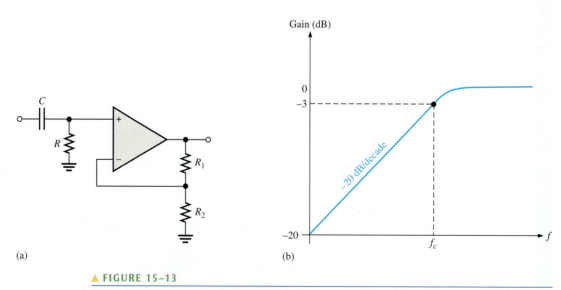

(a)   (b)

▲ **FIGURE 15–13**

Single-pole active high-pass filter and response curve.

Ideally, a high-pass filter passes all frequencies above f_c without limit, as indicated in Figure 15–14(a), although in practice, this is not the case. As you have learned, all op-amps inherently have internal *RC* circuits that limit the amplifier's response at high

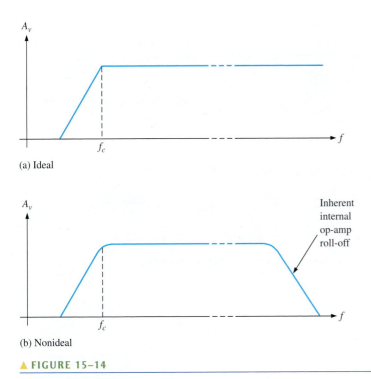

(a) Ideal

(b) Nonideal

▲ FIGURE 15–14

High-pass filter response.

frequencies. Therefore, there is an upper-frequency limit on the high-pass filter's response which, in effect, makes it a band-pass filter with a very wide bandwidth. In the majority of applications, the internal high-frequency limitation is selected to be much greater than that of the filter's critical frequency and the limitation can be neglected. In some very high-frequency applications, discrete transistors or specialized super-fast op-amps can be used for the gain element to increase the high-frequency limitation beyond that realizable with standard op-amps.

The Sallen-Key High-Pass Filter

A high-pass Sallen-Key configuration is shown in Figure 15–15. The components R_A, C_A, R_B, and C_B form the two-pole frequency-selective circuit. Notice that the positions of the resistors and capacitors in the frequency-selective circuit are opposite to those in the low-pass configuration. As with the other filters, the response characteristic can be optimized by proper selection of the feedback resistors, R_1 and R_2.

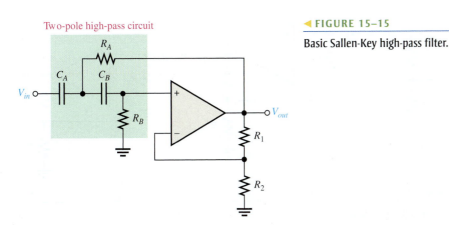

◀ FIGURE 15–15

Basic Sallen-Key high-pass filter.

EXAMPLE 15–5

Choose values for the Sallen-Key high-pass filter in Figure 15–15 to implement an equal-value second-order Butterworth response with a critical frequency of approximately 10 kHz.

Solution

Start by selecting a value for R_A and R_B (R_1 or R_2 can also be the same value as R_A and R_B for simplicity).

$$R = R_A = R_B = R_2 = \mathbf{3.3\ k\Omega}\ \text{(an arbitrary selection)}$$

Next, calculate the capacitance value from $f_c = 1/(2\pi RC)$.

$$C = C_A = C_B = \frac{1}{2\pi R f_c} = \frac{1}{2\pi(3.3\ \text{k}\Omega)(10\ \text{kHz})} = \mathbf{0.0048\ \mu F}$$

For a Butterworth response, the damping factor must be 1.414 and $R_1/R_2 = 0.586$.

$$R_1 = 0.586 R_2 = 0.586(3.3\ \text{k}\Omega) = \mathbf{1.93\ k\Omega}$$

If you had chosen $R_1 = 3.3\ \text{k}\Omega$, then

$$R_2 = \frac{R_1}{0.586} = \frac{3.3\ \text{k}\Omega}{0.586} = 5.63\ \text{k}\Omega$$

Either way, an approximate Butterworth response is realized by choosing the nearest standard values.

Related Problem

Select values for all the components in the high-pass filter of Figure 15–15 to obtain an $f_c = 300$ Hz. Use equal-value components with $R = 10\ \text{k}\Omega$ and optimize for a Butterworth response.

Cascading High-Pass Filters

As with the low-pass configuration, first- and second-order high-pass filters can be cascaded to provide three or more poles and thereby create faster roll-off rates. Figure 15–16 shows a six-pole high-pass filter consisting of three Sallen-Key two-pole stages. With this configuration optimized for a Butterworth response, a roll-off of -120 dB/decade is achieved.

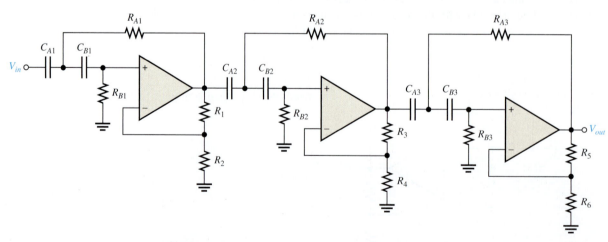

▲ **FIGURE 15–16**

Sixth-order high-pass filter.

SECTION 15–4 CHECKUP	1. How does a high-pass Sallen-Key filter differ from the low-pass configuration?
	2. To increase the critical frequency of a high-pass filter, would you increase or decrease the resistor values?
	3. If three two-pole high-pass filters and one single-pole high-pass filter are cascaded, what is the resulting roll-off?

15–5 ACTIVE BAND-PASS FILTERS

As mentioned, band-pass filters pass all frequencies bounded by a lower-frequency limit and an upper-frequency limit and reject all others lying outside this specified band. A band-pass response can be thought of as the overlapping of a low-frequency response curve and a high-frequency response curve.

After completing this section, you should be able to

❑ **Analyze basic types of active band-pass filters**
❑ Describe how to cascade low-pass and high-pass filters to create a band-pass filter
  ◆ Calculate the critical frequencies and the center frequency
❑ Identify and analyze a multiple-feedback band-pass filter
  ◆ Determine the center frequency, quality factor (Q), and bandwidth
  ◆ Calculate the voltage gain
❑ Identify and describe the state-variable filter
  ◆ Explain the basic filter operation ◆ Determine the Q
❑ Identify and discuss the biquad filter

Cascaded Low-Pass and High-Pass Filters

One way to implement a band-pass filter is a cascaded arrangement of a high-pass filter and a low-pass filter, as shown in Figure 15–17(a), as long as the critical frequencies are sufficiently separated. Each of the filters shown is a Sallen-Key Butterworth configuration so that the roll-off rates are −40 dB/decade, indicated in the composite response curve of Figure 15–17(b). The critical frequency of each filter is chosen so that the response curves overlap sufficiently, as indicated. The critical frequency of the high-pass filter must be sufficiently lower than that of the low-pass stage. This filter is generally limited to wide bandwidth applications.

The lower frequency f_{c1} of the passband is the critical frequency of the high-pass filter. The upper frequency f_{c2} is the critical frequency of the low-pass filter. Ideally, as discussed earlier, the center frequency f_0 of the passband is the geometric mean of f_{c1} and f_{c2}. The following formulas express the three frequencies of the band-pass filter in Figure 15–17.

$$f_{c1} = \frac{1}{2\pi\sqrt{R_{A1}R_{B1}C_{A1}C_{B1}}}$$

$$f_{c2} = \frac{1}{2\pi\sqrt{R_{A2}R_{B2}C_{A2}C_{B2}}}$$

$$f_0 = \sqrt{f_{c1}f_{c2}}$$

Of course, if equal-value components are used in implementing each filter, the critical frequency equations simplify to the form $f_c = 1/(2\pi RC)$.

▶ FIGURE 15–17

Band-pass filter formed by cascading a two-pole high-pass and a two-pole low-pass filter (it does not matter in which order the filters are cascaded).

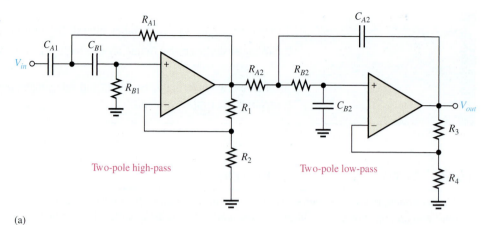

(a)

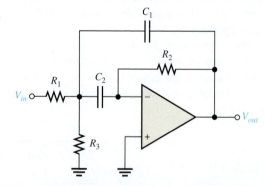

(b)

Multiple-Feedback Band-Pass Filter

Another type of filter configuration, shown in Figure 15–18, is a multiple-feedback band-pass filter. The two feedback paths are through R_2 and C_1. Components R_1 and C_1 provide the low-pass response, and R_2 and C_2 provide the high-pass response. The maximum gain, A_0, occurs at the center frequency. Q values of less than 10 are typical in this type of filter.

▶ FIGURE 15–18

Multiple-feedback band-pass filter.

An expression for the center frequency is developed as follows, recognizing that R_1 and R_3 appear in parallel as viewed from the C_1 feedback path (with the V_{in} source replaced by a short).

$$f_0 = \frac{1}{2\pi\sqrt{(R_1 \parallel R_3)R_2C_1C_2}}$$

Making $C_1 = C_2 = C$ yields

$$f_0 = \frac{1}{2\pi\sqrt{(R_1 \| R_3)R_2C^2}} = \frac{1}{2\pi C\sqrt{(R_1 \| R_3)R_2}}$$

$$= \frac{1}{2\pi C}\sqrt{\frac{1}{R_2(R_1 \| R_3)}} = \frac{1}{2\pi C}\sqrt{\left(\frac{1}{R_2}\right)\left(\frac{1}{R_1R_3/R_1 + R_3}\right)}$$

$$f_0 = \frac{1}{2\pi C}\sqrt{\frac{R_1 + R_3}{R_1R_2R_3}}$$

Equation 15–8

A value for the capacitors is chosen and then the three resistor values are calculated to achieve the desired values for f_0, BW, and A_0. As you know, the Q can be determined from the relation $Q = f_0/BW$. The resistor values can be found using the following formulas (stated without derivation):

$$R_1 = \frac{Q}{2\pi f_0 C A_0}$$

$$R_2 = \frac{Q}{\pi f_0 C}$$

$$R_3 = \frac{Q}{2\pi f_0 C(2Q^2 - A_0)}$$

To develop a gain expression, solve for Q in the R_1 and R_2 formulas as follows:

$$Q = 2\pi f_0 A_0 C R_1$$
$$Q = \pi f_0 C R_2$$

Then,

$$2\pi f_0 A_0 C R_1 = \pi f_0 C R_2$$

Cancelling yields

$$2A_0 R_1 = R_2$$

$$A_0 = \frac{R_2}{2R_1}$$

Equation 15–9

In order for the denominator of the equation $R_3 = Q/[2\pi f_0 C(2Q^2 - A_0)]$ to be positive, $A_0 < 2Q^2$, which imposes a limitation on the gain.

EXAMPLE 15–6

Determine the center frequency, maximum gain, and bandwidth for the filter in Figure 15–19.

▶ **FIGURE 15–19**

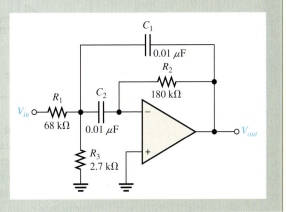

The frequency responses of the band-pass filter and the low-pass filter are shown on the Bode plotters in Figure 15–33. As you can see, the peak response of the band-pass filter is approximately 125 kHz and the critical frequency of the low-pass filter is approximately 16 kHz.

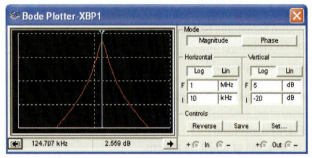

(a) Band-pass response

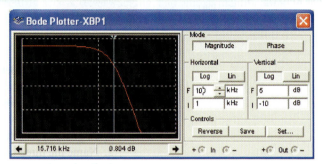

(b) Low-pass response

▲ FIGURE 15–33

Bode plots for the RFID reader filters.

5. What is the purpose of the band-pass filter in the RFID reader?
6. What is the purpose of the low-pass filter in the RFID reader?
7. Calculate the gain of each amplifier in the reader in Figure 15–32.
8. Use the formula for a multiple-feedback band-pass filter to verify the center frequency of the band-pass filter in the reader.
9. What type of response characteristic is the low-pass filter set up for?
10. Calculate the critical frequency of the low-pass filter and compare to the measured value.
11. Calculate the reference voltage for the comparator and explain why a reference above ground is necessary.

Measurements at points on the reader circuit are shown on the oscilloscope in Figure 15–34. The top waveform is the modulated carrier at the output of amplifier U3. The second waveform is the output of the rectifier D1. The third waveform is the output of the low-pass filter (notice that the carrier frequency has been removed by the filter). The bottom waveform is the output of the comparator and represents the digital data sent to the processor.

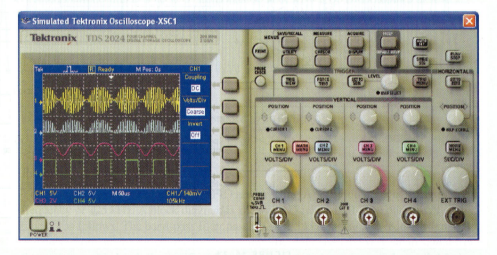

▲ FIGURE 15–34

RFID reader waveforms.

Simulate the RFID reader circuit using your Multisim or LT Spice software. Observe the operation with the oscilloscope and Bode plotter.

Prototyping and Testing

Now that the circuit has been simulated, the prototype circuit is constructed and tested. After the circuit is successfully tested on a protoboard, it is ready to be finalized on a printed circuit board.

Lab Experiment

To build and test a low-pass filter similar to one used in the RFID reader, go to Experiment 15–A in your lab manual (*Laboratory Exercises for Electronic Devices* by David Buchla and Steven Wetterling).

Circuit Board

The RFID reader circuit is implemented on a printed circuit board as shown in Figure 15–35. The dark gray lines represent backside traces.

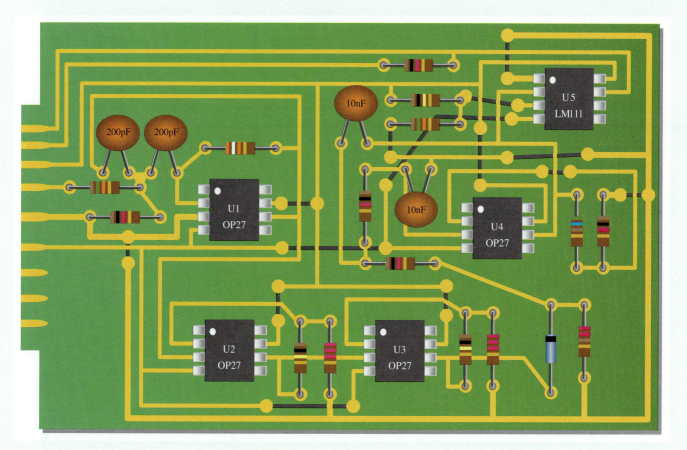

▲ FIGURE 15–35

RFID reader board.

12. Check the printed circuit board and verify that it agrees with the simulation schematic in Figure 15–32.

13. Label each input and output pin according to function.

Programmable Analog Technology

The material you have learned in this chapter is necessary to give you a basic understanding of active filters. However, filter design can be quite complex mathematically. To avoid tedious calculations and trial-and-error breadboarding, the preferred method for development of many filters is to use computer software and then download the design to a programmable analog array. AnadigmDesigner2 software is used in this section to illustrate the ease with which active filters can be developed and implemented in hardware. If you have checked out the optional *Programmable Analog Technology* feature, which appeared first in Chapter 12, you are aware that this software is available and can be downloaded free from www.anadigm.com. You can easily implement a filter design in an FPAA or dpASP chip if you have an evaluation board and interface cable connected to your computer.

Filter Specification

Once you have downloaded the AnadigmDesigner2 software, the first thing you see when opening it is an outline representation of the blank FPAA chip, as shown in Figure 15–36. Under the Tools menu, select *AnadigmFilter*, as shown, and you will get the screen shown in Figure 15–37.

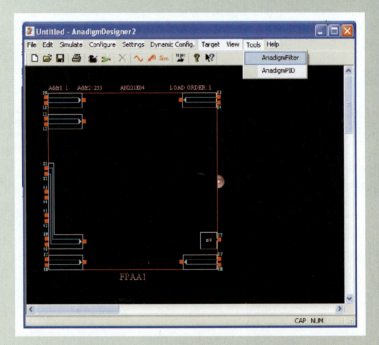

▲ **FIGURE 15–36**

FPAA chip outline screen showing AnadigmFilter selection.

You are now ready to specify a filter. For example, select a filter type and approximation and enter the desired parameters, as shown in Figure 15–38, for a band-pass Butterworth filter. Note that you can use your mouse to drag the limits, shown in red and blue on the screen, to set the desired response.

When the filter has been completely specified, click on "To AnadigmDesigner2" and the filter components will be placed in the FPAA chip screen, as shown in Figure 15–39(a). Notice that the filter consists of three stages in this case. Now use the connection tool to connect the filter to an input and output, as shown in part (b).

▶ FIGURE 15–37

Filter default screen.

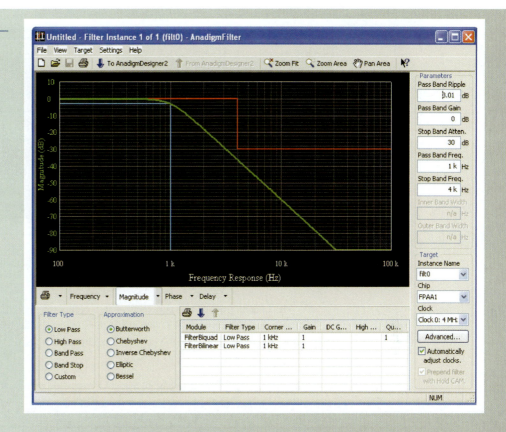

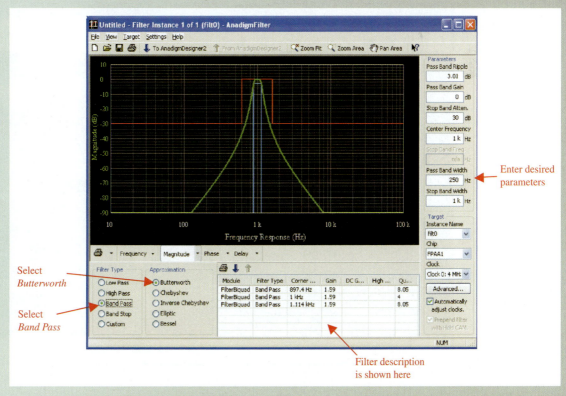

Select *Butterworth*

Select *Band Pass*

Enter desired parameters

Filter description is shown here

▲ FIGURE 15–38

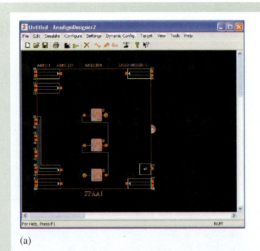 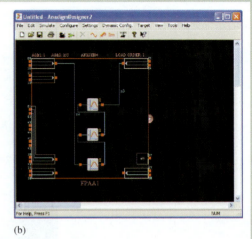

(a)   (b)

▲ **FIGURE 15–39**

By attaching actual signal generators and oscilloscope probes to the board, you can verify that the downloaded circuit is behaving just as the simulator indicated it would. Note that an FPAA or dpASP is reprogrammable so you can make circuit changes, download, and test indefinitely.

Design Assignment

Implement the RFID reader circuit using AnadigmDesigner2 software.

Procedure: Figure 15–40 shows a version of the circuit implemented in FPAA1. Because of limitations on implementing the ASK input signal, modifications have been made. Since the input

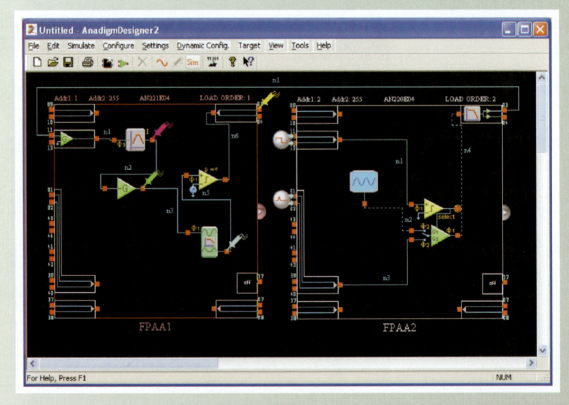

▲ **FIGURE 15–40**

Design screen showing the RFID reader in FPAA1 and an ASK generator representing the RFID tag in FPAA2.

cell contains an amplifier with gain, the amplifier in the RFID reader circuit has less gain than if a 1 mV ASK signal were available. Also, the rectifier and low-pass filter are combined in one CAM. FPAA2 is used as a signal source to replicate a 125 kHz carrier modulated with a 10 kHz square wave. This chip is for test purposes only and is not part of the RFID reader.

Analysis: The simulation of the RFID reader is shown in Figure 15–41. The top waveform is the output of the 125 kHz band-pass filter CAM and is an ASK input signal representing a digital 1 followed by a 0. The second waveform is the output of the inverting gain stage CAM with a unity gain. The third waveform is the output of the half-wave rectifier/low-pass filter CAM. The bottom output is the digital signal from the comparator.

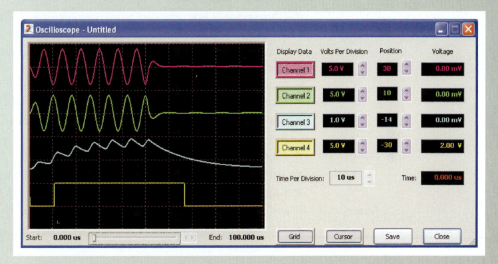

▲ **FIGURE 15–41**

Simulation waveforms for the RFID reader.

Programming Exercises

1. Why is a software program the best way to specify and implement active filters?
2. List the filter types available in the AnadigmFilter software.
3. List the filter approximations available in the AnadigmFilter software.

PAM Experiment

To program, download, and test a circuit using AnadigmDesigner2 software and the programmable analog module (PAM) board, go to Experiment 15–B in *Laboratory Exercises for Electronic Devices* by David Buchla and Steven Wetterling.

SUMMARY

Section 15–1   ◆ In filter terminology, a one-pole filter can be constructed from one resistor and one capacitor.

◆ The bandwidth in a low-pass filter equals the critical frequency because the response extends to 0 Hz.

◆ The passband of a high-pass filter extends above the critical frequency and is limited only by the inherent frequency limitation of the active circuit.

◆ A band-pass filter passes all frequencies within a band between a lower and an upper critical frequency and rejects all others outside this band.

◆ The bandwidth of a band-pass filter is the difference between the upper critical frequency and the lower critical frequency.

◆ The quality factor Q of a band-pass filter determines the filter's selectivity. The higher the Q, the narrower the bandwidth and the better the selectivity.

◆ A band-stop filter rejects all frequencies within a specified band and passes all those outside this band.

Section 15–2
◆ Filters with the Butterworth response characteristic have a very flat response in the passband, exhibit a roll-off of -20 dB/decade/pole, and are used when all the frequencies in the passband must have the same gain.

◆ Filters with the Chebyshev characteristic have ripples or overshoot in the passband and exhibit a faster roll-off per pole than filters with the Butterworth characteristic.

◆ Filters with the Bessel characteristic are used for filtering pulse waveforms. Their linear phase characteristic results in minimal waveshape distortion. The roll-off rate per pole is slower than for the Butterworth.

◆ Each pole in a Butterworth filter causes the output to roll-off at a rate of -20 dB/decade.

◆ The damping factor determines the filter response characteristic (Butterworth, Chebyshev, or Bessel).

Section 15–3
◆ Single-pole low-pass filters have a -20 dB/decade roll-off.

◆ The Sallen-Key low-pass filter has two poles (second order) and has a -40 dB/decade roll-off.

◆ Each additional filter in a cascaded arrangement adds -20 dB to the roll-off rate.

Section 15–4
◆ Single-pole high-pass filters have a -20 dB/decade roll-off.

◆ The Sallen-Key high-pass filter has two poles (second order) and has a -40 dB/decade roll-off.

◆ Each additional filter in a cascaded arrangement adds -20 dB to the roll-off rate.

◆ The upper frequency response of an active high-pass filter is limited by the internal op-amp roll-off.

Section 15–5
◆ Band-pass filters pass a specified band of frequencies.

◆ A band-pass filter can be achieved by cascading a low-pass and a high-pass filter.

◆ The multiple-feedback band-pass filter uses two feedback paths to achieve its response characteristic.

◆ The state-variable band-pass filter uses a summing amplifier and two integrators.

◆ The biquad filter consists of an integrator followed by an inverting amplifier and a second integrator.

Section 15–6
◆ Band-stop filters reject a specified band of frequencies.

◆ Multiple-feedback and state-variable are common types of band-stop filters.

Section 15–7
◆ Filter response can be measured using discrete-point measurement or swept-frequency measurement.

KEY TERMS

Key terms and other bold terms in the chapter are defined in the end-of-book glossary.

Band-pass filter A type of filter that passes a range of frequencies lying between a certain lower frequency and a certain higher frequency.

Band-stop filter A type of filter that blocks or rejects a range of frequencies lying between a certain lower frequency and a certain higher frequency.

Damping factor A filter characteristic that determines the type of response.

Filter A circuit that passes certain frequencies and attenuates or rejects all other frequencies.

High-pass filter A type of filter that passes frequencies above a certain frequency while rejecting lower frequencies.

Low-pass filter A type of filter that passes frequencies below a certain frequency while rejecting higher frequencies.

Pole In electronic filter circuits, a circuit containing one resistor and one capacitor that contributes -20 dB/decade to a filter's roll-off rate.

Roll-off The rate of decrease in gain, below or above the critical frequencies of a filter.

KEY FORMULAS

15–1	$BW = f_c$	Low-pass bandwidth
15–2	$BW = f_{c2} - f_{c1}$	Filter bandwidth of a band-pass filter
15–3	$f_0 = \sqrt{f_{c1} f_{c2}}$	Center frequency of a band-pass filter
15–4	$Q = \dfrac{f_0}{BW}$	Quality factor of a band-pass filter
15–5	$DF = 2 - \dfrac{R_1}{R_2}$	Damping factor
15–6	$A_{cl(NI)} = \dfrac{R_1}{R_2} + 1$	Closed-loop voltage gain
15–7	$f_c = \dfrac{1}{2\pi \sqrt{R_A R_B C_A C_B}}$	Critical frequency for a second-order Sallen-Key filter
15–8	$f_0 = \dfrac{1}{2\pi C} \sqrt{\dfrac{R_1 + R_3}{R_1 R_2 R_3}}$	Center frequency of a multiple-feedback filter
15–9	$A_0 = \dfrac{R_2}{2R_1}$	Gain of a multiple-feedback filter

TRUE/FALSE QUIZ

Answers can be found at www.pearsonhighered.com/floyd.

1. The response of a filter can be identified by its passband.
2. A filter pole is the cutoff frequency of a filter.
3. A single-pole filter has one RC circuit.
4. A single-pole filter produces a roll-off of -25 dB/decade.
5. A low-pass filter can pass a dc voltage.
6. A high-pass filter passes any frequency above dc.
7. The critical frequency of a filter depends only on R and C values.
8. The band-pass filter has two critical frequencies.
9. The quality factor of a band-pass filter is the ratio of bandwidth to the center frequency.
10. The higher the Q, the narrower the bandwidth of a band-pass filter.
11. The Butterworth characteristic provides a flat response in the passband.
12. Filters with a Chebyshev response have a slow roll-off.
13. A Chebyshev response has ripples in the passband.
14. Bessel filters are useful in filtering pulse waveforms.
15. The order of a filter is the number of poles it contains.
16. A Sallen-Key filter is also known as a VCVS filter.
17. Multiple feedback is used in low-pass filters.
18. A state-variable filter uses differentiators.
19. A band-stop filter rejects certain frequencies.
20. Filter response can be measured using a sweep generator.

CIRCUIT-ACTION QUIZ

Answers can be found at www.pearsonhighered.com/floyd.

1. If the critical frequency of a low-pass filter is increased, the bandwidth will

   (a) increase   (b) decrease   (c) not change

2. If the critical frequency of a high-pass filter is increased, the bandwidth will

   (a) increase   (b) decrease   (c) not change

3. If the Q of a band-pass filter is increased, the bandwidth will

   (a) increase   (b) decrease   (c) not change

4. If the values of C_A and C_B in Figure 15–11 are increased by the same amount, the critical frequency will

   (a) increase   (b) decrease   (c) not change

5. If the the value of R_2 in Figure 15–11 is increased, the bandwidth will

   (a) increase   (b) decrease   (c) not change

6. If two filters like the one in Figure 15–15 are cascaded, the roll-off rate of the frequency response will

   (a) increase   (b) decrease   (c) not change

7. If the value of R_2 in Figure 15–19 is decreased, the Q will

   (a) increase   (b) decrease   (c) not change

8. If the capacitors in Figure 15–19 are changed to 0.022 μF, the center frequency will

   (a) increase   (b) decrease   (c) not change

SELF-TEST

Answers can be found at www.pearsonhighered.com/floyd.

Section 15–1

1. The term *pole* in filter terminology refers to

   (a) a high-gain op-amp   (b) one complete active filter

   (c) a single *RC* circuit   (d) the feedback circuit

2. A single resistor and a single capacitor can be connected to form a filter with a roll-off rate of

   (a) −20 dB/decade   (b) −40 dB/decade

   (c) −6 dB/octave   (d) answers (a) and (c)

3. A band-pass response has

   (a) two critical frequencies   (b) one critical frequency

   (c) a flat curve in the passband   (d) a notch in the bandwidth

4. The lowest frequency passed by a low-pass filter is

   (a) 1 Hz   (b) 0 Hz   (c) 10 Hz   (d) dependent on the critical frequency

5. The quality factor (Q) of a band-pass filter depends on

   (a) the critical frequencies   (b) only the bandwidth

   (c) the center frequency and the bandwidth   (d) only the center frequency

Section 15–2

6. The damping factor of an active filter determines

   (a) the voltage gain   (b) the critical frequency

   (c) the response characteristic   (d) the roll-off rate

7. A maximally flat frequency response is known as

   (a) Chebyshev   (b) Butterworth   (c) Bessel   (d) Colpitts

8. The damping factor of a filter is set by

   (a) the negative feedback circuit   (b) the positive feedback circuit

   (c) the frequency-selective circuit   (d) the gain of the op-amp

9. The number of poles in a filter affect the

   (a) voltage gain   (b) bandwidth

   (c) center frequency   (d) roll-off rate

Section 15–3 **10.** Sallen-Key low-pass filters are

    (a) single-pole filters   **(b)** second-order filters

    (c) Butterworth filters   **(d)** band-pass filters

11. When low-pass filters are cascaded, the roll-off rate

    (a) increases   **(b)** decreases   **(c)** does not change

Section 15–4 **12.** In a high-pass filter, the roll-off occurs

    (a) above the critical frequency   **(b)** below the critical frequency

    (c) during the mid range   **(d)** at the center frequency

13. A two-pole Sallen-Key high-pass filter contains

    (a) one capacitor and two resistors   **(b)** two capacitors and two resistors

    (c) a feedback circuit   **(d)** answers (b) and (c)

Section 15–5 **14.** When a low-pass and a high-pass filter are cascaded to get a band-pass filter, the critical frequency of the low-pass filter must be

    (a) equal to the critical frequency of the high-pass filter

    (b) less than the critical frequency of the high-pass filter

    (c) greater than the critical frequency of the high-pass filter

15. A state-variable filter consists of

    (a) one op-amp with multiple-feedback paths

    (b) a summing amplifier and two integrators

    (c) a summing amplifier and two differentiators

    (d) three Butterworth stages

Section 15–6 **16.** When the gain of a filter is minimum at its center frequency, it is

    (a) a band-pass filter   **(b)** a band-stop filter

    (c) a notch filter   **(d)** answers (b) and (c)

PROBLEMS

Answers to all odd-numbered problems are at the end of the book.

BASIC PROBLEMS

Section 15–1 **Basic Filter Responses**

1. Identify each type of filter response (low-pass, high-pass, band-pass, or band-stop) in Figure 15–42.

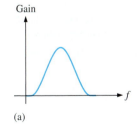
(a)

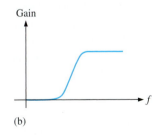
(b)

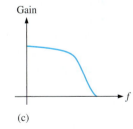
(c)

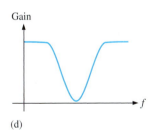
(d)

▲ **FIGURE 15–42**

2. A certain low-pass filter has a critical frequency of 800 Hz. What is its bandwidth?

3. A single-pole high-pass filter has a frequency-selective circuit with $R = 2.2 \text{ k}\Omega$ and $C = 0.0015 \text{ }\mu\text{F}$. What is the critical frequency? Can you determine the bandwidth from the available information?

4. What is the roll-off rate of the filter described in Problem 3?

5. What is the bandwidth of a band-pass filter whose critical frequencies are 3.2 kHz and 3.9 kHz? What is the Q of this filter?

6. What is the center frequency of a filter with a Q of 15 and a bandwidth of 1 kHz?

Section 15–2   Filter Response Characteristics

7. What is the damping factor in each active filter shown in Figure 15–43? Which filters are approximately optimized for a Butterworth response characteristic?

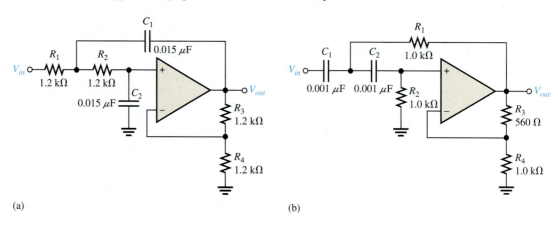

(a)

(b)

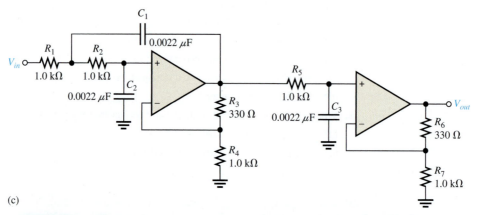

(c)

▲ **FIGURE 15–43**

Multisim and LT Spice file circuits are identified with a logo and are in the Problems folder on the website. Filenames correspond to figure numbers (e.g., FGM15-43 or FGS15-43).

8. Identify each filter type in Figure 15–43 (low-pass, high-pass, band-pass or band-stop).

9. For each filter in Figure 15–43, state the number of poles and the approximate roll-off rate.

10. For the filters in Figure 15–43 that do not have a Butterworth response, specify the changes necessary to convert them to Butterworth responses. (Use nearest standard values.)

11. Response curves for second-order filters are shown in Figure 15–44. Identify each as Butterworth, Chebyshev, or Bessel.

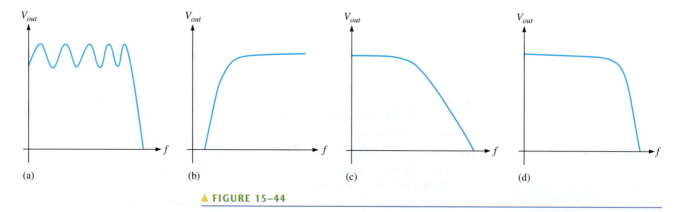

(a)   (b)   (c)   (d)

▲ **FIGURE 15–44**

Section 15–3 Active Low-Pass Filters

12. Is the four-pole filter in Figure 15–45 approximately optimized for a Butterworth response? What is the roll-off rate?

13. Determine the critical frequency in Figure 15–45.

14. Without changing the response curve, adjust the component values in the filter of Figure 15–45 to make it an equal-value filter. Select $C = 0.22 \ \mu F$ for both stages.

15. Modify the filter in Figure 15–45 to increase the roll-off rate to -120 dB/decade while maintaining an approximate Butterworth response.

16. Using a block diagram format, show how to implement the following roll-off rates using single-pole and two-pole low-pass filters with Butterworth responses.

   (a) -40 dB/decade   (b) -20 dB/decade

   (c) -60 dB/decade   (d) -100 dB/decade

   (e) -120 dB/decade

▶ FIGURE 15–45

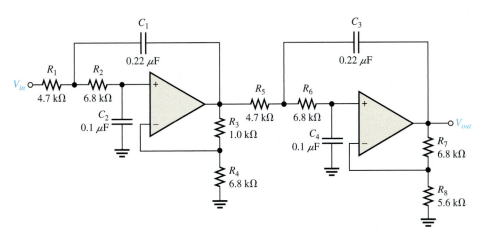

Section 15–4 Active High-Pass Filters

17. Convert the filter in Problem 14 to a high-pass with the same critical frequency and response characteristic.

18. Make the necessary circuit modification to reduce by half the critical frequency in Problem 17.

19. For the filter in Figure 15–46, (a) how would you increase the critical frequency? (b) How would you increase the gain?

▶ FIGURE 15–46

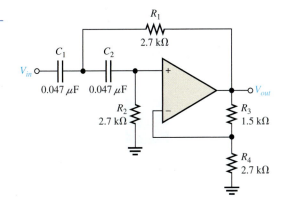

Section 15–5 **Active Band-Pass Filters**

20. Identify each band-pass filter configuration in Figure 15–47.

21. Determine the center frequency and bandwidth for each filter in Figure 15–47.

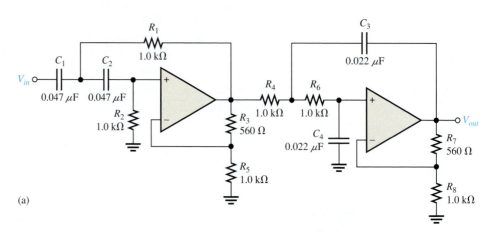

(a)

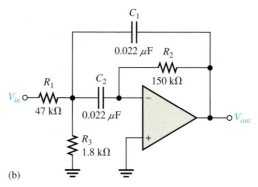

(b)

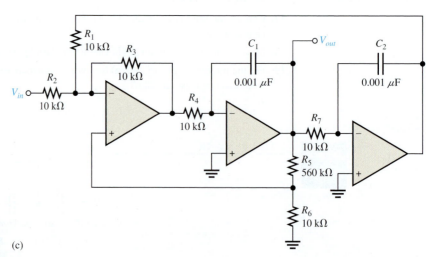

(c)

▲ **FIGURE 15–47**

▶ FIGURE 15–48

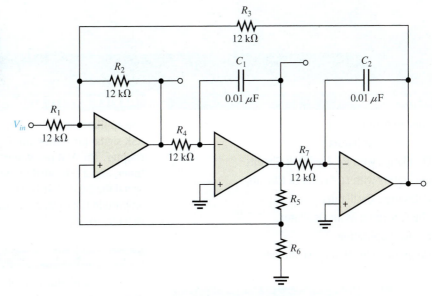

22. Optimize the state-variable filter in Figure 15–48 for $Q = 50$. What bandwidth is achieved?

Section 15–6 **Active Band-Stop Filters**

23. Show how to make a notch (band-stop) filter using the basic circuit in Figure 15–48.

24. Modify the band-stop filter in Problem 23 for a center frequency of 120 Hz.

MULTISIM TROUBLESHOOTING PROBLEMS

These file circuits are in the Troubleshooting Problems folder on the website.

25. Open file TPM-25 and determine the fault.

26. Open file TPM-26 and determine the fault.

27. Open file TPM-27 and determine the fault.

28. Open file TPM-28 and determine the fault.

29. Open file TPM-29 and determine the fault.

30. Open file TPM-30 and determine the fault.

31. Open file TPM-31 and determine the fault.

32. Open file TPM-32 and determine the fault.

33. Open file TPM-33 and determine the fault.

The Wien-bridge oscillator circuit can be viewed as a noninverting amplifier configuration with the input signal fed back from the output through the lead-lag circuit. Recall that the voltage divider determines the closed-loop gain of the amplifier.

$$A_{cl} = \frac{1}{B} = \frac{1}{R_2/(R_1 + R_2)} = \frac{R_1 + R_2}{R_2}$$

The circuit is redrawn in Figure 16–7(b) to show that the op-amp is connected across the bridge circuit. One leg of the bridge is the lead-lag circuit, and the other is the voltage divider.

Positive Feedback Conditions for Oscillation As you know, for the circuit to produce a sustained sinusoidal output (oscillate), the phase shift around the positive feedback loop must be 0° and the gain around the loop must equal unity (1). The 0° phase-shift condition is met when the frequency is f_r because the phase shift through the lead-lag circuit is 0° and there is no inversion from the noninverting (+) input of the op-amp to the output. This is shown in Figure 16–8(a).

▶ FIGURE 16–8

Conditions for a sustained sine wave output.

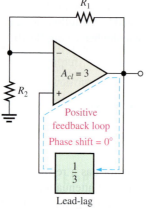

(a) The phase shift around the loop is 0°.

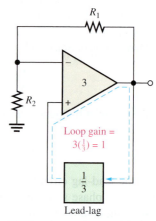

(b) The voltage gain around the loop is 1.

The unity-gain condition in the feedback loop is met when

$$A_{cl} = 3$$

This offsets the 1/3 attenuation of the lead-lag circuit, thus making the total gain around the positive feedback loop equal to 1, as depicted in Figure 16–8(b). To achieve a closed-loop gain of 3,

$$R_1 = 2R_2$$

Then

$$A_{cl} = \frac{R_1 + R_2}{R_2} = \frac{2R_2 + R_2}{R_2} = \frac{3R_2}{R_2} = 3$$

Start-Up Conditions Initially, the closed-loop gain of the amplifier itself must be more than 3 ($A_{cl} > 3$) until the output signal builds up to a desired level. Ideally, the gain of the amplifier must then decrease to 3 so that the total gain around the loop is 1 and the output signal stays at the desired level, thus sustaining oscillation. This is illustrated in Figure 16–9.

Although the Wien bridge was developed conceptually before it was practical, it was Bill Hewlett that found a solution to stabilizing the bridge using a lamp. He built his first stable oscillator in 1938 based on the fact that as it warms up, the resistance of a tungsten

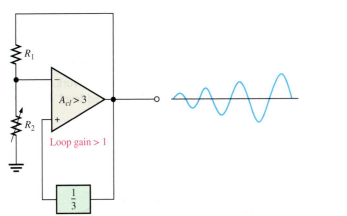

Loop gain > 1

(a) Loop gain greater than 1 causes output to build up.

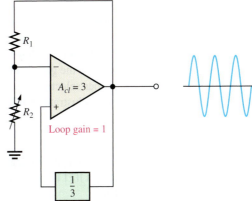

Loop gain = 1

(b) Loop gain of 1 causes a sustained sine wave output.

▲ **FIGURE 16–9**

Conditions for start-up and a sustained sine wave output.

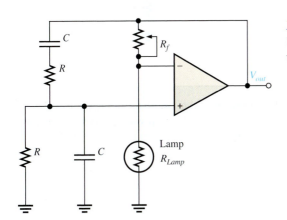

◄ **FIGURE 16–10**

Basic Wien-bridge oscillator with tungsten lamp for stability.

lamp decreases. The lamp lasts a very long time because it is used below its incandescent point and is specially selected to have a cold resistance of several hundred ohms. Many oscillators were built with this principle because a low-distortion output could be achieved. The basic circuit is shown in Figure 16–10 using equal values of *R* and equal values of *C*.

Initially, the feedback resistor, R_f, is set to a resistance that is slightly more than twice the lamp's cold resistance. This means the gain of the noninverting amplifier will be >3 as required for startup. As current warms the lamp, its resistance increases until the lamp resistance is exactly one-half R_f, producing a gain of exactly 3 and the output is stable. The circuit will oscillate with a frequency given by:

$$f = \frac{1}{2\pi RC}$$

Another method to control the gain uses a JFET as a voltage-controlled resistor in a negative feedback path. This method can produce an excellent sinusoidal waveform that is stable. A JFET operating with a small or zero V_{DS} is operating in the ohmic region. As the gate voltage increases, the drain-source resistance increases. If the JFET is placed in the negative feedback path, automatic gain control can be achieved because of this voltage-controlled resistance.

A JFET stabilized Wien bridge is shown in Figure 16–11. The gain of the op-amp is controlled by the components shown in the green box, which include the JFET. The JFET's drain-source resistance depends on the gate voltage. With no output signal, the gate is at

A FET can be used in place of a BJT, as shown in Figure 16–19, to minimize the load-ing effect of the transistor's input impedance. Recall that FETs have much higher input impedances than do bipolar junction transistors. Also, when an external load is connected to the oscillator output, as shown in Figure 16–20(a), f_r may decrease, again because of a reduction in Q. This happens if the load resistance is too small. In some cases, one way to eliminate the effects of a load resistance is by transformer coupling, as indicated in Figure 16–20(b).

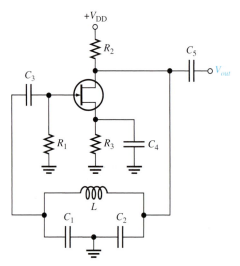

▲ **FIGURE 16–19**

A basic FET Colpitts oscillator.

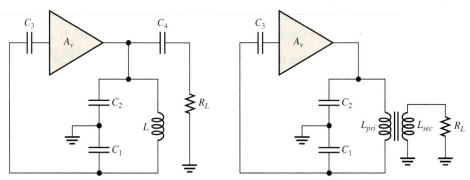

(a) A load capacitively coupled to oscillator output can reduce circuit Q and f_r.

(b) Transformer coupling of load can reduce loading effect by impedance transformation.

▲ **FIGURE 16–20**

Oscillator loading.

EXAMPLE 16–3

(a) Determine the frequency for the oscillator in Figure 16–21. Assume there is negligible loading on the feedback circuit and that its Q is greater than 10.

(b) Find the frequency if the oscillator is loaded to a point where the Q drops to 8.

▶ **FIGURE 16–21**

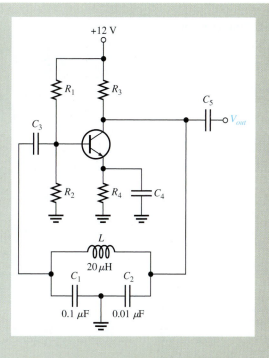

Solution (a) $C_T = \dfrac{C_1 C_2}{C_1 + C_2} = \dfrac{(0.1\ \mu F)(0.01\ \mu F)}{0.11\ \mu F} = 0.0091\ \mu F$

$f_r \cong \dfrac{1}{2\pi\sqrt{LC_T}} = \dfrac{1}{2\pi\sqrt{(20\ \mu H)(0.0091\ \mu F)}} = \textbf{373 kHz}$

(b) $f_r = \dfrac{1}{2\pi\sqrt{LC_T}}\sqrt{\dfrac{Q^2}{Q^2 + 1}} = (373\ \text{kHz})(0.9923) = \textbf{370 kHz}$

Related Problem What frequency does the oscillator in Figure 16–21 produce if it is loaded to a point where $Q = 4$?

The Clapp Oscillator

The Clapp oscillator is a variation of the Colpitts. The basic difference is an additional capacitor, C_3, in series with the inductor in the resonant feedback circuit, as shown in Figure 16–22. Since C_3 is in series with C_1 and C_2 around the tank circuit, the total capacitance is

$$C_T = \cfrac{1}{\dfrac{1}{C_1} + \dfrac{1}{C_2} + \dfrac{1}{C_3}}$$

and the approximate frequency of oscillation ($Q > 10$) is

$$f_r \cong \dfrac{1}{2\pi\sqrt{LC_T}}$$

If C_3 is much smaller than C_1 and C_2, then C_3 almost entirely controls the resonant frequency [$f_r \cong 1/(2\pi\sqrt{LC_3})$]. Since C_1 and C_2 are both connected to ground at one end, the junction capacitance of the transistor and other stray capacitances appear in parallel with C_1 and C_2 to ground, altering their effective values. C_3 is not affected, however, and thus provides a more accurate and stable frequency of oscillation.

A basic Clapp oscillator.

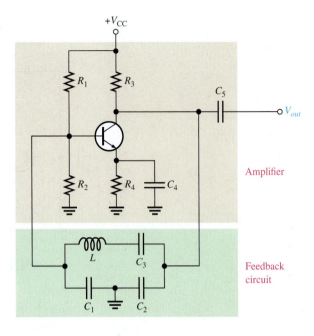

The Hartley Oscillator

The Hartley oscillator is similar to the Colpitts except that the feedback circuit consists of two series inductors and a parallel capacitor as shown in Figure 16–23.

► FIGURE 16–23

A basic Hartley oscillator.

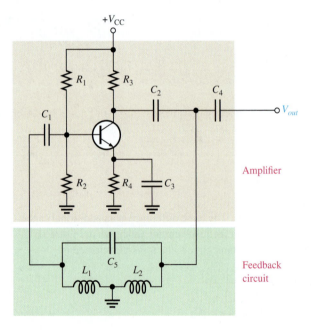

In this circuit, the frequency of oscillation for $Q > 10$ is

$$f_r \cong \frac{1}{2\pi\sqrt{L_T C}}$$

where $L_T = L_1 + L_2$. The inductors act in a role similar to C_1 and C_2 in the Colpitts to determine the attenuation, B, of the feedback circuit.

$$B \cong \frac{L_1}{L_2}$$

To assure start-up of oscillation, A_v must be greater than $1/B$.

$$A_v > \frac{L_2}{L_1}$$

Equation 16–8

Loading of the tank circuit has the same effect in the Hartley as in the Colpitts; that is, the Q is decreased and thus f_r decreases.

The Armstrong Oscillator

This type of *LC* feedback oscillator uses transformer coupling to feed back a portion of the signal voltage, as shown in Figure 16–24. It is sometimes called a "tickler" oscillator in reference to the transformer secondary or "tickler coil" that provides the feedback to keep the oscillation going. The Armstrong is less common than the Colpitts, Clapp, and Hartley, mainly because of the disadvantage of transformer size and cost. The frequency of oscillation is set by the inductance of the primary winding (L_{pri}) in parallel with C_1.

$$f_r = \frac{1}{2\pi\sqrt{L_{pri}C_1}}$$

Equation 16–9

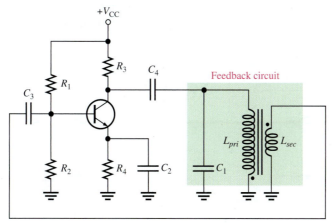

◀ FIGURE 16–24

A basic Armstrong oscillator.

Crystal-Controlled Oscillators

The most stable and accurate type of feedback oscillator uses a piezoelectric **crystal** in the feedback loop to control the frequency.

The Piezoelectric Effect Quartz is one type of crystalline substance found in nature that exhibits a property called the **piezoelectric effect**. When a changing mechanical stress is applied across the crystal to cause it to vibrate, a voltage develops at the frequency of mechanical vibration. Conversely, when an ac voltage is applied across the crystal, it vibrates at the frequency of the applied voltage. The greatest vibration occurs at the crystal's natural resonant frequency, which is determined by the physical dimensions and by the way the crystal is cut.

Crystals used in electronic applications typically consist of a quartz wafer mounted between two electrodes and enclosed in a protective "can" as shown in Figure 16–25(a) and (b). A schematic symbol for a crystal is shown in Figure 16–25(c), and an equivalent *RLC* circuit for the crystal appears in Figure 16–25(d). As you can see, the crystal's equivalent circuit is a series-parallel *RLC* circuit and can operate in either series resonance or parallel resonance. At the series resonant frequency, the inductive reactance is cancelled by the reactance of C_s. The remaining series resistor, R_s, determines the impedance of the crystal. Parallel resonance occurs when the inductive reactance and the reactance of the parallel capacitance, C_p, are equal. The parallel resonant frequency is usually at least 1 kHz higher

A quartz crystal.

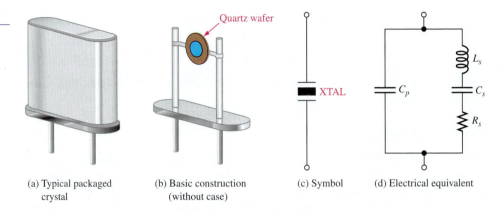

(a) Typical packaged crystal

(b) Basic construction (without case)

(c) Symbol

(d) Electrical equivalent

than the series resonant frequency. A great advantage of the crystal is that it exhibits a very high Q (Qs with values of several thousand are typical). In critical applications, the crystal is mounted in a shock-proof enclosure, and temperature is controlled to avoid frequency drift.

An oscillator that uses a crystal as a series resonant tank circuit is shown in Figure 16–26(a). The impedance of the crystal is minimum at the series resonant frequency, thus providing maximum feedback. The crystal tuning capacitor, C_C, is used to "fine tune" the oscillator frequency by "pulling" the resonant frequency of the crystal slightly up or down.

Basic crystal oscillators.

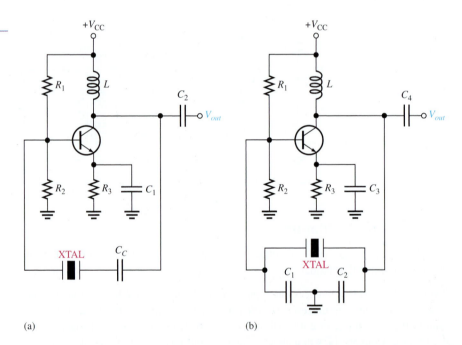

(a)

(b)

A modified Colpitts configuration is shown in Figure 16–26(b) with a crystal acting as a parallel resonant tank circuit. The impedance of the crystal is maximum at parallel resonance, thus developing the maximum voltage across the capacitors. The voltage across C_1 is fed back to the input.

Modes of Oscillation in the Crystal   Piezoelectric crystals can oscillate in either of two modes—fundamental or overtone. The fundamental frequency of a crystal is the lowest frequency at which it is naturally resonant. The fundamental frequency depends on the crystal's mechanical dimensions, type of cut, and other factors, and is inversely proportional to the thickness of the crystal slab. Because a slab of crystal cannot be cut too thin without fracturing, there is an upper limit on the fundamental frequency. For most crystals, this upper limit is less than 20 MHz. For higher frequencies, the crystal must be operated in the overtone mode. Overtones are approximate integer multiples of the fundamental frequency. The overtone frequencies are usually, but not always, odd multiples (3, 5, 7, . . .) of the fundamental. Many crystal oscillators are available in integrated circuit packages.

SECTION 16–4 CHECKUP	1. What is the basic difference between the Colpitts and the Hartley oscillators?
	2. What is the advantage of a FET amplifier in a Colpitts or Hartley oscillator?
	3. How can you distinguish a Colpitts oscillator from a Clapp oscillator?

16–5   RELAXATION OSCILLATORS

The second major category of oscillators is the relaxation oscillator. Relaxation oscillators use an *RC* timing circuit and a device that changes states to generate a periodic waveform. In this section, you will learn about several circuits that are used to produce nonsinusoidal waveforms.

After completing this section, you should be able to

❑ **Describe and analyze the operation of relaxation oscillators**
❑ Describe the operation of a triangular-wave oscillator
   ◆ Discuss a practical triangular-wave oscillator   ◆ Define *function generator*
   ◆ Determine the UTP, LTP, and frequency of oscillation
❑ Describe a sawtooth voltage-controlled oscillator (VCO)
   ◆ Explain the purpose of the PUT in this circuit   ◆ Determine the frequency of oscillation
❑ Describe a square-wave oscillator

A Triangular-Wave Oscillator

The op-amp integrator covered in Chapter 13 can be used as the basis for a triangular-wave oscillator. The basic idea is illustrated in Figure 16–27(a) where a dual-polarity, switched input is used. We use the switch only to introduce the concept; it is not a practical way to implement this circuit. When the switch is in position 1, the negative voltage is applied, and the output is a positive-going ramp. When the switch is thrown into position 2, a negative-going ramp is produced. If the switch is thrown back and forth at fixed intervals, the output is a triangular wave consisting of alternating positive-going and negative-going ramps, as shown in Figure 16–27(b).

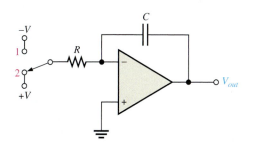

(a)

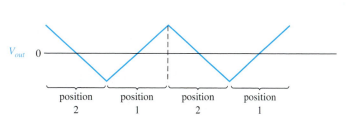

(b) Output voltage as the switch is thrown back and forth at regular intervals

▲ FIGURE 16–27

Basic triangular-wave oscillator.

A Practical Triangular-Wave Oscillator   One practical implementation of a triangular-wave oscillator utilizes an op-amp comparator with hysteresis to perform the switching function, as shown in Figure 16–28. The operation is as follows. To begin, assume that the output voltage of the comparator is at its maximum negative level. This output is connected to the inverting input of the integrator through R_1, producing a positive-going ramp on the output of the integrator. When the ramp voltage reaches the upper trigger point (UTP), the comparator switches to its maximum positive level. This positive level causes the integrator ramp to change to a negative-going direction. The ramp continues in this direction until the lower trigger point (LTP) of the comparator is reached. At this point, the comparator output switches back to the maximum negative level and the cycle repeats. This action is illustrated in Figure 16–29.

▶ **FIGURE 16–28**

A triangular-wave oscillator using two op-amps.

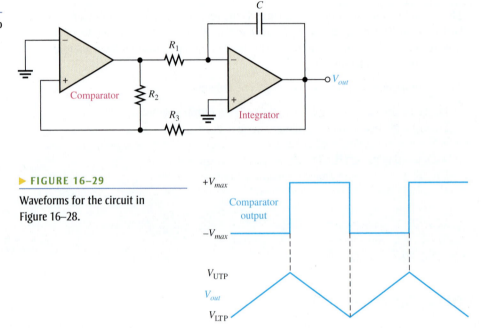

▶ **FIGURE 16–29**

Waveforms for the circuit in Figure 16–28.

Since the comparator produces a square-wave output, the circuit in Figure 16–28 can be used as both a triangular-wave oscillator and a square-wave oscillator. Devices of this type are commonly known as **function generators** because they produce more than one output function. The output amplitude of the square wave is set by the output swing of the comparator, and the resistors R_2 and R_3 set the amplitude of the triangular output by establishing the UTP and LTP voltages according to the following formulas:

$$V_{\text{UTP}} = +V_{max}\left(\frac{R_3}{R_2}\right)$$

$$V_{\text{LTP}} = -V_{max}\left(\frac{R_3}{R_2}\right)$$

where the comparator output levels, $+V_{max}$ and $-V_{max}$, are equal. The frequency of both waveforms depends on the R_1C time constant as well as the amplitude-setting resistors, R_2 and R_3. By varying R_1, the frequency of oscillation can be adjusted without changing the output amplitude.

Equation 16–10

$$f_r = \frac{1}{4R_1C}\left(\frac{R_2}{R_3}\right)$$

EXAMPLE 16–4

Determine the frequency of oscillation of the circuit in Figure 16–30. To what value must R_1 be changed to make the frequency 5.0 kHz?

◢ **FIGURE 16–30**

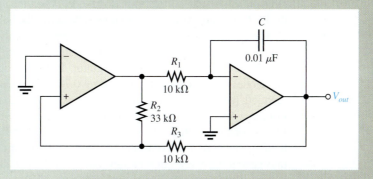

Solution

$$f_r = \frac{1}{4R_1C}\left(\frac{R_2}{R_3}\right) = \left(\frac{1}{4(10\ k\Omega)(0.01\ \mu F)}\right)\left(\frac{33\ k\Omega}{10\ k\Omega}\right) = \textbf{8.25 kHz}$$

To make f = 5 kHz,

$$R_1 = \frac{1}{4fC}\left(\frac{R_2}{R_3}\right) = \left(\frac{1}{4(5\ kHz)(0.01\ \mu F)}\right)\left(\frac{33\ k\Omega}{10\ k\Omega}\right) = \textbf{16.5 k}\Omega$$

Related Problem

What is the amplitude of the triangular wave in Figure 16–30 if the comparator output is ± 10 V?

MultiSim
LT Spice

Open the Multisim file EXM16-04 or the LT Spice file EXS16-04 and observe the operation.

A Sawtooth Voltage-Controlled Oscillator (VCO)

The **voltage-controlled oscillator (VCO)** is a relaxation oscillator whose frequency can be changed by a variable dc control voltage. VCOs can be either sinusoidal or nonsinusoidal. One way to build a sawtooth VCO is with an op-amp integrator that uses a switching device (PUT) in parallel with the feedback capacitor to terminate each ramp at a prescribed level and effectively "reset" the circuit. Figure 16–31(a) shows the implementation.

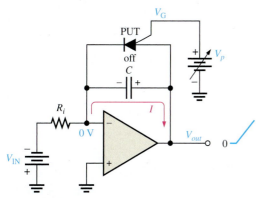

(a) Initially, the capacitor charges, the output ramp begins, and the PUT is off.

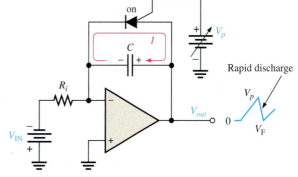

(b) The capacitor rapidly discharges when the PUT momentarily turns on.

▲ **FIGURE 16–31**

Sawtooth VCO operation.

As you learned in Chapter 11, the PUT is a programmable unijunction transistor with an anode, a cathode, and a gate terminal. The gate is always biased positively with respect to the cathode. When the anode voltage exceeds the gate voltage by approximately 0.7 V, the PUT turns on and acts as a forward-biased diode. When the anode voltage falls below this level, the PUT turns off. Also, the current must be above the holding value to maintain conduction.

The operation of the sawtooth VCO begins when the negative dc input voltage, $-V_{IN}$, produces a positive-going ramp on the output. During the time that the ramp is increasing, the circuit acts as a regular integrator. The PUT triggers on when the output ramp (at the anode) exceeds the gate voltage by 0.7 V. The gate is set to the approximate desired sawtooth peak voltage. When the PUT turns on, the capacitor rapidly discharges, as shown in Figure 16–31(b). The capacitor does not discharge completely to zero because of the PUT's forward voltage, V_F. Discharge continues until the PUT current falls below the holding value. At this point, the PUT turns off and the capacitor begins to charge again, thus generating a new output ramp. The cycle continually repeats, and the resulting output is a repetitive sawtooth waveform, as shown. The sawtooth amplitude and period can be adjusted by varying the PUT gate voltage.

The frequency of oscillation is determined by the R_iC time constant of the integrator and the peak voltage set by the PUT. Recall that the charging rate of a capacitor is V_{IN}/R_iC. The time it takes a capacitor to charge from V_F to V_p is the period, T, of the sawtooth waveform (neglecting the rapid discharge time).

$$T = \frac{V_p - V_F}{|V_{IN}|/R_iC}$$

From $f = 1/T$,

Equation 16–11

$$f = \frac{|V_{IN}|}{R_iC}\left(\frac{1}{V_p - V_F}\right)$$

EXAMPLE 16–5

(a) Find the amplitude and frequency of the sawtooth output in Figure 16–32. Assume that the forward PUT voltage, V_F, is approximately 1 V.

(b) Sketch the output waveform.

▶ **FIGURE 16–32**

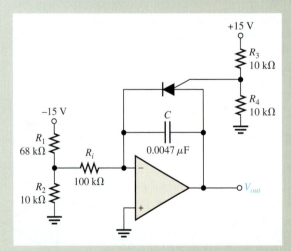

Solution (a) First, find the gate voltage in order to establish the approximate voltage at which the PUT turns on.

$$V_G = \frac{R_4}{R_3 + R_4}(+V) = \frac{10\ k\Omega}{20\ k\Omega}(15\ V) = 7.5\ V$$

This voltage sets the approximate maximum peak value of the sawtooth output (neglecting the 0.7 V).

$$V_p \cong 7.5 \text{ V}$$

The minimum peak value (low point) is

$$V_F \cong 1 \text{ V}$$

So the peak-to-peak amplitude is

$$V_{pp} = V_p - V_F = 7.5 \text{ V} - 1 \text{ V} = \mathbf{6.5 \text{ V}}$$

Determine the frequency as follows:

$$V_{IN} = \frac{R_2}{R_1 + R_2}(-V) = \frac{10 \text{ k}\Omega}{78 \text{ k}\Omega}(-15 \text{ V}) = -1.92 \text{ V}$$

$$f = \frac{|V_{IN}|}{R_i C}\left(\frac{1}{V_p - V_F}\right) = \left(\frac{1.92 \text{ V}}{(100 \text{ k}\Omega)(0.0047\mu\text{F})}\right)\left(\frac{1}{7.5 \text{ V} - 1 \text{ V}}\right) = \mathbf{628 \text{ Hz}}$$

(b) The output waveform is shown in Figure 16–33, where the period is determined as follows:

$$T = \frac{1}{f} = \frac{1}{628 \text{ Hz}} = 1.59 \text{ ms}$$

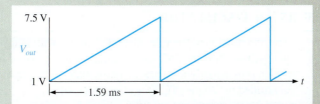

▲ **FIGURE 16–33**

Output of the circuit in Figure 16–32.

Related Problem   If R_i is changed to 56 kΩ in Figure 16–32, what is the frequency?

A Square-Wave Oscillator

The basic square-wave oscillator shown in Figure 16–34 is a type of relaxation oscillator because its operation is based on the charging and discharging of a capacitor. Notice that the op-amp's inverting input is the capacitor voltage and the noninverting input is a portion of the output fed back through resistors R_2 and R_3 to provide hysteresis. When the circuit is first turned on, the capacitor is uncharged, and thus the inverting input is at 0 V. This makes the output a positive maximum, and the capacitor begins to charge toward V_{out} through R_1. When the capacitor voltage (V_C) reaches a value equal to the feedback voltage (V_f) on the noninverting input, the op-amp switches to the maximum negative state. At this point, the capacitor begins to discharge from $+V_f$ toward $-V_f$. When the capacitor voltage reaches $-V_f$, the op-amp switches back to the maximum positive state. This action repeats, as shown in Figure 16–35, and a square-wave output voltage is obtained.

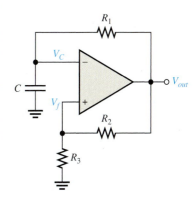

▲ **FIGURE 16–34**

A square-wave relaxation oscillator.

The Phase-Locked Loop

An integrated circuit that contains a voltage-controlled oscillator (VCO) is the phase-locked loop (PLL). The complete **phase-locked loop (PLL)** is an integrated circuit with

◀ **FIGURE 16–35**

Waveforms for the square-wave relaxation oscillator.

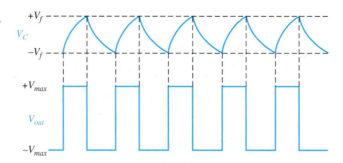

external components that form the entire circuit, which consists of a phase detector, a low-pass filter, and a voltage-controlled oscillator (VCO). If you only need a VCO, it can be used separately without using other circuitry in a PLL. The basic free running frequency of the VCO is configured by the user with just two external components—a resistor and a capacitor; a voltage is sent to a separate pin to change the frequency. PLLs are widely used in communication systems, so details will be covered in Section 18–8.

SECTION 16–5 CHECKUP	1. What is a VCO, and basically, what does it do?
	2. Upon what principle does a relaxation oscillator operate?

16–6 THE 555 TIMER AS AN OSCILLATOR

The 555 timer is a versatile integrated circuit with many applications. In this section, you will see how the 555 is configured as an astable or free-running multivibrator, which is essentially a square-wave oscillator. The use of the 555 timer as a voltage-controlled oscillator (VCO) is also discussed.

After completing this section, you should be able to

❏ **Discuss and analyze the 555 timer and use it in oscillator applications**
❏ Describe the astable operation of a 555 timer
  ◆ Determine the frequency of oscillation ◆ Determine the duty cycle
❏ Discuss the 555 timer as a voltage-controlled oscillator
  ◆ Describe the connections

The 555 timer consists basically of two comparators, a flip-flop, a discharge transistor, and a resistive voltage divider, as shown in Figure 16–36. The flip-flop (bistable multivibrator) is a digital device that may be unfamiliar to you at this point unless you already have taken a digital fundamentals course. Briefly, it is a two-state device whose output can be at either a high voltage level (set, S) or a low voltage level (reset, R). The state of the output can be changed with proper input signals.

The resistive voltage divider is used to set the voltage comparator levels. All three resistors are of equal value; therefore, the upper comparator has a reference of $\frac{2}{3}V_{CC}$, and the lower comparator has a reference of $\frac{1}{3}V_{CC}$. The comparators' outputs control the state of the flip-flop. When the trigger voltage goes below $\frac{1}{3}V_{CC}$, the flip-flop sets and the output jumps to its high level. The threshold input is normally connected to an external RC timing circuit. When the external capacitor voltage exceeds $\frac{2}{3}V_{CC}$, the upper comparator resets the flip-flop, which in turn switches the output back to its low level. When the device output is low, the discharge transistor (Q_d) is turned on and provides a path for rapid discharge of the external timing capacitor. This basic operation allows the timer to be configured with external components as an oscillator, a one-shot, or a time-delay element.

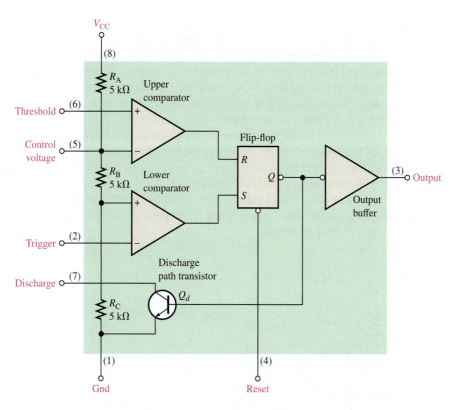

Internal diagram of a 555 integrated circuit timer. (IC pin numbers are in parentheses.)

Astable Operation

A 555 timer connected to operate in the **astable** mode as a free-running relaxation oscillator (astable multivibrator) is shown in Figure 16–37. Notice that the threshold input (THRESH) is now connected to the trigger input (TRIG). The external components R_1, R_2, and C_{ext} form the timing circuit that sets the frequency of oscillation. The 0.01 μF capacitor connected to the control (CONT) input is strictly for decoupling and has no effect on the operation.

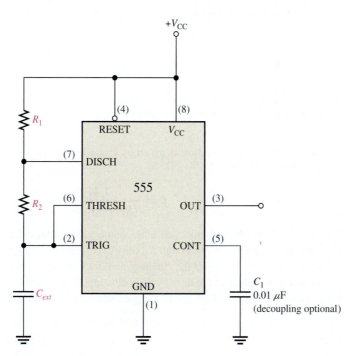

The 555 timer connected as an astable multivibrator.

Initially, when the power is turned on, the capacitor C_{ext} is uncharged and thus the trigger voltage (pin 2) is at 0 V. This causes the output of the lower comparator to be high and the output of the upper comparator to be low, forcing the output of the flip-flop, and thus the base of Q_d, low and keeping the transistor off. Now, C_{ext} begins charging through R_1 and R_2 as indicated in Figure 16–38. When the capacitor voltage reaches $\frac{1}{3}V_{CC}$, the lower comparator switches to its low output state, and when the capacitor voltage reaches $\frac{2}{3}V_{CC}$, the upper comparator switches to its high output state. This resets the flip-flop, causes the base of Q_d to go high, and turns on the transistor. This sequence creates a discharge path for the capacitor through R_2 and the transistor, as indicated. The capacitor now begins to discharge, causing the upper comparator to go low. At the point where the capacitor discharges down to $\frac{1}{3}V_{CC}$, the lower comparator switches high, setting the flip-flop, which makes the base of Q_d low and turns off the transistor. Another charging cycle begins, and the entire process repeats. The result is a rectangular wave output whose duty cycle depends on the values of R_1 and R_2.

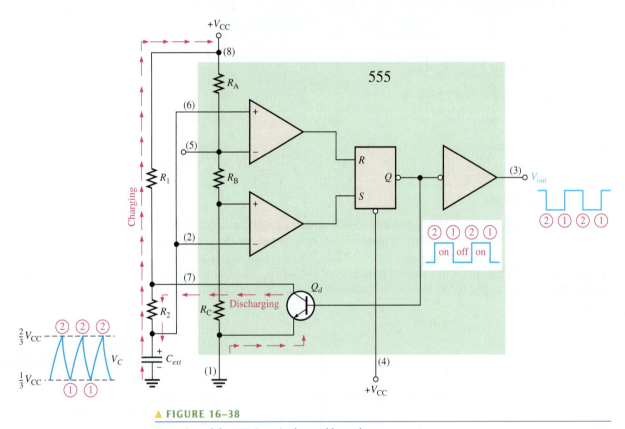

▲ **FIGURE 16–38**

Operation of the 555 timer in the astable mode.

The frequency of oscillation is given by Equation 16–12, or it can be found using the graph in Figure 16–39.

Equation 16–12

$$f_r = \frac{1.44}{(R_1 + 2R_2)C_{ext}}$$

By selecting R_1 and R_2, the duty cycle of the output can be adjusted. Since C_{ext} charges through $R_1 + R_2$ and discharges only through R_2, duty cycles approaching a minimum of 50% can be achieved if $R_2 >> R_1$ so that the charging and discharging times are approximately equal.

A formula to calculate the duty cycle is developed as follows. The time that the output is high (t_H) is how long it takes C_{ext} to charge from $\frac{1}{3}V_{CC}$ to $\frac{2}{3}V_{CC}$. It is expressed as

$$t_H = 0.694(R_1 + R_2)C_{ext}$$

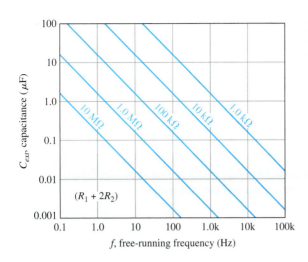

Frequency of oscillation (free-running frequency) of a 555 timer in the astable mode as a function of C_{ext} and $R_1 + 2R_2$. The sloped lines are values of $R_1 + 2R_2$.

The time that the output is low (t_L) is how long it takes C_{ext} to discharge from $\frac{2}{3}V_{CC}$ to $\frac{1}{3}V_{CC}$. It is expressed as

$$t_L = 0.694R_2C_{ext}$$

The period, T, of the output waveform is the sum of t_H and t_L. The following formula for T is the reciprocal of f in Equation 16–12.

$$T = t_H + t_L = 0.694(R_1 + 2R_2)C_{ext}$$

Finally, the percent duty cycle is

$$\text{Duty cycle} = \left(\frac{t_H}{T}\right)100\% = \left(\frac{t_H}{t_H + t_L}\right)100\%$$

$$\textbf{Duty cycle} = \left(\frac{R_1 + R_2}{R_1 + 2R_2}\right)\textbf{100\%}$$

Equation 16–13

To achieve duty cycles of less than 50%, the circuit in Figure 16–37 can be modified so that C_{ext} charges through only R_1 and discharges through R_2. This is achieved with a diode, D_1, placed as shown in Figure 16–40. The duty cycle can be made less than 50% by

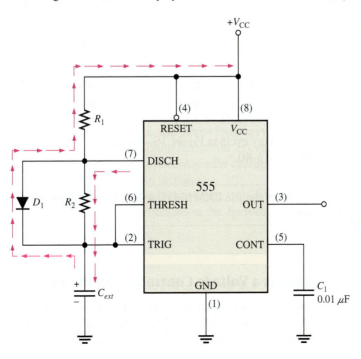

◀ **FIGURE 16–40**

The addition of diode D_1 allows the duty cycle of the output to be adjusted to less than 50% by making $R_1 < R_2$.

Lab Experiment

> To build and test a circuit similar to one used in the ASK test generator, go to Experiment 16–A in your lab manual (*Laboratory Exercises for Electronic Devices* by David Buchla and Steven Wetterling).

Circuit Board

The ASK test generator is implemented on a printed circuit board as shown in Figure 16–51 and will be housed in a unit for use in testing RFID readers on the assembly line. The dark gray lines represent backside connections.

▶ **FIGURE 16–51**

ASK test generator board.

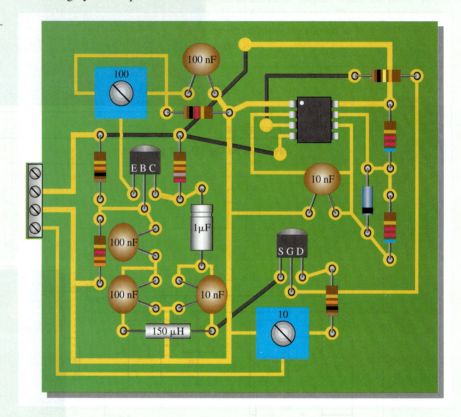

7. Check the printed circuit board and verify that it agrees with the simulation schematic in Figure 16–48.
8. Label each input and output pin according to function.

Programmable Analog Technology

Oscillators with various types of outputs can be programmed into an FPAA or a dpASP. These are described as follows.

Sine-Wave Oscillator

Oscillators can be implemented in programmable analog arrays using software. A sine-wave oscillator is shown in Figure 16–52 using AnadigmDesigner2.

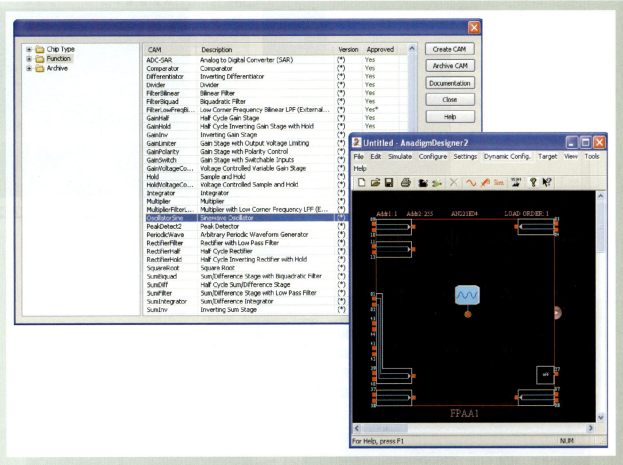

▲ FIGURE 16–52

Selection and placement of the sine-wave oscillator CAM.

The frequency and peak amplitude of the oscillator can be programmed, and the oscillator CAM can be connected to an output as shown in Figure 16–53(a). Running the simulation produces the results shown in part (b).

Square-Wave Oscillator

A square-wave oscillator can be programmed using the sine-wave oscillator CAM and the comparator CAM, as illustrated in Figure 16–54. The frequency of the square wave can be changed by reprogramming the frequency of the sine-wave oscillator.

Variable Duty Cycle Pulse Oscillator

By adding a variable reference to the comparator CAM and by changing its value, the duty cycle as well as the frequency of the pulse waveform can be varied, as shown in Figure 16–55.

Triangular-Wave Oscillator

One way to program a triangular-wave oscillator is shown in Figure 16–56. A sine-wave oscillator is used to drive an inverting gain stage into nonlinear operation. This is followed by an integrator with a properly selected integration constant. A comparator could have been used instead of the over-driven gain stage except that Designer2 does not allow the output of a comparator to be connected to anything but a chip output.

Programming Exercises

1. How do you adjust the duty cycle of the variable duty cycle pulse oscillator?
2. To change the frequency of the triangular-wave oscillator, what parameters must be changed by programming?

SUMMARY

Section 16–1
- ◆ Sinusoidal feedback oscillators operate with positive feedback; after startup, the loop gain must be exactly 1.
- ◆ Relaxation oscillators use an RC timing circuit.

Section 16–2
- ◆ The two conditions for positive feedback are the phase shift around the feedback loop must be 0° and the voltage gain around the feedback loop must equal or greater than 1.
- ◆ For initial start-up, the voltage gain around the feedback loop must be greater than 1.

Section 16–3
- ◆ Sinusoidal RC oscillators include the Wien bridge, phase-shift, and twin-T.

Section 16–4
- ◆ Sinusoidal LC oscillators include the Colpitts, Clapp, Hartley, Armstrong, and crystal-controlled.
- ◆ The feedback signal in a Colpitts oscillator is derived from a capacitive voltage divider in the LC circuit.
- ◆ The Clapp oscillator is a variation of the Colpitts with a capacitor added in series with the inductor.
- ◆ The feedback signal in a Hartley oscillator is derived from an inductive voltage divider in the LC circuit.
- ◆ The feedback signal in an Armstrong oscillator is derived by transformer coupling.
- ◆ Crystal oscillators are the most stable type of feedback oscillator.

Section 16–5
- ◆ A relaxation oscillator uses an RC timing circuit and a device that changes states to generate a periodic waveform.
- ◆ The frequency in a voltage-controlled oscillator (VCO) can be varied with a dc control voltage.

Section 16–6
- ◆ The 555 timer is an integrated circuit that can be used as an oscillator, in addition to many other applications.

KEY TERMS

Key terms and other bold terms in the chapter are defined in the end-of-book glossary.

Astable   Characterized by having no stable states.

Oscillator   An electronic circuit that produces a periodic waveform on its output with only the dc supply voltage as an input.

Phase-locked loop (PLL)   An integrated circuit consisting of a phase detector, a low-pass filter, and a voltage-controlled oscillator.

Positive feedback   The return of a portion of the output signal to the input such that it reinforces and sustains the output.

Voltage-controlled oscillator (VCO)   A type of relaxation oscillator whose frequency can be varied by a dc control voltage.

KEY FORMULAS

16–1 $\dfrac{V_{out}}{V_{in}} = \dfrac{1}{3}$ Wien-bridge positive feedback attenuation

16–2 $f_r = \dfrac{1}{2\pi RC}$ Wien-bridge resonant frequency

16–3 $B = \dfrac{1}{29}$ Phase-shift feedback attenuation

16–4 $f_r = \dfrac{1}{2\pi\sqrt{6}RC}$ Phase-shift oscillator frequency

16–5 $f_r \cong \dfrac{1}{2\pi\sqrt{LC_T}}$ Colpitts, Clapp, and Hartley approximate resonant frequency

16–6 $A_v = \dfrac{C_1}{C_2}$ Colpitts amplifier gain

16–7 $\quad f_r = \dfrac{1}{2\pi\sqrt{LC_T}}\sqrt{\dfrac{Q^2}{Q^2 + 1}}$ $\qquad$ Colpitts resonant frequency

16–8 $\quad A_v > \dfrac{L_2}{L_1}$ $\qquad$ Hartley self-starting gain

16–9 $\quad f_r = \dfrac{1}{2\pi\sqrt{L_{pri}C_1}}$ $\qquad$ Armstrong resonant frequency

16–10 $\quad f_r = \dfrac{1}{4R_1C}\left(\dfrac{R_2}{R_3}\right)$ $\qquad$ Triangular-wave oscillator frequency

16–11 $\quad f = \dfrac{|V_{IN}|}{R_iC}\left(\dfrac{1}{V_p - V_F}\right)$ $\qquad$ Sawtooth VCO frequency

16–12 $\quad f_r = \dfrac{1.44}{(R_1 + 2R_2)C_{ext}}$ $\qquad$ 555 astable frequency

16–13 $\quad$ **Duty cycle** $= \left(\dfrac{R_1 + R_2}{R_1 + 2R_2}\right)$**100%** $\qquad$ 555 astable

TRUE/FALSE QUIZ
Answers can be found at www.pearsonhighered.com/floyd.

1. Two categories of oscillators are feedback and relaxation.
2. A feedback oscillator uses only negative feedback.
3. Positive feedback is never used in an oscillator.
4. The net phase shift around the oscillator feedback loop must be zero.
5. The voltage gain around the closed feedback loop must be greater than 1 to sustain oscillations.
6. For start-up, the loop gain must be greater than 1.
7. A Wien-bridge oscillator uses an RC circuit in the positive feedback loop.
8. The phase-shift oscillator utilizes RC circuits.
9. The twin-T oscillator contains an LC feedback circuit.
10. Colpitts, Clapp, Hartley, and Armstrong are examples of LC oscillators.
11. The crystal oscillator is based on the photoelectric effect.
12. A relaxation oscillator uses no positive feedback.
13. Most relaxation oscillators produce sinusoidal outputs.
14. VCO stands for variable-capacitance oscillator.
15. The 555 timer can be used as an oscillator.

CIRCUIT-ACTION QUIZ
Answers can be found at www.pearsonhighered.com/floyd.

1. If R_1 and R_2 are increased to 18 kΩ in Figure 16–12, the frequency of oscillation will
   (a) increase   (b) decrease   (c) not change
2. If the feedback potentiometer R_f is adjusted to a higher value, the voltage gain in Figure 16–12 will
   (a) increase   (b) decrease   (c) not change
3. In Figure 16–14, if the R_f is decreased, the feedback attenuation will
   (a) increase   (b) decrease   (c) not change
4. If the capacitors in Figure 16–14 are increased to 0.01 μF, the frequency of oscillation will
   (a) increase   (b) decrease   (c) not change
5. In order to increase V_{UTP} in Figure 16–30, R_3 must
   (a) increase   (b) decrease   (c) not change
6. If the capacitor in Figure 16–30 opens, the frequency of oscillation will
   (a) increase   (b) decrease   (c) not change

7. If the value of R_1 in Figure 16–32 is decreased, the peak value of the sawtooth output will

(a) increase   (b) decrease   (c) not change

8. If the diode in Figure 16–40 opens, the duty cycle will

(a) increase   (b) decrease   (c) not change

SELF-TEST

Answers can be found at www.pearsonhighered.com/floyd.

Section 16–1

1. An oscillator differs from an amplifier because the oscillator

(a) has more gain   (b) requires no input signal

(c) requires no dc supply   (d) always has the same output

Section 16–2

2. One condition for oscillation is

(a) a phase shift around the feedback loop of 180°

(b) a gain around the feedback loop of one-third

(c) a phase shift around the feedback loop of 0°

(d) a gain around the feedback loop of less than 1

3. A second condition for oscillation is

(a) no gain around the feedback loop

(b) a gain of 1 or more around the feedback loop

(c) the attenuation of the feedback circuit must be one-third

(d) the feedback circuit must be capacitive

4. In a certain sine wave oscillator, $A_v = 50$. The attenuation of the feedback circuit must be

(a) 1   (b) 0.01   (c) 10   (d) 0.02

5. For an oscillator to properly start, the gain around the feedback loop must initially be

(a) 1   (b) less than 1   (c) greater than 1   (d) equal to B

Section 16–3

6. Wien-bridge oscillators are based on

(a) positive feedback   (b) negative feedback

(c) the piezoelectric effect   (d) high gain

7. In a Wien-bridge oscillator, if the resistances in the positive feedback circuit are decreased, the frequency

(a) decreases   (b) increases   (c) remains the same

8. The Wien-bridge oscillator's positive feedback circuit is

(a) an RL circuit   (b) an LC circuit

(c) a voltage divider   (d) a lead-lag circuit

9. A phase-shift oscillator has

(a) three RC circuits   (b) three LC circuits

(c) a T-type circuit   (d) a π-type circuit

Section 16–4

10. Colpitts, Clapp, and Hartley are names that refer to

(a) types of RC oscillators   (b) inventors of the transistor

(c) types of LC oscillators   (d) types of filters

11. The main feature of a crystal oscillator is

(a) economy   (b) reliability

(c) stability   (d) high frequency

Section 16–5

12. An oscillator whose frequency is changed by a variable dc voltage is known as

(a) a crystal oscillator   (b) a VCO

(c) an Armstrong oscillator   (d) a piezoelectric device

13. The operation of a relaxation oscillator is based on

(a) the charging and discharging of a capacitor   (b) a highly selective resonant circuit

(c) a very stable supply voltage   (d) low power consumption

14. Which one of the following is *not* an input or output of the 555 timer?

      (a) Threshold   **(b)** Control voltage   **(c)** Clock

      (d) Trigger   **(e)** Discharge   **(f)** Reset

PROBLEMS

Answers to odd-numbered problems are at the end of the book.

BASIC PROBLEMS

Section 16–1 **The Oscillator**

    1. What type of input is required for an oscillator?

    2. What are the basic components of an oscillator circuit?

Section 16–2 **Feedback Oscillators**

    3. If the attenuation of a sine wave feedback oscillator is $1/B$, what should be the gain to maintain an undistorted sine wave output?

    4. Explain why sine wave feedback oscillators need some form of automatic gain control.

Section 16–3 **Oscillators with *RC* Feedback Circuits**

    5. A certain lead-lag circuit has a resonant frequency of 3.5 kHz and equal resistors with equal capacitors. What is the rms output voltage if an input signal with a frequency equal to f_r and with an rms value of 2.2 V is applied to the input?

    6. Calculate the resonant frequency of a lead-lag circuit with the following values: $R_1 = R_2 = 6.2 \text{ k}\Omega$, and $C_1 = C_2 = 0.02 \ \mu\text{F}$.

    7. For the circuit in Figure 16–57, what is the range of output frequencies?

    8. Assume the Wien bridge in Figure 16–57 is oscillating with an undistorted sine wave output. To what value must R_f be set to maintain oscillations if the lamp has a resistance of 160 Ω?

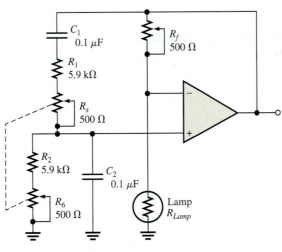

▲ **FIGURE 16–57**

    9. For the Wien-bridge oscillator in Figure 16–58, calculate the setting for R_f, assuming the internal drain-source resistance, r'_{ds}, of the JFET is 350 Ω when oscillations are stable.

   10. Find the frequency of oscillation for the Wien-bridge oscillator in Figure 16–58.

▶ FIGURE 16–58

Multisim and LT Spice file circuits are identified with a logo and are in the Problems folder on the website. Filenames correspond to figure numbers (e.g., FGM16-58 or FGS16-58).

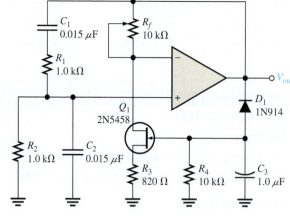

11. What value of R_f is required for the phase-shift oscillator in Figure 16–59? What is f_r?

▶ FIGURE 16–59

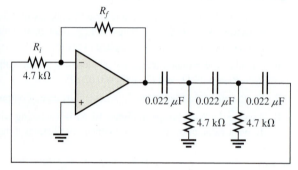

Section 16–4   **Oscillators with *LC* Feedback Circuits**

12. Calculate the frequency of oscillation for each circuit in Figure 16–60 and identify the type of oscillator. Assume $Q > 10$ in each case.

▶ FIGURE 16–60

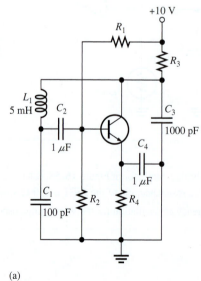

(a)

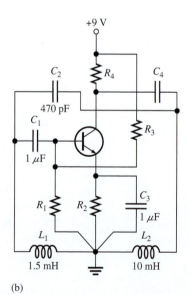

(b)

13. Determine what the gain of the amplifier stage must be in Figure 16–61 in order to have sustained oscillation.

▶ FIGURE 16–61

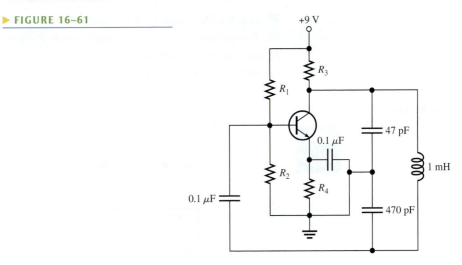

Section 16–5   **Relaxation Oscillators**

14. What type of output does the circuit in Figure 16–62 produce? Determine the frequency of the output.

15. Show how to change the frequency of oscillation in Figure 16–62 to 10 kHz.

16. Determine the amplitude and frequency of the output voltage in Figure 16–63. Use 1 V as the forward PUT voltage.

17. Modify the sawtooth generator in Figure 16–63 so that its peak-to-peak output is 4 V.

▶ FIGURE 16–62

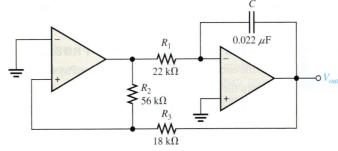

▶ FIGURE 16–63

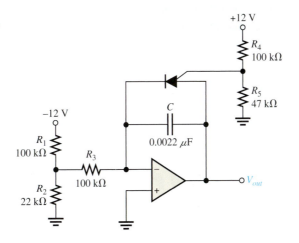

18. A certain sawtooth generator has the following parameter values: $V_{IN} = 3$ V, $R = 4.7$ kΩ, $C = 0.001$ μF. Determine its peak-to-peak output voltage if the period is 10 μs.

EXAMPLE 17–2

A certain voltage regulator has a 12 V output when there is no load ($I_L = 0$). When there is a full-load current of 140 mA, the output voltage is 11.9 V. Express the voltage regulation as a percentage change from no-load to full-load and also as a percentage change for each mA change in load current.

Solution   The no-load output voltage is

$$V_{NL} = 12V$$

The full-load output voltage is

$$V_{FL} = 11.9 \text{ V}$$

The load regulation as a percentage change from no-load to full-load is

$$\text{load regulation} = \left(\frac{V_{NL} - V_{FL}}{V_{FL}}\right)100\% = \left(\frac{12 \text{ V} - 11.9 \text{ V}}{11.9 \text{ V}}\right)100\% = \mathbf{0.840\%}$$

The load regulation can also be expressed as a percentage change per milliamp as

$$\text{load regulation} = \frac{0.840\%}{140 \text{ mA}} = \mathbf{0.006\%/mA}$$

where the change in load current from no-load to full-load is 140 mA.

Related Problem   A regulator has a no-load output voltage of 18 V and a full-load output of 17.8 V at a load current of 500 mA. Determine the voltage regulation as a percentage change from no-load to full-load and also as a percentage change for each mA change in load current.

Sometimes power supply manufacturers specify the equivalent output resistance of a power supply (R_{OUT}) instead of its load regulation. Recall that an equivalent Thevenin circuit can be drawn for any two-terminal linear circuit. Figure 17–3 shows the equivalent Thevenin circuit for a power supply with a load resistor. The Thevenin voltage is the voltage from the supply with no load (V_{NL}), and the Thevenin resistance is the specified output resistance, R_{OUT}. Ideally, R_{OUT} is zero, corresponding to 0% load regulation, but in practical power supplies R_{OUT} is a small value. With the load resistor in place, the output voltage is found by applying the voltage-divider rule:

$$V_{OUT} = V_{NL}\left(\frac{R_L}{R_{OUT} + R_L}\right)$$

▶ **FIGURE 17–3**

Thevenin equivalent circuit for a power supply with a load resistor.

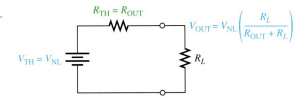

If we let R_{FL} equal the smallest-rated load resistance (largest-rated current), then the full-load output voltage (V_{FL}) is

$$V_{FL} = V_{NL}\left(\frac{R_{FL}}{R_{OUT} + R_{FL}}\right)$$

By rearranging and substituting into Equation 17–3,

$$V_{NL} = V_{FL}\left(\frac{R_{OUT} + R_{FL}}{R_{FL}}\right)$$

$$\text{Load regulation} = \frac{V_{FL}\left(\dfrac{R_{OUT} + R_{FL}}{R_{FL}}\right) - V_{FL}}{V_{FL}} \times 100\%$$

$$= \left(\frac{R_{OUT} + R_{FL}}{R_{FL}} - 1\right)100\%$$

$$\textbf{Load regulation} = \left(\frac{R_{OUT}}{R_{FL}}\right)\textbf{100}\%$$

Equation 17–4

Equation 17–4 is a useful way of finding the percent load regulation when the output resistance and minimum load resistance are specified.

SECTION 17–1 CHECKUP
Answers can be found at www
.pearsonhighered.com/floyd.

1. Define *line regulation*.
2. Define *load regulation*.
3. The input of a certain regulator increases by 3.5 V. As a result, the output voltage increases by 0.042 V. The nominal output is 20 V. Determine the line regulation in both % and in %/V.
4. If a 5.0 V power supply has an output resistance of 80 mΩ and a specified maximum output current of 1.0 A, what is the load regulation? Give the result as a % and as a %/mA.

17–2 BASIC LINEAR SERIES REGULATORS

The fundamental classes of voltage regulators are linear regulators and switching regulators. Both of these are available in integrated circuit form. Two basic types of linear regulator are the series regulator and the shunt regulator.

After completing this section, you should be able to

❑ **Describe and analyze the operation of linear series regulators**
❑ Explain regulating action
  ◆ Determine the closed-loop gain ◆ Determine the output voltage
❑ Discuss overload protection
  ◆ Explain constant-current limiting ◆ Determine the maximum load current
❑ Discuss fold-back current limiting

A simple representation of a series type of **linear regulator** is shown in Figure 17–4(a), and the basic components are shown in the block diagram in Figure 17–4(b). The control element is a pass transistor in series with the load between the input and output. The output sample circuit senses a change in the output voltage. The error detector compares the sample voltage with a reference voltage and causes the control element to

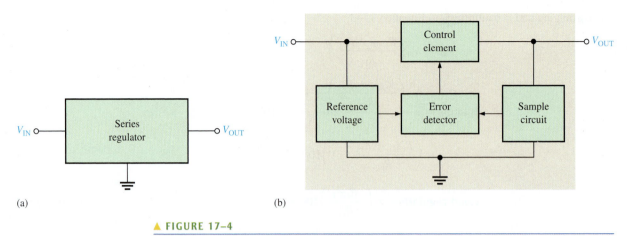

(a)

(b)

▲ **FIGURE 17–4**

Simple series voltage regulator and block diagram.

▶ **FIGURE 17–5**

Basic op-amp series regulator.

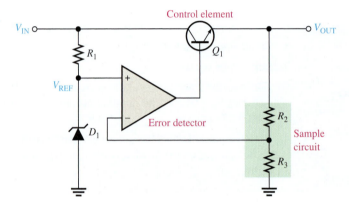

compensate in order to maintain a constant output voltage. A basic op-amp series regulator is shown in Figure 17–5.

Regulating Action

The operation of the series regulator is illustrated in Figure 17–6 and is as follows. The resistive voltage divider formed by R_2 and R_3 senses any change in the output voltage. When the output tries to decrease, as indicated in Figure 17–6(a), because of a decrease in V_{IN} or because of an increase in I_L caused by a decrease in R_L, a proportional voltage decrease is applied to the op-amp's inverting input by the voltage divider. Since the zener diode (D_1) holds the other op-amp input at a nearly constant reference voltage, V_{REF}, a small difference voltage (error voltage) is developed across the op-amp's inputs. This difference voltage is amplified, and the op-amp's output voltage, V_B, increases. This increase is applied to the base of Q_1, causing the emitter voltage V_{OUT} to increase until the voltage to the inverting input again equals the reference (zener) voltage. This action offsets the attempted decrease in output voltage, thus keeping it nearly constant. The power transistor, Q_1, is usually used with a heat sink because it must handle all of the load current.

The opposite action occurs when the output tries to increase, as indicated in Figure 17–6(b). The op-amp in the series regulator is actually connected as a noninverting amplifier where the reference voltage V_{REF} is the input at the noninverting terminal, and the R_2/R_3 voltage divider forms the negative feedback circuit. The closed-loop voltage gain is

$$A_{cl} = 1 + \frac{R_2}{R_3}$$

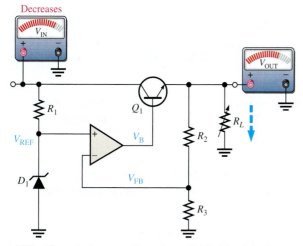

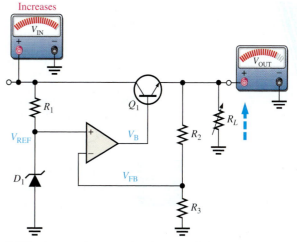

(a) When V_{IN} or R_L decreases, V_{OUT} drops slightly. The slight drop is sampled by the feedback voltage, V_{FB}, and the op-amp immediately increases V_B. This increase causes the output to remain nearly the same as before the original drop. Changes in V_{OUT} are exaggerated for illustration.

When V_{IN} (or R_L) stabilizes at its new lower value, the voltages return to their original values, thus keeping V_{OUT} nearly constant as a result of the negative feedback.

(b) When V_{IN} or R_L increases, V_{OUT} increases slightly. The slight increase is sampled by the feedback voltage, V_{FB}, and the op-amp immediately decreases V_B. This decrease causes the output to remain nearly the same as before the original increase.

When V_{IN} (or R_L) stabilizes at its new higher value, the voltages return to their original values, thus keeping V_{OUT} nearly constant as a result of the negative feedback.

▲ FIGURE 17–6

Illustration of series regulator action that keeps V_{OUT} constant when V_{IN} or R_L changes.

Therefore, the regulated output voltage of the series regulator (neglecting the base-emitter voltage of Q_1) is

$$V_{OUT} \cong \left(1 + \frac{R_2}{R_3}\right) V_{REF}$$

Equation 17–5

From this analysis, you can see that the output voltage is determined by the zener voltage and the resistors R_2 and R_3. It is relatively independent of the input voltage, and therefore, regulation is achieved (as long as the input voltage and load current are within specified limits).

EXAMPLE 17–3   Determine the output voltage for the regulator in Figure 17–7.

▶ FIGURE 17–7

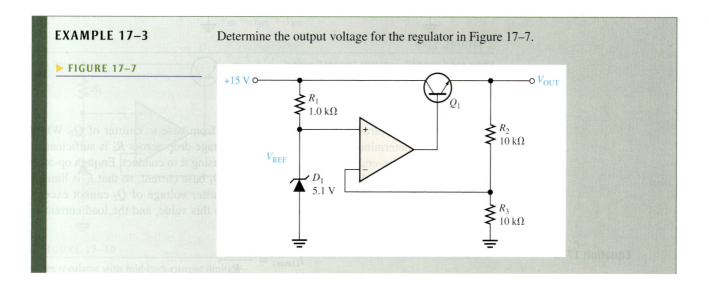

Simplified step-down regulator.

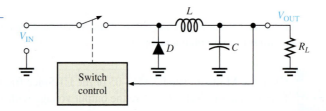

opens and closes rapidly from a control circuit that senses the output, and it adjusts the on-time and the off-time to keep the desired output. When the switch is closed, the diode is *off* and the magnetic field of the inductor builds, storing energy. When the switch opens, the magnetic field collapses, keeping nearly constant current in the load. A path for the load current is provided through the forward-biased diode (as long as the load resistance is not too large). The capacitor smoothes the dc to a nearly constant level.

Let's look at the circuit, including the switching device, in more detail. The switch turns the input voltage on and off at a rapid rate and with a duty cycle that is based on the regulator's load requirement. Figure 17–17 shows a basic step-down switching regulator using an E-MOSFET switching transistor. MOSFET transistors can switch faster than BJTs and have become the preferred type of switching device, provided that the off-state voltage is not too high. As in most electronic devices, there are trade-offs for designers in choosing a switching device. Differences in breakdown voltage, on-state resistance, and switching time must all be considered for a given design. In addition to transistor switches, you may see thyristors used occasionally.

A basic step-down switching regulator.

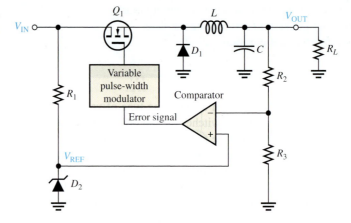

The pulsed current from the transistor switch is smoothed by an *LC* filter. The inductor tries to keep current constant, and the capacitor tends to keep voltage constant. Ideally, these components do not dissipate power, but in practice some loss is encountered due to various factors. To avoid requiring large (and expensive) inductors and capacitors, the switching frequency is selected to be much higher than the utility frequency; 20 kHz is common. The drawback to higher frequencies is electrical noise. Switching power supplies can radiate harmonic frequency noise to nearby circuits, so they need to be well shielded and frequently require EMI (electromagnetic interference) filters. Since the switching device spends most of its time either in cutoff or saturation, the power lost in the control element is usually relatively small (although instantaneous power dissipated in the switching device can be large).

The *on* and *off* intervals of Q_1 are shown in the waveform of Figure 17–18(a). For an *n*-channel E-MOSFET, the control voltage swings below and above the threshold voltage (*off* and *on* states). The capacitor charges during the on-time (t_{on}) and discharges during the off-time (t_{off}). When the on-time is increased relative to the off-time, the capacitor charges

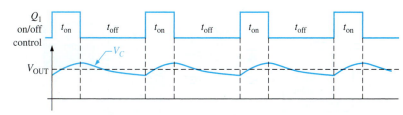

(a) V_{OUT} depends on the duty cycle.

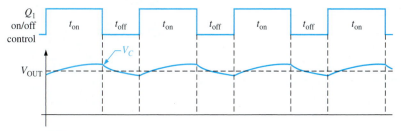

(b) Increase the duty cycle and V_{OUT} increases.

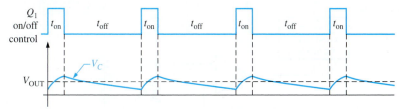

(c) Decrease the duty cycle and V_{OUT} decreases.

▲ **FIGURE 17–18**

Switching regulator waveforms. The V_C waveform is shown for no inductive filtering to illustrate the charge and discharge action (ripple). L and C smooth V_C to a nearly constant level, as indicated by the dashed line for V_{OUT}.

more, thus increasing the output voltage, as indicated in Figure 17–18(b). When the on-time is decreased relative to the off-time, the capacitor discharges more, thus decreasing the output voltage, as in Figure 17–18(c). The inductor further smoothes the fluctuations of the output voltage caused by the charging and discharging action.

Ideally, the output voltage is expressed as

$$V_{OUT} = \left(\frac{t_{on}}{T} \right) V_{IN}$$

Equation 17–8

T is the period of the on-off cycle of Q_1 and is related to the frequency by $T = 1/f$. The period is the sum of the on-time and the off-time.

$$T = t_{on} + t_{off}$$

As you know, the ratio t_{on}/T is called the *duty cycle*.

The regulating action is as follows and is illustrated in Figure 17–19. When V_{OUT} tries to decrease, the on-time of Q_1 is increased, causing an additional charge on C to offset the attempted decrease. When V_{OUT} tries to increase, the on-time of Q_1 is decreased, causing the capacitor to discharge enough to offset the attempted increase.

Step-Up Configuration

A basic step-up type of switching regulator (sometimes called a *boost converter*) is shown in Figure 17–20, where transistor Q_1 operates as a switch to ground.

Basic regulating action of a step-down switching regulator.

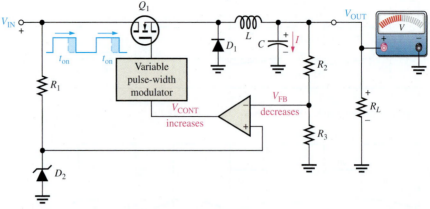

(a) When V_{OUT} attempts to decrease, the on-time of Q_1 increases.

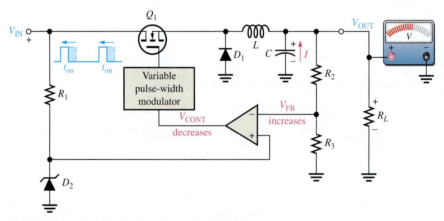

(b) When V_{OUT} attempts to increase, the on-time of Q_1 decreases.

Basic step-up switching regulator.

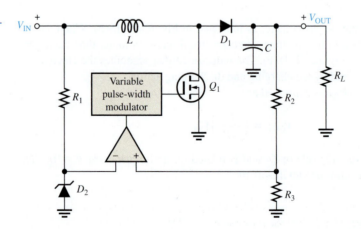

The switching action is illustrated in Figures 17–21 and 17–22. When Q_1 turns on, a voltage equal to approximately V_{IN} is induced across the inductor with a polarity as indicated in Figure 17–21. During the on-time (t_{on}) of Q_1, the inductor voltage, V_L, decreases from its initial maximum and diode D_1 is reverse-biased. The longer Q_1 is on, the smaller V_L becomes. During the on-time, the capacitor only discharges an extremely small amount through the load.

When Q_1 turns off, as indicated in Figure 17–22, the inductor voltage suddenly reverses polarity and adds to V_{IN}, forward-biasing diode D_1 and allowing the capacitor to charge.

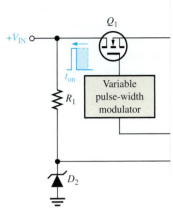

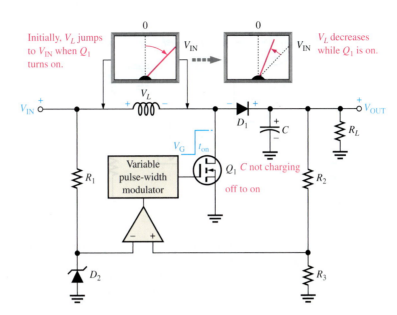

► FIGURE 17–21

Basic action of a step-up regulator when Q_1 is on.

(a) When $-V_{OUT}$ tries to decrease, t_{on} de decrease in $-V_{OUT}$.

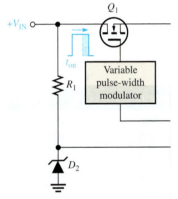

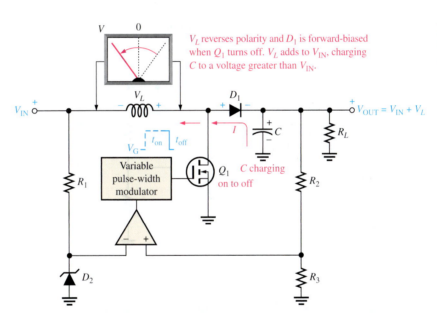

◄ FIGURE 17–22

Basic switching action of a step-up regulator when Q_1 turns off.

(b) When $-V_{OUT}$ tries to increase, t_{on} inc increase in $-V_{OUT}$.

SECTION 17–4 CHECKUP

17–5 INTEGRATED C

In the previous sections, the l
Several types of both linear a
(IC) form. Generally, the line
either positive or negative ou
section, typical linear and sw

After completing this section

❑ **Discuss integrated circu**
❑ Discuss fixed positive line
    • Describe the 78XX reg

The output voltage is equal to the capacitor voltage and can be larger than V_{IN} because the capacitor is charged to V_{IN} plus the voltage induced across the inductor during the off-time of Q_1. The output voltage is dependent on both the inductor's magnetic field action (determined by t_{on}) and the charging of the capacitor (determined by t_{off}).

Voltage regulation is achieved by the variation of the on-time of Q_1 (within certain limits) as related to changes in V_{OUT} due to changing load or input voltage. If V_{OUT} tries to increase, the on-time of Q_1 will decrease, resulting in a decrease in the amount that C will charge. If V_{OUT} tries to decrease, the on-time of Q_1 will increase, resulting in an increase in the amount that C will charge. This regulating action maintains V_{OUT} at an essentially constant level.

Voltage-Inverter Configuration

A third type of switching regulator produces an output voltage that is opposite in polarity to the input. A basic diagram is shown in Figure 17–23. This is sometimes called a *buck-boost converter*.

▶ FIGURE 17–23

Basic inverting switching regulat

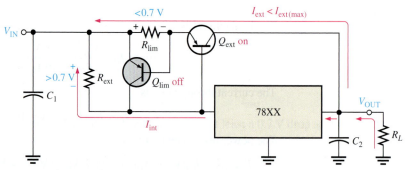

(a) During normal operation, when the load current is not excessive, Q_{lim} is off.

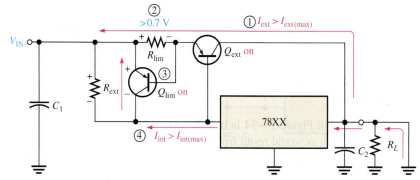

(b) When short occurs ① , the external current becomes excessive and the voltage across R_{lim} increases ② and turns on Q_{lim} ③ , which then routes current through the regulator and conducts it away from Q_{ext} , causing the internal regulator current to become excessive ④ which forces the regulator into thermal shutdown.

▲ FIGURE 17–37

The current-limiting action of the regulator circuit.

A Current Regulator

The three-terminal regulator can be used as a current source when an application requires that a constant current be supplied to a variable load. The basic circuit is shown in Figure 17–38 where R_1 is the current-setting resistor. The regulator provides a fixed constant voltage, V_{OUT}, between the ground terminal (not connected to ground in this case) and the output terminal. This determines the constant current supplied to the load.

Equation 17–11

$$I_L = \frac{V_{OUT}}{R_1} + I_G$$

The current, I_G, to the ground pin is very small compared to the output current and can often be neglected.

▶ FIGURE 17–38

The three-terminal regulator as a current source.

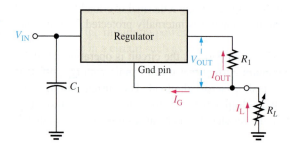

EXAMPLE 17–9

What value of R_1 is necessary in a 7805 regulator to provide a constant current of 0.5 A to a variable load that can be adjusted from 1 Ω to 10 Ω?

Solution

The 7805 produces 5 V between its ground terminal and its output terminal. Therefore, if you want 0.5 A of current, the current-setting resistor must be (neglecting I_G)

$$R_1 = \frac{V_{\text{OUT}}}{I_L} = \frac{5\text{ V}}{0.5\text{ A}} = 10\text{ }\Omega$$

The circuit is shown in Figure 17–39.

▶ **FIGURE 17–39**

A constant-current source of 0.5 A.

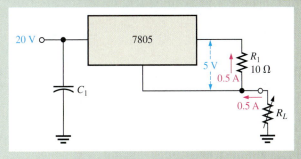

Related Problem

If a 7808 regulator is used instead of the 7805, to what value would you change R_1 to maintain a constant current of 0.5 A?

SECTION 17–6 CHECKUP

1. What is the purpose of using an external pass transistor with an IC voltage regulator?
2. What is the advantage of current limiting in a voltage regulator?
3. What does *thermal overload* mean?

Device Application: *Variable DC Power Supply*

A regulated power supply with a fixed output voltage of +12 V was developed in Chapter 3. The company that manufactures this power supply plans to offer a new line of variable power supplies for which a specified voltage can be preset at the factory or can be adjusted by the user. In this application, the power supply with a variable regulator is developed to provide an output voltage from +9 V to +30 V and a maximum load current of 250 mA.

The Circuit

Recall that in the original power supply, a 7812 provided a +12 V regulated output. In this new power supply, a 7809 is used to produce that variable output voltage. As in the earlier design, it is recommended by the manufacturer that a 0.33 μF capacitor be connected from the input terminal to ground and a 0.1 μF capacitor be connected from the output terminal to ground, as shown in Figure 17–40, to prevent high frequency oscillations and improve the performance. The reason for a small-value capacitor in parallel with a large one is that the large filter capacitor has an internal equivalent series resistance, which affects the high frequency response of the system. The effect is cancelled with the small capacitor.

The Transformer The transformer must convert the 120 V rms line voltage to an ac voltage that will result in a rectified voltage that will produce 34 V ± 10% when filtered.

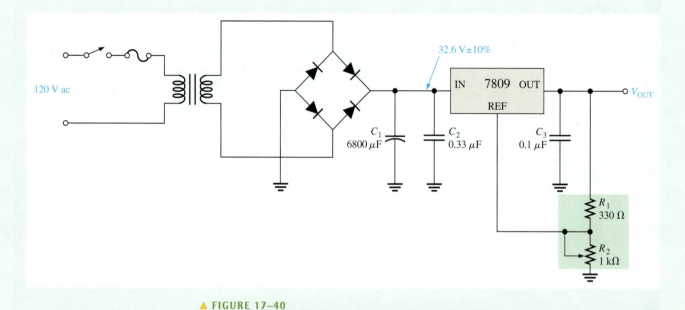

▲ FIGURE 17–40

Variable output power supply.

The Voltage Regulator A partial datasheet for a 7809 is shown in Figure 17–41. Notice that there is a range of nominal output voltages, but it is typically 9.0 V. The line and load regulations specify how much the output can vary about the nominal output value. For example, the typical 9.0 V output will change no more than 12 mV (typical) as the load current changes from 5 mA to 1.5 A. The output voltage of the regulator is the voltage between the output (OUT) terminal and the reference (REF) terminal. The voltage divider formed by R_1 and R_2 provides a reference voltage other than ground and increases the output voltage with respect to ground above the 9 V nominal regulator output by an amount equal to the voltage across R_2.

1. What are the minimum and maximum nominal output voltages specified on the datasheet when I_O is 500 mA.
2. From the datasheet, determine the maximum change in the output voltage when the load current changes from 5 mA to 1.5 A.

▶ FIGURE 17–41

Partial datasheet for a 7809 regulator. Copyright Fairchild Semiconductor Corporation. Used by permission.

Electrical Characteristics (LM7809) (Continued)

Refer to the test circuits. -40°C < T_J < 125°C, I_O = 500mA, V_I = 15V, C_I = 0.33μF, C_O = 0.1μF, unless otherwise specified.

Symbol	Parameter	Conditions		Min.	Typ.	Max.	Unit
V_O	Output Voltage	T_J = +25°C		8.65	9.0	9.35	V
		5mA ≤ I_O ≤ 1A, P_O ≤ 15W, V_I = 11.5V to 24V		8.6	9.0	9.4	
Regline	Line Regulation[7]	T_J = +25°C	V_I = 11.5V to 25V	–	6.0	180	mV
			V_I = 12V to 17V	–	2.0	90.0	
Regload	Load Regulation[7]	T_J = +25°C	I_O = 5mA to 1.5A	–	12.0	180	mV
			I_O = 250mA to 750mA	–	4.0	90.0	
I_Q	Quiescent Current	T_J = +25°C		–	5.0	8.0	mA
ΔI_Q	Quiescent Current Change	I_O = 5mA to 1A		–	–	0.5	mA
		V_I = 11.5V to 26V		–	–	1.3	
$\Delta V_O/\Delta T$	Output Voltage Drift[8]	I_O = 5mA		–	-1.0	–	mV/°C
V_N	Output Noise Voltage	f = 10Hz to 100kHz, T_A = +25°C		–	58.0	–	μV/V_O
RR	Ripple Rejection[8]	f = 120Hz, V_O = 13V to 23V		56.0	71.0	–	dB
V_{DROP}	Dropout Voltage	I_O = 1A, T_J = +25°C		–	2.0	–	V
r_O	Output Resistance[8]	f = 1kHz		–	17.0	–	mΩ
I_{SC}	Short Circuit Current	V_I = 35V, T_A = +25°C		–	250	–	mA
I_{PK}	Peak Current[8]	T_J = +25°C		–	2.2	–	A

3. Calculate the maximum power dissipation in R_1.
4. Calculate the maximum power dissipation in R_2.

The Fuse   The fuse will be in series with the primary winding of the transformer, as shown in Figure 17–40. The fuse should be calculated based on the maximum allowable primary current. Recall from your dc/ac circuits course that if the voltage is stepped down, the current is stepped up. From the specifications for the unregulated power supply, the maximum load current is 100 mA.

5. Calculate the primary current and use this value to select a fuse rating for the circuit in Figure 17–40.

Simulation

Multisim is used to simulate this power supply circuit. Figure 17–42 shows the simulated regulated power supply circuit adjusted to show that it meets or exceeds the specified minimum and maximum output voltages.

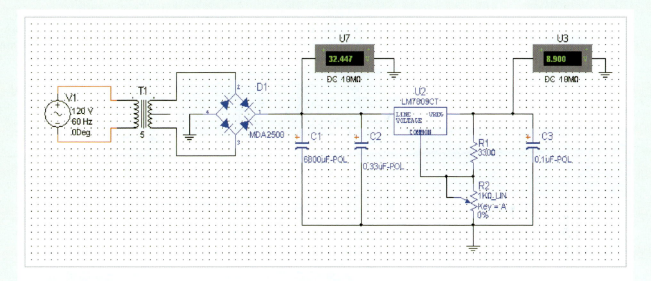

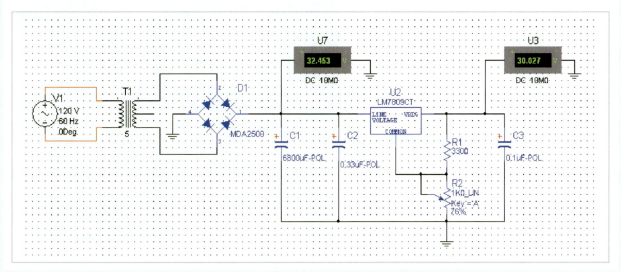

▲ **FIGURE 17–42**

Simulation of the regulated variable power supply circuit at the minimum and maximum specified output voltages.

Build and simulate the circuit using your Multisim or LT Spice software. Verify the operation.

Prototyping and Testing

Now that all the component s have been selected and the circuit has been simulated, the circuit is breadboarded and tested.

Lab Experiment

To build and test a similar circuit, go to Experiment 17 in your lab manual (*Laboratory Exercises for Electronic Devices* by David Buchla and Steven Wetterling).

Printed Circuit Board

The variable regulated power supply prototype has been built and tested. It is now committed to a printed circuit layout, as shown in Figure 17–43. Notice that a heat sink is used with the regulator IC to increase its ability to dissipate power. The output voltage is measured at the potentiometer.

6. Compare the printed circuit board to the schematic in Figure 17–40.
7. Calculate the power dissipated by the regulator for an output of 9 V and $I_L = 100$ mA.
8. Calculate the power dissipated by the regulator for an output of 30 V and $I_L = 100$ mA.

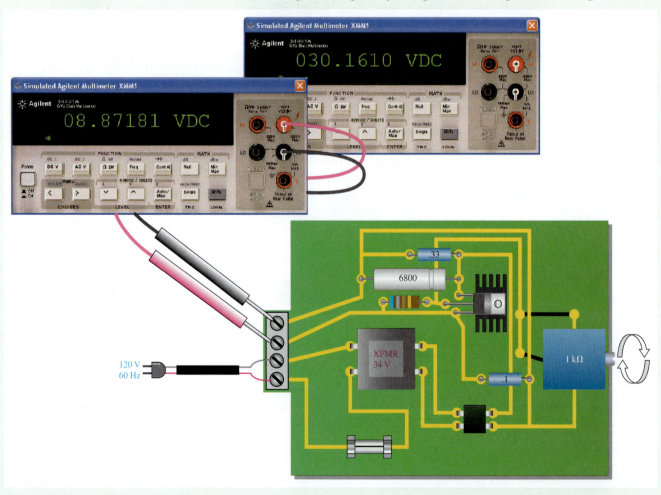

▲ **FIGURE 17–43**

Regulated power supply PC board adjusted for output voltages that meet the minimum and maximum specifications.

SUMMARY

Section 17–1 ◆ Voltage regulators keep an essentially constant dc output voltage when the input or load varies within limits.

◆ Line regulation is the percentage change in the output voltage for a given change in the input voltage of a regulator.

◆ Load regulation is the percentage change in output voltage for a given change in load current.

Section 17–2 ◆ A basic voltage regulator consists of a reference voltage source, an error detector, a sampling element, and a control device. Protection circuitry is also found in most regulators.

◆ Two basic categories of voltage regulators are linear and switching.

◆ Two basic types of linear regulators are series and shunt.

◆ In a linear series regulator, the control element is a transistor in series with the load.

Section 17–3 ◆ In a linear shunt regulator, the control element is a transistor in parallel with the load.

Section 17–4 ◆ Three configurations for switching regulators are step-down, step-up, and inverting.

◆ Switching regulators are more efficient than linear regulators and are particularly useful in low-voltage, high-current applications.

Section 17–5 ◆ Three-terminal linear IC regulators are available for either fixed output or variable output voltages of positive or negative polarities.

◆ The 78XX series are three-terminal IC regulators with fixed positive output voltage.

◆ The 79XX series are three-terminal IC regulators with fixed negative output voltage.

◆ The LM317 is a three-terminal IC regulator with a positive variable output voltage.

◆ The LM337 is a three-terminal IC regulator with a negative variable output voltage.

Section 17–6 ◆ An external pass transistor increases the current capability of a regulator.

KEY TERMS

Key terms and other bold terms in the chapter are defined in the end-of-book glossary.

Linear regulator   A voltage regulator in which the control element operates in the linear region.

Line regulation   The percentage change in output voltage for a given change in input (line) voltage.

Load regulation   The percentage change in output voltage for a given change in load current from no load to full load.

Regulator   An electronic circuit that maintains an essentially constant output voltage with a changing input voltage or load current.

Switching regulator   A voltage regulator in which the control element operates as a switch.

Thermal overload   A condition in a rectifier where the internal power dissipation of the circuit exceeds a certain maximum due to excessive current.

KEY FORMULAS

Voltage Regulation

17–1   **Line regulation** $= \left(\dfrac{\Delta V_{OUT}}{\Delta V_{IN}} \right) 100\%$ Line regulation as a percentage

17–2   **Line regulation** $= \dfrac{(\Delta V_{OUT}/V_{OUT})100\%}{\Delta V_{IN}}$ Line regulation in %/V

17–3   **Load regulation** $= \left(\dfrac{V_{NL} - V_{FL}}{V_{FL}} \right) 100\%$ Percent load regulation

17–4   **Load regulation** $= \left(\dfrac{R_{OUT}}{R_{FL}} \right) 100\%$ Load regulation in terms of output resistance and full-load resistance

Basic Series Regulator

17–5 $\quad V_{OUT} \cong \left(1 + \dfrac{R_2}{R_3}\right) V_{REF}$ $\qquad$ Regulator output

17–6 $\quad I_{L(max)} = \dfrac{0.7\ V}{R_4}$ $\qquad$ For constant-current limiting (silicon)

Basic Shunt Regulator

17–7 $\quad I_{L(max)} = \dfrac{V_{IN}}{R_1}$ $\qquad$ Maximum load current

Basic Switching Regulators

17–8 $\quad V_{OUT} = \left(\dfrac{t_{on}}{T}\right) V_{IN}$ $\qquad$ For step-down switching regulator

Integrated Circuit Voltage Regulators

17–9 $\quad V_{OUT} = V_{REF}\left(1 + \dfrac{R_2}{R_1}\right) + I_{ADJ}\ R_2$ $\qquad$ IC regulator

17–10 $\quad R_{ext} = \dfrac{0.7\ V}{I_{max}}$ $\qquad$ For external pass circuit

17–11 $\quad I_L = \dfrac{V_{OUT}}{R_1} + I_G$ $\qquad$ Regulator as a current source

TRUE/FALSE QUIZ

Answers can be found at www.pearsonhighered.com/floyd.

1. Line regulation is a measure of how constant the output voltage is for a given change in the input voltage.
2. Load regulation depends on the amount of power dissipated in the load.
3. Linear and switching are two main categories of voltage regulators.
4. Two types of linear regulator are series and bypass.
5. Three types of switching regulator are step-down, step-up, and inverting.
6. The three terminals of a 78XX series regulator are input, output, and control.
7. An external bypass transistor is sometimes used to increase the current capability of a regulator.
8. Current limiting is used to protect the external bypass transistor.
9. The purpose of a heat sink is to help the regulator dissipate excessive heat.
10. A variable pulse-width modulator is part of a linear voltage regulator.

CIRCUIT-ACTION QUIZ

Answers can be found at www.pearsonhighered.com/floyd.

1. If the input voltage in Figure 17–7 is increased by 1 V, the output voltage will
   (a) increase (b) decrease (c) not change

2. If the zener diode in Figure 17–7 is changed to one with a zener voltage of 6.8 V, the output voltage will
   (a) increase (b) decrease (c) not change

3. If R_3 in Figure 17–7 is increased in value, the output voltage will
   (a) increase (b) decrease (c) not change

4. If R_4 in Figure 17–9 is reduced, the amount of current that the regulator can supply to the load will
   (a) increase (b) decrease (c) not change

5. If R_2 in Figure 17–15 is increased, the power dissipation in R_1 will

   (a) increase   (b) decrease   (b) not change

6. If the duty cycle of the variable pulse-width modulator in Figure 17–17 is increased, the output voltage will

   (a) increase   (b) decrease   (c) not change

7. If R_2 in Figure 17–30 is adjusted to a lower value, the output voltage will

   (a) increase   (b) decrease   (c) not change

8. To increase the maximum current that the regulator in Figure 17–35 can supply, the value of R_{ext} must

   (a) increase   (b) decrease   (c) not change

SELF-TEST   Answers can be found at www.pearsonhighered.com/floyd.

Section 17–1   1. In the case of line regulation,

   (a) when the temperature varies, the output voltage stays constant

   (b) when the output voltage changes, the load current stays constant

   (c) when the input voltage changes, the output voltage stays constant

   (d) when the load changes, the output voltage stays constant

2. In the case of load regulation,

   (a) when the temperature varies, the output voltage stays constant

   (b) when the input voltage changes, the load current stays constant

   (c) when the load changes, the load current stays constant

   (d) when the load changes, the output voltage stays constant

3. All of the following are parts of a basic voltage regulator *except*

   (a) control element   (b) sampling circuit

   (c) voltage-follower   (d) error detector   (e) reference voltage

Section 17–2   4. The basic difference between a series regulator and a shunt regulator is

   (a) the amount of current that can be handled   (b) the position of the control element

   (c) the type of sample circuit   (d) the type of error detector

5. In a basic series regulator, V_{OUT} is determined by

   (a) the control element   (b) the sample circuit

   (c) the reference voltage   (d) answers (b) and (c)

6. The main purpose of current limiting in a regulator is

   (a) protection of the regulator from excessive current

   (b) protection of the load from excessive current

   (c) to keep the power supply transformer from burning up

   (d) to maintain a constant output voltage

7. In a linear regulator, the control transistor is conducting

   (a) a small part of the time   (b) half the time

   (c) all of the time   (d) only when the load current is excessive

Section 17–3   8. In a basic shunt regulator, V_{OUT} is determined by

   (a) the control element   (b) the sample circuit

   (c) the reference voltage   (d) answers (b) and (c)

Section 17–4   9. In a switching regulator, the control transistor is conducting

   (a) part of the time   (b) all of the time

   (c) only when the input voltage exceeds a set limit   (d) only when there is an overload

Section 17–5   **10.** The LM317 is an example of an IC

      (a) three-terminal negative voltage regulator   **(b)** fixed positive voltage regulator

      (c) switching regulator   **(d)** linear regulator

      (e) variable positive voltage regulator   **(f)** answers (b) and (d) only

      (g) answers (d) and (e) only

Section 17–6   **11.** An external pass transistor is used for

      (a) increasing the output voltage   **(b)** improving the regulation

      (c) increasing the current that the regulator can handle   **(d)** short-circuit protection

---

PROBLEMS

Answers to all odd-numbered problems are at the end of the book.

BASIC PROBLEMS

Section 17–1   **Voltage Regulation**

    1. The nominal output voltage of a certain regulator is 8 V. The output changes 2 mV when the input voltage goes from 12 V to 18 V. Determine the line regulation and express it as a percentage change over the entire range of V_{IN}.

    2. Express the line regulation found in Problem 1 in units of %/V.

    3. A certain regulator has a no-load output voltage of 10 V and a full-load output voltage of 9.90 V. What is the percent load regulation?

    4. In Problem 3, if the full-load current is 250 mA, express the load regulation in %/mA.

Section 17–2   **Basic Linear Series Regulators**

    5. Label the functional blocks for the voltage regulator in Figure 17–44.

▶ **FIGURE 17–44**

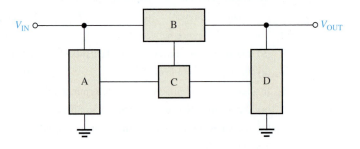

    6. Determine the output voltage for the regulator in Figure 17–45.

▶ **FIGURE 17–45**

Multisim and LT Spice file circuits are identified with a logo and are in the Problems folder on the website. Filenames correspond to figure numbers (e.g., FGM17-45 or FGS17-45).

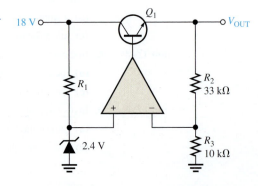

7. Determine the output voltage for the series regulator in Figure 17–46.

◆ FIGURE 17–46

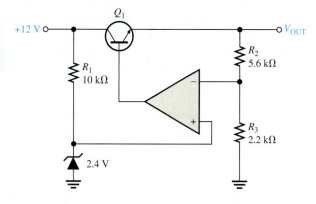

8. If R_3 in Figure 17–46 is increased to 4.7 kΩ, what happens to the output voltage?

9. If the zener voltage is 2.7 V instead of 2.4 V in Figure 17–46, what is the output voltage?

10. A series voltage regulator with constant-current limiting is shown in Figure 17–47. Determine the value of R_4 if the load current is to be limited to a maximum value of 250 mA. What power rating must R_4 have?

11. If the R_4 determined in Problem 10 is halved, what is the maximum load current?

◆ FIGURE 17–47

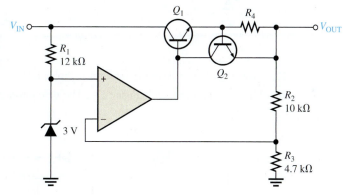

Section 17–3   Basic Linear Shunt Regulators

12. In the shunt regulator of Figure 17–48, when the load current increases, does Q_1 conduct more or less? Why?

◆ FIGURE 17–48

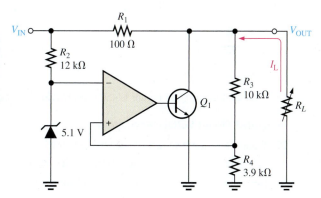

13. Assume I_L remains constant and V_{IN} changes by 1 V in Figure 17–48. What is the change in the collector current of Q_1?

14. With a constant input voltage of 17 V, the load resistance in Figure 17–48 is varied from 1 kΩ to 1.2 kΩ. Neglecting any change in output voltage, how much does the shunt current through Q_1 change?

15. If the maximum allowable input voltage in Figure 17–48 is 25 V, what is the maximum possible output current when the output is short-circuited? What power rating should R_1 have?

Section 17–4   Basic Switching Regulators

16. A basic switching regulator is shown in Figure 17–49. If the switching frequency of the transistor is 10 kHz with an off-time of 60 μs, what is the output voltage?

► FIGURE 17–49

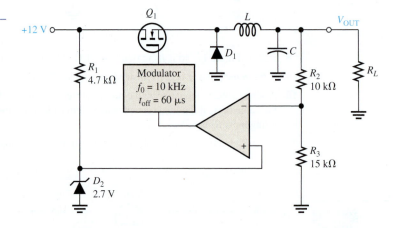

17. What is the duty cycle of the transistor in Problem 16?

18. When does the diode D_1 in Figure 17–50 become forward-biased?

► FIGURE 17–50

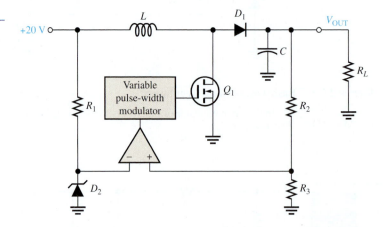

19. If the on-time of Q_1 in Figure 17–50 is decreased, does the output voltage increase or decrease?

Section 17–5   Integrated Circuit Voltage Regulators

20. What is the output voltage of each of the following IC regulators?

    (a) 7806   **(b)** 7905   **(c)** 7818   **(d)** 7924

21. Determine the output voltage of the regulator in Figure 17–51. $I_{ADJ} = 50$ μA.

◄ **FIGURE 17–51**

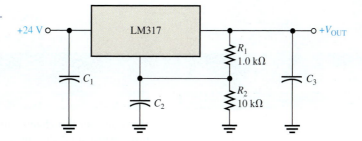

22. Determine the minimum and maximum output voltages for the circuit in Figure 17–52. $I_{ADJ} = 50\ \mu A$.

◄ **FIGURE 17–52**

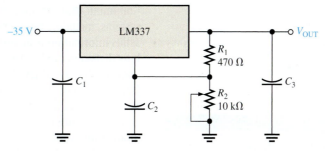

23. With no load connected, how much current is there through the regulator in Figure 17–51? Neglect the adjustment terminal current.

24. Select the values for the external resistors to be used in an LM317 circuit that is required to produce an output voltage of 12 V with an input of 18 V. The maximum regulator current with no load is to be 2 mA. There is no external pass transistor.

Section 17–6 **Integrated Circuit Voltage Regulator Configurations**

25. In the regulator circuit of Figure 17–53, determine R_{ext} if the maximum internal regulator current is to be 250 mA.

◄ **FIGURE 17–53**

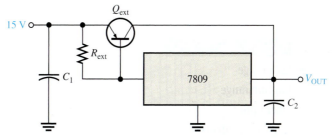

26. Using a 7812 voltage regulator and a 10 Ω load in Figure 17–53, how much power will the external pass transistor have to dissipate? The maximum internal regulator current is set at 500 mA by R_{ext}.

27. Show how to include current limiting in the circuit of Figure 17–53. What should the value of the limiting resistor be if the external current is to be limited to 2 A?

28. Using an LM317, design a circuit that will provide a constant current of 500 mA to a load.

29. Repeat Problem 28 using a 7908.

30. How do you set up the ADP1612/1613 switching regulator for a switching frequency of 1.3 MHz?

MULTISIM TROUBLESHOOTING PROBLEMS

These file circuits are in the Troubleshooting Problems folder on the website.

31. Open file TPM17-31 and determine the fault.

32. Open file TPM17-32 and determine the fault.

33. Open file TPM17-33 and determine the fault.

34. Open file TPM17-34 and determine the fault.

The Superheterodyne AM Receiver

A block diagram of a superheterodyne AM receiver is shown in Figure 18–2. The receiver shown consists of an antenna, an RF (radio frequency) amplifier, a mixer, a local oscillator (LO), an IF (intermediate frequency) amplifier, a detector, an audio amplifier, a power amplifier, and a speaker.

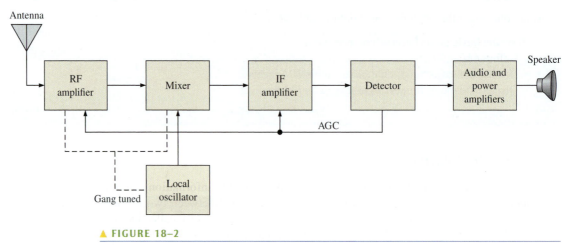

▲ FIGURE 18–2

Superheterodyne AM receiver block diagram.

Antenna   The antenna picks up all radiated signals and feeds them into the RF amplifier. These signals are very small (usually only a few microvolts).

RF Amplifier   This circuit can be adjusted (tuned) to select and amplify any carrier frequency within the AM broadcast band. Only the selected frequency and its two side bands pass through the amplifier. (Some AM receivers do not have a separate RF amplifier stage.)

Local Oscillator   This circuit generates a steady sine wave at a frequency 455 kHz above the selected RF frequency.

Mixer   This circuit accepts two inputs, the amplitude modulated RF signal from the output of the RF amplifier (or the antenna when there is no RF amplifier) and the sinusoidal output of the local oscillator (LO). These two signals are then "mixed" by a nonlinear process called *heterodyning* to produce sum and difference frequencies. For example, if the RF carrier has a frequency of 1000 kHz, the LO frequency is 1455 kHz and the sum and difference frequencies out of the mixer are 2455 kHz and 455 kHz, respectively. The difference frequency is always 455 kHz no matter what the RF carrier frequency.

IF Amplifier   The input to the IF amplifier is the 455 kHz AM signal, a replica of the original AM carrier signal except that the frequency has been lowered to 455 kHz. The IF amplifier significantly increases the level of this signal. The advantage of the IF stage is that it can be designed for a single frequency, simplifying the receiver.

Detector   This circuit recovers the modulating signal (audio signal) from the 455 kHz intermediate frequency (IF). At this point the IF is no longer needed, so the output of the detector consists of only the audio signal.

Audio and Power Amplifiers   This circuit amplifies the detected audio signal and drives the speaker to produce sound.

AGC   The automatic gain control (AGC) provides a dc level out of the detector that is proportional to the strength of the received signal. This level is fed back to the IF amplifier, and sometimes to the mixer and RF amplifier, to adjust the gains so as to maintain constant signal levels throughout the system over a wide range of incoming carrier signal strengths.

Figure 18–3 shows the signal flow through an AM superheterodyne receiver. The receiver can be tuned to accept any frequency in the AM band. The RF amplifier, mixer, and local oscillator are tuned simultaneously so that the LO frequency is always 455 kHz above the incoming RF signal frequency. This is called *gang tuning*.

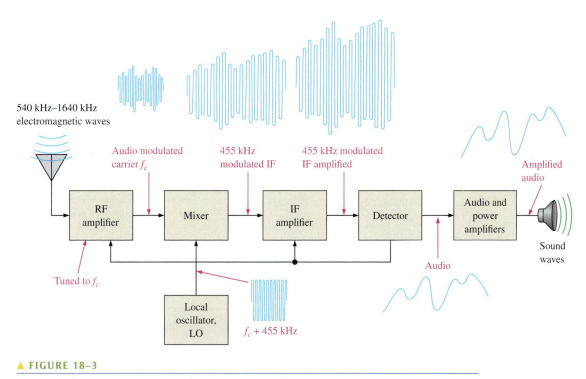

▲ **FIGURE 18–3**

Illustration of signal flow through an AM receiver.

Frequency Modulation

In **frequency modulation (FM)**, the modulating signal (audio) varies the frequency of a carrier as opposed to the amplitude, as in the case of AM. Figure 18–4 illustrates basic frequency modulation. The standard FM broadcast band consists of carrier frequencies from 88 MHz to 108 MHz, which is significantly higher than AM.

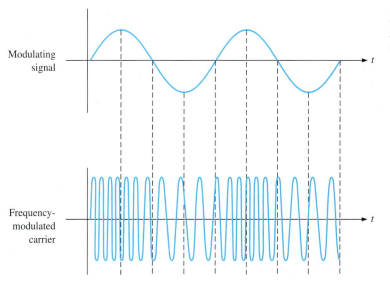

◀ **FIGURE 18–4**

An example of frequency modulation.

The Superheterodyne FM Receiver

The FM receiver is similar to the AM receiver in many ways, but there are several differences. A block diagram of a superheterodyne FM receiver is shown in Figure 18–5. Notice that it includes an RF amplifier, mixer, local oscillator, and IF amplifier just as in the AM receiver. These circuits operate at higher frequencies than in a commercial AM system. A significant difference in FM is the way the audio signal must be recovered from the modulated IF. This is accomplished by the limiter, discriminator, and de-emphasis network. Figure 18–6 depicts the signal flow through an FM receiver.

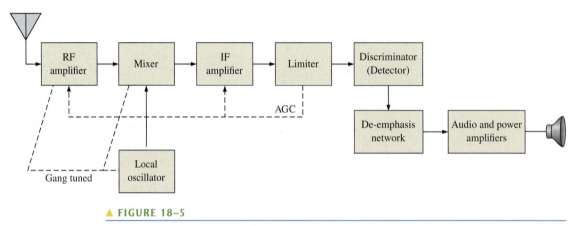

▲ FIGURE 18–5

Superheterodyne FM receiver block diagram.

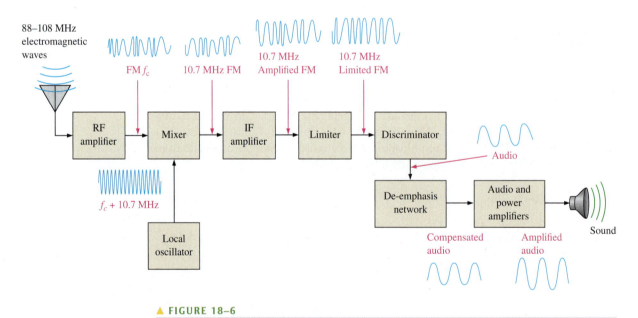

▲ FIGURE 18–6

Example of signal flow through an FM receiver.

RF Amplifier   This circuit must be capable of amplifying any frequency between 88 MHz and 108 MHz. It is highly selective so that it passes only the selected carrier frequency and significant side-band frequencies that contain the audio.

Local Oscillator   This circuit produces a sine wave at a frequency 10.7 MHz above the selected RF frequency.

Mixer   This circuit performs the same function as in the AM receiver, except that its output is a 10.7 MHz FM signal regardless of the RF carrier frequency.

IF Amplifier   This circuit amplifies the 10.7 MHz FM signal.

Limiter   The limiter removes any unwanted variations in the amplitude of the FM signal as it comes out of the IF amplifier and produces a constant amplitude FM output at the 10.7 MHz intermediate frequency.

Discriminator   This circuit performs the equivalent function of the detector in an AM system and is sometimes called a detector rather than a discriminator. The discriminator recovers the audio from the FM signal.

De-emphasis Network   For certain reasons, the higher modulating frequencies are amplified more than the lower frequencies at the transmitting end of an FM system by a process called *preemphasis*. The de-emphasis circuit in the FM receiver brings the high-frequency audio signals back to the proper amplitude relationship with the lower frequencies.

Audio and Power Amplifiers   This circuit is the same as in the AM system and can be shared when there is a dual AM/FM configuration.

SECTION 18–1 CHECKUP Answers are found at the end of the chapter.	1. What do *AM* and *FM* mean? 2. How do *AM* and *FM* differ? 3. What are the standard broadcast frequency bands for AM and FM?

18–2   THE LINEAR MULTIPLIER

The linear multiplier is used in many types of communications systems for diverse applications such as multiplication, division, modulation, demodulation, and voltage-controlled filters and amplifiers. In this section, you will examine the basic principles of linear multipliers and look at a few multiplier configurations that are found in communications as well as in other areas.

After completing this section, you should be able to

❏ **Discuss the function of a linear multiplier**
   ◆ Describe multiplier quadrants and transfer characteristic   ◆ Discuss scale factor
   ◆ Show how to use a multiplier circuit as a multiplier, squaring circuit, divide circuit, square root circuit, and mean square circuit

Multiplier Quadrants

There are one-quadrant, two-quadrant, and four-quadrant multipliers. The quadrant classification indicates the number of input polarity combinations that the multiplier can handle. A graphical representation of the quadrants is shown in Figure 18–7. A **four-quadrant multiplier** can accept any of the four possible input polarity combinations and produce an output with the corresponding polarity.

▶ FIGURE 18–7

Four-quadrant polarities and their products.

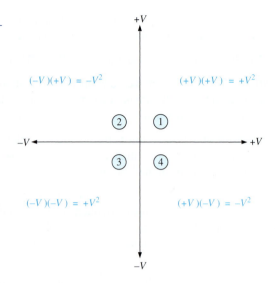

The Multiplier Transfer Characteristic

Figure 18–8 shows the transfer characteristic for a typical linear multiplier of two input voltages V_X and V_Y. Values of V_X run along the horizontal axis and values of V_Y are the sloped lines. To find the output voltage from the transfer characteristic graph, find the intersection of the two input voltages V_X and V_Y. Then find the output voltage by projecting the point of intersection over to the vertical axis. An example will illustrate this.

▶ FIGURE 18–8

A four-quadrant multiplier transfer characteristic.

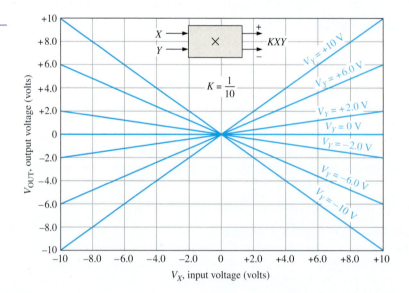

EXAMPLE 18–1

Determine the output voltage for a four-quadrant linear multiplier whose transfer characteristic is given in Figure 18–8. The input voltages are $V_X = -4$ V and $V_Y = +10$ V.

Solution   The output voltage is **−4 V** as illustrated in Figure 18–9. For this transfer characteristic, the output voltage is a factor of ten smaller than the actual product of the two input voltages. This is due to the scale factor of the multiplier, which is discussed next.

▶ FIGURE 18–9

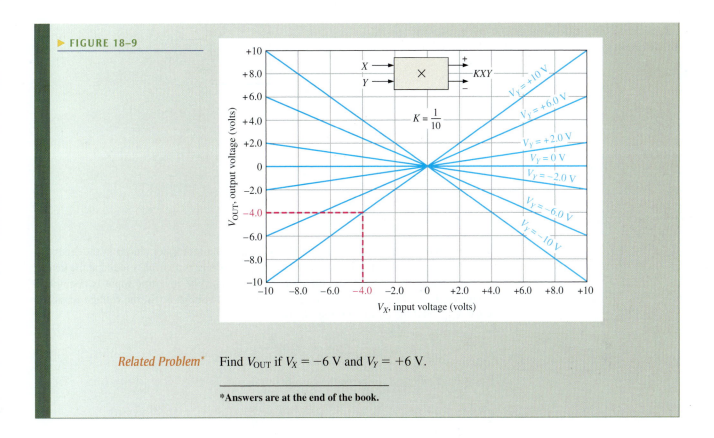

*Related Problem**   Find V_{OUT} if $V_X = -6$ V and $V_Y = +6$ V.

*Answers are at the end of the book.

The Scale Factor, K

The scale factor, K, is basically an internal attenuation that reduces the output by a fixed amount. The scale factor has a typical value of 0.1.

The expression for the output voltage of the linear multiplier includes the scale factor, K, as indicated in Equation 18–1. The symbol is shown in Figure 18–10.

$$V_{OUT} = KV_XV_Y$$

Equation 18–1

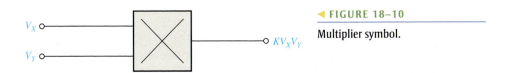

◀ FIGURE 18–10

Multiplier symbol.

Other Multiplier Configurations

Squaring Circuit   A special case of the multiplier is a squaring circuit that is realized by simply applying the same voltage to both inputs by connecting the inputs together as shown in Figure 18–11.

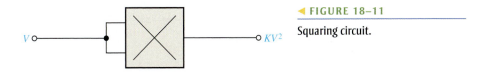

◀ FIGURE 18–11

Squaring circuit.

Divide circuit.

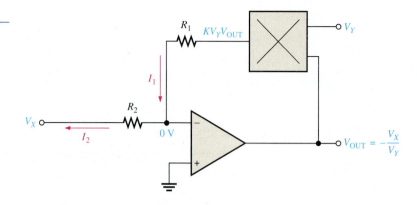

Divide Circuit The circuit in Figure 18–12 shows the multiplier placed in the feedback loop of an op-amp. The basic operation is as follows. There is a virtual ground at the inverting (–) input of the op-amp and therefore the current at the inverting input is negligible. Therefore, I_1 and I_2 are equal. Since the inverting input voltage is 0 V, the voltage across R_1 is KV_YV_{OUT} and the current through R_1 is

$$I_1 = \frac{KV_YV_{OUT}}{R_1}$$

The voltage across R_2 is V_X, so the current through R_2 is

$$I_2 = \frac{V_X}{R_2}$$

Since $I_1 = -I_2$,

$$\frac{KV_YV_{OUT}}{R_1} = -\frac{V_X}{R_2}$$

Solving for V_{OUT},

$$V_{OUT} = -\frac{V_XR_1}{KV_YR_2}$$

If $R_1 = KR_2$,

$$V_{OUT} = -\frac{V_X}{V_Y}$$

Square Root Circuit The square root circuit is a special case of the divide circuit where V_{OUT} is applied to both inputs of the multiplier as shown in Figure 18–13.

Square root circuit.

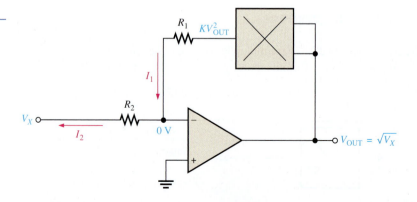

Mean Square Circuit   In this application, the multiplier is used as a squaring circuit with its output connected to an op-amp integrator as shown in Figure 18–14. The integrator produces the average or mean value of the squared input over time, as indicated by the integration sign ($\int$).

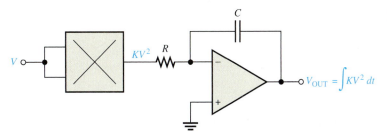

◀ FIGURE 18–14

Mean square circuit.

SECTION 18–2 CHECKUP	1. Compare a four-quadrant multiplier to a one-quadrant multiplier in terms of the inputs that can be handled.
	2. How do you convert a basic multiplier to a squaring circuit?

18–3   AMPLITUDE MODULATION

Amplitude modulation (AM) is an important method for transmitting information. Of course, the AM superheterodyne receiver is designed to receive transmitted AM signals. In this section, we take a further look at amplitude modulation and show how the linear multiplier can be used as an amplitude-modulated device.

After completing this section, you should be able to

❑ **Discuss the fundamentals of amplitude modulation**
   ◆ Explain how AM is basically a multiplication process   ◆ Describe sum and difference frequencies   ◆ Discuss balanced modulation   ◆ Describe the frequency spectra   ◆ Explain standard AM

As you learned in Section 18–1, amplitude modulation is the process of varying the amplitude of a signal of a given frequency (carrier) with another signal of much lower frequency (modulating signal). One reason that the higher-frequency carrier signal is necessary is because audio or other signals with relatively low frequencies cannot be transmitted with antennas of a practical size. The basic concept of standard amplitude modulation is illustrated in Figure 18–15.

A Multiplication Process

If a signal is applied to the input of a variable-gain device, the resulting output is an amplitude-modulated signal because $V_{out} = A_v V_{in}$. The output voltage is the input voltage

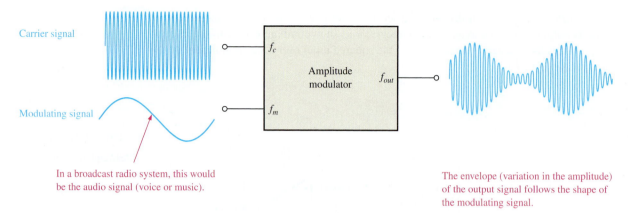

Carrier signal

Modulating signal

In a broadcast radio system, this would be the audio signal (voice or music).

The envelope (variation in the amplitude) of the output signal follows the shape of the modulating signal.

▲ FIGURE 18–15

Basic concept of amplitude modulation.

multiplied by the voltage gain. For example, if the gain of an amplifier is made to vary sinusoidally at a certain frequency and an input signal is applied at a higher frequency, the output signal will have the higher frequency. However, its amplitude will vary according to the variation in gain as illustrated in Figure 18–16. Amplitude modulation is basically a multiplication process (input voltage multiplied by a variable gain).

▶ FIGURE 18–16

The amplitude of the output voltage varies according to the gain and is the product of voltage gain and input voltage.

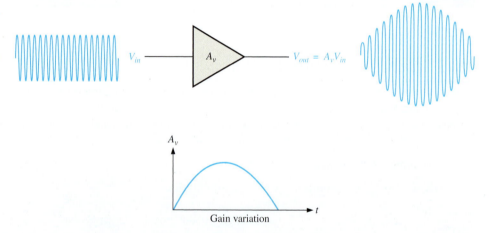

Gain variation

Sum and Difference Frequencies

If the expressions for two sinusoidal signals of different frequencies are multiplied mathematically, a term containing both the difference and the sum of the two frequencies is produced. Recall from ac circuit theory that a sinusoidal voltage can be expressed as

$$v = V_p \sin 2\pi f t$$

where V_p is the peak voltage and f is the frequency. Two different sinusoidal signals can be expressed as follows:

$$v_1 = V_{1(p)} \sin 2\pi f_1 t$$
$$v_2 = V_{2(p)} \sin 2\pi f_2 t$$

Multiplying these two sinusoidal wave terms,

$$v_1 v_2 = (V_{1(p)} \sin 2\pi f_1 t)(V_{2(p)} \sin 2\pi f_2 t) = V_{1(p)} V_{2(p)} (\sin 2\pi f_1 t)(\sin 2\pi f_2 t)$$

The general trigonometric identity for the product of two sinusoidal functions is

$$(\sin A)(\sin B) = \frac{1}{2}[\cos(A - B) - \cos(A + B)]$$

Applying this identity to the previous formula for $v_1 v_2$,

$$v_1 v_2 = \frac{V_{1(p)} V_{2(p)}}{2}[\cos(2\pi f_1 t - 2\pi f_2 t) - \cos(2\pi f_1 t + 2\pi f_2 t)]$$

$$= \frac{V_{1(p)} V_{2(p)}}{2}[\cos 2\pi(f_1 - f_2)t - \cos 2\pi(f_1 + f_2)t]$$

$$v_1 v_2 = \frac{V_{1(p)} V_{2(p)}}{2}\cos 2\pi(f_1 - f_2)t - \frac{V_{1(p)} V_{2(p)}}{2}\cos 2\pi(f_1 + f_2)t$$

Equation 18–2

You can see in Equation 18–2 that the product of the two sinusoidal voltages V_1 and V_2 contains a difference frequency $(f_1 - f_2)$ and a sum frequency $(f_1 + f_2)$. The fact that the product terms are cosine simply indicates a 90° phase shift in the multiplication process.

Analysis of Balanced Modulation

Since amplitude modulation is simply a multiplication process, the preceding analysis is now applied to carrier and modulating signals. The expression for the sinusoidal carrier signal can be written as

$$v_c = V_{c(p)}\sin 2\pi f_c t$$

Assuming a sinusoidal modulating signal, it can be expressed as

$$v_m = V_{m(p)}\sin 2\pi f_m t$$

Substituting these two signals in Equation 18–2,

$$v_c v_m = \frac{V_{c(p)} V_{m(p)}}{2}\cos 2\pi(f_c - f_m)t - \frac{V_{c(p)} V_{m(p)}}{2}\cos 2\pi(f_c + f_m)t$$

An output signal described by this expression for the product of two sinusoidal signals is produced by a linear multiplier. Notice that there is a difference frequency term $(f_c - f_m)$ and a sum frequency term $(f_c + f_m)$, but the original frequencies, f_c and f_m, do not appear alone in the expression. Thus, the product of two sinusoidal signals contains no signal with the carrier frequency, f_c, or with the modulating frequency, f_m. This form of amplitude modulation is called **balanced modulation** because there is no carrier frequency in the output. The carrier frequency is "balanced out."

The Frequency Spectra of a Balanced Modulator

A graphical picture of the frequency content of a signal is called its frequency spectrum. A frequency spectrum shows voltage on a frequency base rather than on a time base as a waveform diagram does. The frequency spectra of the product of two sinusoidal signals are shown in Figure 18–17. Part (a) shows the two input frequencies and part (b) shows the output frequencies. In communications terminology, the sum frequency is called the *upper-side frequency* and the difference frequency is called the *lower-side frequency* because the frequencies appear on each side of the missing carrier frequency.

The Linear Multiplier as a Balanced Modulator

As mentioned, the linear multiplier acts as a balanced modulator when a carrier signal and a modulating signal are applied to its inputs, as illustrated in Figure 18–18. A balanced modulator produces an upper-side frequency and a lower-side frequency, but it does not

▶ FIGURE 18–17

Illustration of the input and output frequency spectra for a linear multiplier.

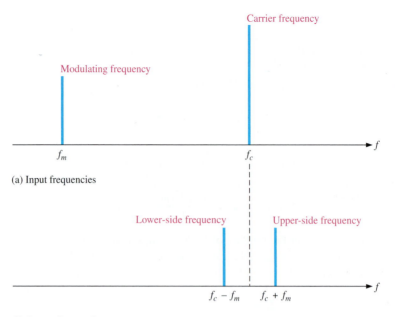

(a) Input frequencies

(b) Output frequencies

▶ FIGURE 18–18

The linear multiplier as a balanced modulator.

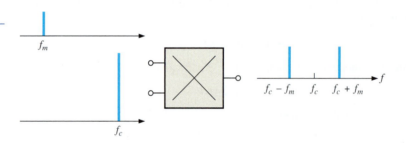

produce a carrier frequency. Since there is no carrier signal, balanced modulation is sometimes known as *suppressed-carrier modulation*. Balanced modulation is used in certain types of communications systems. For example, in single sideband communication systems, the carrier is suppressed as well as one of the sidebands, which results in a smaller bandwidth and higher efficiency; this is an advantage for many voice communication systems. No information is lost by transmitting using single sideband, but the recovery of the signal requires a more complex receiver, and the receiver must be tuned to exactly the same frequency for full fidelity. For this reason, it is not used in standard AM broadcast systems.

EXAMPLE 18–2

Determine the frequencies contained in the output signal of the balanced modulator in Figure 18–19.

▶ FIGURE 18–19

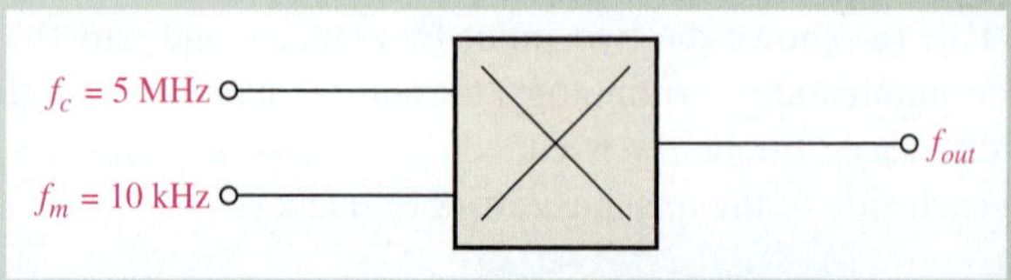

$f_c = 5$ MHz

$f_m = 10$ kHz

f_{out}

Solution   The upper-side frequency is

$$f_c + f_m = 5 \text{ MHz} + 10 \text{ kHz} = \textbf{5.01 MHz}$$

The lower-side frequency is

$$f_c - f_m = 5 \text{ MHz} - 10 \text{ kHz} = \mathbf{4.99 \, MHz}$$

Related Problem   Explain how the separation between the side frequencies can be increased using the same carrier frequency.

Standard Amplitude Modulation (AM)

In standard AM systems, the output signal contains the carrier frequency as well as the sum and difference side frequencies. The frequency spectrum in Figure 18–20 illustrates standard amplitude modulation.

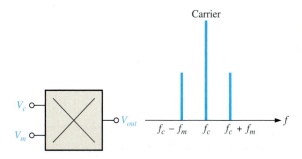

◀ **FIGURE 18–20**

The output frequency spectrum of a standard amplitude modulator.

The expression for a standard amplitude-modulated signal is

$$V_{out} = V_{c(p)}^2 \sin 2\pi f_c t + \frac{V_{c(p)} V_{m(p)}}{2} \cos 2\pi (f_c - f_m)t - \frac{V_{c(p)} V_{m(p)}}{2} \cos 2\pi (f_c + f_m)t$$

Equation 18–3

Notice in Equation 18–3 that the first term is for the carrier frequency and the other two terms are for the side frequencies. Let's see how the carrier-frequency term gets into the equation.

If a dc voltage equal to the peak of the carrier voltage is added to the modulating signal before the modulating signal is multiplied by the carrier signal, a carrier-signal term appears in the final result as shown in the following steps. Add the peak carrier voltage to the modulating signal, and you get the following expression:

$$V_{c(p)} + V_{m(p)} \sin 2\pi f_m t$$

Multiply by the carrier signal.

$$V_{out} = (V_{c(p)} \sin 2\pi f_c t)(V_{c(p)} + V_{m(p)} \sin 2\pi f_m t)$$

$$= \underbrace{V_{c(p)}^2 \sin 2\pi f_c t}_{\text{carrier term}} + \underbrace{V_{c(p)} V_{m(p)} (\sin 2\pi f_c t)(\sin 2\pi f_m t)}_{\text{product term}}$$

Apply the basic trigonometric identity to the product term.

$$V_{out} = V_{c(p)}^2 \sin 2\pi f_c t + \frac{V_{c(p)} V_{m(p)}}{2} \cos 2\pi (f_c - f_m)t - \frac{V_{c(p)} V_{m(p)}}{2} \cos 2\pi (f_c + f_m)t$$

This result shows that the output of the multiplier contains a carrier term and two side-frequency terms. Figure 18–21 illustrates how a standard amplitude modulator can be implemented by a summing circuit followed by a linear multiplier. Figure 18–22 shows a possible implementation of the summing circuit.

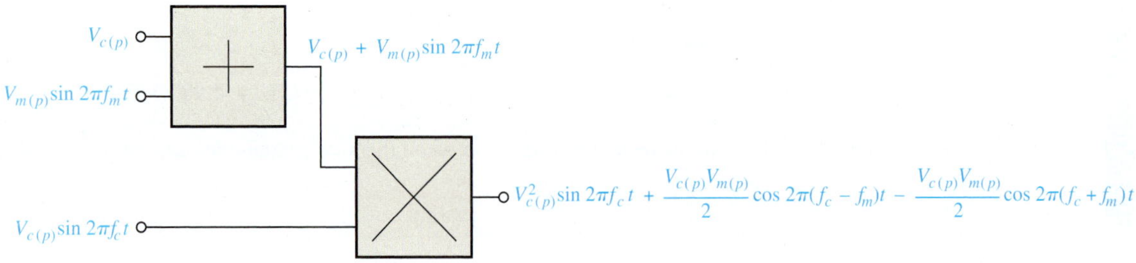

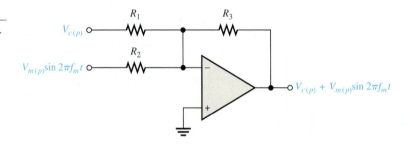

▲ FIGURE 18–21

Basic block diagram of an amplitude modulator.

▶ FIGURE 18–22

Implementation of the summing circuit in the amplitude modulator.

EXAMPLE 18–3

A carrier frequency of 1200 kHz is modulated by a sinusoidal wave with a frequency of 3 kHz by a standard amplitude modulator. Determine the output frequency spectrum.

Solution   The lower-side frequency is

$$f_c - f_m = 1200 \text{ kHz} - 3 \text{ kHz} = \textbf{1197 kHz}$$

The upper-side frequency is

$$f_c + f_m = 1200 \text{ kHz} + 3 \text{ kHz} = \textbf{1203 kHz}$$

The output contains the carrier frequency and the two side frequencies as shown in Figure 18–23.

▶ FIGURE 18–23

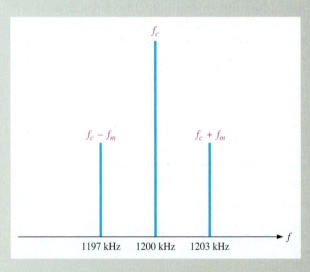

Related Problem   Compare the output frequency spectrum in this example to that of a balanced modulator having the same inputs.

Amplitude Modulation with Voice or Music

To this point in our discussion, we have considered the modulating signal to be a pure sinusoidal signal just to keep things fairly simple. If you receive an AM signal modulated by a pure sinusoidal signal in the audio frequency range, you will hear a single tone from the receiver's speaker.

An audio signal (voice or music) consists of many sinusoidal components within a range of frequencies from about 20 Hz to 20 kHz. For example, if a carrier frequency is amplitude modulated with voice or music with frequencies from 100 Hz to 10 kHz, the frequency spectrum is as shown in Figure 18–24. Instead of one lower-side and one upper-side frequency as in the case of a single-frequency modulating signal, a band of lower-side frequencies and a band of upper-side frequencies correspond to the sum and difference frequencies of each sinusoidal component of the voice or music signal.

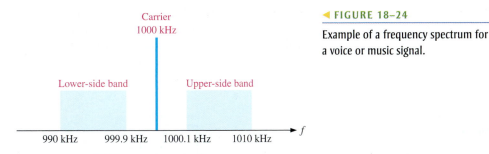

◀ FIGURE 18–24

Example of a frequency spectrum for a voice or music signal.

SECTION 18–3 CHECKUP	1. What is amplitude modulation?

1. What is amplitude modulation?
2. What is the difference between balanced modulation and standard AM?
3. What two input signals are used in amplitude modulation? Explain the purpose of each signal.
4. What are the upper-side frequency and the lower-side frequency?
5. How can a balanced modulator be changed to a standard amplitude modulator?

18–4 THE MIXER

The mixer in the receiver system discussed in Section 18–1 can be implemented with a linear multiplier as you will see in this section. The basic principles of linear multiplication of sinusoidal signals are covered, and you will see how sum and difference frequencies are produced. The difference frequency is a critical part of the operation of many types of receiver systems.

After completing this section, you should be able to

❑ **Discuss the basic function of a mixer**
   ◆ Explain why a mixer is a linear multiplier ◆ Describe the frequencies in the mixer and IF portion of a receiver

The **mixer** is basically a frequency converter because it changes the frequency of a signal to another value. The mixer in a receiver system takes the incoming modulated RF signal (which is sometimes amplified by an RF amplifier and sometimes not) along with the signal from the local oscillator and produces a modulated signal with a frequency equal to the difference of its two input frequencies (RF and LO). The mixer also produces a frequency equal to the sum of the input frequencies. The mixer function is illustrated in Figure 18–25.

▶ FIGURE 18–25

The mixer function.

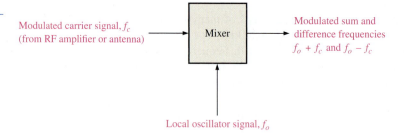

In the case of receiver applications, the mixer must produce an output that has a frequency component equal to the difference of its input frequencies. From the mathematical analysis in Section 18–3, you can see that if two sinusoidal signals are multiplied, the product contains the difference frequency and the sum frequency. Thus, the mixer is actually a linear multiplier as indicated in Figure 18–26.

▶ FIGURE 18–26

The mixer as a linear multiplier.

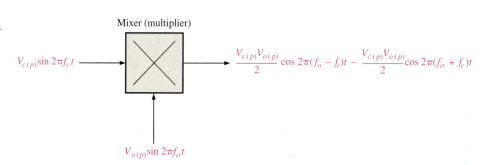

EXAMPLE 18–4

Determine the output expression for a multiplier with one sinusoidal input having a peak voltage of 5 mV and a frequency of 1200 kHz and the other input having a peak voltage of 10 mV and a frequency of 1655 kHz.

Solution The two input expressions are

$$v_1 = (5 \text{ mV})\sin 2\pi(1200 \text{ kHz})t$$
$$v_2 = (10 \text{ mV})\sin 2\pi(1655 \text{ kHz})t$$

Multiplying,

$$v_1 v_2 = (5 \text{ mV})(10 \text{ mV})[\sin 2\pi(1200 \text{ kHz})t][\sin 2\pi(1655 \text{ kHz})t]$$

Applying the trigonometric identity, $(\sin A)(\sin B) = \frac{1}{2}[\cos(A - B) - \cos(A + B)]$,

$$V_{out} = \frac{(5 \text{ mV})(10 \text{ mV})}{2}\cos 2\pi(1655 \text{ kHz} - 1200 \text{ kHz})t$$

$$- \frac{(5 \text{ mV})(10 \text{ mV})}{2}\cos 2\pi(1655 \text{ kHz} + 1200 \text{ kHz})t$$

$$V_{out} = (25 \text{ } \mu\text{V})\cos 2\pi(455 \text{ kHz})t - (25 \text{ } \mu\text{V})\cos 2\pi(2855 \text{ kHz})t$$

Related Problem What is the value of the peak amplitude and frequency of the difference frequency component in this example?

In the receiver system, both the sum and difference frequencies from the mixer are applied to the IF (intermediate frequency) amplifier. The IF amplifier is actually a tuned amplifier that is designed to respond to the difference frequency while rejecting the sum frequency. You can think of the IF amplifier section of a receiver as a band-pass filter plus an amplifier because it uses resonant circuits to provide the frequency selectivity. This is illustrated in Figure 18–27.

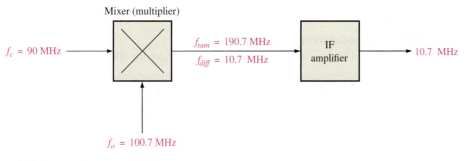

▲ FIGURE 18–27

Example of frequencies in the mixer and IF portion of a receiver.

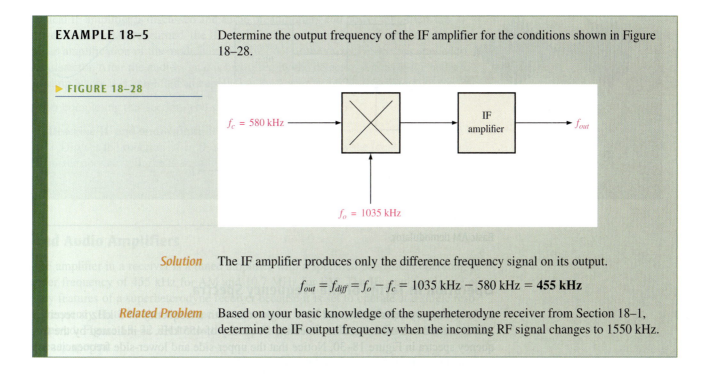

EXAMPLE 18–5   Determine the output frequency of the IF amplifier for the conditions shown in Figure 18–28.

▶ FIGURE 18–28

Solution   The IF amplifier produces only the difference frequency signal on its output.

$$f_{out} = f_{diff} = f_o - f_c = 1035 \text{ kHz} - 580 \text{ kHz} = \textbf{455 kHz}$$

Related Problem   Based on your basic knowledge of the superheterodyne receiver from Section 18–1, determine the IF output frequency when the incoming RF signal changes to 1550 kHz.

SECTION 18–4 CHECKUP

1. What is the purpose of the mixer in a superheterodyne receiver?
2. How does the mixer produce its output?
3. If a mixer has 1000 kHz on one input and 350 kHz on the other, what frequencies appear on the output?

These formulas show that frequency increases as capacitance decreases and vice versa.

Capacitance decreases as reverse voltage (control voltage) increases. Therefore, an increase in control voltage to the VCO causes an increase in frequency and vice versa. Basic VCO operation is illustrated in Figure 18–40. The graph in part (b) shows that at the nominal control voltage, $V_{c(nom)}$, the oscillator is running at its nominal or free-running frequency, $f_{o(nom)}$. An increase in V_c above the nominal value forces the oscillator frequency to increase, and a decrease in V_c below the nominal value forces the oscillator frequency to decrease. There are, of course, limits on the operation as indicated by the minimum and maximum points. The transfer function or conversion gain, K, of the VCO is normally expressed as a certain frequency deviation per unit change in control voltage.

$$K = \frac{\Delta f_o}{\Delta V_c}$$

▶ FIGURE 18–40

Basic VCO operation.

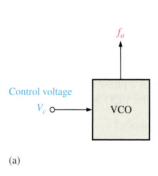

(a)

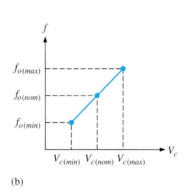

(b)

EXAMPLE 18–7

The output frequency of a certain VCO changes from 50 kHz to 65 kHz when the control voltage increases from 0.5 V to 1 V. What is the conversion gain, K?

Solution

$$K = \frac{\Delta f_o}{\Delta V_c} = \frac{65\ \text{kHz} - 50\ \text{kHz}}{1\ \text{V} - 0.5\ \text{V}} = \frac{15\ \text{kHz}}{0.5\ \text{V}} = \textbf{30 kHz/V}$$

Related Problem

If the conversion gain of a certain VCO is 20 kHz/V, how much frequency deviation does a change in control voltage from 0.8 V to 0.5 V produce? If the VCO frequency is 250 kHz at 0.8 V, what is the frequency at 0.5 V?

Basic PLL Operation

When the PLL is locked, the incoming frequency, f_i, and the VCO frequency, f_o, are equal. However, there is always a phase difference between them called the *static phase error*. The phase error, θ_e, is the parameter that keeps the PLL locked in. As you have seen, the filtered voltage from the phase detector is proportional to θ_e (Equation 18–4). This voltage controls the VCO frequency and is always just enough to keep $f_o = f_i$.

Figure 18–41 shows the PLL and two sinusoidal signals of the same frequency but with a phase difference, θ_e. For this condition the PLL is in lock and the VCO control voltage is constant. If f_i decreases, θ_e increases to θ_{e1} as illustrated in Figure 18–42. This increase in θ_e is sensed by the phase detector causing the VCO control voltage to decrease, thus decreasing f_o until $f_o = f_i$ and keeping the PLL in lock. If f_i increases, θ_e decreases to θ_{e1} as

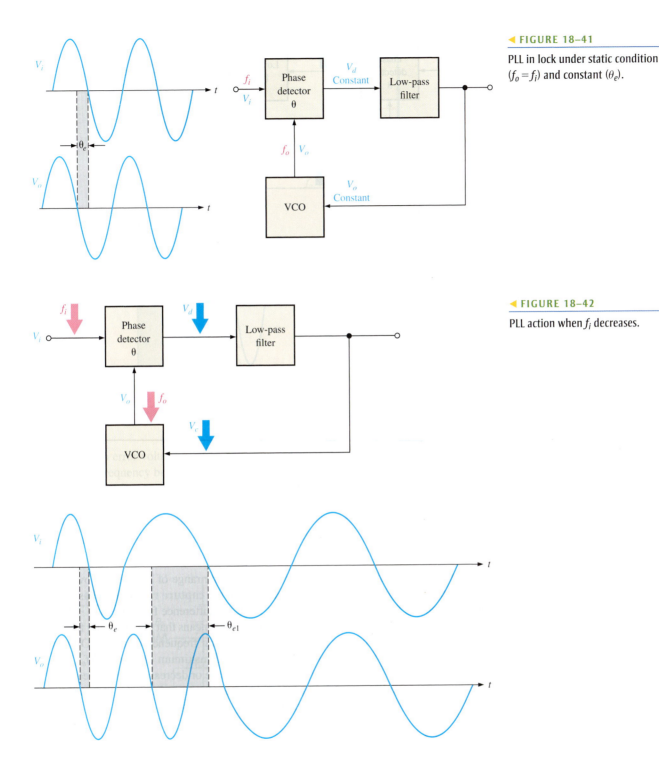

PLL in lock under static condition $(f_o = f_i)$ and constant (θ_e).

PLL action when f_i decreases.

illustrated in Figure 18–43. This decrease in θ_e causes the VCO control voltage to increase, thus increasing f_o until $f_o = f_i$ and keeping the PLL in lock.

Lock Range   Once the PLL is locked, it will track frequency changes in the incoming signal. The range of frequencies over which the PLL can acquire lock is called the **lock range** or tracking range. The hold-in range is the range of frequencies over which the PLL can remain locked once it has acquired lock. Limitations on the hold-in range are the maximum frequency deviations of the VCO and the output limits of the phase detector. The hold-in range is independent of the bandwidth of the low-pass filter because, when the PLL is in lock, the difference frequency $(f_i - f_o)$ is zero or a very low instantaneous value that falls

Section 18–8 ◆ A phase-locked loop (PLL) is an integrated circuit consisting of a phase detector, a low-pass filter, a VCO, and sometimes an amplifier.

◆ The purpose of a PLL is to lock onto and track incoming frequencies. There are a variety of applications for PLLs including as an FM demodulator.

◆ A linear multiplier can be used as a phase detector.

Section 18–9 ◆ Fiber optics provides a light path from a light-emitting device to a light-activated device.

◆ The three basic parts of a fiber-optic cable are the core, the cladding, and the jacket.

◆ Light rays must bounce off the core boundary at an angle (angle of incidence) greater than the critical angle in order to be reflected.

◆ Light rays that strike the core boundary at an angle less than the critical angle are refracted into the cladding, resulting in attenuation of the light.

◆ The angle of reflection is equal to the angle of incidence.

◆ Three types of fiber-optic cable are multimode step index, single-mode step index, and multimode graded index.

KEY TERMS

Amplitude modulation (AM)   A communication method in which a lower-frequency signal modulates (varies) the amplitude of a higher-frequency signal (carrier).

Angle of incidence   The angle at which a light ray strikes a surface.

Balanced modulation   A form of amplitude modulation in which the carrier is suppressed; sometimes known as *suppressed-carrier modulation.*

Capture range   The range of frequencies over which a PLL can acquire lock.

Critical angle   The angle that defines whether a light ray will be reflected or refracted as it strikes a surface.

Fiber optics   The use of light for the transmission of information through tiny fiber cables.

Four-quadrant multiplier   A linear device that produces an output voltage proportional to the product of two input voltages.

Frequency modulation (FM)   A communication method in which a lower-frequency intelligence-carrying signal modulates (varies) the frequency of a higher-frequency signal.

Index of refraction   An optical characteristic of a material that determines the critical angle.

Lock range   The range of frequencies over which a PLL can maintain lock.

Mixer   A device for down-converting frequencies in a receiver system.

Phase-locked loop (PLL)   A device for locking onto and tracking the frequency of an incoming signal.

KEY FORMULAS

18–1 $V_{\text{OUT}} = KV_XV_Y$ Multiplier output voltage

18–2 $v_1v_2 = \dfrac{V_{1(p)}V_{2(p)}}{2} \cos 2\pi(f_1 - f_2)t$ Sum and difference frequencies

$\qquad\quad -\dfrac{V_{1(p)}V_{2(p)}}{2} \cos 2\pi(f_1 + f_2)t$

18–3 $V_{out} = V_{c(p)}^2 \sin 2\pi f_c t$ Standard AM

$\qquad\quad +\dfrac{V_{c(p)}V_{m(p)}}{2} \cos 2\pi(f_c - f_m)t$

$\qquad\quad -\dfrac{V_{c(p)}V_{m(p)}}{2} \cos 2\pi(f_c + f_m)t$

$$18\text{--}4 \qquad V_c = \frac{V_i V_o}{2} \cos \theta_e \qquad\qquad\qquad \text{PLL control voltage}$$

$$18\text{--}5 \qquad \theta_c = \cos^{-1}\left(\frac{n_2}{n_1}\right) \qquad\qquad\qquad \text{Critical angle}$$

TRUE/FALSE QUIZ

Answers are at the end of the chapter.

1. In amplitude modulation, the amplitude of a higher-frequency carrier signal varies according to a lower-frequency modulating signal.
2. The AM broadcast band is from 1 MHz to 540 MHz.
3. In frequency modulation, the frequency of a higher-frequency carrier signal varies according to a lower-frequency modulating signal.
4. In an AM receiver, the detector recovers the modulating signal from the RF carrier signal.
5. In an FM receiver, the discriminator recovers the modulating signal from the IF signal.
6. The linear multiplier is a key circuit in many types of communications systems.
7. Amplitude modulation is a division process.
8. There is no carrier signal in balanced modulation.
9. IF stands for interference frequency.
10. PLL stands for phase-locked loop.
11. A PLL consists of a phase detector, a low-pass filter, and a VCO.
12. A VCO is a variable capacitance oscillator.

CIRCUIT-ACTION QUIZ

Answers are at the end of the chapter.

1. If R_2 in Figure 18–12 is increased in value, the output voltage will
   (a) increase   (b) decrease   (c) not change
2. If R_1 in Figure 18–12 is increased in value, the output voltage will
   (a) increase   (b) decrease   (c) not change
3. Refer to Figure 18–23. If the amplitude modulating frequency is increased, the lower-side frequency will
   (a) increase   (b) decrease   (c) not change
4. Refer to Figure 18–23. If the carrier frequency is decreased, the upper-side frequency will
   (a) increase   (b) decrease   (c) not change
5. In amplitude modulation, if the amplitude of the modulating signal increases, the carrier frequency will
   (a) increase   (b) decrease   (c) not change

SELF-TEST

Answers are at the end of the chapter.

Section 18–1

1. In amplitude modulation, the pattern produced by the peaks of the carrier signal is called the
   (a) index   (b) envelope
   (c) audio signal   (d) upper-side frequency
2. Which of the following is not a part of an AM superheterodyne receiver?
   (a) Mixer   (b) IF amplifier
   (c) DC restorer   (d) Detector
   (e) Audio amplifier   (f) Local oscillator
3. In an AM receiver, the local oscillator always produces a frequency that is above the incoming RF by
   (a) 10.7 kHz   (b) 455 MHz   (c) 10.7 MHz   (d) 455 kHz

4. An FM receiver has an intermediate frequency that is
   (a) in the 88 MHz to 108 MHz range
   (b) in the 540 kHz to 1640 kHz range
   (c) 455 kHz
   (d) greater than the IF in an AM receiver

5. The detector or discriminator in an AM or an FM receiver
   (a) detects the difference frequency from the mixer
   (b) changes the RF to IF
   (c) recovers the audio signal
   (d) maintains a constant IF amplitude

6. The IF in a receiver is the
   (a) sum of the local oscillator frequency and the RF carrier frequency
   (b) local oscillator frequency
   (c) difference of the local oscillator frequency and the RF carrier frequency
   (d) difference of the carrier frequency and the audio frequency

7. When a receiver is tuned from one RF frequency to another,
   (a) the IF changes by an amount equal to the LO (local oscillator) frequency
   (b) the IF stays the same
   (c) the LO frequency changes by an amount equal to the audio frequency
   (d) both the LO and the IF frequencies change

8. The output of the AM detector goes directly to the
   (a) IF amplifier   (b) mixer
   (c) audio amplifier   (d) speaker

Section 18–2 9. In order to handle all combinations of input voltage polarities, a multiplier must have
   (a) four-quadrant capability
   (b) three-quadrant capability
   (c) four inputs
   (d) dual-supply voltages

10. The internal attenuation of a multiplier is called the
   (a) transconductance   (b) scale factor   (c) reduction factor

11. When the two inputs of a multiplier are connected together, the device operates as a
   (a) voltage doubler   (b) square root circuit
   (c) squaring circuit   (d) averaging circuit

Section 18–3 12. Amplitude modulation is basically a
   (a) summing of two signals
   (b) multiplication of two signals
   (c) subtraction of two signals
   (d) nonlinear process

13. The frequency spectrum of a balanced modulator contains
   (a) a sum frequency   (b) a difference frequency
   (c) a carrier frequency   (d) answers (a), (b), and (c)
   (e) answers (a) and (b)   (f) answers (b) and (c)

14. Balanced modulation is sometimes known as
   (a) sum and difference modulation
   (b) carrier modulation
   (c) suppressed-carrier modulation
   (d) standard modulation

Section 18–4   **15.** The mixer in a receiver system

     (a) can be implemented with a linear multiplier

     (b) is basically a frequency converter

     (c) produces sum and difference frequencies

     (d) all of the above

   16. The two inputs to a mixer are

     (a) the modulating signal and the carrier signal

     (b) the modulated carrier signal and the local oscillator signal

     (c) the IF signal and the RF signal

     (d) none of the above

Section 18–5   **17.** An AM demodulator is basically a

     (a) linear multiplier   **(b)** low-pass filter

     (c) rectifier   **(d)** linear multiplier followed by a low-pass filter

   18. The final output of an AM demodulator is the

     (a) IF signal   **(b)** audio signal   **(c)** carrier signal   **(d)** side-band frequencies

Section 18–6   **19.** In an AM receiver, the IF amplifier is tuned to

     (a) the carrier frequency   **(b)** the local oscillator frequency

     (c) 10.7 MHz   **(d)** 455 kHz

   20. In an AM receiver system, the audio amplifier

     (a) drives the speaker(s)

     (b) follows the detector

     (c) has a typical bandwidth of 3 kHz to 15 kHz

     (d) all of the above

Section 18–7   **21.** In FM, modulation is achieved by

     (a) varying the frequency of an oscillator

     (b) varying the amplitude of an oscillator

     (c) mixing the local oscillator signal with the IF

     (d) mixing the carrier signal with the modulating signal

   22. A method for FM demodulation is

     (a) slope detection   **(b)** phase-shift discrimination   **(c)** ratio detection

     (d) quadrature detection   **(e)** phase-locked loop demodulation   **(f)** all of the above

Section 18–8   **23.** If the control voltage to a VCO increases, the output frequency

     (a) decreases   **(b)** does not change   **(c)** increases

   24. A PLL maintains lock by comparing

     (a) the phase of two signals

     (b) the frequency of two signals

     (c) the amplitude of two signals

Section 18–9   **25.** In a fiber-optic cable, the light travels through the

     (a) core   **(b)** cladding   **(c)** shell   **(d)** jacket

   26. If the angle of incidence of a light ray is greater than the critical angle, the light will be

     (a) absorbed   **(b)** reflected   **(c)** amplified   **(d)** refracted

   27. The critical angle of a reflective material is determined by the

     (a) absorption

     (b) amount of scattering

     (c) index of refraction

     (d) attenuation

PROBLEMS

Answers to all odd-numbered problems are at the end of the book.

BASIC PROBLEMS

Section 18–1 **Basic Receivers**

1. Label each block in the AM receiver in Figure 18–52.

▶ FIGURE 18–52

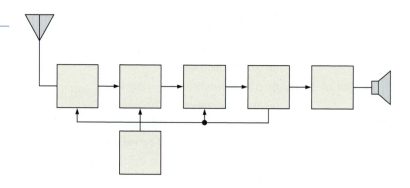

2. Label each block in the FM receiver in Figure 18–53.

▶ FIGURE 18–53

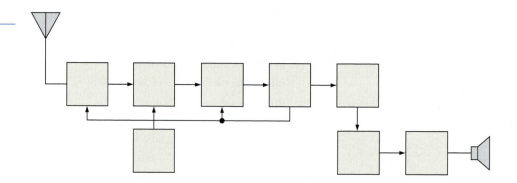

3. An AM receiver is tuned to a transmitted frequency of 680 kHz. What is the local oscillator (LO) frequency?

4. An FM receiver is tuned to a transmitted frequency of 97.2 MHz. What is the LO frequency?

5. The LO in an FM receiver is running at 101.9 MHz. What is the incoming RF? What is the IF?

Section 18–2 **The Linear Multiplier**

6. From the graph in Figure 18–54, determine the multiplier output voltage for each of the following pairs of input voltages.
   (a) $V_X = -4$ V, $V_Y = +6$ V
   (b) $V_X = +8$ V, $V_Y = -2$ V
   (c) $V_X = -5$ V, $V_Y = -2$ V
   (d) $V_X = +10$ V, $V_Y = +10$ V

▶ FIGURE 18–54

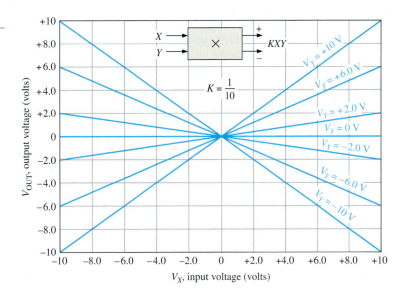

7. If a certain multiplier has a scale factor of 0.125 and the inputs are +3.5 V and −2.9 V, what is the output voltage?

8. Explain how to use a linear multiplier as a squaring circuit.

9. Determine the output voltage for each circuit in Figure 18–55.

▶ FIGURE 18–55

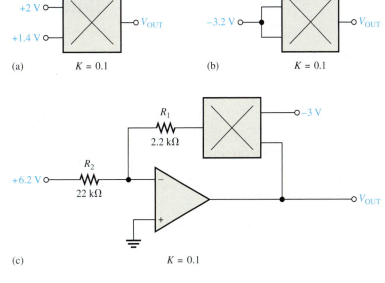

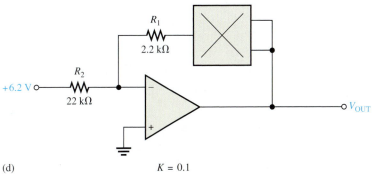

Section 18–3   **Amplitude Modulation**

10. If a 100 kHz signal and a 30 kHz signal are applied to a balanced modulator, what frequencies will appear on the output?

11. What are the frequencies on the output of the balanced modulator in Figure 18–56?

▶ FIGURE 18–56

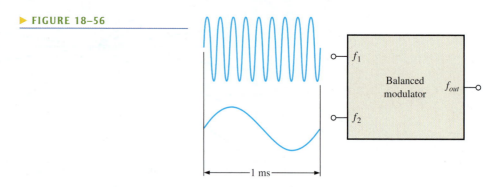

12. If a 1000 kHz signal and a 3 kHz signal are applied to a standard amplitude modulator, what frequencies will appear on the output?

13. What are the frequencies on the output of the standard amplitude modulator in Figure 18–57?

▶ FIGURE 18–57

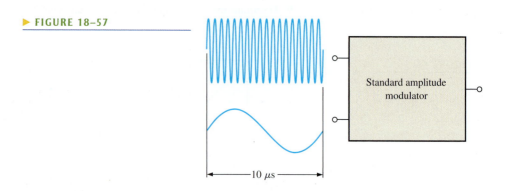

14. The frequency spectrum in Figure 18–58 is for the output of a standard amplitude modulator. Determine the carrier frequency and the modulating frequency.

▶ FIGURE 18–58

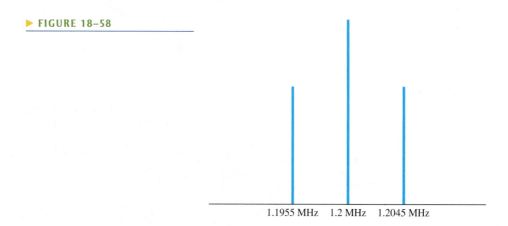

1.1955 MHz   1.2 MHz   1.2045 MHz

15. The frequency spectrum in Figure 18–59 is for the output of a balanced modulator. Determine the carrier frequency and the modulating frequency.

▶ **FIGURE 18–59**

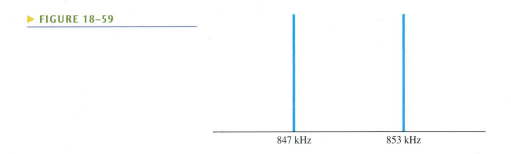

847 kHz   853 kHz

16. A voice signal ranging from 300 Hz to 3 kHz amplitude modulates a 600 kHz carrier. Develop the frequency spectrum.

Section 18–4   The Mixer

17. Determine the output expression for a multiplier with one sinusoidal input having a peak voltage of 0.2 V and a frequency of 2200 kHz and the other input having a peak voltage of 0.15 V and a frequency of 3300 kHz.

18. Determine the output frequency of the IF amplifier for the frequencies shown in Figure 18–60.

▶ **FIGURE 18–60**

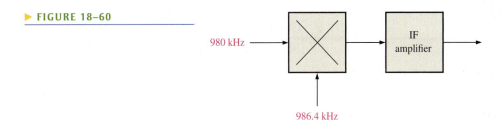

980 kHz

IF amplifier

986.4 kHz

Section 18–5   AM Demodulation

19. The input to a certain AM receiver consists of a 1500 kHz carrier and two side frequencies separated from the carrier by 2 kHz. Determine the frequency spectrum at the output of the mixer amplifier.

20. For the same conditions stated in Problem 19, determine the frequency spectrum at the output of the IF amplifier.

21. For the same conditions stated in Problem 19, determine the frequency spectrum at the output of the AM detector (demodulator).

Section 18–6   IF and Audio Amplifiers

22. For a carrier frequency of 1.2 MHz and a modulating frequency of 8.5 kHz, list all of the frequencies on the output of the mixer in an AM receiver.

23. In a certain AM receiver, one amplifier has a passband from 450 kHz to 460 kHz and another has a passband from 10 Hz to 5 kHz. Identify these amplifiers.

Section 18–7   Frequency Modulation

24. How does an FM signal differ from an AM signal?

25. What is the variable reactance element shown in Figure 18–37?

Section 18–8   **The Phase-Locked Loop (PLL)**

26. Label each block in the PLL diagram of Figure 18–61.

▶ **FIGURE 18–61**

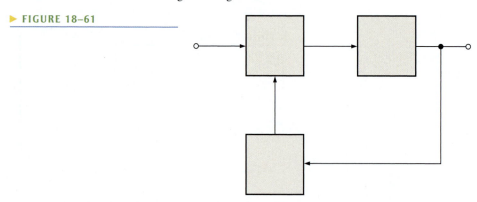

27. A PLL is locked onto an incoming signal with a peak amplitude of 250 mV and a frequency of 10 MHz at a phase angle of 30°. The 400 mV peak VCO signal is at a phase angle of 15°.

   (a) What is the VCO frequency?

   (b) What is the value of the control voltage being fed back to the VCO at this point?

28. What is the conversion gain of a VCO if a 0.5 V increase in the control voltage causes the output frequency to increase by 3.6 kHz?

29. If the conversion gain of a certain VCO is 1.5 kHz per volt, how much does the frequency change if the control voltage increases 0.67 V?

Section 18–9   **Fiber Optics**

30. A light ray strikes the core of a fiber-optic cable at 30° angle of incidence. If the critical angle of the core is 15°, will the light ray be reflected or refracted?

31. Determine the critical angle of a fiber-optic cable if the core has an index of refraction of 1.55 and the cladding has an index of refraction of 1.25.

Answers to Odd-Numbered Problems

Chapter 1

1. Atoms have a planetary type of structure that consists of a central nucleus surrounded by orbiting electrons. The **nucleus** consists of positively charged particles called **protons** and uncharged particles called **neutrons**.

3. 6 electrons; 6 protons

5. **(a)** insulator   **(b)** semiconductor   **(c)** conductor

7. Four

9. Conduction band and valence band

11. A type of chemical bond found in metal solids in which fixed positive ion cores are held together in a lattice by mobile electrons.

13. Antimony is a pentavalent material. Boron is a trivalent material. Both are used for doping.

15. No. The barrier potential is a voltage drop.

Chapter 2

1. p region

3. Reverse bias can be applied not to exceed reverse breakdown value of the diode.

5. To generate the forward bias portion of the characteristic curve, connect a voltage source across the diode for forward bias and place an ammeter in series with the diode and a voltmeter across the diode. Slowly increase the voltage from zero and plot the forward voltage versus the current.

7. **(a)** reversed-biased   **(b)** forward-biased

    (c) forward-biased   **(d)** forward-biased

9. **(a)** -3 V   **(b)** 0 V   **(c)** 0 V   **(d)** 0 V

11. See Figure ANS–1.

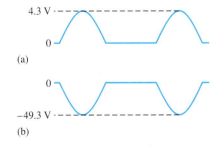

▲ FIGURE ANS–1

13. 63.7 V

15. 24 V rms

17. **(a)** 1.59 V   **(b)** 63.7 V   **(c)** 16.4 V   **(d)** 10.5 V

19. 186 V

21. 78.5 V

23. See Figure ANS–2.

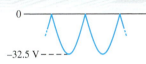

▲ FIGURE ANS–2

25. $V_r = 8.33$ V; $V_{DC} = 25.8$ V

27. 556 μF

29. $V_{r(pp)} = 1.25$ V; $V_{DC} = 48.9$ V

31. 4%

33. See Figure ANS–3.

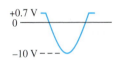

▲ FIGURE ANS–3

35. See Figure ANS–4.

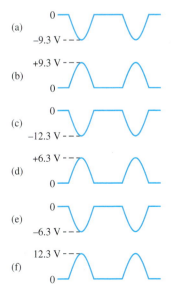

▲ FIGURE ANS–4

37. See Figure ANS–5.

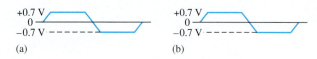

(a)
(b)

▲ **FIGURE ANS–5**

39. **(a)** 7.86 mA   **(b)** 8.5 mA

   (c) 18.8 mA   **(d)** 19.4 mA

41. **(a)** A sine wave with a positive peak at +0.7 V, a negative peak at −7.3 V, and a dc value of −3.3 V.

   (b) A sine wave with a positive peak at +29.3 V, a negative peak at −0.7 V, and a dc value of +14.3 V.

   (c) A square wave varying from +0.7 V down to −15.3 V, with a dc value of −7.3 V.

   (d) A square wave varying from +1.3 V down to −0.7 V, with a dc value of +0.3 V.

43. 56.6 V

45. 100 V

47. 50 Ω

49. $V_A = +25$ V; $V_B = +24.3$ V; $V_C = +8.7$ V; $V_D = +8.0$ V

51. R_{surge} is open. Capacitor is shorted.

53. The circuit should not fail because the diode ratings exceed the actual PIV and maximum current.

55. The rectifier must be connected backwards.

57. 177 μF

59. 651 mΩ (nearest standard 0.68 Ω)

61. See Figure ANS–6.

63. $V_{C1} = 170$ V; $V_{C2} = 338$ V

65. Diode open

67. Diode shorted

69. Diode shorted

71. Diode open

73. Diode shorted

75. Diode open

77. Reduced transformer turns ratio

79. Diode leaky

81. Load resistor open

Chapter 3

1. See Figure ANS–7.

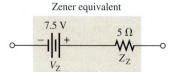

Zener equivalent

▲ **FIGURE ANS–7**

3. 5 Ω

5. 6.92 V

7. **(a)** 36 V   **(b)** 37.8 V   **(c)** 0.25 mA

   (d) 6.67 mW/°C   **(e)** 50°C

9. 14.3 V

11. See Figure ANS–8.

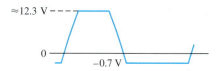

▲ **FIGURE ANS–8**

13. 10.3%

15. 3.13%

17. 5.88%

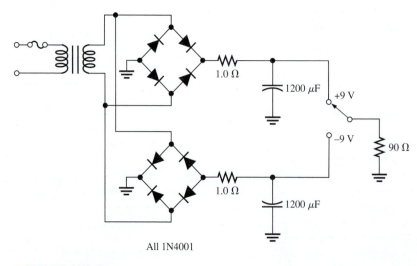

All 1N4001

▲ **FIGURE ANS–6**

19. 3 V

21. $V_R \cong 3$ V

23. See Figure ANS–9.

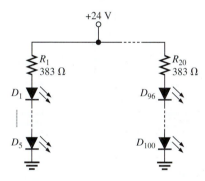

▲ **FIGURE ANS–9**

25. See Figure ANS–10.

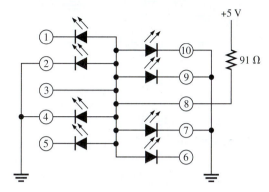

▲ **FIGURE ANS–10**

27. (a) 30 kΩ (b) 8.57 kΩ (c) 5.88 kΩ

29. p-region, n-region, conductive layer, conductive grid, and reflective coating

31. A series connection of 30 cells

33. −750 Ω

35. The reflective ends cause the light to bounce back and forth, thus increasing the intensity of the light. The partially reflective end allows a portion of the reflected light to be emitted.

37. (a) ≈30 V dc

(b) 0 V

(c) Excessive 120 Hz ripple limited to 12 V by zener

(d) Full-wave rectified waveform limited at 12 V by zener

(e) 60 Hz ripple limited to 12 V

(f) 60 Hz ripple limited to 12 V

(g) 0 V

(h) 0 V

39. Incorrect transformer secondary voltage

41. 48 mW

43. (a) 60 pF (b) 20 pF (c) CR = 3

45. (a) 1 μV (b) 940 nm (c) 830 nm

47. $V_{OUT(1)} = 6.8$ V; $V_{OUT(2)} = 24$ V

49. See Figure ANS–11.

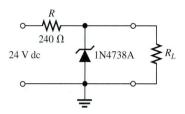

▲ **FIGURE ANS–11**

51. See Figure ANS–12.

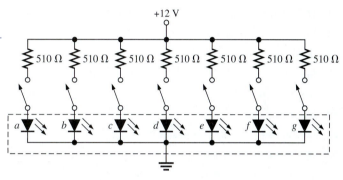

▲ **FIGURE ANS–12**

53. Zener diode open

55. Zener diode shorted

Chapter 4

1. The *npn* type consists of two *n* regions separated by a *p* region, and the *pnp* type consists of two *p* regions separated by an *n* region.

3. Holes

5. The base is narrow and lightly doped so that a small recombination (base) current is generated compared to the collector current.

7. Negative, positive

9. 0.947

11. 101.5

13. 8.98 mA

15. 0.99

17. 5.3 V increase

19. (a) $V_{BE} = 0.7$ V, $V_{CE} = 5.10$ V, $V_{CB} = 4.40$ V

(b) $V_{BE} = -0.7$ V, $V_{CE} = -3.83$ V, $V_{CB} = -3.13$ V

21. $I_B = 30$ μA, $I_E = 1.3$ mA, $I_C = 1.27$ mA

23. 3 μA

25. 425 mW

27. 33.3

29. 1.1 kΩ

31. 500 μA, 3.33 μA, 4.03 V

33. 20 mA

35. 1.45 V

37. 30 mA

39. See Figure ANS–13.

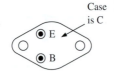

▲ **FIGURE ANS–13**

41. Open, low resistance

43. **(a)** 27.8   **(b)** 109

45. 60 Ω

47. **(a)** 40 V   **(b)** 200 mA dc   **(c)** 625 mW

(d) 500 mW   **(e)** 70

49. 840 mW

51. **(a)** Saturated   **(b)** Not saturated

53. **(a)** No parameters are exceeded.

(b) No parameters are exceeded.

55. Yes, marginally; $V_{CE} = 1.5$ V; $I_C = 75$ mA

57. See Figure ANS–14.

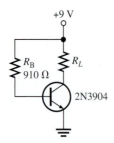

▲ **FIGURE ANS–14**

59. R_B shorted

61. Collector-emitter shorted

63. R_E leaky

65. R_B open

Chapter 5

1. Apply proper dc bias voltage for the maximum input signal.

3. Saturation

5. 18 mA

7. $V_{CE} = 20$ V; $I_{C(sat)} = 2$ mA

9. See Figure ANS–15.

11. **(a)** $I_{C(sat)} = 50$ mA

(b) $V_{CE(CUTOFF)} = 10$ V

(c) $I_B = 250$ μA; $I_C = 25$ mA; $V_{CE} = 5$ V

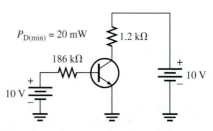

▲ **FIGURE ANS–15**

13. 63.2

15. $I_C \cong 809$ μA; $V_{CE} = 13.2$ V

17. See Figure ANS–16.

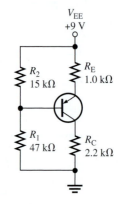

▲ **FIGURE ANS–16**

19. **(a)** −1.63 mA, −8.16 V

(b) 13.3 mW

21. $V_B = -186$ mV; $V_E = -0.886$ V; $V_C = 3.14$ V

23. 0.09 mA

25. $I_C = 16.3$ mA; $V_{CE} = -6.95$ V

27. 2.53 kΩ

29. 7.87 mA; 2.56 V

31. $I_{CQ} = 92.5$ mA; $V_{CEQ} = 2.75$ V

33. 27.7 mA to 69.2 mA; 6.23 V to 2.08 V; Yes

35. $V_1 = 0$ V, $V_2 = 0$ V, $V_3 = 8$ V

37. **(a)** Open collector

(b) No problems

(c) Transistor shorted collector-to-emitter

(d) Open emitter

39. **(a)** 1: 10 V,   2: float,   3: −3.59 V,   4: 10 V

(b) 1: 10 V,   2: 4.05 V,   3: 4.75 V,   4: 4.05 V

(c) 1: 10 V,   2: 0 V,   3: 0 V,   4: 10 V

(d) 1: 10 V,   2: 570 mV,   3: 1.27 V,   4: float

(e) 1: 10 V,   2: 0 V,   3: 0.7 V,   4: 0 V

(d) 1: 10 V,   2: 0 V,   3: 3.59 V,   4: 10 V

41. R_1 open, R_2 shorted, BE junction open

43. $V_C = V_{CC} = 9.1$ V, V_B normal, $V_E = 0$ V

45. None are exceeded.

47. 457 mW

49. See Figure ANS–17.

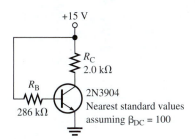

▲ **FIGURE ANS–17**

51. See Figure ANS–18.

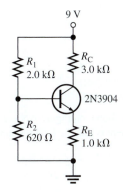

▲ **FIGURE ANS–18**

53. Yes

55. V_{CEQ} will be less, causing the transistor to saturate at a slightly higher temperature, thus limiting the low temperature response.

57. R_C open

59. R_2 open

61. R_C shorted

Chapter 6

1. Slightly greater than 1 mA min.

3. The end points of an ac load line are $I_{c(sat)}$ and $V_{ce(cutoff)}$.

5. 8.33 Ω

7. $r'_e \cong 19$ Ω

9. See Figure ANS–19.

11. 37.5 mW

13. (a) 1.29 kΩ (b) 968 Ω (c) 171

15. (a) $I_E = 2.63$ mA (b) $V_E = 2.63$ V
(c) $V_B = 3.76$ V (d) $I_C \cong 2.63$ mA
(e) $V_C = 9.32$ V (f) $V_{CE} = 6.69$ V

17. $A'_v = 131; \theta = 180°$

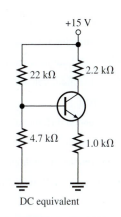

DC equivalent

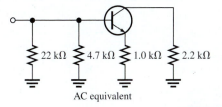

AC equivalent

▲ **FIGURE ANS–19**

19. $A_{v(max)} = 74.7, A_{v(min)} = 2.07$

21. A_v is reduced to 30.1. See Figure ANS–20.

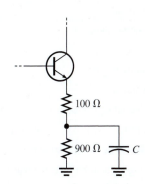

▲ **FIGURE ANS–20**

23. $R_{in(tot)} = 3.1$ kΩ; $V_{OUT} = 1.06$ V

25. 270 Ω

27. 8.8

29. $R_{in(emitter)} = 2.28$ Ω; $A_v = 526$; $A_i \cong 1$; $A_p = 526$

31. 400

33. (a) $A_{v1} = 93.6, A_{v2} = 303$
(b) $A'_v = 28,361$
(c) $A_{v1(dB)} = 39.4$ dB, $A_{v2(dB)} = 49.6$ dB, $A'_{v(dB)} = 89.1$ dB

35. $V_{B1} = 2.16$ V, $V_{E1} = 1.46$ V, $V_{C1} \cong 5.16$ V, $V_{B2} = 5.16$ V, $V_{E2} = 4.46$ V, $V_{C2} \cong 7.54$ V, $A_{v1} = 66, A_{v2} = 179$, $A'_v = 11,814$

37. (a) 1.41 (b) 2.00 (c) 3.16
(d) 10.0 (e) 100

39. V_1: differential output voltage

V_2: noninverting input voltage

V_3: single-ended output voltage

V_4: differential input voltage

I_1: bias current

41. **(a)** Single-ended differential input; differential output

(b) Single-ended differential input; single-ended output

(c) Double-ended differential input; single-ended output

(d) Double-ended differential input; differential output

43. Cutoff, 10 V

45.

TEST POINT	DC VOLTS	AC VOLTS (RMS)
Input	0 V	25 μV
Q_1 base	2.99 V	20.8 μV
Q_1 emitter	2.29 V	0 V
Q_1 collector	7.44 V	1.95 mV
Q_2 base	2.99 V	1.95 mV
Q_2 emitter	2.29 V	0 V
Q_2 collector	7.44 V	589 mV
Output	0 V	589 mV

47. **(a)** No output signal

(b) Reduced output signal

(c) No output signal

(d) Reduced output signal

(e) No output signal

(f) Increased output signal (perhaps with distortion)

49. **(a)** Q_1 is in cutoff   **(b)** V_{EE} **(c)** Unchanged

51. **(a)** 700   **(b)** 40 Ω **(c)** 20 kΩ

53. A leaky coupling capacitor affects the bias voltages and attenuates the ac voltage.

55. Make $R_9 = 69.1\ \Omega$.

57. See Figure ANS–21.

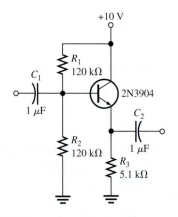

+10 V

▲ **FIGURE ANS–21**

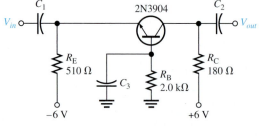

▲ **FIGURE ANS–22**

59. See Figure ANS–22.

61. $A_v = R_C/r_e'$

$A_v \cong (V_{R_C}/I_C)/(0.025\text{ V}/I_C) = V_{R_C}/0.025 = 40V_{R_C}$

63. C_2 shorted

65. C_1 open

67. C_3 open

Chapter 7

1. **(a)** $I_{CQ} = 68.4$ mA; $V_{CEQ} = 5.14$ V

(b) $A_v = 11.7$; $A_p = 263$

3. The changes are shown on Figure ANS–23. The advantage of this arrangement is that the load resistor is referenced to ground.

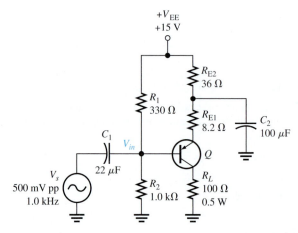

▲ **FIGURE ANS–23**

5. **(a)** $I_C = 54$ mA; $V_{CE} = 2.3$ V

(b) $I_C = 15.7$ mA; $V_{CE} = 2.39$ V

7. For Figure 7–42(a): 46 mA, 2.3 V; For Figure 7–42(b): 10.2 mA, 2.39 V

9. $V_{CE} = 3.5$ V, $I_C = 26.9$ mA

11. 169 mW

13. **(a)** $V_{B(Q1)} = 0.7$ V; $V_{B(Q2)} = -0.7$ V; $V_E = 0$ V;

$V_{CEQ(Q1)} = 9$ V; $V_{CEQ(Q2)} = -9$ V; $I_{CQ} = 8.3$ mA

(b) $P_L = 0.5$ W

15. 457 Ω

17. **(a)** $V_{B(Q1)} = 8.2$ V; $V_{B(Q2)} = 6.8$ V;

$V_E = 7.5$ V;

$I_{CQ} = 6.8$ mA; $V_{CEQ(Q1)} = 7.5$ V;

$V_{CEQ(Q2)} = -7.5$ V

(b) $P_L = 167$ mW

19. **(a)** Power supply off, open R_1, Q_1 base shorted to ground

(b) Q_1 has collector-to-emitter short

(c) One or both diodes shorted

21. 450 μW

23. 24 V

25. Negative half of input cycle

27. **(a)** No dc supply voltage or R_1 open

(b) R_1 is shorted.

(c) No fault

(d) Q_1 emitter open

29. -15 V dc, output signal same as input signal

31. The vertically oriented diode is connected backwards.

33. 10 W

35. Gain increases, then decreases at a certain value of I_C.

37. T_C is much closer to the actual junction temperature than T_A. In a given operating environment, T_A is always less than T_C.

39. See Figure ANS–24.

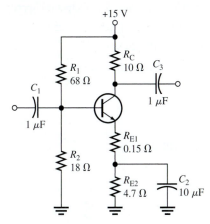

▲ **FIGURE ANS–24**

41. C_{in} open

43. Q_1 collector-emitter open

45. Q_2 drain-source open

Chapter 8

1. **(a)** Narrows   **(b)** Increases

3. See Figure ANS–25.

5. 5 V

7. 10 mA

9. 4 V

n–channel p–channel

▲ **FIGURE ANS–25**

11. -2.63 V

13. $g_m = 1429$ μS, $g_{fs} = 1429$ μS

15. $V_{GS} = 0$ V, $I_D = 8$ mA
$V_{GS} = -1$ V, $I_D = 5.12$ mA
$V_{GS} = -2$ V, $I_D = 2.88$ mA
$V_{GS} = -3$ V, $I_D = 1.28$ mA
$V_{GS} = -4$ V, $I_D = 0.320$ mA
$V_{GS} = -5$ V, $I_D = 0$ mA

17. See Figure ANS–26.

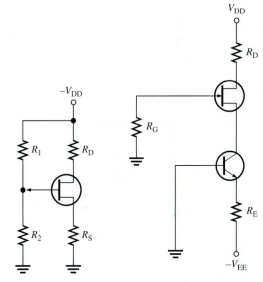

▲ **FIGURE ANS–26**

19. 800 Ω

21. **(a)** 20 mA   **(b)** 0 A   **(c)** Increases

22. 211 Ω

25. 9.80 MΩ

27. $I_D \cong 5.3$ mA, $V_{GS} \cong 2.1$ V

29. $I_D \cong 1.9$ mA, $V_{GS} \cong -1.5$ V

31. From 1.33 kΩ to 2.67 kΩ

33. 935 Ω

35. The enhancement mode

37. The gate is insulated from the channel.

39. 4.69 mA

41. **(a)** Depletion   **(b)** Enhancement

(c) Zero bias   **(d)** Depletion

43. (a) 4 V   (b) 5.4 V   (c) −4.52 V

45. (a) 5 V, 3.18 mA   (b) 3.2 V, 1.02 mA

47. The input resistance of an IGBT is very high because of the insulated gate structure.

49. R_D or R_S open, JFET open D-to-S, $V_{DD} = 0$ V, or ground connection open.

51. Essentially no change

53. The 1.0 MΩ bias resistor is open.

55. $V_{G2S} = 6$ V, $I_D \cong 10$ mA; $V_{G2S} = 1$ V, $I_D \cong 5$ mA

57. 3.04 V

59. (a) −0.5 V   (b) 25 V   (c) 310 mW   (d) −25 V

61. 2000 μS

63. $I_D \cong 1.4$ mA

65. $I_D \cong 13$ mA when $V_{GS} = +3$ V, $I_D \cong 0.4$ mA when $V_{GS} = -2$ V.

67. −3.0 V

69. $I_D = 3.58$ mA; $V_{GS} = -4.21$ V

71. 6.01 V

73. See Figure ANS–27.

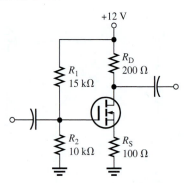

+12 V

R_1 15 kΩ

R_D 200 Ω

R_2 10 kΩ

R_S 100 Ω

▲ **FIGURE ANS–27**

75. R_D shorted

77. R_1 open

79. R_D open

81. Drain-source shorted

Chapter 9

1. dc analysis and ac analysis

3. (a) 60 μA   (b) 900 μA   (c) 3.6 mA   (d) 6 mA

5. 14.2

7. (a) n-channel D-MOSFET with zero-bias; $V_{GS} = 0$

   (b) p-channel JFET with self-bias; $V_{GS} = -0.99$ V

   (c) n-channel E-MOSFET with voltage-divider bias; $V_{GS} = 3.84$ V

9. (a) n-channel D-MOSFET

   (b) n-channel JFET

   (c) p-channel E-MOSFET

11. Figure 9–16(b): approximately 4 mA
    Figure 9–16(c): approximately 3.2 mA

13. 920 mV

15. (a) 4.32   (b) 9.92

17. 7.5 mA

19. 2.54

21. 33.6 mV rms

23. 9.84 MΩ

25. $V_{GS} = 9$ V; $I_D = 3.13$ mA; $V_{DS} = 13.3$ V; $V_{ds} = 675$ mV

27. $R_{in} \cong 10$ MΩ; $A_v = 0.620$

29. (a) 0.906   (b) 0.299

31. $R_{in} = R_{IN(gate)} \| (R_3 + R_1 \| R_2)$

33. 250 Ω

35. $A_v = 2640$; $R_{in} = 14.6$ MΩ

37. 0.95

39. 30 kHz

41. 40 kΩ

43. (a) 3.3 V   (b) 3.3 V   (c) 3.3 V   (d) 0 V

45. The MOSFET has lower on-state resistance and can turn off faster.

47. (a) $V_{out} = 0$ V if C_1 is open

   (b) $A_{v1} = 7.5$, $A_{v2} = 2.24$, $A_v = 16.8$, $V_{out} = 168$ mV

   (c) $V_{GS} = 0$ V for Q_2. $I_D = I_{DSS}$. Output is clipped.

   (d) No V_{out} because no signal at Q_2 gate.

49. (a) −3.0 V   (b) 20 V dc

   (c) 200 mW   (d) ± 10 V dc

51. 900 μS

53. 1.5 mA

55. 2.0; 6.82

57. See Figure ANS–28.

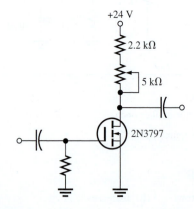

+24 V

2.2 kΩ

5 kΩ

2N3797

▲ **FIGURE ANS–28**

59. C_2 open

61. R_S shorted

63. R_1 open

65. R_2 open

Chapter 10

1. **(a)** Parasitic capacitance affects the high-frequency response.

(b) A designer can choose a transistor with a lower internal capacitance, lower the gain to reduce the Miller effect, or change the circuit to use a noninverting amplifier.

3. BJT: C_{be}, C_{bc}, C_{ce}; FET: C_{gs}, C_{gd}, C_{ds}

5. 812 pF

7. $C_{in(miller)} = 6.95$ pF; $C_{out(miller)} \cong 5.28$ pF

9. 24 mV rms; 34 dB

11. **(a)** 3.01 dBm   **(b)** 0 dBm

(c) 6.02 dBm   **(d)** −6.02 dBm

13. **(a)** 318 Hz   **(b)** 1.59 kHz

15. At $0.1 f_c$: $A_v = 18.8$ dB

At f_c: $A_v = 35.8$ dB

At $10 f_c$: $A_v = 38.8$ dB

17. Input RC circuit: $f_c = 3.34$ Hz

Output RC circuit: $f_c = 3.01$ kHz

Output f_c is dominant.

19. Input circuit: $f_c = 4.32$ MHz

Output circuit: $f_c = 94.9$ MHz

Input f_c is dominant.

21. Input circuit: $f_c = 12.9$ MHz

Output circuit: $f_c = 54.5$ MHz

Input f_c is dominant.

23. $f_{cl} = 136$ Hz, $f_{cu} = 8$ kHz

25. $BW = 5.26$ MHz, $f_{cu} \cong 5.26$ MHz

27. 230 Hz; 1.2 MHz

29. 514 kHz

31. ≈ 2.5 MHz

33. Increase the frequency until the output voltage drops to 3.54 V rms. This is f_{cu}.

35. 15.9 Hz

37. No effect

39. 112 pF

41. $C_{gd} = 1.3$ pF; $C_{gs} = 3.7$ pF; $C_{ds} = 3.7$ pF

43. ≈ 10.9 MHz

45. R_C open

47. R_2 open

Chapter 11

1. $I_A = 24.1$ mA

3. See "Turning the SCR On" in Section 11–2.

5. When the switch is closed, the battery V_2 causes illumination of the lamp. The light energy causes the LASCR to conduct and thus energize the relay. When the relay is energized, the contacts close and 115 V ac are applied to the motor.

7. Add a transistor to provide inversion of negative half-cycle in order to obtain a positive gate trigger.

9. See Figure ANS–29.

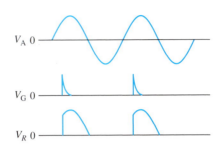

▲ **FIGURE ANS–29**

11. See Figure ANS–30.

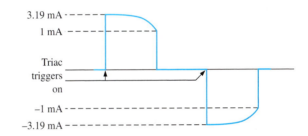

▲ **FIGURE ANS–30**

13. Anode, cathode, anode gate, cathode gate

15. 6.48 V

17. **(a)** 9.79 V   **(b)** 5.2 V

19. See Figure ANS–31.

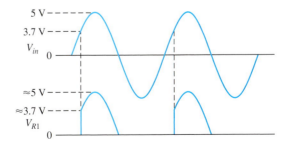

▲ **FIGURE ANS–31**

21. 0 V

23. As the PUT gate voltage increases, the PUT triggers on later to the ac cycle causing the SCR to fire later in the cycle, conduct for a shorter time, and decrease power to the motor.

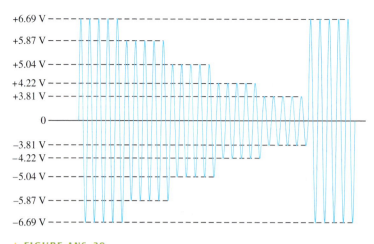

▲ **FIGURE ANS–39**

37. R_f open

39. C open

Chapter 14

1. $A_{v(1)} = A_{v(2)} = 101$

3. 1.005 V

5. 51.5

7. Change R_G to 2.2 kΩ.

9. 300

11. Change the 18 kΩ resistor to 68 kΩ.

13. Connect pin 6 directly to pin 10, and connect pin 14 directly to pin 15 to make $R_f = 0$.

15. 500 μA, 5 V

17. $A_v \cong 11.6$

19. See Figure ANS–39.

21. See Figure ANS–40.

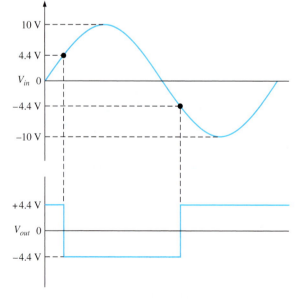

▲ **FIGURE ANS–40**

23. **(a)** −0.301 **(b)** 0.301 **(c)** 1.70 **(d)** 2.11

25. The output of a log amplifier is limited to 0.7 V because of the transistor's *pn* junction.

27. −157 mV

29. $V_{out(max)} = -147$ mV, $V_{out(min)} = -89.2$ mV; the 1 V input peak is reduced 85% whereas the 100 mV input peak is reduced only 10%.

31. See Figure ANS–41.

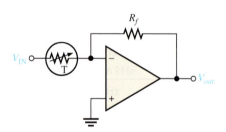

▲ **FIGURE ANS–41**

33. R open

35. Zener diode open

Chapter 15

1. **(a)** Band-pass **(b)** High-pass
(c) Low-pass **(d)** Band-stop

3. 48.2 kHz, No

5. 700 Hz, 5.04

7. **(a)** 1, not Butterworth
(b) 1.44, approximate Butterworth
(c) 1st stage: 1.67; 2nd stage: 1.67; Not Butterworth

9. (a) and (b) are two-pole filters with approximately a –40 dB/decade roll-off. Filter (c) is a three-pole filter with approximately a –60 dB/decade roll-off rate.

11. **(a)** Chebyshev **(b)** Butterworth
(c) Bessel **(d)** Butterworth

13. 190 Hz

15. Add another identical stage and change the ratio of the feedback resistors to 0.068 for first stage, 0.586 for second stage, and 1.482 for third stage.

17. Exchange positions of resistors and capacitors.

19. (a) Decrease R_1 and R_2 or C_1 and C_2.

    (b) Increase R_3 or decrease R_4.

21. (a) $f_0 = 4.95$ kHz, $BW = 3.84$ kHz

    (b) $f_0 = 449$ Hz, $BW = 96.4$ Hz

    (c) $f_0 = 15.9$ kHz, $BW = 838$ Hz

23. Sum the low-pass and high-pass outputs with a two-input adder.

25. R_4 shorted

27. C_3 shorted

29. R_1 open

31. R_1 open

33. R_7 open

Chapter 16

1. An oscillator requires no input (other than dc power).

3. Gain should be B.

5. 733 mV

7. 249 Hz to 270 Hz.

9. 2.34 kΩ

11. 136 kΩ, 628 Hz

13. 10

15. Change R_1 to 3.54 kΩ

17. $R_4 = 65.8$ kΩ, $R_5 = 47$ kΩ

19. 3.33 V, 6.67 V

21. 0.0076 μF

23. The VCO provides a variable frequency to lock onto the incoming signal.

25. Drain-to-source shorted

27. Collector-to-emitter shorted

29. R_2 open

Chapter 17

1. 0.0333%

3. 1.01%

5. A: Reference voltage, B: Control element, C: Error detector, D: Sampling circuit

7. 8.51 V

9. 9.57 V

11. 500 mA

13. 10 mA

15. $I_{L(max)} = 250$ mA, $P_{R1} = 6.25$ W

17. 40%

19. V_{OUT} decreases

21. 14.3 V

23. 1.3 mA

25. 2.8 Ω

27. $R_{limit} = 0.35$ Ω

29. See Figure ANS–42.

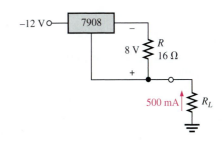

▲ **FIGURE ANS–42**

31. R_2 leaky

33. Q_2 collector-to-emitter open

Chapter 18

1. See Figure ANS–43.

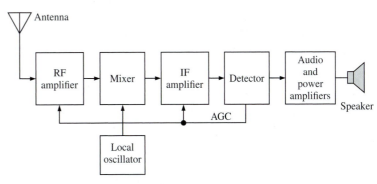

▲ **FIGURE ANS–43**

3. 1135 kHz

5. $f_{RF} = 91.2$ MHz; $f_{IF} = 10.7$ MHz

7. −1.27 V

9. (a) +0.28 V   (b) +1.024 V

    (c) +2.07 V   (d) +2.49 V

11. $f_{diff} = 8$ kHz; $f_{sum} = 10$ kHz

13. $f_{diff} = 1.7$ MHz; $f_{sum} = 1.9$ MHz; $f_c = 1.8$ MHz

15. $f_c = 850$ kHz; $f_m = 3$ kHz

17. $V_{out} = 15$ mV $\cos[2\pi(1100$ kHz$)t] -$

               15 mV $\cos[2\pi(5500$ kHz$)t]$

19. See Figure ANS–44.

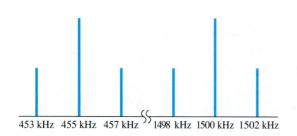

453 kHz 455 kHz 457 kHz 1498 kHz 1500 kHz 1502 kHz

▲ FIGURE ANS–44

21. See Figure ANS–45.

23. 450 kHz − 460 kHz: If amplifier; 10 Hz − 5 kHz: Audio/Power amplifiers

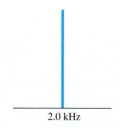

2.0 kHz

▲ FIGURE ANS–45

25. Varactor

27. **(a)** 10 MHz **(b)** 48.3 mV

29. 1005 Hz

31. 36.2°

GLOSSARY

ac ground A point in a circuit that appears as ground to ac signals only.

active filter A frequency-selective circuit consisting of active devices such as transistors or op-amps coupled with reactive components.

A/D conversion A process whereby information in analog form is converted into digital form.

alpha (α) The ratio of dc collector current to dc emitter current in a bipolar junction transistor.

amplification The process of increasing the power, voltage, or current by electronic means.

amplifier An electronic circuit having the capability to amplify power, voltage, or current.

amplitude modulation A method of modulation in which the amplitude of a carrier signal is varied by a lower frequency information signal.

analog Characterized by a linear process in which a variable takes on a continuous set of values.

analog switch A device that switches an analog signal on and off.

AND gate A digital circuit in which the output is at a high-level voltage when all of the inputs are at a high-level voltage.

angle of incidence The angle at which a light ray strikes the reflective surface of an optical cable on entry.

anode The *p* region of a diode.

antilogarithm The result obtained when the base of a number is raised to a power equal to the logarithm of that number.

assembly language A low-level programming language that represents each machine language instruction with an English-like instruction that is easier to remember than groups of 0s and 1s.

astable Characterized by having no stable states.

atom The smallest particle of an element that possesses the unique characteristics of that element.

atomic number The number of protons in an atom.

attenuation The reduction in the level of power, current, or voltage.

automated test system A system that operates under the control of an automated controller to conduct tests on a component, circuit, or system.

avalanche breakdown The higher voltage breakdown in a zener diode.

avalanche effect The rapid buildup of conduction electrons due to excessive reverse-bias voltage.

band gap The difference in energy between energy levels in an atom.

band-pass filter A type of filter that passes a range of frequencies lying between a certain lower frequency and a certain higher frequency.

band-stop filter A type of filter that blocks or rejects a range of frequencies lying between a certain lower frequency and a certain higher frequency.

bandwidth The characteristic of certain types of electronic circuits that specifies the usable range of frequencies that pass from input to output.

barrier potential The amount of energy required to produce full conduction across the *pn* junction in forward bias.

base One of the semiconductor regions in a BJT. The base is very thin and lightly doped compared to the other regions.

Bessel A type of filter response having a linear phase characteristic and less than -20 dB/decade/pole roll-off.

beta (β) The ratio of dc collector current to dc base current in a BJT; current gain from base to collector.

bias The application of a dc voltage to a diode, transistor, or other device to produce a desired mode of operation.

bipolar Characterized by both free electrons and holes as current carriers.

BJT Bipolar junction transistor; a transistor constructed with three doped semiconductor regions separated by two *pn* junctions.

blocking voltage (BV_{DSS}) Maximum drain-to-source voltage that can be applied to the MOSFET.

Bode plot An idealized graph of the gain in dB versus frequency used to graphically illustrate the response of an amplifier or filter.

bounding The process of limiting the output range of an amplifier or other circuit.

breakdown The phenomenon of a sudden and drastic increase when a certain voltage is reached across a device.

bridge rectifier A type of full-wave rectifier consisting of diodes arranged in a four-cornered configuration.

Butterworth A type of filter response characterized by flatness in the passband and a -20 dB/decade/pole roll-off.

bypass capacitor A capacitor placed across the emitter resistor of an amplifier.

CAM Configurable analog module; a predesigned analog circuit used in an FPAA or dpASP for which some of its parameters can be selectively programmed.

capacitance ratio The ratio of varactor capacitances at minimum and at maximum reverse voltages.

capture range The range of frequencies over which a PLL can acquire lock.

cascade An arrangement of circuits in which the output of one circuit becomes the input to the next.

cascode A FET amplifier configuration in which a common-source amplifier and a common-gate amplifier are connected in a series arrangement.

cathode The *n* region of a diode.

center-tapped rectifier A type of full-wave rectifier consisting of a center-tapped transformer and two diodes.

channel The conductive path between the drain and source in a FET.

Chebyshev A type of filter response characterized by ripples in the passband and a greater than −20 dB/decade/pole roll-off.

clamper A circuit that adds a dc level to an ac voltage using a diode and a capacitor.

class A A type of amplifier that operates entirely in its linear (active) region.

class AB A type of amplifier that is biased into slight conduction.

class B A type of amplifier that operates in the linear region for 180° of the input cycle because it is biased at cutoff.

class C A type of amplifier that operates only for a small portion of the input cycle.

class D A nonlinear amplifier in which the transistors are operated as switches.

clipper See Limiter.

closed-loop An op-amp configuration in which the output is connected back to the input through a feedback circuit.

closed-loop voltage gain (A_{cl}) The voltage gain of an op-amp with external feedback.

CMOS Complementary MOS.

CMRR Common-mode rejection ratio; the ratio of open-loop gain to common-mode gain; a measure of an op-amp's ability to reject common-mode signals.

coherent light Light having only one wavelength.

cohesion An indication of how well a procedure or program keeps together the code associated with a specific task.

collector The largest of the three semiconductor regions of a BJT.

common-base (CB) A BJT amplifier configuration in which the base is the common terminal to an ac signal or ground.

common-collector (CC) A BJT amplifier configuration in which the collector is the common terminal to an ac signal or ground.

common-drain (CD) A FET amplifier configuration in which the drain is the grounded terminal.

common-emitter (CE) A BJT amplifier configuration in which the emitter is the common terminal to an ac signal or ground.

common-gate (CG) A FET amplifier configuration in which the gate is the grounded terminal.

common mode A condition where two signals applied to differential inputs are of the same phase, frequency, and amplitude.

common-source (CS) A FET amplifier configuration in which the source is the grounded terminal.

comparator A circuit which compares two input voltages and produces an output in either of two states indicating the greater or less than relationship of the inputs.

complementary symmetry transistors Two transistors, one *npn*, and one *pnp*, having matched characteristics.

conditional execution The selective processing of program instructions based upon the validity of some condition.

conduction electron A free electron.

conductor A material that conducts electrical current very well.

continuous drain current (I_D) The maximum current that can safely be carried by a FET continuously.

core The central part of an atom, includes the nucleus and all but the valence electrons.

coupling An indication of how much one part of a program interacts with or potentially affects another part of the program.

covalent Related to the bonding of two or more atoms by the interaction of their valence electrons.

covalent bond A bond between atoms created by the sharing of a valence electron.

critical angle The value of the angle of incidence at which a light ray is reflected back into the core of an optical cable.

critical frequency The frequency at which the response of an amplifier or filter is 3 dB less than at midrange.

crossover distortion Distortion in the output of a class B push-pull amplifier at the point where each transistor changes from the cutoff state to the *on* state.

crystal A solid material that is a three-dimensional symmetrical arrangement of atoms.

CTR Current transfer ratio. An indication of how efficiently a signal is coupled from input to output.

current The rate of flow of electrical charge.

current mirror A circuit that uses matching diode junctions to form a current source. The current in a diode junction is reflected as a matching current in the other junction (which is typically the base-emitter junction of a transistor). Current mirrors are commonly used to bias a push-pull amplifier.

cutoff The nonconducting state of a transistor.

cutoff frequency Another term for critical frequency.

cutoff voltage The value of the gate-to-source voltage that makes the drain current approximately zero.

D/A conversion The process of converting a sequence of digital codes to an analog form.

damping factor A filter characteristic that determines the type of response.

dark current The amount of thermally generated reverse current in a photodiode in the absence of light.

Darlington pair A configuration of two transistors in which the collectors are connected and the emitter of the first drives the base of the second to achieve beta multiplication.

dBm A unit for measuring power levels referenced to 1 mW.

dc load line A straight line plot of I_C and V_{CE} for a transistor circuit.

dc power supply The dc power of an amplifier with no input signal.

dc quiescent power The maximum power that a class A amplifier must handle.

decade A ten-times increase or decrease in the value of a quantity such as frequency.

decibel (dB) A logarithmic measure of the ratio of one power to another or one voltage to another.

depletion In a MOSFET, the process of removing or depleting the channel of charge carriers and thus decreasing the channel conductivity.

depletion region The area near a *pn* junction on both sides that has no majority carriers.

diac A two-terminal four-layer semiconductor device (thyristor) that can conduct current in either direction when properly activated.

differential amplifier (diff-amp) An amplifier in which the output is a function of the difference between two input voltages, used as the input stage of an op-amp.

differential mode A mode of op-amp operation in which two opposite polarity signal voltages are applied to two inputs (double-ended) or in which a signal is applied to one input and ground to the other (single-ended).

differentiator A circuit that produces an output which approximates the instantaneous rate of change of the input function.

digital Characterized by a process in which a variable takes on either of two values.

diode A semiconductor device with a single *pn* junction that conducts current in only one direction.

diode drop The voltage across the diode when it is forward-biased; approximately the same as the barrier potential and typically 0.7 V for silicon.

doping The process of imparting impurities to an intrinsic semiconductive material in order to control its conduction characteristics.

downloading The process of implementing the software description of a circuit in an FPAA.

drain One of the three terminals of a FET analogous to the collector of a BJT.

dynamic reconfiguration The process of downloading a design modification or new design in an FPAA while it is operating in a system without the need to power down or reset the system; also known as "on-the-fly" reprogramming.

dynamic resistance The nonlinear internal resistance of a semiconductive material.

efficiency The ratio of the signal power delivered to a load to the power from the power supply of an amplifier.

electroluminescence The process of releasing light energy by the recombination of electrons in a semiconductor.

electron The basic particle of negative electrical charge.

electron cloud In the quantum model, the area around an atom's nucleus where an electron can probably be found.

electron-hole pair The conduction electron and the hole created when the electron leaves the valence band.

electrostatic discharge (ESD) The discharge of a high voltage through an insulating path that can destroy an electronic device.

emitter The most heavily doped of the three semiconductor regions of a BJT.

emitter-follower A popular term for a common-collector amplifier.

enhancement In a MOSFET, the process of creating a channel or increasing the conductivity of the channel by the addition of charge carriers.

feedback The process of returning a portion of a circuit's output back to the input in such a way as to oppose or aid a change in the output.

feedback oscillator An electronic circuit that operates with positive feedback and produces a time-varying output signal without an external input signal.

FET Field-effect transistor; a type of unipolar, voltage-controlled transistor that uses an induced electric field to control current.

fiber optics A system for transmitting information using light through an optical cable.

filter In a power supply, a capacitor used to reduce the variation of the output voltage from a rectifier; a type of circuit that passes or blocks certain frequencies to the exclusion of all others.

floating point A point in the circuit that is not electrically connected to ground or a "solid" voltage.

flowchart A graphical means of representing the organization and process flow of a program using distinctively-shaped interconnected blocks.

fold-back current limiting A method of current limiting in voltage regulators.

forced commutation A method of turning off an SCR.

forward bias The condition in which a diode conducts current.

forward-breakover voltage ($V_{BR(F)}$) The voltage at which a device enters the forward-blocking region.

four-layer diode The type of two-terminal thyristor that conducts current when the anode-to-cathode voltage reaches a specified "breakover" value.

four-quadrant multiplier A device that accepts any of four input polarity combinations and produces an output with the corresponding polarity.

FPAA Field-programmable analog array; an integrated circuit that can be programmed for implementation of an analog circuit design.

free electron An electron that has acquired enough energy to break away from the valence band of the parent atom; also called a conduction electron.

frequency modulation A method of modulation in which the frequency of a carrier signal is varied by the information-carrying signal.

frequency response The change in gain or phase shift over a specified range of input signal frequencies.

full-wave rectifier A circuit that converts an ac sinusoidal input voltage into a pulsating dc voltage with two output pulses occurring for each input cycle.

fuse A protective device that burns open when the current exceeds a rated limit.

programming language A set of instructions and rules for their use that allow programmers to provide a processor with the necessary information to accomplish some specific task.

proton The basic particle of positive charge.

pulse-width modulation A process in which a signal is converted to a series of pulses with widths that vary proportionally to the signal amplitude.

push-pull A type of class B amplifier with two transistors in which one transistor conducts for one half-cycle and the other conducts for the other half-cycle.

PUT Programmable unijunction transistor; a type of three-terminal thyristor (more like an SCR than a UJT) that is triggered into conduction when the voltage at the anode exceeds the voltage at the gate.

PV cell Photovoltaic cell or solar cell.

Q-point The dc operating (bias) point of an amplifier specified by voltage and current values.

quality factor (Q) For a reactive component, a figure of merit which is the ratio of energy stored and returned by the component to the energy dissipated; for a band-pass filter, the ratio of the center frequency to its bandwidth.

quantum dots A form of nanocrystals made from semiconductor material such as silicon, germanium, cadmium sulfide, cadmium selenid, and indium phosphide.

radiant intensity (I_0) The output power of an LED per steradian in units of mW/sr.

radiation The process of emitting electromagnetic or light energy.

recombination The process of a free (conduction band) electron falling into a hole in the valence band of an atom.

rectifier An electronic circuit that converts ac into pulsating dc; one part of a power supply.

regulator An electronic device or circuit that maintains an essentially constant output voltage for a range of input voltage or load values; one part of a power supply.

relaxation oscillator An electronic circuit that uses an RC timing circuit to generate a nonsinusoidal waveform without an external input signal.

reverse bias The condition in which a diode prevents current.

ripple factor A measure of effectiveness of a power supply filter in reducing the ripple voltage; ratio of the ripple voltage to the dc output voltage.

ripple voltage The small variation in the dc output voltage of a filtered rectifier caused by the charging and discharging of the filter capacitor.

r parameter One of a set of BJT characteristic parameters that include α_{DC}, β_{DC}, r_e', r_b', and r_c'.

roll-off The rate of decrease in the gain above or below the critical frequencies of a filter; usually it is specified in dB/decade.

safe operating area (SOA) A set of curves drawn on a log-log plot that define the maximum value of drain-source voltage as a function of drain current, which guarantees safe operation when the device is forward biased.

saturation The state of a BJT in which the collector current has reached a maximum and is independent of the base current.

schematic A symbolized diagram representing an electrical or electronic circuit.

Schmitt trigger A comparator with built-in hysteresis.

SCR Silicon-controlled rectifier; a type of three-terminal thyristor that conducts current when triggered on by a voltage at the single gate terminal and remains on until the anode current falls below a specified value.

SCS Silicon-controlled switch; a type of four-terminal thyristor that has two gate terminals that are used to trigger the device on and off.

semiconductor A material that lies between conductors and insulators in its conductive properties.

sequential programming Programming in which instructions execute in the order in which they appear in the program.

shell An energy band in which electrons orbit the nucleus of an atom.

signal compression The process of scaling down the amplitude of a signal voltage.

silicon A semiconductive material.

slew rate The rate of change of the output voltage of an op-amp in response to a step input.

source One of the three terminals of a FET analogous to the emitter of a BJT.

source-follower The common-drain amplifier.

spectral Pertaining to a range of frequencies.

stability A measure of how well an amplifier maintains its design values (Q-point, gain, etc.) over changes in beta and temperature.

stage One of the amplifier circuits in a multistage configuration.

standoff ratio The characteristic of a UJT that determines its turn-on point.

stiff voltage divider A voltage divider for which loading effects can be neglected.

summing amplifier An op-amp configuration with two or more inputs that produces an output voltage that is proportional to the negative of the algebraic sum of its input voltages.

switched-capacitor circuit A combination of a capacitor and transistor switches used in programmable analog devices to emulate resistors.

switching current (I_S) The value of anode current at the point where the device switches from the forward-blocking region to the forward-conduction region.

switching regulator A voltage regulator in which the control element operates as a switch.

Sziklai pair A complementary Darlington arrangement.

test equipment The components in an automated test system that provide the voltages, signals, and currents for the unit under test.

test fixture The component in an automated test system that selectively connects the test equipment and instrumentation to the unit under test.

test instrumentation The components in an automated test system that measure and record the response of the unit under test to the test equipment.

thermal overload A condition in a rectifier where the internal power dissipation of the circuit exceeds a certain maximum due to excessive current.

thermistor A temperature-sensitive resistor with a negative temperature coefficient.

thyristor A class of four-layer (*pnpn*) semiconductor devices.

transconductance (g_m) The ratio of a change in drain current to a change in gate-to-source voltage in a FET; in general, the ratio of the output current to the input voltage.

transformer An electrical device constructed of two or more coils (windings) that are electromagnetically coupled to each other to provide a transfer of power from one coil to another.

transistor A semiconductive device used for amplification and switching applications.

triac A three-terminal thyristor that can conduct current in either direction when properly activated.

trigger The activating input of some electronic devices and circuits.

trivalent Describes an atom with three valence electrons.

troubleshooting A systematic process of isolating, identifying, and correcting a fault in a circuit or system.

turns ratio The number of turns in the secondary of a transformer divided by the number of turns in the primary.

UJT Unijunction transistor; a three-terminal single *pn* junction device that exhibits a negative resistance characteristic.

unit under test (UUT) The component, circuit, or system to be tested in a test system. The UUT is sometimes referred to as a device under test (DUT).

valence Related to the outer shell of an atom.

varactor A variable capacitance diode.

V-I characteristic A curve showing the relationship of diode voltage and current.

visual programming Programming that uses graphical objects rather than textual instructions to create the final program.

voltage-controlled oscillator (VCO) A type of relaxation oscillator whose frequency can be varied by a dc control voltage; an oscillator for which the output frequency is dependent on a controlling input voltage.

voltage-follower A closed-loop, noninverting op-amp with a voltage gain of 1.

voltage multiplier A circuit using diodes and capacitors that increases the input voltage by two, three, or four times.

wavelength The distance in space occupied by one cycle of an electromagnetic or light wave.

Wien bridge oscillator A type of feedback oscillator that is characterized by an *RC* lead-lag circuit in the positive feedback loop.

zener breakdown The lower voltage breakdown in a zener diode.

zener diode A diode designed for limiting the voltage across its terminals in reverse bias.